Lecture Notes in Physics

325

E. Brändas N. Elander (Eds.)

Resonances

The Unifying Route Towards the Formulation of Dynamical Processes

Foundations and Applications in Nuclear, Atomic and Molecular Physics

Proceedings of a Symposium Held at Lertorpet, Värmland, Sweden, August 19–26, 1987

Springer-Verlag
Berlin Heidelberg New York London Paris Tokyo Hong Kong

Editors

Erkki Brändas
Quantum Chemistry Group for
Atomic, Molecular and Solid State Physics
University of Uppsala, S-751 20 Uppsala, Sweden

Nils Elander
Manne Siegbahn Institute of Physics
Frescativägen 24, S-104 05 Stockholm, Sweden

ISBN 3-540-50994-1 Springer-Verlag Berlin Heidelberg New York
ISBN 0-387-50994-1 Springer-Verlag New York Berlin Heidelberg

Printed in Germany

Printing: Druckhaus Beltz, Hemsbach/Bergstr.
Bookbinding: J. Schäffer GmbH & Co. KG., Grünstadt
2153/3140-543210 – Printed on acid-free paper

INTRODUCTION

Let any mountain be submerged gradually, and coral grow in the sea in which it is sinking, and there will be a ring of coral, and finally only a lagoon in the centre ... Coral islands are the last efforts of drowning continents to lift their heads above water. Regions of elevation and subsidence in the ocean may be traced by the state of coral reefs. The preceding quotation appears in the autobiography of Charles Darwin. The great scientist and universal genius, botanist, zoologist and geologist, Sir Charles Darwin gave birth to theories on most phenomena in nature through his exceptional intellectual and creative power. In a cited letter from Sir Charles Lyell to Sir John Herschel, the present, almost poetic, picture of pattern recognition, connected with Darwin's theory of coral island formation, is presented as an apparitional phenomenon. The quasi-stationary structure of a coral reef is depicted and comprehended by the supposed knowledge of the sinking mountain. In a general sense, this reasoning provides the flavour of pattern recognition as it appeared in the contentious theory of evolution associated with the origin of species.

Even if the above quoted passage refers to a macroscopic phase, and the **Lertorpet Symposium on** *Resonances - the unifying route towards the formulation of dynamical processes*, held in the wilderness of Värmland, Sweden, August 19–26, 1987, in contrast, focused on a microcosmic level, there are some subtle points in common, see further below. The subject of the symposium is to a large extent interdisciplinary and ranges from pure mathematics, via nuclear, atomic and molecular physics, to the study of chemical reactions, with applications reaching into the domains of biology and medicine. So, for instance, this approach involves a long list of topics including the correlation problem of the collision complex constituting the compound nucleus; shape resonances, such as those in α-decay; nuclear reactions and heavy ion resonances; Auger spectra; electron-atom scattering; molecular predissociations; wavepacket propagation; electron transfer; applications to van der Waals complexes; trapping of atoms and molecules on surfaces; far-infrared absorption in polar liquids; photon-counting in steady-state luminescence with possible implications for macromolecular phenomena.

The surprising circumstance that scientists from so many different areas have so much in common rests on their veracity and ability to retrace the successive stages in the respective theoretical development to an equivalent or even identical origin. For instance, consider the picture used by a fusion oriented nuclear physicist who wants to describe how a helium nucleus is created by two colliding deuterons. This visualization will not differ much from the image painted by a theoretical chemist explaining the gaseous reactions of H_2 and O_2 to form H_2O. In both cases, the initial and final states of the participating particles are well defined. However, one does not possess detailed information concerning the connecting paths. The interpretation of these processes is usually in terms of shortlived unstable decaying intermediaries, called resonances. Although these ideas originated more than fifty years ago, the correct mathematical description was not forthcoming until the beginning of the seventies, and the exploitation by theoretical physicists and chemists followed shortly thereafter.

The fundamental concept of an extended spectral classification, which subsequently provided the basis for the popular complex scaling method (CSM), had nevertheless been anticipated and utilized in connection with regularization routines. However, with the pioneering paper of Balslev and Combes, Commun. Math. Phys. **22**, 280 (1971), the unexpected facets of a dilation analytic view gave support to a wealth of new applications.

Since nuclear, atomic and molecular physicists, and chemists and mathematicians, all investigate various aspects of reaction paths and associated resonance phenomena, it was suggested that an interdisciplinary meeting of the **Lertorpet** type would enhance the understanding and development of respective subject areas and would lead to improved ideas and fundamental knowledge concerning the elementary processes governing our natural environment.

Returning to the quote at the beginning, one can draw a specific analogy between Darwin's geological theory and the present microscopic concept of a resonance. In the quantum mechanical formulation, the supposed relationship between motion and force is given by a generator. According to the postulates of quantum theory, this operator is self-adjoint, implying that its spectrum is always to be found on the real axis $\mathbf{R}$. CSM, on the other hand, guarantees the existence of a *generalized spectrum* that is not necessarily real. Although this seems to be contradictory, it is not! Every structure or pattern that corresponds to such a *generalized spectrum* has a projection on $\mathbf{R}$, which during a certain time interval defines the appropriate (projected) generator. In a sense one can say that the resonance is the *sinking mountain* that provides the knowledge of the dynamics of the quasi-stationary state, *the coral island.*

The present book contains the invited talks, and accompanying discussions, that took place at **Lertorpet** during a very intensive week in August, 1987. The number of participants was restricted to 35, the maximum capacity of the field station. In addition to scheduled lectures and extracurricular activities - organized sightseeing tours and study visits to local industries and municipalities - there was also ample time for spontaneous discussions, either privately or in small groups.

To aid the reader of these proceedings, we have tried to structure the arrangement of contributions according to subject area. Although the actual lectures were chronologically scheduled with a certain thought of coherence, it does not follow that they occurred in exactly the same order as here. Nevertheless, we feel that these proceedings correctly reflect the sequence of events as they took place last year in the wilderness of Värmland.

The introductory article is concerned with the notion of a rigged hilbert space. This idea gives a well-known but important extension of traditional L^2-theories. Even if this concept will not give a full characterization of the actual resonances, it will provide a proper imbedding of the theoretical models (see below) into the appropriate mathematical framework. The next two articles deal with a microlocal analysis of partial differential equations, the most modern and sophisticated mathematical theory of resonances, albeit of a physical nature and origin. Furthermore, various resonance formulations are appraised and compared. An interesting theorem on the equivalence between short range potentials and finite rank representations is outlined in the following contribution. Finally, the section on pure mathematics is concluded by one of the founding fathers of CSM; resonances with background potentials is the topic here. In an appendix the resonating group approach as well as the familiar *back-rotation problem* are given proper mathematical attention (see particularly the section on nuclear physics, where these questions are explicitly raised).

The following three articles belong, like the introductory one, to the domain of mathematical physics. The almost periodic Schrödinger equation is revisited in a very careful setting. In subsequent work, variational principles for non-self-adjoint operators are analyzed and applied to model problems. It was demonstrated for the first time that one can determine resonance states with such an accuracy that possible errors will be at most due to the numerical round-off of the computer.

Expansion theorems involving bound states, resonances, etc. are at the center of interest in the next two papers beginning the section on selected applications in nuclear physics. It is surprising to note that the pioneers in this field made their achievements well before the appearance of CSM. The particular applications presented in the ensuing work seem to corroborate the equivalence theorem earlier referred to in connection with finite rank potentials. Moreover, the remarkable accuracy reached here led, as mentioned above, to a re-examination of the *back-rotation problem*.

The transition from nuclear to atomic and molecular physics is made smoothly in these proceedings. Here a different view is advocated; instead of transforming the evolution or propagator to a study of the analytic properties of the associated resolvent, one resorts directly to an explicit time dependent representation. The interplay between classical and quantum formulations is investigated as well as the possibility of treating relevant parts of the dynamics classically. The possibility of elaborating on genuine quantum effects, when necessary, is a very attractive alternative here. Further work pertains to a semi-classical formulation which, as is clearly demonstrated, allows for systematic corrections converging to the exact result.

Returning to the view of the resonance as a Gamow wave or *sinking mountain*, the explicit boundary value problem has to be analyzed (see also appropriate papers in the nuclear physics section). Atomic and molecular applications are here presented in a multichannel version of CSM and/or optical potential models. The multichannel formulation is also presented in a potential model used to reproduce experimental data for scattering of He-atoms off a corrugated copper surface. Although not accounted for here, the somewhat controversial idea of a complex quasi-probability aroused comments. Since no satisfactory interpretation was offered as an alternative, we have decided to leave out this interesting discussion in these proceedings.

In most applications of CSM one more or less automatically assumes complex symmetric representations. Even if this is not always the case, a fundamental theorem tells us that every matrix can be brought to complex symmetric form by a similarity transformation. A very simple construction is offered here as an additional concretizing proof. The form itself is shown to appear in certain extreme situations, i.e. organized forms called coherent-dissipative structures. Finally the complex scaling method is treated as an analytic extension of conventional theories such as those of the S-matrix and Hartree-Fock, as well as properly defined scalar products, etc.

During the symposium it was generally felt that one could really benefit immensely from talking to scientists with motives and interests being fertilized in different fields. In the way science develops today, with increasing specialization, this is not at all a self-evident conclusion. It also needs to be said in this connection that the aim of interdisciplinary work is not to make mathematicians out of theoretically oriented physicists and chemists or to retrain a mathematician to be a physical chemist. Instead, we want to emphasize that the success of a program, like the present one, rests on the the premise that all who are sincerely participating should be deeply rooted in their own subject. This, of course, is easier said than done. However, the organizers of this workshop are convinced that with sufficiently many personal contacts, with the exchange of ideas, with a lack of fear of appearing *stupid* in another territory, with a willingness to appreciate *foreign* viewpoints, to listen to criticism, to debate *confusing terminology*, and to freely share one's brainchild with colleagues *on the other side of the fence*, the present enterprise would lead to something more substantial than just another statement in support of the interdisciplinary belief.

This symposium could not have been held without support from various loyal people and financial sources. First, we would like to thank all our students at that time: Christina Carlsund, who, besides running all kinds of errands also took a large number of photographs which will remind the participants of their good time at Lertorpet in Värmland; Erik Engdahl, who turned out to be an excellent potato peeler, and, together with Mikael Höghede, also helped the participants with changes in their manuscripts, delivered to us on floppy disks written in TEX; the always stable and reliable Peter Krylstedt, who resolved every possible little problem that came his way, and also helped us to the bitter end with the editing of this book.

Of course, this book would not have seen the light of day *without the extraordinary effort of Lisa Mowat.* With the help of her husband, Richard, she learnt TEX and typeset at least half of the contributions in this book. In this context we would also especially like to express our gratitude to the Manne Sigbahn Institute of Physics for its generosity in making available to us its computing facilities and for the assistance of its staff in the editorial work on this book.

We could not have survived as well and happily without our two ladies, Susanne Engström and Ann-Marie Karlstedt, who organized the kitchen and the wonderful lunches.

Behind everything, and wanting everything, to the last detail, to run smoothly, but himself to be invisible, was **Dr Allan Hellstrand**, the Managing Director of the **Albert and Maria Bergström Foundation**, to whom we say *thank you for the wonderful time you arranged for us all.*

To Prof. Ingmar Bergström, the grandson of Albert and Maria Bergström and his wife Britta, who gave us the idea to use **Lertorpet** and then took time to guide us in the land of his childhood summers, we know that your efforts are well remembered and appreciated by all of us.

Lektor Rolf Karlstedt was our local guru who seemed to know everyone in the area. Without his initiative and helpful contacts we would not have had the two nice ladies to help us, the industries to welcome, support and entertain us, and the Ferlin society to sing for us. Thank you again, Rolf.

To Mr Finn Madison, managing director of the world's largest crisp bread factory we say thank you for the lunch and the tour of the plant.

To Mr Svein Kalgraff, managing director of Lesjöfors Industrifjädrar AB, we would like to express our deepest gratitude. Not only did he take us to a pleasant countryside inn at Långban, but he also arranged a guided tour of the old iron mill *Långbanshyttan* and the beautiful surrounding countryside.

To Mr Uno Eriksson, manager of the FIMEK company, we are greatly indebted for providing the final symposium dinner. Typical Swedish Crayfish with all its accessories was served, and the non-educated part of the group was furthermore taught how to eat this crustacean. The dinner was a success and very much appreciated by the participants.

The town of Filipstad is acknowledged with thanks for its practical support with expert advice of all kinds through its different agencies. We appreciated in particular the marvellous dinner at the Kalhyttans Herrgård.

Kalhyttans Herrgård supplied us with the dinners at **LERTORPET**. You really contributed to the success of the symposium by providing excellent propaganda for Swedish *husmanskost* at its best. Many thanks from all of us.

The Swedish National Research Council (NFR) kindly supported the event in two ways. First, a considerable part of the funds which made the symposium possible came from an interdisciplinary program between Mathematics and Physics through the committee of Physics and Mathematics. Funds for this book and its editing were kindly supplied by the publishing committee of NFR.

Finally, the Marianne and Marcus Wallenberg Foundation through a grant made it possible for us to carry out our plans at a late stage when we realised that our funds were not enough to support our wild plans. Their quick and informal support is gratefully acknowledged.

Thank You All !!

Erkki Brändas

Quantum Chemistry Group
for Research in Atomic
Molecular and Solid State
Physics
Uppsala
Sweden

Nils Elander

Manne Siegbahn
Institute of Physics
Stockholm
Sweden

CONTENTS

INTRODUCTION TO RIGGED HILBERT SPACES (RHS)

Bengt Nagel

COMPARISON BETWEEN DIFFERENT NOTIONS OF RESONANCES

Bernard Helffer

RESONANCES AND SEMICLASSICAL ANALYSIS

Johannes Sjöstrand

GENERALIZATION OF MÜLLER'S VARIATIONAL PRINCIPLE

Geert-Ulrich Sölter

RESONANCE STATE EXPANSIONS IN NUCLEAR PHYSICS

Tore Berggren

MITTAG-LEFFLER EXPANSIONS IN NUCLEAR PHYSICS

ON THE ANALYTICAL CONTINUATION OF THE PARTIAL WAVE S-MATRIX USING COMPLEX SCALING TECHNIQUES

Magnus Rittby, Nils Elander and Erkki Brändas

CALCULATION OF RESONANT WAVE FUNCTIONS IN NUCLEAR PHYSICS

Borbála Gyarmati

THE USE OF GAMOW FUNCTIONS IN NUCLEAR PROBLEMS

Tamas Vertse, Patricia Curutchet and Roberto J. Liotta

RESONANCES AS AN EIGENVALUE PROBLEM

Lidia S. Ferreira

A MODEL FOR STUDYING TIME DEPENDENT QUANTUM MECHANICAL PROCESSES AND ITS APPLICATION FOR QUASI-STATIONARY STATES

János Révai

TIME-DEPENDENT DYNAMICS APPLIED TO ELECTRON TRANSFER

Erik Deumens and Yngve Öhrn

SEMICLASSICAL DESCRIPTION OF RESONANCES

H. Jürgen Korsch

REGGE POLES AND ATOM-MOLECULE DIFFRACTION

Karl-Erik Thylwe

ENERGY SHIFTS AND WIDTHS IN ATOMIC AND MOLECULAR PHYSICS: MULTICHANNEL APPROACH

Roland Lefebvre

TWO STUDY CASES IN THE CALCULATION OF RESONANCES USING THE MULTICHANNEL SCHRÖDINGER EQUATION

Mario A. Natiello and Alejandro R. Engelmann

MULTICHANNEL COMPLEX SCALED TITCHMARSH WEYL THEORY. A MODEL FOR DIATOMIC FRAGMENTATION

Erkki Brändas, Magnus Rittby and Nils Elander

ON THE WAY TO A MULTICONFIGURATION TREATMENT OF RESONANCE PHENOMENA IN ATOMS

Nils Elander, Christina Carlsund, Peter Krylstedt and Peter Winkler

COMPLEX SCALING APPLIED TO TRAPPING OF ATOMS AND MOLECULES ON SOLID SURFACES

Nimrod Moiseyev

ON A THEOREM FOR COMPLEX SYMMETRIC MATRICES AND ITS RELEVANCE IN THE STUDY OF DECAY PHENOMENA

Charles E. Reid and Erkki Brändas

CREATION OF LONG RANGE ORDER IN AMORPHOUS CONDENSED SYSTEMS

Erkki Brändas and C. Aris Chatzidimitriou-Dreismann

THE LERTORPET SYMPOSIUM VIEW ON A GENERALIZED INNER PRODUC

Edited by Nils Elander and Erkki Brändas

INTRODUCTION
TO
RIGGED HILBERT SPACES (RHS)

Bengt Nagel

Department of Theoretical Physics
Royal Institute of Technology
S-100 44
Stockholm
SWEDEN

Abstract

The purpose of this talk is to give a simple introduction to ideas related to the concept of Rigged Hilbert Space (or Gelfand Triplet). This concept has sometimes been used in defining resonances, but time does not permit us to go into this application. However, some references are given.

1. Background and Introduction

The basic Hilbert Space (HS) structure of Quantum Mechanics (QM), as worked out especially by Johann von Neumann around 1930, is based on the concepts of observable and state, defined as follows:

Observables are self-adjoint operators in a complex Hilbert space $\mathcal{H}$. To such an operator A we can associate a projection valued spectral family $\{E_\lambda : -\infty < \lambda < \infty\}$.

States (also called pure states) are unit rays (i.e. normalized states up to a phase factor) or equivalently 1-dimensional subspaces (or 1-dimensional projectors) of $\mathcal{H}$. Call a state φ.

Mixed States (also called states) are density operators (self-adjoint positive operators ρ with trace Tr $\rho = 1$).

To every pair (A,φ) (or (A,ρ)) of an observable A and a state $\varphi(\rho)$ we can associate a real-valued random variable $A_\varphi(A_\rho)$ with distribution function

$$F_{A_\varphi}(\lambda) = \text{Prob } (A_\varphi \leq \lambda) = (\varphi, E_\lambda \varphi)$$

$$\left(\text{or } F_{A_\rho}(\lambda) = Tr(\rho E_\lambda)\right)$$

A_φ with distribution function F_{A_φ} describes the outcome of a measurement of the observale A if the system is in state φ. Compatible (i.e. mutually commuting) observables (and only those) have simultaneous joint probability distributions.

This basic structure has been abstracted and generalized in various directions. We can roughly classify these extensions in two groups:

I: Extensions that don't start from the Hilbert space structure;
II: Those that do.

In group I we can mention

I 1: Algebraic Q M (Quantum Mechanics)
I 2: Convex Q M
I 3: Lattice Q M,

abstracting from the algebra of Hilbert space operators, the convex set of density operators, and the lattice of orthogonal projections, respectively.

An advantage of these approaches is that one can give, up to a certain point, a unified treatment of classical mechanics and quantum mechanics, and after that clarify the crucial structural differences between them. If one wants to apply the formalism to concrete problems, one generally specializes to the Hilbert space formulation, however.

In group II

II 1: generalizes the concept of observable, introduces generalized observables (also called measurements), with a more general family of positive operator valued measures, still enabling a probability interpretation. This formalism can e.g. describe unsharp measurements.

Our main interest is in what could be called the most conservative approach:

II 2: Rigged Hilbert Space (RHS) approach, primarily designed to handle generalized (= non normalizable) eigenvectors of ordinary Hilbert space observables. It can be said to give a rigorous foundation for Dirac's bra-ket-formalism (but without the full algorithmic power of that formalism).

The motto for the RHS formalism could be taken to be the French "Il faut reculer pour mieux sauter."

In short, a RHS (or essentially a Gelfand Triplet) is a sequence of linear spaces

ϕ	$\subset$	$\mathcal{H}$	$\subset$	ϕ'
set of "nice" states		set of normalizable states		set of generalized states

Here the scalar product, well-defined for vectors in $\mathcal{H}$ and ϕ, can be naturally extended to the pair ϕ, ϕ', so that the vectors in ϕ' can be considered to be (continuous) linear functionals on ϕ.

For the triplet above to be a Gelfand Triplet one usually requires ϕ to be equipped with a topology making it a nuclear space (we explain what this means in simple cases later; generally one can say that an infinite-dimensional nuclear space shares many nice properties with finite-dimensional linear spaces - these latter ones are trivially nuclear).

A simple example of the triplet above appears in distribution theory (or theory of generalized functions):

$\mathcal{D}$ or $\mathcal{S}$	$\subset$	$L^2(\mathbf{R}^n)$	$\subset$	$\mathcal{D}'$ or $\mathcal{S}'$
test functions		Square integrable functions of n variables		distributions

Here $\mathcal{D}$ is the set of C^∞ (infinitely differentiable) functions with compact support (i.e. vanishing outside a bounded region of $\mathbf{R}^n$), and $\mathcal{S}$ the (larger) set of C^∞ functions which, together with all derivatives, vanish faster than any inverse power of $r = \left(x_1^2 + ...x_n^2\right)^{1/2}$ as $r \to \infty$. Each of these spaces is equipped with a 'natural' topology (which we don't specify here; see however later for the topology of $\mathcal{S}$), a topology finer (i.e. demanding more for convergence of a sequence) than the Hilbert space norm topology given in $L^2(\mathbf{R}^n)$, and making $\mathcal{D}$ (or $\mathcal{S}$) a nuclear space (see later for a definition what this means in the topology defined on $\mathcal{S}$). $\mathcal{D}$ is a linear subset of $\mathcal{S}$ and $L^2(\mathbf{R}^n)$ and is dense in each of them, and even in $\mathcal{D}'$ (meaning that every element in $\mathcal{D}'$ is the limit, in the topology of $\mathcal{D}'$, of some sequence of elements in $\mathcal{D}$; an example is that the Dirac δ-function can be considered the limit of functions in $\mathcal{D}$, or (more common) of e.g. Gaussians, that is functions in $\mathcal{S}$). $\mathcal{D}'(\mathcal{S}')$ is the linear space of continuous linear (or in this construction anti-linear, see below) forms (also called functionals) on $\mathcal{D}(\mathcal{S})$, i.e. the set of distributions (tempered distributions) in $\mathbf{R}^n$.

2. Examples of Generalized Eigenfunctions

Taking the simple example of a particle moving in one dimension, we have in the ordinary configuration space representation

$$\mathcal{H} = L^2(\mathbf{R}) = \left\{ \varphi(x) : \int |\varphi(x)|^2 dx < \infty \right\}.$$

The position and momentum operators $\hat{x}$ and $\hat{p}$, defined by $(\hat{x}\varphi)(x) = x\varphi(x)$ and $(\hat{p}\varphi)(x) = -i\frac{d}{dx}\varphi(x)$, both have the full line $\mathbf{R}$ as their spectra: Sp $\hat{x}$ = Sp $\hat{p}$ = $\mathbf{R}$. The corresponding sets of generalized eigenfunctions can "obviously" be taken as $\{\delta(x - x_o) : x_o \in \mathbf{R}\}$ and $\left\{e^{ip_o x} : p_o \in \mathbf{R}\right\}$, since we have $\hat{x}\delta(x - x_o) = x\delta(x - x_o) = x_o\delta(x - x_o)$ and $\hat{p}e^{ip_o x} = p_o e^{ip_o x}$. The first relation is valid in the distribution sense: for a test function $\varphi(x)$ we have $\int \varphi(x)x\delta(x - x_o)dx = \varphi(x_o)x_o = \int \varphi(x)x_o\delta(x-x_o)dx$. The second relation is valid as a function relation, but both are equations "outside Hilbert space $\mathcal{H}$."

We cannot give sense (pointwise, for given x_o or p_o) to '$\int \varphi(x)\delta(x - x_o)dx$' or '$\int \varphi(x)e^{ip_o x}dx$' for <u>all</u> functions $\varphi \in L_2(\mathbf{R})$, but for "nice" functions (test functions) we can.

Since we would like to have a domain of nice functions on which we can apply $\hat{x}$ and $\hat{p}$ freely an arbitrary number of times (i.e. apply an arbitrary linear differential operator with polynomial coefficients) without getting out of the set of nice functions (which should furthermore be a subset of $L^2(\mathbf{R})$) a natural choice is $\mathcal{S}$, the set of C^∞, "rapidly decreasing" functions. $\mathcal{D}$ is also possible, but since $\mathcal{S}$ is larger than $\mathcal{D}$, the dual $\mathcal{S}'$ of $\mathcal{S}$ is smaller than $\mathcal{D}'$, the dual of $\mathcal{D}$, which is sometimes an advantage (see later). Besides $\mathcal{S}$ has the advantage of treating $\hat{x}$ and $\hat{p}$ on an equal footing: the Fourier transform, interchanging $\hat{x}$ and $\hat{p}$, is an isomorphism of $\mathcal{S}$ (i.e. a bi-continuous one-to-one mapping of $\mathcal{S}$ onto itself), whereas the Fourier transform maps $\mathcal{D}$ onto another function space.

$\hat{x}$ and $\hat{p}$ are both continuous linear operators in $\mathcal{S}$; observe that when defined in $L^2(\mathbf{R})$ they are not continuous, but unbounded operators defined only on a dense subset.

It is known that every (continuous) antilinear form on $\mathcal{H}$ (antilinear means $L(a_1\varphi_1 + a_2\varphi_2) = a_1^* L(\varphi_1) + a_2^* L(\varphi_2)$) can be represented by a unique vector φ_L in $\mathcal{H} : L(\varphi) = (\varphi, \varphi_L)$. The linear set of these antilinear forms can hence be identified with $\mathcal{H}$ (i.e. $\mathcal{H}$ is self-anti-dual).

Restricting φ to the smaller set $\mathcal{S}$ with a finer topology, the set of possible φ_L:s is increased to be $\mathcal{S}'$, the set of tempered distributions. The operators $\hat{x}$ and $\hat{p}$, continuous on $\mathcal{S}$ and selfadjoint there (i.e. $(\varphi, \hat{x}\psi) = (\hat{x}\varphi, \psi)$ for $\varphi, \psi \in \mathcal{S}$) can be extended "by duality" (or more precisely anti-duality) to continuous operators defined in all $\mathcal{S}'$:

$$(\hat{x}\varphi, \varphi') = (\varphi, \hat{x}'\varphi'), (\hat{p}\varphi, \varphi') = (\varphi, \hat{p}'\varphi'), \varphi \in \mathcal{S}, \varphi' \in \mathcal{S}'.$$

Evidently both $\delta(x - x_o)$ and $e^{ip_o x}$ belong to $\mathcal{S}'$, and the "formal" eigenvalue equations they satisfy are mathematically well-defined equations in $\mathcal{S}'$.

Hence the operators $\hat{x}$ and $\hat{p}$, selfadjoint in $\mathcal{H}$, and continuous in $\mathcal{S}$, have a complete set of eigenfunctions in $\mathcal{S}'$, proper solutions of the eigenvalue equations, considered as equations for the extended operators $\hat{x}'$ and $\hat{p}'$ acting as continuous operators in $\mathcal{S}'$.

3. Definition of Gelfand Triplet

Generalizing the example above, let us call Rigged Hilbert Space (RHS) a triplet

$$\phi \subset \mathcal{H} \subset \phi'$$

where ϕ is a dense linear subset of $\mathcal{H}$ with a finer topology than $\mathcal{H}$, and ϕ' (containing ϕ and $\mathcal{H}$ as dense linear subspaces) is the set of antilinear forms on ϕ'. If - as is generally desirable - the topology on ϕ makes it a nuclear space, the RHS is said to form a Gelfand Triplet. See Section 4 for an explanation of what "nuclear" means for a space of a type similar to $\mathcal{S}$.

Assume given a linear operator A, selfadjoint (but in general unbounded, hence not defined on all of $\mathcal{H}$), and such that, restricted to ϕ, A is a continuous operator in ϕ. Then A can be extended by duality to an operator A$'$, defined and continuous in ϕ', using

$$(A\varphi, \varphi') = (\varphi, A'\varphi'), \text{all } \varphi \in \phi, \varphi' \in \phi'.$$

A$'$ can also be considered to be the closure of A (in the operator topology defined in ϕ') from ϕ to ϕ', meaning: if $\{\varphi_n\}$ is a sequence in ϕ converging to φ' in ϕ' - such sequences always exist since ϕ is dense in ϕ' - then $A\varphi_n \to A'\varphi'$.

The generalized spectrum GenSp$_\phi$ A of A is then the set of possible eigenvalues (with multiplicity) of the eigenvalue equation $A'\varphi'_\lambda = \lambda\varphi'_\lambda$ in ϕ'. $\{\varphi'_\lambda\}$ is the set of generalized eigenfunctions.

Obvious questions arising are:

1. What is the connection between Sp A and GenSp$_\phi$ A? We would of course like to have GenSp$_\phi$ A $\supset$ Sp A and maybe even equality.

2. If $\text{GenSp}_\phi \text{A} \supset \text{Sp A}$, can we expand (in some sense) all elements in $\mathcal{H}$, or at least in ϕ, in generalized eigenfunctions?

For $\hat{x}$ we have $\text{GenSp}_\phi \hat{x} = \mathbf{R} = \text{Sp}\ \hat{x}$ with generalized eigenfunctins $\delta(x - x_o)$ belonging to both $\mathcal{D}'$ and $\mathcal{S}'$, and for φ in $\mathcal{D}$ or $\mathcal{S}$ we can write (trivially)

$$\varphi(x) = \int \varphi(x_o)\delta(x - x_o)dx_o$$

For $\hat{p}$ $\text{GenSp}_\mathcal{S}\hat{p} = \mathbf{R} = Sp\ \hat{p}$, with generalized eigenfunctions $\left\{e^{ip_o x} : p_o \in \mathbf{R}\right\}$ and the expansion $\varphi(x) = \int \hat{\varphi}(p_o)e^{ip_o x}dp_o$. But in $\mathcal{D}'$ $e^{ip_o x}$ makes sense also for all complex p_o, so $\text{GenSp}_\mathcal{D}\ \hat{p} = \mathbf{C}$; the expansion of course only involves eigenfunctions connected with $\text{Sp}\hat{p} = \mathbf{R}$.

Another example is the harmonic oscillator Hamiltonian $H = \hat{p}^2/2 + \hat{x}^2/2$, which can be seen to have $\text{GenSp}_\mathcal{S}$ H = Sp H = $\{n + 1/2 : n = 0, 1, 2, ..\}$ but $\text{GenSp}_\mathcal{D}$ H = 2 x C, since the equation $H\varphi(x) = \lambda\varphi(x)$ has for every complex λ two linearly independent solutions, both in general behaving as exp $(x^2/2)$ for large x.

This situation with redundant generalized eigenvalues is in general not desired, hence $\mathcal{S}$ is a better space to use for the "nice" functions than $\mathcal{D}$ in the cases above. But in some situations complex generalized eigenvalues are wanted or forced on us. An example of this enforcement appears when one wants to derive unitary representations for non-compact Lie groups (as e.g. SU(1,1) or, essentially equivalent, SO(2,1)) working with the commutation relations of the infinitesimal generators. Although these generators are selfadjoint operators and hence have a real spectrum, working without restriction with the commutation relations, well-defined in ϕ', we are forced to accept non-real generalized eigenvalues for some of the generators (see Reference 1 for SU(1,1)). Also when using the RHS concept to define resonances as non-real eigenvalues of the Hamiltonian a similar situation should appear.

4. Countably-Hilbert Spaces

The harmonic oscillator Hamiltonian can be used to define the subspace $\mathcal{S}$ of $L^2(\mathbf{R})$ and its "natural" topology by a construction giving a somewhat extended version of a Gelfand Triplet, called a Countably-Hilbert Space, at the same time enabling us to define the concept "nuclear" in this particular case. We generalize somewhat and assume

A is a selfadjoint positive operator in $\mathcal{H}$, bounded below by the identity operator, A $\geq$ I (for convenience). Define, for n = 1,2,... Hilbert spaces (linear subsets of $\mathcal{H}$) $\mathcal{H}_n = \mathcal{D}(A^n)$ (we assume A to be unbounded above, so that the domains of definition of $A^n, \mathcal{D}(A^n)$, form linear subsets of $\mathcal{H}$ which get successively smaller as n increases). The scalar product in $\mathcal{H}_n$ is defined as $(\varphi, \psi)_n = (A^n\varphi, A^n\psi)$.

Similarly, define for positive n Hilbert spaces $\mathcal{H}_{-n}$ which are the completions of $\mathcal{H}$ with respect to the norm $\|\varphi\|_{-n} = \|A^{-n}\varphi\|$. It is not hard to see that $\mathcal{H}_{-n}$ is the antidual of $\mathcal{H}_n$ (under the scalar product of $\mathcal{H}$).

$$\text{Define } \phi = \overset{\infty}{\cap} \mathcal{H}_n \quad \text{(mathematically called projective limit)}$$

$$\text{and } \phi' = \overset{\infty}{\cup} \mathcal{H}_{-n} \quad \text{(inductive limit).}$$

We then get a Countably-Hilbert Space

$$\phi \subset \mathcal{H}_n \subset \mathcal{H} \subset \mathcal{H}_{-n} \subset \phi'.$$

The topology of ϕ is defined with the sequence of norms $\{\|\cdot\|_n : n = 0, 1, 2, ...\}$. ϕ', of course, turns out to be its antidual. One should observe that all spaces involved except the extremes ϕ and ϕ' are Hilbert spaces. In fact, ϕ will be what is called in topology a Fréchet space, which has a lot of nice properties.

As a simple example of this construction, we can choose A $=$ $H_{osc} + 1/2$, with $\{e_\nu(x)\}$ the ON basis of normalized Hermite functions in $\mathcal{H} = L^2(\mathbf{R}) : Ae_\nu = (\nu+1)e_\nu, \ \nu = 0, 1, ...$

Then we can characterize the various spaces in the corresponding Countably-Hilbert space by their expansions in e_ν:

$$\begin{aligned}
\mathcal{H} &= \{\varphi = \sum a_\nu e_\nu : \sum |a_\nu|^2 < \infty\} \\
\mathcal{H}_n &= \{\varphi = \sum a_\nu e_\nu : \sum (\nu+1)^{2n} |a_\nu|^2 < \infty\}, \quad n > 0 \text{ or } n < 0 \\
\phi &= \{\varphi = \sum a_\nu e_\nu : |\nu^M a_\nu| \le C_M, \text{ all } M\} \\
\phi' &= \{\varphi' \sim \sum a'_\nu e_\nu : |a'_\nu| \le C\nu^M, \quad \text{for some } M = M(\varphi')\}
\end{aligned}$$

It is not hard to see that if $\varphi = \sum a_\nu e_\nu \in \mathcal{H}_n$ (or ϕ) and $\varphi' \sim \sum a'_\nu e_\nu \in \mathcal{H}_{-n}$ (or ϕ'), then the "scalar product" $(\varphi, \varphi') = \sum a^*_\nu a'_\nu$ is finite and well-defined.

Projective and inductive limits imply: an element in ϕ lies in every $\mathcal{H}_n$ and an element in ϕ' lies in all $\mathcal{H}_{-n}$ from some n onwards, is hence an antilinear form on $\mathcal{H}_n$ (and of course on all smaller Hilbert spaces).

5. Nuclear Space (special case)

If we take A $=$ $H_{osc} + 1/2$ it turns out that the resulting projective limit, with the topology defined above, is just the space $\mathcal{S}$ with its standard topology. In fact the resulting definition of the topology of S with the sequence of Hilbert space norms $\|\cdot\|_n$ is very convenient. Using the fact that the Fourier transform of a Hermite function is up to a phase factor the same Hermite function implies directly that the Fourier transform is an isomorphism of ϕ : the basis $\{e_\nu\}$ diagonalizes the Fourier transformation.

Looking at the inverse A^{-1} it is evidently (as operator in $\mathcal{H}$) a bounded selfadjoint operator with Sp $A^{-1} = \{1/(n+1)\}$. This means that the sum of the squared eigenvalues is convergent, so that A^{-1} is a Hilbert-Schmidt operator.

This means that the ON basis $\{A^{-n}e_\nu\}$ in $\mathcal{H}_n$ (ON basis since $(A^{-n}e_\mu, A^{-n}e_\nu)_n = (e_\mu, e_\nu) = \delta_{\mu\nu}$), considered as elements in the larger Hilbert space $\mathcal{H}_{n-1}$ has a finite norm-squared sum

$$\sum_\nu (A^{-n}e_\nu, A^{-n}e_\nu)_{n-1} = \sum_\nu (A^{-1}e_\nu, A^{-1}e_\nu) = \sum_\nu 1/(v+1)^2 < \infty.$$

In other words, the unit sphere in $\mathcal{H}_n$ is mapped by the natural embedding onto an ellipsoid with a finite sum of the squares of the half-axis in $\mathcal{H}_{n-1}$. One then says that this map $\mathcal{H}_n \to \mathcal{H}_{n-1}$ is a Hilbert-Schmidt map.

Generally, in the Countably-Hilbert case defined above, the resulting ϕ is a <u>nuclear space</u> if and only if for every m there is an $n > m$ such that the embedding $\mathcal{H}_n \to \mathcal{H}_m$ is a Hilbert-Schmidt map.

Expressed in terms of the operator A used above this means that Sp $A = \{\lambda_\nu\}$ must be discrete and $\lambda_\nu \to \infty$ fast enough so that $\sum \lambda_\nu^{-2M} < \infty$ for some M.

In a qualitative sense: the topology should get rapidly finer as we move along to larger n in the sequence $\mathcal{H}_n$.

Nuclear Spaces have nice properties. Of special interest is the

6. Nuclear Spectral Theorem

We can state this theorem in a form somewhat more general than the case of a Countably-Hilbert space.

Assume given a quintuplet of spaces

$$\phi \subset \mathcal{H}_{(1)} \subset \mathcal{H} \subset \mathcal{H}'_{(1)} \subset \phi'$$

where ϕ is nuclear, continuously embedded in $\mathcal{H}_{(1)}$, and the embedding of $\mathcal{H}_{(1)}$ in $\mathcal{H}$ is Hilbert-Schmidt (we can take $\mathcal{H}_{(1)} = \mathcal{H}_M$ with notations as above).

Assume B is a selfadjoint operator in $\mathcal{H}$, continuous in ϕ. Extend B to B′, continuous operator in ϕ'.

The Nuclear Spectral Theorem then states:

$\mathcal{H}'_{(1)}$ contains a complete set of generalized eigenvectors of B, i.e. a set $\{e'(\lambda, i_\lambda) : \lambda \in Sp\ B, i_\lambda = 1, ... n_\lambda \leq \infty$, multiplicity of $\lambda\}$, such that

1) $B'e'(\lambda, i_\lambda) = \lambda e'(\lambda, i_\lambda), \{e'(\lambda, i_\lambda)\}_1^n$ linearly independent.

2) For some positive measure μ on Sp B and any pair x, y $\in \mathcal{H}_{(1)}$ and spectral projector E(Δ) of B, we have the completeness relation

$$(x, E(\Delta)y) = \int_\Delta \sum_{i_\lambda=1}^{n_\lambda} (x, e'(\lambda, i_\lambda)) \cdot (y, e'(\lambda, i_\lambda))^* d\mu(\lambda).$$

This then implies a similar relation for an arbitrary function of B:

$$(x, f(B)y) = \int_{\text{Sp B}} f(\ ) \cdot \sum (...) d\mu(\lambda).$$

There are essentially two different ways of deriving the Nuclear Spectral Theorem:

1) differentiation of the spectral measure of B (Gelfand, Kostiuchenko, Berezanskij and others); see Reference 5.

2) starting from von Neumann's direct integral form of the spectral theorem for B (Gårding, Maurin and others); see Reference 6.

7. References to Applications to Resonances and Virtual States

Here I can only refer to References 2-4 below, which contain also references to earlier work in this area, especially contributions by Böhm. Reference 4 is a short conference summary.

References

1. G. Lindblad and B. Nagel, Ann. Inst. H. Poincaré, **13**, 27 (1970).

2. M. Gadella, J. Math. Phys., **24**, 1462 (1983).

3. M. Gadella, J. Math. Phys., **24**, 2142 (1983).

4. M. Gadella, Physica, **124A**, 317 (1984).

5. Yu. M. Berezanskij, *Expansions in Eigenfunction of Self-adjoint Operators. Translations of Mathematical Monographs* , Ch. V., Am. Math. Soc., Providence, Rhode Island (1968).

6. K. Maurin, *General Eigenfunction Expansions and Unitary Representations of Topological Groups*, Ch. II., Polish Scient. Publ., Warsaw (1968).

COMPARISON BETWEEN DIFFERENT NOTIONS OF RESONANCES

Bernard Helffer

Department of Mathematics
University of Nantes
2, rue de la Houssinière
F-44072 Nantes Cédex 03
FRANCE

Abstract

We compare different definitions of resonances given by different mathematicians or physicists, in particular these of Hunziker and Helffer-Sjöstrand and we describe recent results by Hunziker and by Helffer-Martinez permitting to identify the various definitions.

1. Introduction

In the last years, the shape resonances have been studied in the semi-classical context by different methods. For a first "wave" of results, we will start from the basic papers by Aguilar and Combes [1] and Balslev and Combes [3]. We shall then refer to the survey of Simon [14] where it is possible to find a first discussion about the different notions of resonances. The second "wave" corresponds in the semi-classical context to the contributions by Helffer and Sjöstrand [9]; Combes, Duclos, Klein, and Seiler [5]; Hislop and Sigal [10]; and in the general context by contributions of Sigal [13], Cycon [6], and quite recently by W. Hunziker [11] whose paper was our motivation. In each of these papers a notion of resonances is introduced adapted to a particular problem. It was then natural to prove that for a given problem where different techniques can be used the notions of resonances are the same. For this purpose, the paper by Hunziker was quite enlightening for the analysis of the different techniques of analytic distortions. Our contribution is to compare the definition of Hunziker with the definition of Helffer and Sjöstrand [9] in the case of the Schrödinger operator

$$P(h) = -h^2\Delta + V$$

where $V \in C^\infty(\mathbf{R}^n, \mathbf{R})$ and admits a holomorphic extension in some domain D of the form

$$D = \{x \in C^n; |\Re x| \geq M, |\Im x| \leq \varepsilon_0 |\Re x|\} \tag{1}$$

where it satisfies: $\lim_{\substack{|x| \to \infty \\ x \in D}} V(x) = 0$

In this context, different notions of resonances have been introduced near a level $E_0 > 0$ (as h tends to 0) and the result is that they all coincide. In this lecture we will just sketch the general ideas and refer you to the proof in the recent paper written in collaboration with Martinez [8] for the details.

2. A Brief Review On The Analytic Distortion Or Dilation Techniques

We essentially recall here the analysis by Hunziker. The techniques introduced in References 1,3,10,6, and 11 can all be put in the following general procedure used to define the resonance near some real $E_0 > 0$.

We start from a family U_λ of unitary operators defined for λ real, $\lambda \in [-\lambda_0, +\lambda_0]$ with $\lambda_0 > 0$ and with $U_0 = I$. This family is chosen such that the family $(U_\lambda P U_\lambda^{-1})$ admits a natural analytic extension as an analytic family $P(\lambda)$ of non-bounded operators for $\Im\lambda \geq 0$ s.t. The resolvent $(P(\lambda) - z)^{-1}$ is always well defined for $\Im z > 0$, and $|\lambda| < \lambda_0$, $\Im\lambda \geq 0$, but extends when $\Im\lambda > 0$ as a meromorphic function in some domain $G(\lambda)$ containing E_0. The resonances are then the poles of the resolvent $R(\lambda, z) \equiv (P(\lambda) - z)^{-1}$ for some value of λ . This definition does not show clearly that on $G(\lambda) \cap G(\lambda')$, poles of $R(\lambda, z)$ and $R(\lambda', z)$ are the same. In this approach, there is always an identification of this notion of resonance with a second one we shall now explain.

Let $\mathcal{A}$ be a space of functions s.t.

$$\forall \Psi \in \mathcal{A} \quad \lambda \to U_\lambda \Psi \quad \text{is analytic in } |\lambda| < \lambda_0 \quad \text{with value in } L^2 \tag{2}$$

$$U_\lambda \mathcal{A} \text{ is dense in } L^2, \ \forall \lambda \text{ with } |\lambda| < \lambda_o \tag{3}$$

Then for all $u_1, u_2 \in \mathcal{A}$ and $z \in C$ with $\Im z > 0$ we have

$$z \to \left((P - z)^{-1} u_1 | u_2\right)_{L_2} = \left((P(\lambda) - z)^{-1} U_\lambda u_1 | U_{\bar\lambda} u_2\right)$$

for all λ with $\Im\lambda \geq 0$.

This is in fact true for λ real and then follows for other λ by analytic extension (with the help of (2)).

This permits a definition of the meromorphic extension of $z \to \big((P-z)^{-1}u_1|u_2\big)$ in $\bigcup_{\substack{|\lambda|<\lambda_0 \\ \Im\lambda \geq 0}} G(\lambda)$ and then a resonance becomes a pole of the meromorphic extension for some $u_1, u_2 \in \mathcal{A}$.

The first and second definitions can be identified by using (3). Then we get:

From the second definition

The set of resonances is independent of U_λ in the sense that once we have a way to define the meromorphic extension for $u_1, u_2 \in \mathcal{A}$ this extension is unique. In particular, it is independent of λ.

From the first definition

We get also that for some U_λ the definition does not depend on the choice of $\mathcal{A}$ s.t. (2) and (3) are satisfied.

Then to identify two definitions of resonances related to U_λ, $\tilde{U}_\lambda$ it is sufficient to find a set $\mathcal{B}$ satisfying to (2) and (3) simultaneously for U_λ and $\tilde{U}_\lambda$. Let us consider the examples considered in the literature.

Case a. The analytic dilation (or complex scaling) [1,3]

You start from

$$\theta \to U_0^\theta \text{ with } U_0^\theta f = e^{n\theta/2} f(e^\theta x)$$

(Here you can consider that λ is the parameter $(e^\theta - 1)$). $\mathcal{A}$ can be taken as the set of the holomorphic vectors for the group U_0^θ for $|\Im\theta| > \alpha_0$.

Case b The exterior scaling

This technique was introduced by B. Simon [14] and is used in Reference 5.

Let us introduce : $|x| = r$.

$$f_\theta(r) = \begin{cases} 1 \text{ for} & 0 \leq r \leq R \\ r^{-1}\big(R + e^\theta(r-R)\big); & \text{R} \leq r < \infty \end{cases}$$

for some $R > 0$.

And let us consider the dilation

$$x \to \phi_\theta(x) = x f_\theta(r)$$

and

$$(U_1^\theta u)(x) = J(\theta)^{1/2} u[\phi_\theta(x)]$$

where $J(\theta)$ is the Jacobian of ϕ_θ.

Then under condition (1), and for R big enough, you can define $P_\theta = U_1^\theta P (U_1^\theta)^{-1}$ (for θ real) and extend it holomorphically for (Im θ) small enough.

Note here that the domain of P_θ depends on θ (you introduce boundary conditions on $|x| = R$); so we do not have an analytic family of type A in the sense of Kato. But what is important is to have a good analytic extension for $(P_\theta - z)^{-1}$ as an operator in $\mathcal{L}(L^2)$ (see Appendix 2 in Reference 5). As in case a, $\mathcal{A}$ can be taken as the set of the holomorphic vectors for U_1^θ for $|\Im\theta| < \alpha_1$.

Case c. The "regularized" exterior scaling (Hislop and Sigal [10])

We have seen that the exterior scaling introduces some "singularity" at $|x| = R$. To avoid this problem, Hislop and Sigal consider the following notion of distortion.

They consider ϕ_θ as the flow associated with the following radial vector fields:

$$F(x) = g(r)\frac{x}{r} (r = |x|)$$

where g is a C^k-function ($k \geq 2$) s.t.

(α) g is non decreasing and has bounded derivatives

(β) $$\lim_{r \to \infty} g(r)(r - R)^{-1} = 1$$

(γ) $$g(r) = \begin{cases} 0 & 0 < \mathrm{r} \leq R \\ \tilde{g}(r) & \mathrm{r} \geq R \end{cases}$$

and $\tilde{g}$ is the restriction to $r > R$ of an analytic function defined in a sector S_0 with vertex at $R - \delta_0 (\delta_0 > 0)$, and angle $\theta_0 > 0$ satisfying the following technical conditions at ∞ in S_0 :

$$|\tilde{g}(z)| \leq [|x-R|+|y|][1+O(|x-R|^{-2})]$$

$$\text{for } x > R, \quad z = (x+iy) \in S_0$$

$$\left(\text{example } \ \tilde{g}(z) = (z-R)\left(1-e^{-(z-R)^2}\right)\right.$$

U^θ is defined in the same way as before. The progress is here that the family you get by extending holomorphically $U^\theta P(U^\theta)^{-1}$ is now an analytic family of Type A in the sense of Kato of fixed domain $H^2(R^n)$. The authors then take $\{\mathcal{A} = \{f \in L^2(\mathbf{R}^n) \text{ s.t. } \hat{f} \in C_0^\infty(\mathbf{R}^n)\}$

Case d. A "linearized" analytic distortion [6,11]

In the first 3 examples $\theta \to U^\theta$ was a group and the parameter λ was $(e^\theta - 1)$. Hunziker introduces the following simple class of distortions.

Let $v \in C^\infty(\mathbf{R}^n; \mathbf{R}^n)$ verifying :

$$|v(x)-v(y)| \leq |x-y|$$

$$v(x) = x \ \text{ for } |x| \geq 3R$$

$$v(x) = 0 \quad |x| \leq 2R$$

U_λ is then defined by

$$(U_\lambda u)(x) = J(x,\lambda)^{1/2} \quad u\,(x+\lambda v(x))$$

where $J(x,\lambda)$ is the Jacobian of the diffeomorphism $x \to x+\lambda v(x)$. As $\mathcal{A}$, Hunziker takes the space

$$\mathcal{A} = \{f \text{ entire on } C^n \text{ and } \forall \varepsilon > 0,\ \forall N > 0 \qquad \lim_{\substack{|x|\to\infty \\ |\Im x| \leq (1-\varepsilon)|\Re x|}} (\Re x)^N f(x) = 0\}$$

One can then see that if we take

$$\mathcal{A}' = \{f \text{ entire in } C^n, \text{ s.t. there exists } \varepsilon_2, \varepsilon_3 > 0, \qquad \lim_{\substack{|x|\to\infty \\ |\Im x| \leq \varepsilon_3 |\Re x|}} \exp(\varepsilon_2 x^2) f(x) = 0\}$$

this space satisfies (2) and (3) for the different U^θ. We then see that all definitions of resonances for the operators introduced in section 1 coincide in some ball $B(E_0, \delta)$ in C for δ small enough and $h \in [O, h_0]$, h_0 small enough.

3. Comparison to the definition of resonances by B. Helffer and J. Sjöstrand [9]

It is not possible to recall the definition of the resonances in this lecture which is sketched in the talk by J. Sjöstrand in these proceedings. Let us just establish briefly the properties we need from Reference 9 to identify this definition with the other definitions from Section 2. In Reference 9, a family of Hilbert spaces ($t \in [-t_0, t_0], k \in \mathbf{Z}$). $H^k(\Lambda_t)$ is constructed with the following properties. (In the lecture of Sjöstrand $H^k(\Lambda_t)$ is denoted by $H(\Lambda_{tG}, r^k)$.)

$$H^k(\Lambda_0) = H^k(\mathbf{R}^n) \tag{4}$$

$$C_0^\infty(\mathbf{R}^n) \text{ is dense in } H^k(\Lambda_t)$$

$$(\psi, \Psi) \to \int \psi(x)\bar{\Psi}(x)dx \equiv (\psi|\Psi) \tag{6}$$

extends naturally as a continuous map from $H^k(\Lambda_t) \times H^{-k}(\Lambda_{-t})$ into C and gives a natural duality between $H^k(\Lambda_t)$ and $H^{-k}(\Lambda_{-t})$.

$$H^{k+1}(\Lambda_t) \subset H^k(\Lambda_t) \tag{7}$$

This family of spaces is constructed s.t. for some $\varepsilon_0 > 0$. For each $t > 0$ there exists some domain $D_t = [E_0 - \varepsilon_0, E_0 + \varepsilon_0] + i[-\varepsilon_0 t, \varepsilon_0]$ s.t. for $z \in D_t$ and $h \in]0, h_0(t)]$, the operator $(P_t - z)$ defined on $C_0^\infty(\mathbf{R}^n)$ by $(P(h) - z)$, extends continuously as a Fredholm operator from $H^0(\Lambda_t)$ into $H^{-2}(\Lambda_t)$ with index 0.

Then the resonances $z(h)$ appear to be the values of $z \in D_t$ s.t. $(P_t - z)$ is not bijective for some t and some $h \leq h(t)$.

It is then not too difficult to establish that $z(h)$ is a resonance if there exists ψ and $\Psi \in C_0^\infty(\mathbf{R}^n)$ s.t.

$$\left((P_t - z)^{-1}\psi|\Psi\right)_{H^0(\Lambda_t)\times H^0(\Lambda_{-t})}$$

has a pole as a meromorphic function.

As in Section 2, we need to see that the resonances don't depend on t, which is obtained by proving that for $\Im z > \frac{\varepsilon_0}{2}$.

$$\left((P_t - z)^{-1}\psi|\Psi\right)_{H^0(\Lambda_t)\times H^0(\Lambda_{-t})}$$

is in fact independent of t and then equal to $\left((P - z)^{-1}\psi|\Psi\right)_{L^2\times L^2}$

This fact, which replaces the argument of analyticity with respect to θ, in the preceding section is a consequence of the following regularity theorem in Reference 9.

$$u \in H^0(\Lambda_t), (P - z)u \in \cap_{|s|<s_0} H^{-2}(\Lambda_s) \Rightarrow u \in \cap_{|s|<s_0} H^0(\Lambda_s)$$

It is now not too difficult to compare the definitions of Sections 2 and 3. It is sufficient to remark that the space $\mathcal{A}'$ introduced in Section 2 is in fact in $H^k(\Lambda_t)$, $\forall k \in Z$ and that $\mathcal{A}'$ is dense in $H^k(\Lambda_t)$ which is relatively easy to prove (see Reference 8 with our choice of $H^k(\Lambda_t)$).

In fact, once we have proved this fact, the resonances coincide because the meromorphic extension of $\left((P - z)^{-1}\psi|\Psi\right)_{L^2\times L^2}$ is unique where it exists.

4. Final remarks

1 - In the definition of Reference 9, a resonant state is an element $u \in H^0(\Lambda_t)$ s.t. $(*)(P - z)^k u = 0$ for some $k \in N$, some $z \in D_t$.

We give, in Reference 8, the following link between this definition and definitions resulting from the other approachs. Under the hypothesis on P (see (1)), u admits, because it is a solution of the partial differential equation (*), an holomorphic extension in a domain s.t. we can define $(U_\lambda u)$ quite explicitly.

Then it is possible to prove that for $z \in D_t \cap G(\lambda)$, $u \in H^0(\Lambda_t)$ is a resonant state if and only if $(U_\lambda u)$ is in L^2 and a generalized eigenvector of $P_\lambda = U_\lambda P U_\lambda^{-1}$.

2 - In the case where $n = 1$, and when V is with compact support, this enables us to identify the preceding definitions with the definition used for example in Ashbaugh-Harrell [2]

E is a resonance if $\exists u \in C^\infty$ s.t.

$$Pu = Eu$$

$$u = \begin{cases} \alpha e^{i\sqrt{E}x/h} & \text{for x big enough} \\ \beta e^{-i\sqrt{E}x/h} & \text{for -x big enough} \end{cases}$$

3 - It has been a problem to define the C-product $\int_{R^n} u^2 dx$ for a resonant state (or Gamow state) in $H^0(\Lambda_t)$. (See other lectures in this conference, or Reference 4). In the definition of Reference 9, we can take the definition:

$$\int_{R^n} u^2 dx \equiv (u|\bar{u})_{H^0(\Lambda_t)\times H^0(\Lambda_{-t})}$$

It is sufficient to observe that, in the case we consider here, $u \to \bar{u}$ is continuous from $H^0(\Lambda_t)$ onto $H^0(\Lambda_{-t})$. It is also possible to verify that $e^{-\varepsilon x^2} u \underset{\varepsilon\to 0}{\longrightarrow} u$ in $H^0(\Lambda_t)$. Then we get

$$(u|\bar{u})_{H^0(\Lambda_t)\times H^0(\Lambda_{-t})} = \lim_{\varepsilon\to 0} \int e^{-2\varepsilon x^2} u(x)^2 dx$$

References

1. J. Aguilar and J. M. Combes, Comm. Math. Phys., **22**, 269 (1971).

2. M. Ashbaugh and E. M. Harrell, Comm. Math. Phys., **83**, 151 (1982).

3. E. Balslev and J. M. Combes, Comm. Math. Phys., **22**, 280 (1971).

4. E. Brändas, *The Method of Complex Scaling*, Int. J. Quant. Chem., **S20**, 119 (1986).

5. J. M. Combes, P. Duclos, M. Klein and R. Seiler, Comm. Math. Phys., **110**, (1987).

6. H. Cycon, Helv. Phys. Acta, **58**, 969 (1985).

7. E. M. Harrell and B. Simon, Duke Math. J., **47**, (1980).

8. B. Helffer and A. Martinez, to appear in Helv. Phys. Acta.

9. B. Helffer and J. Sjöstrand, Mémoires de la S.M.F., **114**, (1986).

10. P. Hislop and I. M. Sigal, *Shape resonances in quantum mechanics*, Proceedings of the International Conferences on Differential Equations and Math. Phys., Birmingham (Alabama), March 1986, Preprint.

11. W. Hunziker, Ann. de l'I.H.P., (Section Physique Théorique), **45**, 339 (1986).

12. M. Reed and B. Simon, *Methods of modern mathematical physics*, Vol. IV, Academic Press, New-York (1978).

13. I. M. Sigal, Ann. I.H.P., **41**, 103 (1984).

14. B. Simon, Ann. Math., **97**, 247 (1973); Int. J. Quant. Chem., **14**, 529 (1978); Phys. Lett., **71A**, 211 (1979).

15. J. Sjöstrand, These Proceedings, E. Brändas and N. Elander, Eds., Lecture Notes in Phys., Springer-Verlag, Berlin (1988).

RESONANCES AND SEMICLASSICAL ANALYSIS

Johannes Sjöstrand

Department of Mathematics
Université de Paris Sud
and CNRS
F-91405 Orsay
FRANCE

Abstract

Recent work on resonances by the author and collaborators is reviewed. Rigorous mathematical results are obtained on the asymptotic behaviour of resonances as Planck's constant tends to zero. Some works together with Gérard are presented for the first time. The problem of a potential well in an island is treated with the general theory presented. Some results on the estimates of the number of resonances are finally described

1. Introduction

This is a survey of some fairly recent works by Helffer, Gérard, and the author. We have tried to obtain rigorous mathematical results on the asymptotic behaviour of resonances, as Planck's constant tends to zero, by means of modern microlocal analysis. The basic theory, developed with Helffer [10], is intuitively rather natural but technically more elaborate than the classical approaches of complex scaling [1,2,13]. On the other hand, our definition often applies where the classical definitions don't; in the sense that our potentials are allowed to be quite large near infinity. Helffer and Martinez [9] have showed, also using a work of Hunziker [13], essentially that the definition of Helffer and the author coincides with any other definition in cases when this other definition also applies. The main advantage of our approach is, however, that it is resolutely microlocal (i.e. based on some phase space analysis) from the beginning. This is of great help in all concrete applications that we have encountered so far. This has permitted us to obtain results in situations where traditional methods seem unsuitable (or at least would require an a posteriori addition of microlocal analysis) [7,8,23]; as well as more precise and more complete results in other situations. Comparing for instance, Reference 10 with Reference 4 and the later work of Hislop and Sigal [12], it seems clear that the actual asymptotic expansion of the imaginary part of the shape resonance and more generally its bounds from below, require a deeper understanding of the microlocal nature of this problem. Unfortunately, in this lecture we cannot go into the mathematical details of the various results. We can, however, refer to several other surveys by the author (mostly addressed to mathematicians), References 24-26, which, on some points, are more detailed. The work with Gérard [8] is described here for the first time.

In Section 2, we describe the general theory of Reference 10 with application to the problem of a potential well in an island. In Section 3 we describe two other simple cases, where it is possible to determine the complete asymptotics of resonances. Finally, in Section 4 we describe some results on estimates on the number of resonances and on bounds from below on the imaginary part of resonances. These results are (hopefully) only a first step in the study of some fascinating problems linking resonances to the behavior of the corresponding dynamical systems.

2. A General Theory

We are interested in spectral theory near the energy level 0 for the Schrödinger operator on $\mathbf{R}^n$; $P = -h^2\Delta + V(x)$, when h tends to 0. Let $p(x,\xi) = \xi^2 + V(x)$ be the associated symbol. Here we use the notation $\xi = (\xi_1, .., \xi_n)$, $\xi^2 = \xi_1^2 + .. + \xi_n^2$. In order to define resonances near 0, we make the following assumptions:

(1) There exist smooth real-valued functions r, R on $\mathbf{R}^n$ such that $r \geq 1$, $rR \geq 1$, $\partial^\alpha r = \mathcal{O}\left(rR^{-|\alpha|}\right)$, $\partial^\alpha R = \mathcal{O}\left(R^{1-|\alpha|}\right)$, uniformly on $\mathbf{R}^n$ for every multi-index $\alpha = (\alpha, ..., \alpha_n) \in \mathbf{N}^n$.
Here $\mathbf{N} = \{0,1,2,...\}$, $|\alpha| = \alpha_1 + ... + \alpha_n$ and $\partial^\alpha r$ denotes the partial derivative $(\partial/\partial x_1)^{\alpha_1} ... (\partial/\partial x_n)^{\alpha_n} r$.

(2) V is a smooth real-valued function, which extends to a holomorphic function on $\{x \in \mathbf{C}^n; |\Im x| \leq C^{-1} R(\Re x), |\Re(x)| \geq C\}$. Here $C > 0$ is some constant.

(3) There exists a smooth real-valued 'escape' function $G(x,\xi)$ on $\mathbf{R}^{2n}$ with $\partial_x^\alpha \partial_\xi^\beta G = \mathcal{O}(\mathrm{r}^{1-|\beta|} R^{1-|\alpha|})$ for all multi-indices α and β with $|\alpha| + |\beta| \geq 1$, such that $H_p G \geq r^2/C$ outside a bounded set in $p^{-1}(0) = \{(x,\xi) \in \mathbf{R}^{2n}; p(x,\xi) = 0\}$. Here $C > 0$ is some constant and $\mathrm{r}(x,\xi) = (r(x)^2 + \xi^2)^{1/2}$. Finally H_p denotes the Hamilton vector field; $\sum \partial_{\xi_j} p \partial_{x_j} - \partial_{x_j} p \partial_{\xi_j}$.

After a suitable modification of G in the region where $|\xi| >> r(x)$, we can define certain weighted function spaces $H(\Lambda_{tG}, m)$, when $t > 0$ and $h > 0$ are sufficiently small, see Reference 10 for details. Very roughly, a function $u(x)$ belongs to $H(\Lambda_{tG}, m)$ if and only if $B_t u(x,\xi) m(x,\xi) e^{-tG(x,\xi)/h}$ is square integrable on $\mathbf{R}^{2n}$. Here B_t is a suitable Bargmann type transformation adapted to the scale functions r, R, and $m(x,\xi)$ is a certain weight function which typically could be 1 or a power of $\mathbf{r}$. Thus microlocally, the elements of $H(\Lambda_{tG}, m)$ are allowed to be very large where $G > 0$, but are required to be very small when $G < 0$. Since the function G is increasing along the classical trajectories (i.e. the integral curves of H_p),we see that it is very natural to try to define resonances for P as eigenvalues of P, acting as a non selfadjoint operator on these spaces. In Reference 10, we obtained the following basic result:

Theorem 2.1. *For $t > 0$ sufficiently small, there exists $h_0 > 0$ and a neighborhood $\Omega \subset \mathbf{C}$ of 0, such that for $0 < h < h_0$: For all $z \in \Omega$, the operator $P - z : H(\Lambda_{tG}, \mathbf{r}^2) \to H(\Lambda_{tG}, 1)$ is a Fredholm operator of index 0. Moreover, there is a set $\Gamma(h) \subset \Omega$ consisting only of isolated points, such that $P - z$ is bijective for $z \in \Omega\backslash\Gamma(h)$ and such that $P - z$ splits in a natural way into a direct sum of one bijective operator and one nilpotent operator $F_z \to F_z$, when $z \in \Gamma(h)$. Here $F_z \subset H(\Lambda_{tG}, \mathbf{r}^2) \subset H(\Lambda_{tG}, 1)$ is a non-trivial finite- dimensional vector space (and 'nilpotent' means that $(P-z)^{\dim(F_z)}$ vanishes on F_z).*

The elements of $\Gamma(h)$ are, by definition, the resonances of P in Ω. If z is such a resonance, then $\dim(F_z)$ is, by definition, the corresponding algebraic multiplicity. In Reference 10, we showed that a different choice of $t > 0$ or of G gives rise to the same resonances if Ω is sufficiently small , and the corresponding spaces F_z are the same. We also showed that the resonances belong to the closed lower halfplane. Our theory is intrinsic in the sense that it does not make use of any particular self-adjoint realization of P. (See also Helffer-Martinez[9] for the equivalence with more classical definitions of resonances.)

Let us next discuss the case of a potential well in an island, which was also studied in Reference 10. Let $U \subset \ddot{O} \subset \mathbf{R}^n$, where U is a compact set, $\ddot{O}$ open and connected. We suppose that V is analytic everywhere; and, in addition to the general hypothesis above, we assume

(4) $$V \leq 0 \text{ on } U, \ > 0 \text{ on } \ddot{O}\backslash U, < 0 \text{ on} \mathbf{R}^n\backslash\ddot{O}\ .$$

(5) There are no trapped classical trajectories in $p^{-1}(0)$ over the complement of $\ddot{O}$. (See below for a precise definition).

(6) For every $\varepsilon > 0$, there is a $C_0(\varepsilon) > 0$ such that $V(x) \geq C_0(\varepsilon)^{-1} r(x)^2$, on $\ddot{O}\backslash \cup_{y \in \partial\ddot{O} \cup U} B(y, \varepsilon R(y))$, where $B(y, \varepsilon R)$ denotes the open ball of center y and radius εR.

Here '∂' means 'boundary of'.

Let d denote the Agmon distance on $\ddot{O}$ associated to the metric $\max(V(x), 0) dx^2$, and put $B_d(U, r) = \{x \in \ddot{O}; d(x, U) < r\}$. We assume that

(7) the d-diameter of U is 0.

Let $S_0 = d(U, \partial\ddot{O})$. One can then show that the closure of $B_d(U, S_0)$ is compact. Let $\eta > 0$ be small and independent of h, and let M_0 be the closure of $B_d(U, S_0 - \eta)$. Let P_0 be the Dirichlet realization of P in $L^2(M_0)$. Let $I(h)$ be a compact interval tending to $\{0\}$ when h tends to 0, and let $a(h) > 0$ be a function which tends to 0 with h, such that $a(h) \geq C(\varepsilon)^{-1}e^{-\varepsilon/h}$ for every $\varepsilon > 0$. We then suppose that

$$\sigma(P_0) \cap ((I(h) + [-2a(h), 2a(h)]) \backslash I(h)) = \phi \tag{8}$$

where $\sigma(P_0)$ denotes the spectrum of P_0. Let $b(h) \geq a(h)$ be a second function, which tends to 0 with h and let $\Omega(h) \subset \mathbf{C}$ be the rectangle with vertices at sup $(I(h)) + a(h) \pm ib(h)$ and inf $(I(h)) - a(h) \pm ib(h)$. Then, in Reference 10, we showed,

Theorem 2.2. *For h sufficiently small, there is a bijection b: $\sigma(P_0) \cap I(h) \to \gamma(h) \cap \Omega(h)$, where the elements in the two sets are counted with their natural multiplicity, such that $b(\mu) - \mu = \mathcal{O}\left(e^{-(2S_0 - \varepsilon(\eta))/h}\right)$. Here $\varepsilon(\eta) \to 0$, $h \to 0$.*

For related results, also see References 4 and 12. Theorem 2.2 (and the results of References 4 and 12) states that the 'shape' resonances generated by a potential well in an island are exponentially close to certain eigenvalues of a selfadjoint operator, but they do not give any minorations or asymptotic expansions of $-\Im z$, when z is a resonance. Such results require, apparently, more precise information about the resonant states outside the island Ö, that the microlocal techniques of Reference 10 enabled us to obtain. Let us assume that we study resonances generated by low eigenvalues of P_0. More precisely, we shall assume that

(9) $U = \{x_0\}$ and x_0 is a non-degenerate local minimum of V.

It was shown in References 21 and 11, that the eigenvalues of P_0 in every interval $[0, Ch]$ have the asymptotic expansion:

$$hE(h) = h\left(\sum_{0 \leq j < \infty} E_j h^j\right), \tag{10}$$

where, in general, we have to sum over all half integers. Here E_0 is an eigenvalue of the natural (non-isotropic) harmonic oscillator, that one obtains by quadratic approximation of P_0 at x_0. Let us now concentrate (for simplicity) on an eigenvalue μ of P_0 with the asymptotic expansion (10), where E_0 is a simple eigenvalue of the non-isotropic harmonic oscillator. Then μ is simple for small h and we can apply the theorem above with $I(h) = \{\mu\}$ and $a(h) = C^{-1}h$, if $C > 0$ is large enough, so we have precisely one resonance, $z(h)$; with $|z(h) - \mu| \leq C^{-1}h$, which is in fact exponentially close to μ. It is then possible to obtain the precise asymptotic behavior of the corresponding resonant state in certain crucial regions outside $\ddot{\mathrm{O}}$. Formally this is done by WKB-constructions, involving also Airy-integrals near the boundary. The mathematical justification of the fact that the exact resonant state coincides with the WKB-one, requires delicate estimates, some of which are microlocal. Using this, a simple Green's formula, and some arguments of Martinez [16], we then obtained in Reference 10

Theorem 2.3. *Under the assumptions above, there is a constant $C > 0$, such that for $h > 0$ sufficiently small, we have*

$$C^{-1}h^{1/2}e^{-2S_0/h} \leq -\Im(z(h)) \leq Ch^{1-n/2-2m}e^{-2S_0/h} \tag{11}$$

where $2m$ is the degree of the non-isotropic hermite polynomial associated with E_0. Under an additional non-degeneracy assumption about the minimal Agmon geodesics, joining U to $\partial\ddot{\mathrm{O}}$, we also have

$$-\Im(z(h)) = h^{1/2-2m}f(h)e^{-2S_0/h}, f(h) \sim f_0 + f_1 h + f_2 h^2 + ..., f_0 > 0 \tag{12}$$

Gérard and Martinez[6] have studied the spectral density of P when V is a long range potential. Some of their results concern the case when the assumptions above are also satisfied and, in the situation of Theorem 2.3, they have established mathematically the existence of a sharp Breit-Wigner peak.

3. Results When The Classical Dynamics Is Simple

We keep the general assumptions (1)-(3) and we shall also from now on make the assumption that V is analytic everywhere on $\mathbf{R}^n$. We choose some sufficiently small $\varepsilon_0 > 0$. For $\varrho \in p^{-1}([-\varepsilon_0, \varepsilon_0])$, let $t \to \phi_t(\varrho) = \exp(tH_p)(\varrho)$ denote the classical trajectory through ϱ, defined for t in the maximal interval of definition $]T_-(\varrho), T_+(\varrho)[$,
$T_\pm \in [0, \pm\infty]$. We then have the outgoing (+) and incoming (-) tails, defined by:

$$\Gamma_+ = \{\varrho \in p^{-1}([-\varepsilon_0, \varepsilon_0])\,;\ \phi_t(\varrho) \text{does not tend to } \infty, \text{ when } t \text{ tends to } T_-(\varrho)\},$$

$$\Gamma_- = \{\varrho \in p^{-1}([-\varepsilon_0, \varepsilon_0])\,;\ \phi_t(\varrho) \text{does not tend to } \infty, \text{ when } t \text{ tends to } T_+(\varrho)\},$$

We then have the following properties (see appendix in Reference 7) closely related and inspired from the 'geometric scattering theory,' see Reed-Simon [20].

1. For some T_0, we have $\Gamma_+ \subset \{G \geq -T_0\}$, $\Gamma_- \subset \{G \leq T_0\}$,
2. The union of trapped trajectories, $K = \Gamma_+ \cap \Gamma_-$ is a bounded set,
3. If Γ_- or Γ_+ is non-empty, then so is K,
4. If we define the true tails $\mathbf{T}_\pm$ to be the complement of K in $\Gamma_\pm$, then the volume of $T_\pm$ vanishes.

We also introduce $\Gamma_{\pm,0} = \Gamma_\pm \cap p^{-1}(0), K_0 = K \cap p^{-1}(0)$. The properties 1-3 then remain valid if we replace $K, \Gamma_\pm$, by $K_0, \Gamma_{\pm,0}$. We have the following unwritten theorem of Reference 10 (See also Klein [15] for a similar result.)

Theorem 3.1. *If K_0 is empty, then there is a neighborhood W of 0, independent of h, such that for $h > 0$ sufficiently small, there are no resonances in W.*

We notice that K_0 is the union of trapped trajectories in the energy surface $p^{-1}(0)$, and the last theorem tells us essentially that, in order to have something interesting to study, we must assume that such trapped trajectories exist. In this section we give results about the easiest cases, namely when K_0 is a closed trajectory and when K_0 is a point.

To start with the first case, we shall follow Reference 7. We then assume:

(13) K_0 is the image of a simple closed trajectory $\gamma_0(t) = \phi_t(\varrho_0)$ of minimal period $T_0 > 0$.

Let $H \subset p^{-1}(0)$ be a hypersurface which intersects γ_0 transversally at ϱ_0. We then have the Poincaré map $H \to H$, defined by following the classical flow of H_p from a point of H until it hits H again at a time approximately equal to T_0. ϱ_0 is then a fixed point of the Poincaré map, and we denote by p_0 the differential at ϱ_0. We assume,

(14) p_0 is of hyperbolic type.

By this we mean that ϱ_0 is a hyperbolic fixed point for the Poincaré map, or in other words, that p_0 has no eigenvalues of modulus 1. It is easy to see that the whole situation remains unchanged, if we replace $p^{-1}(0)$ by $p^{-1}(\varepsilon)$, for $|\varepsilon| \leq \varepsilon_0$, if $\varepsilon_0 > 0$ is small enough. Let then $\gamma_\varepsilon, T_\varepsilon, \varrho_\varepsilon, p_\varepsilon$, be corresponding quantities, and let $\theta_1(\varepsilon), .., \theta_{n-1}(\varepsilon), 1/\theta_1(\varepsilon), .., 1/\theta_{n-1}(\varepsilon)$ be the eigenvalues of p_ε with $|\theta_j(\varepsilon)| > 1$. If we introduce the action $C(\varepsilon) = \int_{\gamma_\varepsilon} \xi dx$, then $C'(\varepsilon) = T(\varepsilon)$.

Using an easy WKB construction, Gérard and the author defined [7] a certain analytic function $\varrho(\varepsilon)$, such that $|\varrho(\varepsilon)| = |\theta_1(\varepsilon)..\theta_{n-1}(\varepsilon)|^{-1/2}$. Then we have [7] :

Theorem 3.2. *Let $\varepsilon_0 > 0$ be sufficiently small and fix a $\delta > 0$. Let $C_0 > 0$ and define $\Gamma_0(h) = \{E \in [-\varepsilon_0, \varepsilon_0] - i[0, C_0 h]$; there exist $\alpha \in \mathbf{N}^{n-1}$ and an integer k such that $C(E) = 2\pi kh + ih \log(\varrho(E)) - ih \sum \alpha_j \log(\theta_j(E))\}$. Then if we count the elements of each set with their natural multiplicities, there is, for $h > 0$ sufficiently small, an injective map $b(h) : \Gamma_0(h) \to \{$resonances of $P\}$, such that $b(h)(\mu) - \mu = o(h)$, uniformly in h and μ. The image of $b(h)$ contains all resonances in the slightly smaller rectangle $[-\varepsilon_0 + \delta h, \varepsilon_0 - \delta h] - i[0, (C_0 - \delta)h]$.*

This result is the natural analogue of a similar result of Gérard [5] on accoustical resonances generated by two strictly convex obstacles, improving earlier work of Ikawa.

The assumptions of the theorem are satisfied typically when V has a saddle point and we fix the energy level 0 to be slightly above the value of V at the saddle point. This situation has been studied by quantum chemists such as Pollak [19]. Notice that our theorem gives rise to several pole strings; one for each value of α. There is also a more refined result which determines the resonances, not only up to an error $o(h)$ but up to an error which is $\mathcal{O}(h^N)$ for every $N > 0$. (See Reference 7 for more details.)

We next turn to the case when K_0 is reduced to a point, so instead of (13),(14), we assume

$$K_0 = \{(x_0, \xi_0)\}. \tag{15}$$

Since H_p has to vanish at that point, we have $\xi_0 = 0$, and after a translation, we may also assume that $x_0 = 0$. We also have $\nabla V(0) = 0$, so 0 is a critical point with critical value $V(0) = 0$. We shall assume that this point is non-degenerate,

$$\det V''(0) \neq 0, \tag{16}$$

where $V''(0)$ is the matrix of all second order derivatives at the point 0.

After a linear change of the x-coordinates, we may assume that

$$p(x,\xi) = \sum_{j=1,..,n-d} \lambda_j(\xi_j^2 + x_j^2)/2 + \sum_{j=n-d,..,n} \lambda_j(\xi_j^2 - x_j^2)/2 + \mathcal{O}\left((x,\xi)^3\right), \tag{17}$$

near (0,0), and the eigenvalues of the linearization of H_p at (0,0) are then $\pm z_j, j = 1,..,n$, where $z_j = i\lambda_j$, for $j \leq n-d$, and $z_j = \lambda_j$ for $j > n-d$. In this case it turns out that $\Gamma_{\pm,0}$ is an analytic d-dimensional manifold passing through the origin tangentially to the eigendirections of the linearization of H_p, associated with the eigenvalues $\pm\lambda_{n-d+1},..,\pm\lambda_n$. In Reference 22, we proved:

Theorem 3.3. *Choose $C_0 > 0$ such that no value of the form,*

$$-ih\sum(\alpha_j + 1/2)z_j, \tag{18}$$

with $\alpha \in \mathbf{N}^n$, is on the boundary of the disc $D(0, C_0h)$ in $\mathbf{C}$ with center 0 and radius C_0h. Let $\Gamma_0(h)$ be the set of such values inside the disc. We count the elements with their natural multiplicity. Then for sufficiently small h, there is a bijection b(h) from $\Gamma_0(h)$ onto the set of resonances of P inside $D(0, C_0h)$, such that $b(h)(\mu) - \mu = o(h)$, uniformly with respect to μ and h.

See also Briet, Combes, Duclos [3], for essentially the same result in the case when 0 is a non-degenerate maximum for the potential.

Again, there is a more refined result which determines the resonances up to any power of h. Using the same methods, it is possible to perturb a suitable rotation invariant (radial) potential on $\mathbf{R}^2$, such that for the perturbed (non-radial) potential, there exists a resonance which is a second order pole for the (analytically extended) resolvent. The existence of such resonances seems to have been a general open problem.

4. Estimates On The Resonances In More General Situations

Throughout this section, we keep the general assumptions (1)-(3) and add the assumption that V is analytic on $\mathbf{R}^n$.

We start by describing a result with Gérard [8] which gives resonance- free domains near the real axis. Take ε_0 (in Section 2) sufficiently small and recall the definition of K and K_0. We assume:

(19) In an open neighborhood Ω of K in $\mathbf{R}^{2n}$, there exists a closed symplectic submanifold Σ of regularity class C^1, which contains K_0 and has the property that H_p is tangent to Σ at every point of Σ. Moreover, we assume that there exist two subvectorbundles $N_\pm$ of $T_\Sigma(\mathbf{R}^{2n})$ of regularity class C^0, with the following properties,

(a) $T_\varrho\Sigma^\sigma = N_{+,\varrho} \oplus N_{-,\varrho}$, at every point ϱ of Σ.

(b) $N_\pm|_{K_0}$ are invariant under the maps $d(\phi_t)$ (i.e. the differentials of the maps ϕ_t).

(c) If we choose norms on $N_{\pm,\varrho}$ depending continuously on ϱ, then there exist $C > 1$ and $T > 0$ such that $\|d(\phi_{\pm T})(\nu_\pm)\| \geq C\|\nu_\pm\|$, for all $\nu_\pm \in N_{\pm,\varrho}, \varrho \in K_0$.

Here $T_\varrho\Sigma^\sigma$ denotes the orthogonal space of $T_\varrho\Sigma$ with respect to the standard symplectic form $\sigma(x,\xi;y,\eta) = \xi y - \eta x$, and '$\Sigma$ is symplectic' means that $T_\varrho\Sigma^\sigma$ and $T_\varrho\Sigma$ are transversal at every point $\varrho \in \Sigma$. The property (c) means that the flow is exponentially expanding in the directions of N_+ and exponentially contracting in the directions of N_-, a characteristic feature for hyperbolic dynamical systems. Next, we fix a non-vanishing constant density on each $N_{+,\varrho}$, depending continuously on ϱ. For $\varrho \in K_0$, $t \in \mathbf{R}$, we can then define $D_+(\varrho,t) = \det\left(d\phi_t|_{N_{+,\varrho}}\right)$, and using (19), it is easy to see that there are constants $C > 0$, $d > 0$, such that $D_+(\varrho,t) \geq C^{-1}\exp(dt)$, for all $\varrho \in K$, $t \geq 0$. Let $d_0 > 0$ be the supremum of the set of all d with this property (for some constant C depending on d). It is easy to check that d_0 does not depend on the choice of density on the fibers of N_+ and d_0, is in fact, closely related to certain Lyapunov exponents for the linearized flow. In Reference 8, we proved,

Theorem 4.1. *For every $\delta > 0$, there exists $\varepsilon > 0$ such that P has no resonances in $[-\varepsilon,\varepsilon] - i[0,(1-\delta)d_0h/2]$ when $0 \leq h \leq \varepsilon$.*

It is easy to check that the choice of d_0 is optimal in the cases when Theorem 3.2 or Theorem 3.3 apply. Our result applies also in more general situations; for instance when K_0 is a closed trajectory, which is not necessarily of hyperbolic type, or when K_0 is a point $(x_0, 0)$, where x_0 is a general critical point. Nevertheless we do admit that the assumption about the existence of an invariant symplectic manifold is very restrictive, and it would be of great interest to have mathematical (or numerical) results in cases when this assumption is not satisfied. It would also be extremely interesting to know to what extent resonances appear below the rectangle in the theorem.

We finally describe some estimates on the number of resonances that may appear in certain regions near the real axis [23]. We drop the assumption (19), but keep the assumptions (1)-(3), We notice that if $\varepsilon_0 > 0$ is sufficiently small, then the conclusion (3) is also valid outside a bounded set in $p^{-1}([-\varepsilon_0,\varepsilon_0])$. We shall need some control on G but less regularity also inside a bounded set; for that purpose we introduce the following variant of the assumption (3):

(20) There exist a real-valued function G on $\mathbf{R}^{2n}$ and an open bounded set W in $p^{-1}([-\varepsilon_0,\varepsilon_0])$, and a constant $C > 0$ such that:

(a) G satisfies all the assumptions of (3) outside W and $H_pG \geq r^2/C$ in $p^{-1}([-\varepsilon_0,\varepsilon_0]) \setminus W$.

(b) G and H_pG belong to $C^{1,1}$ in a neighborhood of the closure of W and $H_pG \geq C^{-1}\|\nabla G\|^2$ in $W \cap p^{-1}([-\varepsilon_0,\varepsilon_0])$.

Here we recall that $C^{1,1}$ is the space of continuously differentiable functions with a Lipschitz gradient. Actually the assumption (20) is equivalent to (3), when

(1) and (2) are fulfilled, as assumptions on P; but our main result below will be formulated in terms of a particular function G, which satisfies (20)

For $0 \leq \delta \leq r \leq 1$, $\delta \leq C_O^{-1}$, let $M(r,\delta)$ be the number of resonances of P in the disc $D(ir, r+\delta)$ of center ir and radius $r+\delta$. Put,

$$V(r,\delta) = \text{Volume of } \{\varrho \in \mathbf{R}^{2n}; p^2 + rH_pG < r\delta\}\ . \tag{21}$$

Then our main result is:

Theorem 4.2. *Assume (1)-(3) and let G satisfy assumption (20). Let* $C_0, C > 0$ *be sufficiently large constants. Then for* $C_0h \leq \delta \leq r \leq 1$, $\delta \leq C_0^{-1}$, $r\delta \geq C_0h$, *we have*

$$M(r,\delta) \leq CV(r, C\delta)h^{-n}. \tag{22}$$

In the proof of this theorem, we work in the Hilbert space $H(\Lambda_{tG}, 1)$ and, it turns out, that some classical inequalities of Weyl can be applied which compare eigenvalues of $P - ir$ with those of the self-adjoint operator $((P-ir)^*(P-ir))^{1/2}$. The work is then,

(a) to estimate the lowest eigenvalue of $(P-ir)^*(P-ir)$ from below; and,

(b) to estimate the number of eigenvalues of $(P-ir)^*(P-ir)$, which are smaller than $(r+\delta)^2$.

Using the techniques found in Reference 10, combined with other microlocal arguments developed in connection with the theory of the propagation of analytic singularities for linear partial differential equations, this becomes rather easy.

Depending on the behaviour of the classical flow, one wants of course to choose G so that V becomes as small as possible,. In cases when the set of classically trapped points is not a manifold, it seems essential that we can allow G to be only of class $C^{1,1}$ (see Proposition 4.5 below). Under the general assumptions (1)-(3), it is always possible to find a smooth function G as in (20) (see the appendix in Reference 7); but the set where $H_p(G) = 0$ may be of non-zero volume. We then get the following result, which can be viewed as a semiclassical analogue of a result of Melrose [18], later extended by Intissar [14].

Corollary 4.3. *Under the general assumptions (1)-(3), there are constants* $\delta_0 > 0$ *and* $C > 0$, *such that,*

$$M(1,\delta_0) \leq Ch^{-n}. \tag{23}$$

If $\delta \geq (C_0 h)^{1/2}$ then from Theorem 4.2 we get, with a new constant C:

$$M(\delta,\delta) \leq C\mathbf{V}(C\delta)h^{-n}, \tag{24}$$

where,

$$\mathbf{V}(\delta) = \text{ Volume of } \{\varrho \in \mathbf{R}^{2n}; |p| \leq \delta, H_pG \leq \delta\}. \tag{25}$$

Theorem 4.2 remains uniformly valid if we replace P by $P-\varepsilon, |\varepsilon| \leq \varepsilon_0/2$. If I is a subinterval of $[-\varepsilon_0/2, \varepsilon_0/2]$, and $C_0 h \leq \delta \leq 1/C_0$, we can cover the rectangle $I - i[0,\delta]$ by closed discs of center $\varepsilon_j + 2i\delta$ and radius 4δ, where $\varepsilon_{j+1} - \varepsilon_j = \delta 2\sqrt{3}$. If $M(I,\delta)$ denotes the number of resonances in the rectangle $I - i[0,\delta]$, we obtain quite easily from Theorem 4.2:

Corollary 4.4. *Assume (1),(2),(20) and let C_0, $C > 0$ be sufficiently large. Then for $C_0 h \leq \delta \leq 1/C_0$ and all intervals I as above, we have*

(26) $M(I,\delta) \leq C$ (Volume of $R(I + [-C\delta^*, C\delta^*], C\delta))\, h^{-n}$, *where* $R(J,\varepsilon) = \{\varrho; p(\varrho) \in J, H_pG \leq \varepsilon\}$, *and* $\delta^* = \max\left(\delta, (C_0 h)^{1/2}\right)$.

In the case when the assumptions of Theorem 4.1 are satisfied, we can find a C^2-function G satisfying the condition (4.2) such that $H_pG > 0$ outside Σ and $H_pG(\varrho) \geq \left(\text{dist}\,(\varrho,\Sigma)\right)^2/C$ in a neighborhood of Σ. This implies that we get fairly explicit estimates on the number of resonances in all such cases.

In cases when the Hamilton flow is a more general dynamical system of hyperbolic type (for instance with a Cantor set structure on K, and $\Gamma_\pm$), the following result seems applicable:

Proposition 4.5. *Assume (1)-(3) and that there is a neighborhood Ω of K in $p^{-1}([-\varepsilon_0,\varepsilon_0])$ such that,*

(27) dist (ϱ, K) is uniformly of the same order of magnitude as dist (ϱ,Γ_+) + $\text{dist}(\varrho,\Gamma_-)$, *for all $\varrho \in \Omega$,*

(28) dist $(\phi_{\pm t}(\varrho),\Gamma_\pm) \leq Ce^{-t/C}\text{dist}(\varrho,\Gamma_\pm)$ if $\phi_{\pm s}(\varrho) \in \Omega$ *for all $0 \leq s \leq t$ Then (possibly after decreasing ε_0 slightly), there exist G and W as in (20), such that* $G = \mathcal{O}(\text{dist}(.,K)^2)$, *and H_pG is uniformly of the same order of magnitude as dist$(.,K)^2$ in $W \cap p^{-1}([-\varepsilon_0,\varepsilon_0])$.*

This result, together with Corollary 4.4, gives estimates on the number of resonances depending on a suitable Hausdorff type of dimension of K. In general, we cannot hope for better regularity of G than $C^{1,1}$, and we suspect very strongly that the potential $V = \Re(x_1 + ix_2)^3 - 1$ gives an example for this.

References

1. J. Aguilar and J. M. Combes, Comm. Math. Phys., **22**, 269 (1971).
2. E. Balslev and J. M. Combes, Comm. Math. Phys., **22**, 280 (1971).
3. Briet, J. M. Combes and P. Duclos, *On the location of resonances in the semiclassical limit II*, Preprint Marseille CPT, **86**, 1884 (1987).
4. J. M. Combes, P. Duclos, M. Klein and R. Seiler, Comm. Math. Phys., **110**, 280 (1987).
5. C. Gérard, *Asymptotique des pôles de la matrice de scattering pour deux obstacles strictement convexes*, Preprint.
6. C. Gérard and A. Martinez, *Semiclassical asymptotics for the spectral function of long range Schrödinger operators*, Preprint.
7. C. Gérard and J. Sjöstrand, Comm. Math. Phys., **108**, 391 (1987).
8. C. Gérard and J. Sjöstrand, *Resonances en limite semiclassique et exposants de Lyapunov*, Preprint.
9. B. Helffer and A. Martinez, *Comparaison entre les diverses notions de resonances*, Preprint.
10. B. Helffer and J. Sjöstrand, Bull. de la S.M.F., **24,25**, suppl. du tome 114/ Fasc. 3.
11. B. Helffer and J. Sjöstrand, Comm. Part. Diff. Eqs., **9**, 337 (1984).
12. P. Hislop and I. Sigal, Preprint.
13. W. Hunziker, Ann. de l'I.H.P., (Section Physique Théorique), **45**, 339 (1986).
14. A. Intissar, Comm. Part. Diff. Eqs., **11**, 367 (1986).
15. M. Klein, Comm. Math. Phys., **106**, 485 (1986).
16. A. Martinez, J. Math. Pures et Appl., **66**, 195 (1987).
17. A. Martinez, *Estimations de l'effet tunnel pour le double puits II, etats hautement exités*, Preprint.
18. R. Melrose, Proceedings of the Journées des E.D.P de St Jean de Monts, June 1984, Ecole Polytechnique and Universités de Nantes et de Rennes.
19. E. Pollak, J. Chem. Phys., **76**, 5843 (1982).

20. M. Reed and B. Simon, *Methods of modern mathematical physics*, Acad. Press, New York (1974-1982).

21. B. Simon, Ann. de l'I.H.P., **38**, 295 (1983).

22. J. Sjöstrand, *Semiclassical resonances generated by a non-degenerate critical point.*, Lecture Notes in Math., **1256**, Springer-Verlag, (1987) p. 402-429.

23. J. Sjöstrand, work in preparation, see also Reference 26.

24. J. Sjöstrand, *Effet tunnel pour l'opérateur de Schrödinger semiclassique II, Résonances*, Proceedings of the Nato Meeting on microlocal analysis at 'Il Ciocco', Sept. 1985, Reidel Publ. Co.

25. J. Sjöstrand, *Semiclassical resonances in some simple cases.*, Exposé aux Journées des EDP de Saint Jean de Monts, June 1986, Ecole Polytechnique, Universités de Nantes et de Rennes.

26. J. Sjöstrand, *Estimates on the number of resonances for semi-classical Schrödinger operators*, Proceedings of the 8:th Latin-American School of Mathematics, To appear in Lecture Notes in Math., Springer-Verlag, Berlin (1986).

RESONANCES WITH A BACKGROUND POTENTIAL

Erik Balslev

Matematisk Institut
Aarhus Universitet
Ny Munkegade. Bygning 530
DK-8000 AARHUS C
DENMARK

Abstract

We consider a pair of operators $H_1 = -\Delta + V$ and $H_2 = H_1 + W$, where the "background potential" V is S_b-dilation analytic, $V = O(|x|^{-2-\epsilon})$ and $W = O(e^{-2a|x|})$ for $|x| \to \infty$. The resolvent $R_1(z) = (H_1 - z^2)^{-1}$ is shown to have an analytic continuation $\tilde{R}_1(z)$ from $\mathbf{C}^+$ to the region $S_{a,b} = \{z| \ |\arg z| < b, \ -a < \Im z < 0\}$ as an operator from the exponentially weighted space $L_{2,a}(\mathbf{R}^3)$ to $L_{2,-a}(\mathbf{R}^3)$. Resonances of (H_1, H_2) in $S_{a,b}$ are then defined as poles of $(1 + W\tilde{R}_1(z))^{-1}$ and identified with poles of the S-matrix of (H_1, H_2).

1. Introduction

Resonances with a background potential $V(r) = O(r^{-1-\epsilon})$ or $V(r) = cr^{-1}$ were introduced in [4,5]. In [6] these results were generalized to non-radial potentials $V(x), x \in \mathbf{R}^3$, such that $|V(x)| < c|x|^{-2-\epsilon}$ for large $|x|$. The basic idea is to obtain analytic continuations $\tilde{\Psi}(k,\omega,x)$ to a domain $\mathcal{O} \subset \mathcal{C}^-$ of the generalized eigenfunctions $\Psi(k,\omega,x)$ of $H = -\Delta + V$, such that $|\tilde{\Psi}(k,\omega,x)| < ce^{b(x)}$ for $|\Im k| < b$. From $\tilde{\Psi}(k,\omega,x)$ an analytic continuation of the resolvent $R(k) = (-\Delta + V - k^2)^{-1}$ from $\mathcal{C}^+$ to $\mathbf{O}$ is constructed as a bounded operator $\tilde{R}(k)$ from $\mathcal{H}_a$ to $\mathcal{H}_{-a}$ where $\mathcal{H}_{\mp a} = \{f = e^{\pm a|x|}g | g \in L_2\}$. This makes it possible to study resonances obtained by adding a potential $W = 0\left(e^{-2a|x|}\right)$ to V. By the analytic Fredholm Theorem, $\left(1 + W\tilde{R}(k)\right)^{-1}$ is meromorphic on 0. The poles are called resonances and are proved to be identical with the poles of the scattering matrix of $(H, H + W)$.

In the present paper we give a brief exposition and sketch a proof of these results, emphasizing the analogy with the case of the free resolvent $R_0(k) = (-\Delta - k^2)^{-1}$ as background for the potential W.

The results have been extended in [11] to short-range potentials, i.e. $|V(x)| < c|x|^{-1-\varepsilon}$. In the Appendix it is shown, in the case of a radial dilation-analytic potential $V(r)$, that the Gamow function is obtained from the square-integrable eigenfunction found through complex dilation by simply replacing the argument r by $re^{-i\varphi}$ where φ is the dilation angle (backwards rotation). The question was raised by B. Gyarmati in connection with work presented at this conference [10].

2. Exponentially decaying perturbations

Let H_0 be the selfadjoint operator in $\mathcal{H} = L_2(\mathbf{R}^3)$ defined by $H_0 f = -\Delta f$ for $f \in \mathcal{D}_{H_0} = H^2(\mathbf{R}^3)$ (the Sobolev space of order 2).

The generalized eigenfunctions $\phi(k,\omega,x)$ are defined for $k \in \mathbf{R}, \omega \in S^2,\ x \in \mathbf{R}^3$ by $\phi(k,\omega,x) = e^{ik\omega\cdot x}$. The functions $\phi(k,\omega,x)$ satisfy the equation

$$(-\Delta - k^2)\phi(k,\omega,x) = 0$$

and they yield a representation of H_0 as multiplication by k^2 via the Fourier-Plancherel transform $\mathcal{F}$ defined for $f \in L_2(\mathbf{R}^3)$ by

$$(\mathcal{F}f)(k,\omega) = (2\pi)^{-\frac{3}{2}} \lim_{R\to\infty} \int_{|x|<R} F(x)\phi(k,-\omega,x)dx \tag{1}$$

where lim is the limit in $L_2(\mathbf{R}^3)$.

The resolvent $R_0(z) = (H_0 - z^2)^{-1}$ is given in this representation for $z \in \mathbf{C}^+$ by

$$\mathcal{F}R_0(z)\mathcal{F}^{-1} = (k^2 - z^2)^{-1} . \tag{2}$$

The family of function $\phi(k,\omega,x)$ has, for fixed $\omega \in S^2,\ x \in \mathbf{R}^3$, an entire analytic extension in k, and for $|\Im k| < a$

$$\int_{\mathbf{R}^3} e^{-2a|x|}|\phi(k,\omega,x)|^2 dx < \infty . \tag{3}$$

Introducing the exponentially weighted spaces $\mathcal{H}_a$ defined by

$$\mathcal{H}_a = \{f|\ \|f\|_a^2 = \int e^{2a|x|}\ |f(x)|^2 dx < \infty\} ,$$

we see that (3) means $\phi(k,\omega,\cdot) \in \mathcal{H}_{-a}$ for $|\Im k| < a$, and $\phi(k,\omega,\cdot)$ is for fixed $\omega \in S^2$ an analytic function of k in $\mathcal{T}_a = \{k \in \mathbf{C} | \ |\Im k| < a\}$ with values in $\mathcal{H}_{-a}$. This implies that for $f \in \mathcal{H}_a$, $\omega \in S^2$, the Fourier transform $(\mathcal{F}f)(k,\omega)$ has an analytic extension in k to $\mathcal{T}_a$ given by (1).

This yields an analytic continuation $\tilde{R}_0(z)$ of $R_0(z)$ as follows. For $f, g \in \mathcal{H}_a$ and $k \in \mathbf{C}^+$ the representation (2) gives

$$(f, R_0(z)g) = \int_{S^2} d\omega \int_0^\infty dk k^2 (\overline{\mathcal{F}f})(k,\omega)(k^2 - z^2)^{-1}(\mathcal{F}g)(k,\omega)$$

Due to the analyticity of $\mathcal{F}f$ and $\mathcal{F}g$ in $\mathcal{T}_a$ we can deform the integration path $\mathbf{R}^+$ in the k-plane to a curve Γ contained in $\mathcal{T}_a$ (see Figure 1) to obtain, for $z \in \mathbf{C}^+$

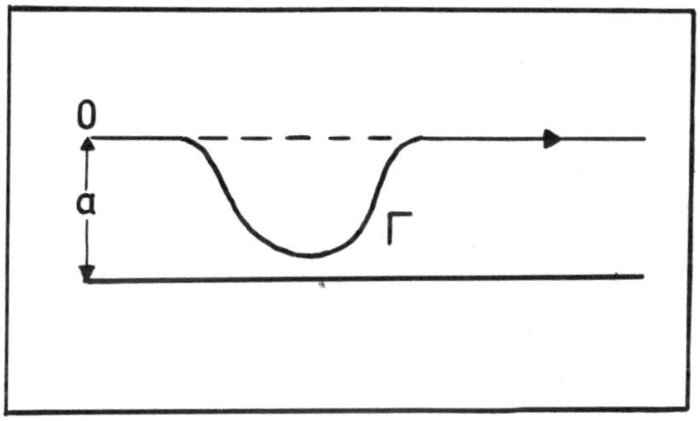

Figure 1

Deformation Γ of integration path in momentum space.

$$(f, R_0(z)g) = \int_{S^2} d\omega \int_\Gamma dk k^2 \left((\overline{\mathcal{F}f})(\bar{k},\omega)(k^2 - z^2)^{-1}(\mathcal{F}g)(k,\omega) \right) . \tag{4}$$

By (4), $(f, R_0(z)g)$ has an analytic continuation $\mathcal{G}_{f,g}(z)$ from $\mathbf{C}^+$ across $\mathbf{R}^+$ to Γ and hence, varying Γ, to $U_a = \{z | \Re z > 0, \ -a < \Im z < 0\}$. It is easy to show that

$$|\mathcal{G}_{f,g}(z)| \le C(z) \|f\|_a \|g\|_a \ ,$$

so that for fixed z, $\mathcal{G}_{f,g}(z)$ is a bounded, sesquilinear form on $\mathcal{H}_a \times \mathcal{H}_a$. This implies that $R_0(z)$ has an analytic continuation $\tilde{R}_0(z)$ to U_a as a bounded operator from $\mathcal{H}_a$ to $\mathcal{H}_{-a}$ defined by

$$(f, \tilde{R}_0(z)g) = \mathcal{G}_{f,g}(z) .$$

Using Cauchy's Theorem, we obtain from (4) the following expression for $\tilde{R}_0(z)$,

$$\tilde{R}_0(z) = R_0(z) + 2\pi i \tilde{T}^*(\bar{z})\tilde{T}(z)$$

where $\tilde{T}(z) \in \mathbf{B}(\mathcal{H}_a, L_2(S^2))$ is given by $\tilde{T}(z)f = \left(\frac{z}{2}\right)^{\frac{1}{2}} (\mathcal{F}f)(z,\cdot)$ and $\tilde{T}^*(z) \in \mathbf{B}(L_2(S^2), \mathcal{H}_{-a})$ is the adjoint of $T(z)$.

It is easy to show that $\tilde{R}_0(z)$ is in fact analytic as a $\mathbf{B}(\mathcal{H}_a, \mathcal{H}^2_{-a})$-valued function, where $\mathcal{H}^2_{-a}$ is the exponentially weighted Sobolev space (see [6] for details).

Let W be an operator of the form

$$W = e^{-a|x|} Q e^{-a|x|} \tag{5}$$

where Q is a symmetric, H_0-compact operator in $\mathcal{H}$. For example, Q could be multiplication by a real-valued, locally square-integrable function going to zero for $|x| \to \infty$.

Then $H = H_0 + W$ is a selfadjoint operator on $\mathcal{D}_H = \mathcal{D}_{H_0}$. Moreover, $WR_0(z)$ has the analytic continuation $W\tilde{R}_0(z)$ from $\mathbf{C}^+$ to U_a as a function with values in $\mathcal{C}(\mathcal{H}_a)$), the space of compact operators on $\mathcal{H}_a$. By the analytic Fredholm Theorem, $\left(1 + W\tilde{R}_0(z)\right)^{-1}$ is meromorphic in $\mathbf{C}^+ \cup \mathbf{R}^+ \cup U_a$, and $R(z) = (H - z^2)^{-1}$ has the meromorphic continuation

$$\tilde{R}(z) = \tilde{R}_0(z)\left(1 + W\tilde{R}_0(z)\right)^{-1}$$

from $\mathbf{C}^+$ to U_a with the same poles as $\left(1 + W\tilde{R}_0(z)\right)^{-1}$. These poles are called resonances of the pair $(H_0, H_0 + W)$, and the set of resonances in U_a is denoted by R_a. It is well known [7] that R_a coincides with the set of poles of the analytically continued scattering matrix $\tilde{S}(z)$, given by

$$\tilde{S}(z) = 1 - \pi i z \tilde{T}(z)(W - W\tilde{R}(z)W)\tilde{T}^*(\bar{z}) . \tag{6}$$

3. Background potentials

The clue to the above treatment of resonances for (H_0, H_0+W) is the existence of an analytic continuation of the family $\phi(k,\omega,x)$ of generalized eigenfunctions of H_0 such that $\phi(k,\omega,\cdot) \in \mathcal{H}_{-a}$ for $|\Im k| < a$. This suggests the possibility of replacing H_0 by an operator $H_1 = H_0 + V$, where the real-valued function V is locally square integrable, and $V(x) = O(|x|^{-1-\varepsilon})$, in which case a family of generalized eigenfunctions is known to exist, cf.[1]. If this family can be extended analytically into the lower half-plane with the same exponential growth properties, a theory of resonances of (H_1, H_1+W)-resonances with the background potential V– can be developed in an analogous way as for (H_0, H_0+W) provided W satisfies (3). We recall the following facts about the generalized eigenfunctions of H_1 [1] under the above assumptions on V.

There exists a family $\Psi(k,\omega,x)$ of generalized eigenfunctions of H_1, defined for $k \in \mathbf{R}/\{0\}$, $\omega \in S^2$, $x \in \mathbf{R}^3$ by

$$\Psi(k,\omega,x) = \left(1 - \lim_{\varepsilon \downarrow 0} R_1(k+i\varepsilon)V\right)\phi(k,\omega,x) \ .$$

The generalized Fourier transform $\mathcal{F}_1$ defined by

$$(\mathcal{F}_1 f)(k,\omega) = (2\pi)^{-\frac{3}{2}} \lim_{R\to\infty} \int\limits_{|x|<R} f(x)\Psi(k,w,x)dx \tag{7}$$

is a unitary operator from $\mathcal{H}_{1,c}$ onto $\mathcal{H}$ such that

$$\mathcal{F}_1 H_{1,c} \mathcal{F}_1^{-1} = k^2$$

where $\mathcal{H}_{1,c} = \mathcal{H} \ominus \mathcal{H}_p$, $H_{1,c} = H\backslash\mathcal{H}_{1,c}$, and $\mathcal{H}_{1,p}$ is the subspace of $\mathcal{H}$ spanned by all the eigenvectors of H_1.

The eigenvalues of H_1 are finite-dimensional and accumulate, at most, at zero. We denote the set of negative eigenvalues of H_1 by $\sigma_d(H_1)$ and set $\sigma_d^{1/2}(H_1) = \{i\beta | -\beta^2 \in \sigma_d(H_1)\}$.

The problem of analytic extension of $\Psi(k,\omega,x)$ gives rise to two very different types of questions

(i) Does $\Psi(k,\omega,x)$, for each $\omega \in S^2$, $x \in \mathbf{R}^3$, have an analytic extension to $k \in \mathbf{C}^+$, such that for $0 < \Im k < a$, $\omega \in S^2$, $\Psi(k,\omega,\cdot) \in \mathcal{H}_{-a}$?

(ii) Does $\Psi(k,\omega,x)$, for each $\omega \in S^2$, $x \in \mathbf{R}^3$, have an analytic continuation $\tilde{\Psi}(k,\omega,x)$ from $\mathbf{C}^+$ across $\mathbf{R}^+$ to a domain $\mathbf{O} \subset \mathbf{C}^-$, such that for $k \in \mathbf{O}_a = \{k \in \mathbf{O} | -a < \Im k < 0\}$, $\tilde{\Psi}(k,\omega,\cdot) \in \mathcal{H}_{-a}$?

A positive answer to (i) and (ii) is given by the following theorem, which is proved in [6]:

Theorem 1. (i) Suppose that V is a real-valued function in $L_2(\mathbf{R}^3)$ such that $|V(x)| < c|x|^{-2-\epsilon}$ for $|x| > R$. Then there exists a family $\Psi(k,\omega,x)$ of generalized eigenfunctions of H, defined for $k \in \overline{\mathbf{C}^+}\backslash\sigma_d^{1/2}(H_1)$, $\omega \in S^2$, $x \in \mathbf{R}^3$, such that for fixed ω and x, $\Psi(k\omega,x)$ is analytic in $\mathbf{C}^+\backslash\sigma_d^{1/2}(H_1)$ and continuous in $\overline{\mathbf{C}^+}$. Moreover, $|\Psi(k,\omega,x)| < ce^{a|x|}$ for $0 \leq \Im k < a$.

(ii) Assume, furthermore, that the S-matrix $S_1(k)$ of (H_0, H_0+V) has an analytic extension $\tilde{S}_1(k)$ to a domain $\mathbf{O} \subset \mathbf{C}^-$. Then $\Psi(k,\cdot,x)$ has for $x \in \mathbf{R}^3$ an $L_2(S^2)$-valued analytic continuation $\tilde{\Psi}(k,\cdot,x)$ to $k \in \mathbf{O}\backslash\sigma_d^{1/2}(H_1)$, given by

$$\tilde{\Psi}(k,-\cdot,x) = \tilde{S}(k)\Psi(-k,\cdot,x) \tag{8}$$

Moreover, $\tilde{\Psi}(k,\cdot,\cdot) \in \mathcal{H}_{-a}(\mathbf{R}^3; L_2(S^2))$ for $k \in \mathbf{O}_a$, and $\tilde{\Psi}(k,\cdot,\cdot)$ is analytic in $\mathbf{O}_a$ with values in $\mathcal{H}_{-a}\left(\mathbf{R}^3; L_2(S^2)\right)$.

Remark 1. The implicit condition on V imposed in (ii), namely that V should be such that the S-matrix of (H_0, H_0+V) has an analytic continuation to a domain $\mathbf{O} \subset \mathbf{C}^-$, is known to hold [3] with $\mathbf{O} = S_b = \{z = \rho e^{i\varphi} | \rho > 0, |\varphi| < b\}$, if $V = O(|x|^{-1-\epsilon})$ is S_b-dilation-analytic. Thus, (i) and (ii) of Theorem 1 are satisfied if $V = V(r,\omega)(r \in \mathbf{R}^+, \omega \in S^2)$ is a real-valued function such that $V(r,\cdot)$ has an analytic $L_2(S^2)$-valued extension $V(z)$ to $z \in S_a$ satisfying for some $R > 0$, $C > 0$

$$\begin{aligned}
&(i)\ |V(r,\omega)| < Cr^{-2-\epsilon} \quad \text{for } r > R,\ \omega \in S^2\\
&(ii)\ |V(re^{i\varphi},\omega)| < Cr^{-1-\epsilon} \quad \text{for } r > R,\ \omega \in S^2,\ |\varphi| < a\\
&(iii)\ \int_0^R r^2 \|V(re^{i\varphi})\|^2_{L_2(S^2)} dr < C \quad \text{for } |\varphi| < a
\end{aligned}$$

Remark 2. It is possible to weaken $|V(x)| < C|x|^{-2-\epsilon}$, the condition in Theorem 1(i), to $|V(x)| < C|x|^{-1-\epsilon}$ for $|x| > R$. The function $\Psi(k,\omega,x)$ is then to be be replaced by an $L_2(S^2)$-valued function $\Psi(k,\cdot,x)$ of $k \in \overline{\mathbf{C}^+}$ and $x \in \mathbf{R}^3$(cf.[8,12]. A different method for obtaining $\Psi(k,\cdot,x)$, utilizing the Green's function, is indicated by Agmon [2], where $V(x) = O(|x|^{-1-\epsilon})$.

Based on Theorem 1, we shall now construct the analytic continuation of the resolvent $R(z) = (H - z^2)^{-1}$ from $\mathbf{C}^+$ to $\mathbf{O}_a$ as a $B(\mathcal{H}_a, \mathcal{H}_{-a})$-valued function. We proceed in the same way as for $R_0(z)$, assuming that V satisfies the conditions of (i) and (ii) of Theorem 1. Then for $f \in \mathcal{H}_a$, $(\mathcal{F}_1 f)(k, \cdot)$ has an $L_2(S^2)$-valued analytic continuation to $\mathbf{O}_a$, given by (7). Moreover, for $f, g \in \mathcal{H}_a \cap \mathcal{H}_c$ we obtain, using the contour deformation Γ of Figure 1

$$\begin{aligned}(f, R_{1,c}(z)g) &= \int_0^\infty dk k^2 \left((\mathcal{F}_1 f)(k, \cdot), (\mathcal{F}_1 g)(k, \cdot)\right)_{L_2(S^2)} (k^2 - z^2)^{-1} \\ &= \int_\Gamma dk k^2 \left((\mathcal{F}_1 f)(\overline{k}, \cdot), (\mathcal{F}_1 g)(k, \cdot)\right)_{L_2(S^2)} (k^2 - z^2)^{-1}\end{aligned}$$

This yields an analytic continuation of $F_{1,f,g}(z) = (f, R_{1,c}(z)g)$ from $\mathbf{C}^+$ across $\mathbf{R}^+$ to Γ and, varying Γ, into $\mathbf{O}_a$. It is easy to show that $F_{1,f,g}(z)$ is a bounded, sesquilinear form on $\mathcal{H}_a \times \mathcal{H}_a$, thus defining an operator $\tilde{R}_{1,c}(z) \in \mathbf{B}(\mathcal{H}_a, \mathcal{H}_{-a})$. In fact, $\tilde{R}_{1,c}(z) \in \mathbf{B}(\mathcal{H}_a, \mathcal{H}^2_{-a})$ and is given by

$$\tilde{R}_{1,c}(z) = R_{1,c}(z) + 2\pi i \tilde{T}_1^*(\overline{z})\tilde{T}_1(z) ,$$

where $\tilde{T}_1(z) \in \mathbf{B}\left(\mathcal{H}_a, L_2(S^2)\right)$ is the generalized trace operator, given by

$$\tilde{T}_1(z)f = \left(\frac{z}{2}\right)^{\frac{1}{2}} (\mathcal{F}_1 f)(z, \cdot) .$$

Since the eigenvalues of H_1 are finite-dimensional and accumulate, at most, at 0, the operator $R_{1,p}(z) = (H_{1,p} - z^2)^{-1}$ has a meromorphic continuation across $\mathbf{R}^+$ as an operator in $\mathbf{B}(\mathcal{H}, H^2)$ and hence in $\mathbf{B}(\mathcal{H}_a, \mathcal{H}^2_{-a})$ with poles at the positive eigenvalues.

We have indicated the proof of the following result.

Theorem 2. Suppose that V satisfies the conditions of (i) and (ii) of Theorem 1. Then the resolvent $R_1(z)$ has a meromorphic continuation $\tilde{R}_1(z)$ across $\mathbf{R}^+$ to $\mathbf{O}_a$ as a $\mathbf{B}(\mathcal{H}_a, \mathcal{H}^2_{-a})$ -valued function with poles contained in $\{\lambda | \lambda^2 \in \sigma_p(H_1)\}$.

Utilizing this result, we obtain the following result on resonances with the background potential V.

Theorem 3. Let V be a potential satisfying the conditions (i) and (ii) of Theorem 1, and assume that W is an operator satisfying (5). Then the resolvent $R_2(z) = (H_1 + W - z^2)^{-1}$ has a meromorphic continuation $\tilde{R}_2(z)$ from $\mathbf{C}^+$ across $\mathbf{R}^+$ to $\mathbf{O}_a$ given by $\tilde{R}_2(z) = \tilde{R}_1(z)\left(1 + W\tilde{R}_1(z)\right)^{-1}$. The poles of $\tilde{R}_2(z)$ in $\mathbf{O}_a$ are the same as the poles of $\left(1 + W\tilde{R}_1(z)\right)^{-1}$. These poles are called resonances of $(H_1, H_1 + W)$

and coincide with the poles of the analytically extended scattering matrix $\tilde{S}_{12}(z)$ of $(H_1, H_1 + W)$, given by

$$\tilde{S}_{12}(z) = 1 - \pi i z \tilde{T}_1(z) \left(W - W\tilde{R}_2(z)w\right) \tilde{T}_1^*(\bar{z}) \,. \tag{9}$$

Proof. It follows from the analytic Fredholm Theorem, since $W\tilde{R}_1(z)$ is an analytic $C(\mathcal{H}_a)$-valued function on $\mathbf{O}_a$, that $\left(1 + W\tilde{R}_1(z)\right)^{-1}$ is meromorphic on $\mathbf{O}_a$. By the second resolvent equation $\tilde{R}_2(z) = \tilde{R}_1(z)(1+W\tilde{R}_1(z))^{-1}$ is a meromorphic $\mathbf{B}(\mathcal{H}_a, \mathcal{H}^2_{-a})$-valued function on $\mathbf{O}_a$ with the same poles. The fact that $\tilde{S}_{12}(z)$, given by (9), has the same poles as $\tilde{R}_2(z)$ is proved in [6].

Corollary. The scattering matrix $S_2(k)$ of the pair $(H_0, H_0 + V + W)$ has a meromorphic extension $\tilde{S}_2(z)$ to $\mathbf{O}_a$ with poles contained in the set of resonances of $(H_1, H_1 + W)$.

Proof. This follows from Theorem 3 and the identity $S_2(k) = S_{12}(k)S_1(k)$, $k > 0$, proved in [6].

We have outlined a theory of resonances with a short-range background potential V produced by an exponentially decaying potential W. Even if these results are extended as indicated in Remark 2 to potentials $V(x) = O(|x|^{-1-\epsilon})$, the Coulomb potential r^{-1}, for example, is not included. This has been treated for fixed angular momentum quantum number ℓ in [5], where the case of radial potentials $V(r) = O(r^{-1-\epsilon})$ is also covered. For perturbations $W(r) = O(e^{-2br})$ the theory of resonances of $(H_1, H_1 + W)$ is developed as in Theorem 3 for fixed ℓ, where $H_1 = -\frac{d^2}{dr^2} + \frac{\ell(\ell+1)}{r^2} + V$ and $V = cr^{-1}$ or $V = O(r^{-1-\epsilon})$.

Remark. The poles of $\tilde{S}_2(z)$ are in fact identical with the resolvent resonances of $H_0 + V + W$, provided V is dilation-analytic, cf.[11].

Appendix

The Gamow function associated with an S-wave resonance

We refer to [4] for background material on resonances and Gamow functions. Let S_a be the angular region defined by

$$S_a = \{z = \rho e^{i\varphi} | \rho > 0,\ |\varphi| < a\} \,,$$

and let $V = V(z)$ be an analytic function on S_a satisfying the following conditions:

(i) $V(z)$ is real for z real

(ii) $\int\limits_0^R |V(\rho e^{i\varphi})|^2 \rho^2 d\rho < C(R) < \infty$ for $|\varphi| < a$

(iii) $\int\limits_R^{R+1} |V(\rho e^{i\varphi})|^2 d\rho \to 0$ for $R \to \infty$, uniformly for $|\varphi| < a$.

(iv) $\int\limits_1^\infty |V(\rho e^{i\varphi})| d\rho < C < \infty$ for $|\varphi| < a$.

Then V is an S_a-dilation-analytic, H_0-compact operator, where H_0 is the selfadjoint operator in $\mathcal{H} = L_2(0, \infty)$ defined by

$$\mathcal{D}_{H_0} = \{u \in \mathcal{H} | u'' \in \mathcal{H},\ u(0) = 0\},\ H_0 u = -u'' \text{ for } u \in \mathcal{D}_{H_0}\ .$$

The Hamiltonian $H = H_0 + V$ is selfadjoint on $\mathcal{D}_H = \mathcal{D}_{H_0}$.

Assume that $k = |k| e^{-i\eta}$ is a resonance of (H_0, H), $0 < \eta < a$, and let Ψ be the Gamow function corresponding to the resonance k, i.e. Ψ is the solution of

$$\left(-\frac{d^2}{dr^2} + V(r) - k^2\right) \Psi(r) = 0,\ 0 < r < \infty \tag{a}$$

$$\Psi(0) = 0, \quad \Psi'(0) = 1 \tag{b}$$

<u>Theorem</u>. The function Ψ has an analytic extension $\Psi(z)$ to $z \in S_a$ with the following properties:

(i) $\Psi(z)$ is the solution of

$$\left(-\frac{d^2}{dz^2} + V(z) - k^2\right) \Psi(z) = 0,\ z \in S_a \tag{a'}$$

$$\Psi(0) = 0, \quad \frac{d\Psi}{dz}|_{z=0} = 1 \tag{b'}$$

where $\frac{d\Psi}{dz}|_{z=0} = \lim_{z \to 0} \frac{\Psi(z)}{z}$.

(ii) $\Psi_\varphi(r) = \Psi(re^{i\varphi})$ satisfies for fixed $\varphi \in (-a, a)$

$$\left(-e^{-2i\varphi}\frac{d^2}{dr^2} + V_\varphi(r) - k^2\right)\Psi_\varphi(r) = 0 \qquad (a_\varphi)$$

where $V_\varphi(r) = V(re^{i\varphi})$

$$\Psi_\varphi(0) = 0, \quad \Psi'_\varphi(0) = e^{-i\varphi} \qquad (b_\varphi)$$

(iii) $\Psi_\varphi(r) \simeq e^{i(ke^{i\varphi})r}$ for $r \to \infty$, $|\varphi| < a$.

In particular, for $\varphi > \eta$, $\Psi_\varphi(\cdot) \in L_2(0,\infty)$, and Ψ_φ is an eigenfunction of $H(e^{i\varphi}) = e^{-2i\varphi}H_0 + V_\varphi$ corresponding to the discrete eigenvalue k^2.

The Gamow function $\Psi(r) = \Psi_0(r)$ is obtained from $\Psi_\varphi(r)$ by analytic continuation.

Proof. By the analytic theory of ordinary differential equations, (a') has a unique solution $\Psi(z)$ analytic in S_a and satisfying (b').

The function $\Psi_0(r)$ is the solution of (a), satisfying (b) (the regular solution of (a)) and hence equal to the Gamow function $\Psi(r)$, which is also equal to $(2ik)^{-1}\mathcal{F}(-k)f_+(k,r)$, where $f_+(k,r)$ is the outgoing solution of (a) at k and $\mathcal{F}$ is the Jost function.

Moreover, by (a') and (b'), keeping φ fixed in $(-a, a)$, we conclude that $\Psi_\varphi(r)$ satisfies (a_φ), (b_φ) or

$$\left(-\frac{d^2}{dr^2} + e^{2i\varphi}V_\varphi(r) - (ke^{i\varphi})^2\right)\Psi_\varphi(r) = 0 \qquad (a'_\varphi)$$

$$\Psi_\varphi(0) = 0, \quad (e^{i\varphi}\Psi_\varphi)'(0) = 1$$

Thus, $e^{i\varphi}\Psi_\varphi(r)$ is the regular solution of (a'_φ). It is well known that this regular solution satisfies, for some $C < \infty$, the estimate

$$|\Psi_\varphi(r)e^{-ike^{i\varphi}}| < C \quad \text{for } 0 < r < \infty, \ -a < \varphi \leq \eta$$

or

$$|\Psi_\varphi(r)e^{-i|k|e^{i(\varphi-\eta)}}| < C \quad \text{for } 0 < r < \infty, \ -a < \varphi \leq \eta \qquad (*)$$

For $\varphi > \eta$ we use the dilation-analytic method. According to this theory, the operator $H(e^{i\varphi}) = -e^{-2i\varphi}\frac{d^2}{dr^2} + V_\varphi(r)$ has k^2 as a discrete eigenvalue for $\varphi > \eta$. The eigenfunction is precisely a multiple of the regular solution of (a'_φ) and hence (choosing the factor $e^{-i\varphi}$) coincides with $\Psi_\varphi(r)$. Being an eigenfunction, $\Psi_\varphi(r)$ also coincides with $e^{-i\varphi}(2ike^{i\varphi})^{-1}\tilde{\mathcal{F}}(-ke^{i\varphi})f_+(k,r)$. Here $\tilde{\mathcal{F}}$ is the analytic continuation of the Jost function, known to exist because V is S_a-dilation-analytic. Thus, for $\varphi > \eta$,

$$\Psi_\varphi(r) \simeq e^{-2i\varphi}(2ik)^{-1}\tilde{\mathcal{F}}(-ke^{i\varphi})e^{i|k|e^{i(\varphi-\eta)}} \quad \text{for } r \to \infty \qquad (**)$$

From (*) and (**) follows, by Vitali's convergence theorem, that for $-a < \varphi < a$

$$\begin{aligned}\Psi_\varphi(r) &\simeq e^{-2i\varphi}(2ik)^{-1}\tilde{\mathcal{F}}(-ke^{i\varphi})e^{i|k|e^{i(\varphi-\eta)}} \\ &= e^{-2i\varphi}(2ik)^{-1}\tilde{\mathcal{F}}(-ke^{i\varphi})e^{ike^{i\varphi}} \,.\end{aligned}$$

Thus, for every $\varphi \in (-a, a)$, $\Psi_\varphi(r)$ is a multiple of the outgoing solution as well as $e^{-i\varphi}$ times the regular solution of (a_φ).

By complex scaling, what is found is for some $\varphi > \eta$ the square-integrable, usual eigenfunctions $\Psi_\varphi(r)$ of $H(e^{i\varphi}) = e^{-2i\varphi}H_0 + V_\varphi$. From this, by the above Theorem, the Gamow function is obtained by analytic continuation of $\Psi_\varphi(r)$ to the positive half-axis. If $\Psi_\varphi(r)$ is given on some explicit form, this analytic continuation can be obtained by replacing the argument by $re^{-i\varphi}$ (backwards rotation).

The same analysis can be carried out also for $\ell > 0$ with the usual modifications. A different proof of the above Theorem was given by E. Skibsted [9], solving the equivalent integral equation for the outgoing solution by iteration and using a contour rotation technique. A corresponding result for non-radial potentials is proved in reference 11.

References

1. S. Agmon, Ann. d. Scuola Norm. Sup. di Pisa, Ser. IV, **2**, 151 (1975).

2. S. Agmon, Comm. Pure and Appl. Math., **39**, 3 (1986).

3. E. Balslev, J. Funct. Anal., **3**, 375 (1978).

4. E. Balslev, *Resonances-Models and Phenomena*, S. Albeverio, L. S. Ferreira and L. Streit, Eds., Lecture Notes in Phys., Springer-Verlag, Berlin (1984) p. 211.

5. E. Balslev, J. Math. Anal. Appl., **2**, 339 (1987).

6. E. Balslev, Comm. Math. Phys., **114**, 599-612 (1988).

7. A. Jensen, J. Math. Anal. Appl., **59**, 503 (1977).

8. Y. Saitō, J.f.d. reine u. angewdt. Math., **343**, 1 (1983).

9. E. Skibsted, *Resonanser for potentialer of typen* $r^{-\alpha}\ e^{-r^{-\beta}}$, *Speciale*, Aarhus Universitet, (1984).

10. B. Gyarmati and A. T. Kruppa, Phys. Rev., **34C**, 95 (1986).

11. E. Balslev and E. Skibstad, *Resonance theory for two-body Schrödinger operators*, Aarhus University preprint (1987).

ON THE GENERAL INVERSION PROBLEM

Anders Melin

Department of Mathematics
University of Lund
Box 118
S-221 00 Lund
Sweden

Abstract

We consider the problem of direct and inverse scattering for the Schrödinger operator $H_v = -\Delta + v(x)$ in odd space dimensions with a short range potential. It will be shown that the wave operators are built up from a family of operators $A_\theta, \theta \in S^{n-1}$, which satisfy the equation $H_v A_\theta = A_\theta H_0$. The corresponding operator kernels are supported in the set where $\langle y - x, \theta \rangle \geq 0$, and they can be described in detail. By introducing polar coordinates for $y - x$ one finds also that these kernels have several properties in common with their one-dimensional analogues. The potential can be easily computed from a special trace of $A_\theta^* A_\theta$, and this operator in turn is given from a factorization of the scattering matrix into upper and lower triangular parts with respect to the direction θ. Finally we give some remarks on the so called miracle, which was introduced by R. G. Newton.

1. Introduction

We shall study the Schrödinger operator

$$H_v = -\Delta + v(x)$$

in $\mathbf{R}^n$, when $n > 1$ is odd. It will be assumed that v is real and of short range in the sense that

$$\int (1+|x|)^{2-n+|\alpha|}|v^{(\alpha)}(x)|dx < \infty \tag{1}$$

for any multi-index α. This condition implies that v is bounded and vanishes at infinity. The restriction of H_v to the Hilbert space $\mathcal{H}_{(2)}$ is a self-adjoint operator for which we use the same notation.

The wave operators

$$W_\pm = \lim_{t\to\pm\infty} e^{itH_v}e^{-itH_0}$$

are isometric on the Hilbert space $L^2(\mathbf{R}^n)$ with the property that

$$H_vW_\pm = W_\pm H_0. \tag{2}$$

We refer to this equation as the *intertwining property* of $W_\pm$. Since H_v may have bound states, the wave operators are in general not unitary. However, they have the same range, and the scattering operator

$$S = W_+^*W_-$$

is therefore unitary.

All objects considered so far are defined with a Hilbert space in the background. In this paper we change the point of view and focus our attention to the distribution kernels of operators instead. In fact, we shall forget about the Hilbert space and identify any operator A on $\mathbf{R}^n$ with its distribution kernel. The intertwining relation then takes the form

$$(\Delta_x - \Delta_y - v(x))\,A(x,y) = 0, \tag{3}$$

where the derivatives should be interpreted in the distribution sense.

2. Construction of Intertwining Operators

If $v(x) = 0$, then $A(x,y) = I = \delta(x-y)$ is a solution of (3). We shall see that for general v there are many solutions to this equation on the form $A(x,y) = \delta(x-y) + U(x,y)$, where U is a function. In the one-dimensional case U is a Volterra operator. In fact, there are two such operators $U_\pm$ corresponding to the choice of the sign of $x-y$ in the support of U. In the multi-dimensional case we therefore expect that there is some interesting class of solutions $A_\theta = I + U_\theta$, which is parametrized by the set S^{n-1} of unit vectors θ in $\mathbf{R}^n$. To describe such operators we have to introduce some definitions.

Definition 1. *$\mathcal{M}$ is the space of all locally integrable functions $U(x,y)$ on $\mathbf{R}^n \times \mathbf{R}^n$ such that*

$$\|U\|_{\mathcal{M}} = \max\left(\sup_x \int |U(x,y)|\,dy\,, \sup_y \int |U(x,y)|\,dx\right) < \infty.$$

If $\theta \in S^{n-1}$ then $\mathcal{M}_\theta$ is the set of all U in $\mathcal{M}$ such that

$$\langle y - x, \theta\rangle \geq 0 \text{ in supp } (U), \tag{4}$$

and

$$\int |U(x,y)|\,dy \to 0 \text{ as } |x| \to \infty,\ x/|x| \to \theta, \tag{5}$$

$$\int |U(x,y)|\,dx \to 0 \text{ as } |y| \to \infty,\ y/|y| \to -\theta. \tag{6}$$

It is clear that $\mathcal{M}$ is a Banach algebra, and $\mathcal{M}_\theta$ is a subalgebra. Moreover, any $U \in \mathcal{M}$ gives rise to a continuous operator which is bounded on L^p when $1 \leq p \leq \infty$. In fact

$$\|U\|_{L^p \to L^p} \leq \|U\|_{\mathcal{M}}.$$

Assume now that we can find a continuous family $A_\theta = I + U_\theta \in I + \mathcal{M}_\theta$, which satisfies (3)-(6). We may formally write any test function u as an integral

$$u = \int_{S^{n-1}} u_\theta\, d\theta,$$

where the spectrum of u_θ is in the direction of θ. This implies that the wave packet $e^{-itH_0}u_\theta$ travels to ∞ in the direction of $\pm\theta$ as $t \to \pm\infty$. It follows from (6) then (still formally) that

$$A_{\mp\theta}e^{-itH_0}u_\theta - e^{-itH_0}u_\theta \to 0$$

as $t \to \pm\infty$. If we multiply by e^{itH_v} and use the intertwining relation we find that

$$A_{\mp\theta}u_\theta = W_\pm u_\theta.$$

This gives us the formula

$$W_\pm u = \int A_{\mp\theta}u_\theta \, d\theta.$$

All these heuristic arguments can be justified. The following result is true, where $\mathcal{F}$ denotes the Fourier transform

$$\mathcal{F}u(\xi) = \hat{u}(\xi) = \int e^{-i\langle x,\xi\rangle}u(x)\,dx.$$

Theorem 2. *Assume that v is small in the sense that $\sum_{|\alpha|=n-2}\int |v^{(\alpha)}(x)|\,dx$ is small. Then there exists a unique continuous family of invertible elements $A_\theta \in I + \mathcal{M}_\theta$ which satisfies (3)-(6). The kernel of $W_\pm\mathcal{F}^*$ is given by*

$$(W_\pm\mathcal{F}^*)(x,\xi) = \int A_\theta(x,y)e^{i\langle y,\xi\rangle}\,dy \quad ,\theta = \mp\xi/|\xi|. \tag{7}$$

There is a simple way to express the scattering operator in terms of the potential and one of the wave operators. This takes the form

$$(\mathcal{F}(S-I)\mathcal{F}^*)(\xi,\eta) = \pi\delta(|\xi|-|\eta|)(\mathcal{F}(vW_-)\mathcal{F}^*)(\xi,\eta)(i|\xi|). \tag{8}$$

Hence, in the case of a small potential, the scattering matrix may be identified with the function

$$(\widehat{vA_\theta})(\xi,\eta)\,,\quad \theta = -\eta/|\eta|,\ |\xi| = |\eta|. \tag{9}$$

It follows from this observation that v can be recovered from the scattering matrix. This is true at least when v is small and integrable, for then the expression in (9) equals $\hat{v}(\xi+\eta)+\varepsilon(\xi,\eta)$, where $\varepsilon(\xi,\eta)$ vanishes at ∞.

In the case of large potentials Theorem 2 is no longer true. However, it can be modified in two ways.

Theorem 2a. *Assume that v is of short range. Then there is a unique A_θ which satisfies (3) such that $e^{-\lambda\langle y-x,\theta\rangle}A_\theta(x,y)$ is an invertible element in $I+\mathcal{M}_\theta$ satisfying (4)-(6) when λ is large enough.*

It follows from Theorem 2a that H_v is conjugated to H_0 in any weighted L^p space

$$L^p_{\lambda\theta} = \{f; e^{\lambda\langle x,\theta\rangle} f(x) \in L^p\},$$

as soon as λ is large enough.

Theorem 2b. *Assume that v is of short range. Then there is a continuous family of invertible elements $B_\theta = I + V_\theta \in I + \mathcal{M}_\theta$ which satisfies (4)-(6) such that*

$$B_\theta^{-1} H_v B_\theta = H_0 + K_\theta, \tag{10}$$

where K_θ is a finite rank operator with all its derivatives in $L^1 \otimes L^1$.

From the proof of this theorem one can also deduce the following result.

Theorem 3. *Assume that v is of short range. Then there is a unitary operator $A \in I + \mathcal{M}$ such that*

$$A^* H_v A = H_0 + K$$

where K is of finite rank and all its derivatives are in $L^1 \otimes L^1$.

The existence of bound states implies that A_θ in Theorem 2a is in general not in $I + \mathcal{M}_\theta$. One can however prove the following result.

Theorem 4. *There exists a continuous family of functions q_θ in $L^1(\mathbf{R}^n)$ with support in the set where $\langle x, \theta\rangle \leq 0$, such that*

$$A_\theta(I - [q_\theta]) \in I + \mathcal{M}_\theta.$$

Here $[q_\theta]$ denotes the convolution operator with kernel $q_\theta(x - y)$.

We shall finally discuss trace formulas and the so called miracle. Before doing that we make some remarks about the construction of the intertwining operators A_θ. If one writes $A_\theta = \sum_0^\infty U_N$ with $U_0(x,y) = \delta(x - y)$, then (3) is fulfilled at least formally if

$$(\Delta_x - \Delta_y) U_{N+1}(x,y) = v(x) U_N(x,y) \tag{11}$$

for any N. To solve these equations one chooses a special fundamental solution $E_\theta(x,y)$ for the ultra-hyperbolic operator :

$$(\Delta_x - \Delta_y) E_\theta(x,y) = \delta(x,y).$$

It turns out that there is a unique solution E_θ if one requires that $\langle y-x,\theta\rangle \geq 0$ in its support and in addition to that introduces some bounds at ∞. In particular, one requires that $E_\theta(x+t\theta, y+t\theta)$ tends to 0 in the distribution sense as $|t| \to \infty$. Having defined $U_N = U_{N,\theta}$ in this way one finds then that the sum $\sum U_N$ is always convergent in the appropriate spaces. (These ideas go back to the constructions of direction dependent Green's functions by Faddeev. [2,3])

3. Trace Formulas

In the one-dimensional case $\theta = \pm 1$, and if $\theta = 1$ then E_θ equals a constant times the characterstic function for the angular domain $x \leq 0, |y| \leq |x|$. It turns out that the formula for E_θ in the n-dimensional case is quite similar once one has introduced the right coordinates in $\mathbf{R}^n \times \mathbf{R}^n$. We have to express points (x, y) there in terms of x and polar coordinates for $y-x$. This leads us to consider the map

$$\gamma : \mathbf{R}^n \times \bar{\mathbf{R}}_+ \times S^{n-1} \ni (x, r, \phi) \to (x, x+r\phi) \in \mathbf{R}^n \times \mathbf{R}^n.$$

We have then a map between distributions

$$\gamma_* : \mathcal{D}'(\mathbf{R}^n \times \bar{\mathbf{R}}_+ \times S^{n-1}) \to \mathcal{D}'(\mathbf{R}^n \times \mathbf{R}^n).$$

It turns out that

$$E_\theta = c_n \int_{\langle\omega,\theta\rangle \geq 0} L_\omega \, d\omega,$$

where L_ω has the simple form

$$(\gamma_*^{-1} L_\omega)(x, r, \phi) = \delta^{(n-2)}(\langle x, \omega\rangle)\delta(\phi - \omega).$$

Definition 5. *We say that $B \in \mathcal{D}'(\mathbf{R}^n \times \mathbf{R}^n)$ is θ-admissible if $B = \gamma_* b$, where b has the form*

$$b(x, r, \phi) = Y_+(\langle\phi, \theta\rangle) b_+(x, r, \phi) + Y_-(\langle\phi, \theta\rangle) b_-(x, r, \phi).$$

Here $Y_\pm$ are the Heaviside functions on the positive and negative half-lines, and we require that the functions $b_\pm$ are continuous in r and smooth in the other variables.

If B is θ-admissible, then we use the notation $\mathrm{tr}_\theta B$ for the function $b(x, 0, \phi)$. If for example $B(x, y) = |x-y|^{1-n}$, then B is θ-admissible, and $\mathrm{tr}_\theta B$ equals a constant. It is also easy to see that if B_1 and B_2 are θ -admissible, then this is also true for $B_1 \circ B_2$ (if the composition is defined) and $\mathrm{tr}_\theta(B_1 \circ B_2) = 0$.

Using explicit formulas for $U_N = U_{N,\theta}$ one finds also that $A_\theta - I$ is θ-admissible, if A_θ is the operator in Theorem 2a. Its trace is given by

$$\mathrm{tr}_\theta(A_\theta - I) = c_n Y_+(\langle \phi, \theta \rangle) \int_{\langle z, \phi \rangle = 0} \langle \phi, \partial_z \rangle^{n-2} v(x+z)\, dz. \tag{12}$$

This formula should be compared with the one-dimensional analogue

$$A_1 - \delta(x-y) = cY_+(y-x) \int_{(x+y)/2}^{\infty} v(z)\, dz + R(x,y),$$

where the remainder term is continuous. (Cf. this with the residu-formulas in Melin. [4])

Set $Q_\theta = A_\theta^* A_\theta = \delta(x-y) + U_{1,\theta} + U_{1,\theta}^* + \cdots$. Then Q_θ is defined when v is small, and when v is large it is defined modulo a smooth kernel. Since the traces of smoothing operators vanish, we can always define $\mathrm{tr}_\theta(Q_\theta - I)$, and one finds that

$$\mathrm{tr}_\theta(Q_\theta - I) = c_n \mathrm{sgn}(\langle \phi, \theta \rangle) \int_{\langle z, \phi \rangle = 0} \langle \phi, \partial_z \rangle^{n-2} v(x+z)\, dz. \tag{13}$$

Theorem 6. *The potential v can be recovered from $Q_\theta = A_\theta^* A_\theta$. In fact,*

$$v(x) = c_n \int_{S^{n-1}} \mathrm{sgn}(\langle \phi, \theta \rangle) \langle \phi, \partial_x \rangle \mathrm{tr}_\theta(Q_\theta - I)(x, \phi)\, d\phi. \tag{14}$$

This result follows from (13) and the formula

$$\begin{aligned}
&\int_{S^{n-1}} \left(\int_{\langle z, \phi \rangle = 0} \langle \phi, \partial_x \rangle \langle \phi, \partial_z \rangle^{n-2} v(x+z)\, dz \right) d\phi \\
&= \int_{S^{n-1}} \left(\int_{\langle z, \phi \rangle = 0} \Delta^{(n-1)} v(x+z)\, dz \right) d\phi \\
&= c_{1,n} \int_{\mathbf{R}^n} |z|^{-1} \Delta^{(n-1)} v(x+z)\, dz = c_{2,n} v(x).
\end{aligned}$$

If v is small and $\theta \in S^{n-1}$, then the scattering matrix is a product $N_+ N_-$, where N_+ and N_- are upper and lower triangular w.r.t. the ordering of the sphere induced by θ. Moreover, Q_θ can be easily computed from $N_+^* N_+$ then.

4. The Miracle

In our discussion of the miracle we refer to Newton [5,6] and Cheney [1] for more details. For the sake of simplicity we shall also assume that v is small. We shall consider the generalized eigenfunctions in the frequency direction θ. Hence we introduce the functions

$$\beta_{\pm}(x,\theta,k) = e^{-ik\langle x,\theta\rangle}(W_{\pm}\mathcal{F}^*)(x,k\theta) - 1.$$

We note that these functions vanish identically if v vanishes. We also introduce their Fourier transforms w.r.t. k and set

$$\eta_{\pm}(x,\theta,s) = (2\pi)^{-1}\int \beta_{\pm}(x,\theta,k)e^{-iks}\,dk.$$

It follows from Theorem 2 then that

$$\eta_{\pm}(x,\theta,s) = \int_{\langle z,\theta\rangle=0} U_{\pm\theta}(x, s\theta + z + x)\,dz.$$

By using the formula for $U_{N,\theta}$ one can prove that the $\eta_{\pm}$ are of the following form

$$\eta_{\pm}(x,\theta,s) = a_{\pm}(x,\theta,s)Y_{\pm}(s),$$

where $a_{\pm}$ are smooth functions. The miracle can then be expressed in the following way.

Theorem 7. *Assume that v is small. Then*

$$\langle\theta,\partial_x\rangle a_{\pm}(x,\theta,0) = \pm c_n v(x),$$

where c_n is a dimension dependent constant.

If v is small, then $W_{\pm}$ are unitary and $W_+S = W_-$. We may write this as

$$(W_+\mathcal{F}^*)\tilde{S} = W_-\mathcal{F}^*,$$

where $\tilde{S}$ denotes the scattering matrix. By taking the Fourier transform of both sides w.r.t. the variable k one finds that the scattering matrix gives an integral equation for $\eta_{\pm}$. See Cheney in Reference 1, Theorem 6.3. Hence Theorem 7 may be viewed as an incomplete characterization of the scattering matrix.

References

1. M. Cheney, J. Math. Phys., **25**, 94 (1984).

2. L. D. Faddeev, J. Math. Phys., **4**, 72 (1963).

3. L. D. Faddeev, J. Sov. Math., **5**, 334 (1976).

4. A. Melin, Comm. Part. Diff. Eqs., **10**, 677 (1985).

5. R. G. Newton, J. Math. Phys., **21**, 1698 (1980).

6. R. G. Newton, SIAM-AMS Proceedings, **14**, 81 (1981).

WEYL'S THEORY STUDIES AND THE ONE-DIMENSIONAL ALMOST PERIODIC SCHRÖDINGER EQUATION

Ladislav Trlifaj

Institute of Physics
Czechoslovak Acad. Sci.,
Na Slovance 2,
180 40 Praha 8,
Czechoslovakia

Abstract

Making use of the Weyl's basis, we derive the modified Milne equation associated with the Schrödinger equation and find it's almost periodic solution, by means of which the Bloch-Weyl solutions can be written. There exist, in general, the generalized Bloch-Weyl functions only within spectral (allowed) bands of the energy parameter and on the negative energy half-axis linking up with the lowest allowed band. One solves the inverse problem in two ways each of them giving distinct formulas for reconstructing a potential. The Darboux transformation proves isospectral.

1. Introduction

If some nonlinear evolution equation has a soliton solution, one can expect that it has a periodic solution, too. Soliton solutions which are constructed by means of the inverse-scattering method use the Schrödinger equation or the allied Zakharov-Shabat AKNS equations. Passing from the soliton solution to the periodic one relies upon replacing a potential of the pulse type by periodic potentials in these equations. The solution of the inverse problem leads naturally to the more general class of almost periodic potentials, if we relax conditions on the boundaries of spectral bands that apply to periodic potentials [1,2]. There is a rich literature that relates to the direct or inverse solutions to the one-dimensional periodic Schrödinger (Hill's) equation. Papers collected and quoted in the books by Levitan [3] and by Eastham [4] are sufficient for our use, so we do not quote them in particular. Much less is done to solve the one-dimensional almost periodic Schrödinger equation and even less written except for its difference form. This is very likely because one cannot transfer methods used in solving periodic equations, which profit by solutions of the Sturm-Liouville problem on a finite segment with periodic or anti-periodic boundary conditions. The only paper we know about which is devoted to the solution of the Schrödinger equation with a general quasi-periodic potential and which employs methods used in physics is a paper by Dinaburg and Sinai [5]. They are especially interested in finding the Bloch solutions, which result from the Floquet theorem for periodic equations and which are important from the physical point of view. They show that there are (not clearly specified) intervals in the asymptotic region of energy, in which the bounded Bloch functions exist. In a recent paper [6] Zheng constructs these functions for a very special potential assuming that they exist for any energy value. Clearing up the existence of the Bloch functions also becomes relevant owing to the discovery of quasi-crystals. It's simplified (single-electron) description just applies the almost periodic Schrödinger equation on the full line (R).

Milne derived a nonlinear equation [7] associated with the Schrödinger equation in the early thirties and used it in calculating the bound-state energies. Recently, Korsch et al. [8] made use of this Milne equation in investigating resonance states. We apply its modified forms to solving the Schrödinger equation with an almost periodic potential and the associated inverse problem. In another context (periodic potentials with a finite number of energy bands), the modified Milne equation was derived and used by McKean and van Moerbeke [2].

We begin by introducing a 3-by-3 matrix Ω in Section 2 of the paper, the matrix elements of which are homogeneous quadratic polynomials in the Weyl's basis ϑ, φ and its derivatives. This matrix is hidden, for example, in some relations of [3]. It has some important properties and represents a fundamental system (monodromy matrix) of solutions to a linear differential equation, the coefficients of which depend on the energy parameter λ and on the potential only. In virtue of this equation, the first-row element $w(t)$ of any combination that we build up from the columns of Ω obeys a nonlinear equation, the modified Milne equation. By means of such $w(t)$, we construct two independent solutions to the primary Schrödinger equation, which resemble the Bloch solutions. We apply derived formulas to periodic potentials in Section 3 in order to find out when these solutions turn into the genuine Bloch-Weyl functions. Since only the periodic solution $w(t) = \varphi(t+c, t, \lambda)$ realizes such functions, we look for the almost periodic solution $w(t, \lambda)$ in Section 4 and show that it actually leads to the Bloch functions at least in some (allowed) energy bands again. These bands (spectrum) are only physically important as the corresponding Bloch functions represent almost periodically distorted waves characterized by a quasi-momentum. In Section 5 we solve, in two ways, the inverse problem of reconstructing a potential from given spectral data. Its solution is based, as usual, on the analycity properties of $w(t, \lambda)$ with respect to λ. The first approach resting upon $w(t, \lambda)$ is a modification of that described in [3]; the second one employs, in an analogous way, another function instead of $w(t,\lambda)$, which obeys a distinct modification of the Milne equation. There exist a variety of potentials which lead to the same spectrum, as necessary and independent spectral data do not consist of it only. Transformations of $w(t,\lambda)$ introduced in Section 6 preserve forms of the modified Milne equation changing only the potential but not the spectral term in it. In other words, these transformations represent the well-known Darboux transformations [9] of the primary Schrödinger equation.

2. Elementary Formulas

We fix zero of the energy scale in the one-dimensional Schrödinger equation with a uniformly almost periodic (u.a.p.) potential $q(t)$

$$\Psi_{tt} + [\lambda - q(t)]\Psi = 0 \quad (t \in R) \tag{1a}$$

by demanding

$$M(q(x)) = \lim_{T\to\infty} \frac{1}{2T} \int_{-T}^{T} dx q(x) = 0. \tag{1b}$$

We are used to studying (1a) by means of the fundamental system $\vartheta(t,t_o,\lambda)$ and $\varphi(t,t_o,\lambda)$, which are defined at a moveable point t_o by the requirements

$$\vartheta(t_o,t_o) = 1 = \varphi_x(t_o,t_o,\lambda), \vartheta_x(t_o,t_o,\lambda) = \varphi(t_o,t_o,\lambda) = 0. \tag{2}$$

Its Wronskian $W(\vartheta,\varphi) = \vartheta\varphi_x - \vartheta_x\varphi = 1$. Derivatives with respect to the first argument (t) are always denoted by the subscript x in what follows. Fundamental systems written in the form of a 2-by-1 matrix $\tilde{S}(t,t_o,\lambda) = (\vartheta(t,t_o,\lambda),\varphi(t,t_o,\lambda))$ and referring to different initial points t_1 and t_o are related as follows

$$S(t,t_1,\lambda) = \begin{pmatrix} \varphi_x(t_1,t_o,\lambda), & -\vartheta_x(t_1,t_o,\lambda) \\ -\varphi(t_1,t_o,\lambda), & \vartheta(t_1,t_o,\lambda) \end{pmatrix} S(t,t_o,\lambda) = w_{t_o}^{t_1}(\lambda) S(t,t_o,\lambda). \tag{3}$$

We have $w_{t_1}^{t}(\lambda) w_{t_o}^{t_1}(\lambda) = w_{t_o}^{t}(\lambda)$ from this definition. The trace $Tr\ w_{t_o}^{t}(\lambda)$ determines eigenvalues of $w_{t_o}^{t}(\lambda)$ as $\det(w_{t_o}^{t}(\lambda)) = 1$.

The 3-by-3 matrix

$$\Omega_{t_o}^t(\lambda) = \begin{pmatrix} \vartheta^2(t,t_o,\lambda), & 2^{\frac{1}{2}}\vartheta(t,t_o,\lambda)\varphi(t,t_o,\lambda), & -\varphi^2(t,t_o,\lambda) \\ 2^{-\frac{1}{2}}[\vartheta^2(t,t_o,\lambda)]_t, & [\vartheta(t,t_o,\lambda)\varphi(t,t_o,\lambda)]_t, & -2^{-\frac{1}{2}}[\varphi^2(t,t_o,\lambda)]_t \\ -\vartheta_x^2(t,t_o,\lambda), & -2^{\frac{1}{2}}\vartheta_x(t,t_o,\lambda)\varphi_x(t,t_o,\lambda), & \varphi_x^2(t,t_o,\lambda) \end{pmatrix} \tag{4}$$

has more interesting properties. Its product $T(t) = \Omega_{t_o}^t(\lambda)T(t_o)$ by a constant 3-by-1 matrix $T(t_o)[\tilde{T}(t_o) = (a,b,c)]$ satisfies the ordinary first-order differential equation

$$T_t(t) = 2^{\frac{1}{2}}Z(t,\lambda)T(t) \tag{5}$$

according to (1a). The only nonzero elements of the singular 3-by-3 matrix $Z(t,\lambda)$ are

$$Z_{12}(t,\lambda) = -Z_{23}(t,\lambda) = 1, \quad -Z_{21}(t,\lambda) = Z_{32}(t,\lambda) = \lambda - q(t), \tag{6}$$

so that

$$-J\tilde{Z}J = Z \quad (J_{ik} = \delta_{i+k,4}; \quad i,k = 1,2,3). \tag{7}$$

Denoting

$$\begin{aligned} T_1(t) = w(t,\lambda) &= (\Omega_{t_o}^t(\lambda)T(t_o))_1 \\ &= a\vartheta^2(t,t_o,\lambda) + 2^{\frac{1}{2}}b\vartheta(t,t_o,\lambda)\varphi(t,t_o,\lambda) - c\varphi^2(t,t_o,\lambda) \end{aligned} \tag{8}$$

we have $T_2(t) = 2^{-\frac{1}{2}}w_t(t,\lambda)$ and $T_3(t,\lambda) = -\frac{1}{2}w_{tt} - [\lambda - q(t)]w$ according to (1a). In these expressions, only the third row of (5) is not an identity giving

$$-\frac{1}{2}w_{ttt} - [\lambda - q(t)]_t w - [\lambda - q(t)]w_t = (\lambda - q(t))w_t. \tag{9}$$

Since it is a linear third-order differential equation, the expression (8) with three independent constants represents its general solution. This connection was first presented by Appell in [10]. (9) turns, however, into the nonlinear second-order differential equation when integrated

$$ww_{tt} - \frac{1}{2}w_t^2 + 2(\lambda - q)w^2 = -\frac{1}{2}\mu^2 = -(b^2 + 2ac). \tag{10}$$

The integration constant $\frac{1}{2}\mu^2$ is equal to the discriminant of the quadratic form (8) and of $T_3(t)$ respectively. (10) is, in fact, the first integral of the system (5) as

$$\tilde{T}(t)JT(t) = \tilde{T}(t_o)JT(t_o) - \frac{1}{2}\mu^2. \tag{11}$$

The function $w(t,\lambda)$ defined in (8) is real for real a, b, c and λ. It is always nonzero provided $\mu^2 < 0$ as the relation $w_t^2(t_2,\lambda) = \mu^2$, which follows from (10) for $w(t_2,\lambda) = 0$ cannot be fulfilled by real $w(t,\lambda)$. We can choose then $w(t,\lambda) > 0$ as both $\pm w(t,\lambda)$ are solutions to (10). There are constant tangents at possible simple zeros t_2 for $\mu^2 > 0$ as

$$w(t,\lambda) = \pm\mu(t - t_2)\,\{1 + const.(t - t_2) + ...\} \tag{12}$$

in the close neighborhood of the chosen zero. The expression (8) turns into the square of a linear combination of ϑ and φ provided $\mu = 0$, so that zeros of $w(t,\lambda)$, if any, are then doubled. Simple and double zeros of $w(t,\lambda)$ never coexist and thus the principle value of integrals from $w^{-1}(t,\lambda)$ always exist for $\mu^2 > 0$. Substituting $f(t) = dw^{\frac{1}{2}}(t,\lambda)$, we transform (10) into the Schrödinger equation (1a) with an added nonlinear term - the famous Milne equation [7]

$$f_{tt} + [\lambda - q(t)]f + \frac{\mu^2 d^4}{4f^3} = 0. \tag{13}$$

Two functions

$$m_\alpha(t,\lambda) = \frac{\alpha\mu + w_t(t,\lambda)}{2w(t,\lambda)} \quad (\alpha = \pm) \tag{14}$$

obey the Riccati equation

$$m_{\alpha,t} + m_\alpha^2 + \lambda - q(t) = 0 \tag{15}$$

in virtue of (10), so that

$$\begin{aligned}\Psi_\alpha(t,t_o,\lambda) &= 2w(t_o\lambda)\eta_\alpha(t)\exp\left\{\int_{t_o}^{t} dx\, m_\alpha(x,\lambda)\right\} \\ &= 2w(t_o\lambda)[\vartheta(t,t_o,\lambda) + m_\alpha(t_o\lambda)\varphi(t,t_o,\lambda)] \quad (\alpha = \pm)\end{aligned} \tag{16}$$

are solutions to (1a). The introduction of the functions $\eta_\alpha(t)(\alpha = \pm)$ is not necessary when $\mu^2 < 0$ i.e. $w(t,\lambda) \neq 0$ as then $\eta_\alpha(t) = 1$. It ensures the continuity of the first derivative $w_t(t,\lambda)$ at zeros of $w(t,\lambda)$ when $\mu^2 > 0$. If there is an infinite number of zeros we denote them by $..., x_{-2}, x_{-1}, x_o, x_1, x_2, ...$ in such a way that $w(t_o,\lambda) > 0$ for $t_o \in (x_o, x_1)$ and $w_t(x_o,\lambda) = \mu = -w_t(x_1,\lambda) > 0$. Then $\eta_+(t) = 1$ for $t \in (x_{4n}, x_{4n+2}), \eta_+(t) = -1$ for $t \in (x_{4n+2}, x_{4n+4})$, $\eta_-(t) = 1$ for $t \in (x_{4n-1}, x_{4n+1}), \eta_-(t) = -1$ for $t \in (x_{4n+1}, x_{4n+3})$ and the relation $\eta_-(t) =$ sign $w(t_o,\lambda)$sign $w(t,\lambda)\eta_+(t)$ applies. Using this relation and the explicit form of $m_\alpha(t,\lambda)$, we have

$$\Psi_+(t,t_o,\lambda)\Psi_-(t,t_o,\lambda) = 4w(t,\lambda)w(t_o,\lambda) \tag{17}$$

in accordance with (8). The discontinuities of $\eta_\alpha(t)$ do not obstruct as they occur at zeros of either $\Psi_\alpha(t,t_o,\lambda)$, which constitute consecutively all zeros of $w(t,\lambda)$ and vice versa.

The matrices $\Omega^t_{t_o}(\lambda)$ represent a fundamental system of solutions to (5) with a group structure $\Omega^{t_o}_{t_o}(\lambda) = \mathbf{1}, \Omega^t_{t_1}(\lambda)\Omega^{t_1}_{t_o}(\lambda) = \Omega^t_{t_o}(\lambda)$. The elementary relations

$$\det(\Omega^t_{t_o}(\lambda)) = 1, \Omega^{t_o}_t(\lambda) = [\Omega^t_{t_o}(\lambda)]^{-1} = J\tilde{\Omega}^t_{t_o}J \tag{18}$$

bring about the equality of the trace

$$Tr\Omega^t_{t_o}(\lambda) = [Trw^t_{t_o}(\lambda)]^2 - 1 = [\vartheta(t,t_o,\lambda) + \varphi_x(t,t_o,\lambda)]^2 - 1 \tag{19}$$

to the sum of the main cofactors of the matrix $\Omega^t_{t_o}(\lambda)$. There exists, therefore, the unit eigenvalue of $\Omega^t_{t_o}(\lambda)$ the corresponding eigenmatrix being

$$\tilde{\Phi}(t,t_o,\lambda) = \left(\varphi(t,t_o,\lambda), 2^{-\frac{1}{2}}[\varphi_x(t,t_o,\lambda) - \vartheta(t,t_o,\lambda)], \vartheta_x(t,t_o,\lambda)\right) \tag{20}$$

apart from an uninteresting proportionality factor. It is a real matrix for real λ.

There is an alternative expression for $\Omega^t_{t_o}(\lambda)$ in terms of the potential $q(t)$. One can transform (5) in the Volterra integral equation by integration, its iterated solution is

$$\begin{aligned}\Omega^t_{t_o}(\lambda) = 1 + \sum_{n=1}^{\infty} 2^{\frac{n}{2}} \int_{t_o}^{t} dt_n \int_{t_o}^{t_n} dt_{n-1} ... \int_{t_o}^{t_2} dt_1 Z(t_n,\lambda)Z(t_{n-1},\lambda)...\\ ...Z(t_2,\lambda)Z(t_1,\lambda) = T \exp 2^{\frac{1}{2}} \int_{t_o}^{t} dx Z(x,\lambda),\end{aligned} \tag{21}$$

where the symbol T requires the usual ordering of matrices $Z(t_i, \lambda)$ in the expansion of the exponential function (see for example, [11,12]). For example

$$\begin{aligned}&\varphi^2(t,t_o,\lambda) = \\ &(t-t_o)^2 + \sum_{n=2}^{\infty}(-1)^{n-1}2^{\frac{n-1}{2}}\int_{t_o}^{t} dt_{2n}\int_{t_o}^{t_{2n}} dt_{2n-1}\dots\int_{t_o}^{t_3} dt_2(t_2-t_o) \\ &[2\lambda - q(t_{2n-2}) - q(t_{2n-1})][2\lambda - q(t_{2n-4}) \\ &- q(t_{2n-3})]\dots[2\lambda - q(t_3) - q(t_2)].\end{aligned} \tag{22}$$

One can show in the analogous way to the convergence proof of the series (21) that $\Omega_t^{t+\tau}(\lambda)$ i.e. $\vartheta(t+\tau,t,\lambda), \varphi(t+\tau,t,\lambda)$ and their x-derivatives are u.a.p. functions of t for any fixed number τ, they are therefore bounded.

3. Periodic Potentials

Continuous periodic (c.p.) potentials constitute a special subclass of u.a.p. potentials. It is worth going through formulas of the preceding Section from the point of view of some c.p. potential $q(t)$ with a period c, as they are simple and instructive.

Substituting $t_\ell = x_\ell + t$ for the integration variable in (21), we immediately see that $\Omega_t^{t+c}(\lambda), \vartheta(t+c,t,\lambda), \varphi(t+c,t,\lambda)$ and their x-derivatives are c.p. functions of t with period c. The trace $\Omega_t^{t+c}(\lambda)$ is equal to $F^2(\lambda) - 1$, where

$$F(\lambda) = \vartheta(t+c,t,\lambda) + \varphi_x(t+c,t,\lambda) = Trw_t^{t+c}(\lambda) \tag{23}$$

is the well-known discriminant of the Hill's equation, which determines its spectrum. The expression (23) along with the Wronskian $W(\vartheta,\varphi)$ are the only invariants with respect to translations of the initial point $t = t_o$. The matrix $\Phi(t+c,t,\lambda)$ defined by (20) is not only a c.p. eigenmatrix of $\Omega_t^{t+c}(\lambda)$ belonging to the unit eigenvalue, but it is also a c.p. solution to (5), so that

$$\begin{aligned}\Phi(t+c,t,\lambda) &= \Omega_{t_o}^t(\lambda)\Phi(t_o+c,t_o,\lambda) = \frac{1}{c}\int_t^{t+c} dx\Omega_x^t\Phi(x+c,x,\lambda) \\ &= M(\Omega_x^t\Phi(x+c,x,\lambda)).\end{aligned} \tag{24}$$

We demonstrate by this mean-value transcription only that it is not easy and simple to solve the corresponding system of the homogeneous Fredholm equations. Thus $w(t,\lambda) = \varphi(t+c,t,\lambda)$ is the only periodic solution to the modified Milne equation (10), in which $\mu^2(\lambda) = F^2(\lambda) - 4$ according to (20), (11) and (23) provided $\mu(\lambda) \neq 0$. Inserting this function into (16) we get the well-known Bloch functions

$$\begin{aligned}\Psi_\alpha(t,t_o,\lambda) &= \exp[\alpha p(\lambda)(t-t_o)]w_\alpha(t,t_o,\lambda)\\ &= 2[|\varphi(t_o+c,t_o,\lambda)\varphi(t+c,t,\lambda)|]^{\frac{1}{2}}\\ &\quad \eta_\alpha(t)\ \exp\left\{\frac{\alpha[F^2(\lambda)-4]^{\frac{1}{2}}}{2}\int_{t_o}^{t}\frac{dx}{\varphi(x+c,x,\lambda)}\right\},\end{aligned} \tag{25}$$

where

$$p(\lambda) = \frac{[F^2(\lambda)-4]^{\frac{1}{2}}}{2c}\int_t^{t+c}\frac{dx}{\varphi(x+c,x,\lambda)} = \frac{[F^2(\lambda)-4]^{\frac{1}{2}}}{2}\,M(\varphi^{-1}(x+c,x,\lambda)) \tag{26}$$

and $\eta_\alpha(t)$ are periodic functions with period c, as $p(\lambda)$ is a well-defined quantity and $w_\alpha(t,t_o,\lambda)$ are c.p. functions with period c. The 'decrement' $p(\lambda)$ is real or purely imaginary depending on whether real values of λ lie within the instability ($|F(\lambda)|>2$) or stability ($|F(\lambda)|<2$) intervals. The formula (26) is identical with that derived in [2] for a potential leading to a finite number of stability intervals, if we realize that the function $m(x,\lambda)$ of the quoted paper equals the function $\varphi(t+c,t,\lambda)$ used here. For λ from the instability interval we employ the Cauchy's principle value and the limit $t\pm\varepsilon(\varepsilon\to 0)$ for contingent zeros t of $\varphi(t+c,t,\lambda)$ respectively in the integrals of (26). The usual definition of $p(\lambda)$, the alternative of which is (26), is given by the equation $\cosh\ cp(\lambda) = F(\lambda)/2$.

Mean value (1b) induces the scalar product $M(f(x)g^*(x))$ for two u.a.p. functions $f(t)$ and $g(t)$ and the norm $M^{\frac{1}{2}}(f(x)f^*(x))$. They are finite for the Bloch functions (25) from the stability intervals. From this point of view, we can interpret the spectrum of (1a), which consists of the closure of the stability intervals, as a continuous point spectrum. We prefer this interpretation to that given in [4] (continuous spectrum), as it corresponds more to its physical rise.

The matrix $\Phi(t+h,t,\lambda)$ defined according to (20) for a fixed $h\neq nc$ is also a c.p. matrix of t; it does not obey,however, (24). Its dependence on h is too complicated it factorizes when h is equal to some periods i.e. $h=\pm nc$ ($n=1,2,...,\infty$). We have then

$$\Phi(t\pm nc,t,\lambda) = \frac{\sinh[\pm ncp(\lambda)]}{\sinh[cp(\lambda)]}\Phi(t+c,t,\lambda). \tag{27}$$

In deriving this relation, we must express in succession $S(t\pm nc,t_o,\lambda)$ by means of $S(t,t_o,\lambda)$. We see that $[\varphi(t_o\pm nc,t_o,\lambda)]^{-1}\Phi(t\pm nc,t,\lambda)$ is independent of n for any n and is a solution of (24). This holds also in the limit $n\to\pm\infty$, which is well defined and finite.

4. The Bloch-Weyl Functions

The preceding consideration indicates that the existence of a u.a.p. solution to (5) and (10) can help us in deciding whether there exists the Bloch-Weyl solution to (1a) or not. If there is a u.a.p. solution to (5) on R, its possible dependence on initial values is factorizable, so hat

$$T(t,\lambda) = \Omega_{t_o}^t(\lambda)T(t_o,\lambda) = \Omega_{t_1}^t T(t_1,\lambda) = M(\Omega_x^t T(x,\lambda)). \tag{28}$$

When solving this equation, we start from the matrix function

$$U(t,\tau,\lambda) = \kappa^{-1}(\tau)\Phi(t+\tau,t,\lambda), \tag{29}$$

where the multiplication constant will be specified later on. The matrix (29) is a u.a.p. function of t for some fixed general τ, but it does not obey (28). In an analogous way to the c.p. potential, we look, however, for particular values of the constant τ, which are now translation numbers $\tau(\varepsilon)$ of the u.a.p. potential $q(t)$, expecting that it will lead us to the u.a.p. solution of (28).

We recall that for any u.a.p. function $f(t)$ $|f(t+\tau(\varepsilon)| - f(t)| < \varepsilon$ uniformly with respect to t, where the translation numbers $\tau(\varepsilon)$ belonging to ε fill in even a subinterval in every inclusion interval of length $\ell(\varepsilon)$ on R [13,14].

Let us develop $U(t_o,\tau(\varepsilon),\lambda)$ according to (28). One obtains

$$\begin{aligned} \Omega_{t_o}^{t+\tau}(\lambda)U(t_o,\tau,\lambda) &= \Omega_{t_o+\tau}^{t+\tau}(\lambda)\ \Omega_{t_o}^{t_o+\tau}(\lambda)U(t_o,\tau(\varepsilon),\lambda) \\ = \Omega_{t_o+\tau}^{t+\tau}(\lambda)U(t_o,\tau,\lambda) &= [\Omega_{t_o}^t + \varepsilon M(t,t_o)]U(t_o,\tau,\lambda) \end{aligned} \tag{30}$$

in virtue of (20) and (21), where the matrix $M(t,t_o)$ is finite. Thus developed values at $t+\tau(\varepsilon)$ and t do not much differ. Hence $\Omega_{t_o}^t(\lambda)U(t_o,\tau(\varepsilon),\lambda)$ remains finite for any t because $\tau(\varepsilon) \to \pm\infty$ as $\varepsilon \to 0^+$ for nonperiodic u.a.p. functions and provided the finite limit $U(t_o,\tau(\varepsilon),\lambda)$ as $\varepsilon \to 0^+$ exists. Choosing t_o and $\tau(\varepsilon)$ such that

$$\kappa(\tau(\varepsilon)) = \varphi(t_o+\tau(\varepsilon),t_o,\lambda) \neq 0, \tag{31}$$

we have $U_1(t_o,\tau(\varepsilon),\lambda) = 1$ and the existence of the finite limits of the other matrix elements follows from the next consideration.

The matrix function $U(t,\tau,\lambda)$ obeys the nonlinear equation

$$U_t = 2^{\frac{1}{2}} ZU + 2^{-\frac{1}{2}} \Delta q W(U), \tag{32}$$

where $\Delta q = q(t + \tau(\varepsilon)) - q(t)$ and the nonlinear term

$$\tilde{W}(U) = (0, U_1, -U_2 \pm [\tilde{U} JU + 2\kappa^{-2}(\tau)]^{\frac{1}{2}}) \tag{33}$$

as

$$\begin{aligned} &\varphi_x(t+\tau, t, \lambda) + \vartheta(t+\tau, t, \lambda) \\ &= [(\varphi_x(t+\tau, t, \lambda) - \theta(t+\tau, t, \lambda))^2 + 4\varphi(t+\tau, t, \lambda)\vartheta_x(t+\tau, t, \lambda) + 4]^{\frac{1}{2}}. \end{aligned} \tag{34}$$

Since $\kappa^{-1}(\tau)$ is bounded and $|\Delta q| < \varepsilon$, one can solve (33) perturbatively $U(t, \tau, \lambda) = T(t, \lambda) + \varepsilon Y_1(t, \tau, \lambda) + ...$ The zero-order term $T(t, \lambda)$ is a solution to (28), so that

$$T(t, \lambda) = \lim_{\varepsilon \to 0+} U(t, \tau(\varepsilon), \lambda) \tag{35}$$

is also a solution of this equation as Δq can be made arbitrarily small. It is the u.a.p. solution to (28) as it is the uniform limit with respect to t of the u.a.p. matrices $U(t, \tau(\varepsilon), \lambda)$. The matrix element $T_1(t, \lambda) = w(t, \lambda)$ is the u.a.p. solution to (10) with $2T(t_o, \lambda) JT_o(t_o, \lambda) = \mu^2(\lambda)$ according to (11). As a u.a.p. function of t, it has an infinite number of simple or double zeros or no zeros at all for real λ's.

No zeros of $w(t, \lambda)$ appear always when $\mu^2(\lambda) < 0$. Then there exists purely imaginary main value

$$p(\lambda) = \frac{1}{2} \mu(\lambda) M(w^{-1}(x, \lambda)) \tag{36}$$

as $w^{-1}(t, \lambda)$ is also a u.a.p. function of t [13, 14]. By means of this formula we rewrite (16) in the form

$$\begin{aligned} &\Psi_\alpha(t, t_o, \lambda) \\ &= 2[w(t_o, \lambda) w(t, \lambda)]^{\frac{1}{2}} \exp \left[\alpha \int_{t_o}^{t} dx \left(\frac{\mu(\lambda)}{2w(x, \lambda)} - p(\lambda) \right) \right] e^{\alpha p(\lambda)(t - t_o)}, \end{aligned} \tag{37}$$

which represents the generalized Bloch-Weyl functions. This is because $w(t,t_o,\lambda) = \int_{t_o}^{t} dx[\frac{1}{2}\mu(\lambda)w^{-1}(x,\lambda) - p(\lambda)]$ is not, in general, a u.a.p. function even though mean value of $w_t(t,t_o\lambda)$ is zero [14]. We stress, however, that on the other hand, norm $M^{\frac{1}{2}}(|\Psi_\alpha(t,t_o,\lambda|^2) = 2w^{\frac{1}{2}}(t_o,\lambda)M^{\frac{1}{2}}(w(x,\lambda))$ always exists and is finite. Those functions $w(t,t_o,\lambda)$, which are not u.a.p., are more or less sophisticated theoretical artefacts [14], and we can expect that they are, for the most part, u.a.p. functions. We presume that this happens when, for example, modules corresponding to the u.a.p. potentials $q(t)$ have no limiting points except infinity-like modules of quasi-periodic potentials.

There can or may not exist mean value (36) for $w(t,\lambda)$ which has an infinite number of simple zeros ($\mu^2(\lambda) > 0$), even though we take the Cauchy's principle value in the integral (36). The only interesting and important property of this integral is that it is independent of t

$$\begin{aligned}&\lim_{\varepsilon\to 0+}\left\{\int_{t_o}^{t_o+\tau(\varepsilon)}\frac{dx}{w(x,\lambda)} - \int_{t}^{t+\tau(\varepsilon)}\frac{dx}{w(x,\lambda)}\right\}\\ &= \lim_{\varepsilon\to 0+}\int_{t_o}^{t} dx\frac{w(x+\tau(\epsilon),\lambda) - w(x,\lambda)}{w(x,\lambda)w(x+\tau(\varepsilon),\lambda)} = 0.\end{aligned} \tag{38}$$

Here, we have replaced the current quantity T of mean value by the translation number $\tau(\varepsilon)$ of $w(t,\lambda)$ belonging to ε and we have also assumed the limits in the integral not to be zeros of $w(t,\lambda)$, which is always possible.

The function $w(t,\lambda)$ need not possess any zero in spite of the inequality $\mu^2(\lambda) > 0$. We shall soon see that there is even an infinite interval of real λ's, for which $|w(t,\lambda)| > 0$. We can then apply formulas (36) and (37) again. Norm of these Bloch functions is not finite, as $p(\lambda)$ is real now.

If $\mu(\lambda) = 0$, the only Bloch function

$$\begin{aligned}\Psi(t,t_o,\lambda) &= 2w(t_o,\lambda)\vartheta(t,t_o,\lambda) + w_t(t_o,\lambda)\varphi(t,t_o,\lambda)\\ &= 2|w(t_o,\lambda)w(t,\lambda)|^{\frac{1}{2}}\eta(t)\end{aligned} \tag{39}$$

exists, according to (16), and has finite norm $2w^{\frac{1}{2}}(t_o,\lambda)M^{\frac{1}{2}}(|w(x,\lambda)|)$.

5. Inverse Problems

We need to know, as usual, analytic properties of $w(t,\lambda)$ as a function of complex λ when solving the inverse problem of (1a). According to (29), (20), (31) and (35),

$w(t,\lambda)$ is a limit ratio of two entire functions of order $\frac{1}{2}$ [15, 16]. Possible poles of this ratio do not depend, however, on t. These poles, as well as the other possible t-independent factors of $w(t,\lambda)$, appear quadratically in $\mu^2(\lambda)$ according to (10) so that we can cancel them in both $w(t,\lambda)$ and $\mu^2(\lambda)$. This means then that we can consider $\mu^2(\lambda)$ as an infinite product of linear terms $\lambda - \Lambda_n$ and assume t-dependent zeros $\lambda = \Lambda(t)$ of $w(t,\lambda)$ only.

The root functions $\Lambda(t)$, which are implicitly defined by $w(t,\Lambda) = 0$, satisfy the differential equation

$$\Lambda_t(t) w_\lambda(t,\Lambda) = -w_t(t,\Lambda) = \pm\mu(\Lambda(t)) \tag{40}$$

according to (10). Both Λ_t and $w_\lambda(t,\Lambda)$ are, therefore, nonzero provided $\Lambda(t)$ does not assume value Λ_n of a root of $\mu^2(\lambda)$. When $\Lambda(t_n) = \Lambda_n$, we have $2w_\lambda(t_n,\Lambda_n)$ $w_{tt}(t_n,\Lambda_n) = -[\frac{\delta}{\delta\lambda}\mu^2(\lambda)]_{\lambda=\Lambda_n} \neq 0$ according to (10), so that $0 \neq w_\lambda(t_n,\Lambda_n)$ again (and $w_{tt}(t_n,\Lambda_n) \neq 0$ as it ought to be). At the same time, $\Lambda_{tt}(t_n) = -w_\lambda^{-1}(t_n,\Lambda_n) w_{tt}(t_n,\Lambda_n) \neq 0$, so that $\Lambda(t)$ returns after catching up the root Λ_n. The derivative $\Lambda_t(t)$ changes its sign accordingly, which is made possible by the alternative signs in (40).

The roots $\Lambda(t)$ are simple, as $w_\lambda(t,\Lambda)$ is always nonzero. They are real, as $w(t,\lambda) = 0$ means that either $\Psi_+(t,t_o,\lambda) = 0$ or $\Psi_-(t,t_o,\lambda) = 0$ according to (17), which can apply only if λ is real [3]. There is, in general, an infinite number of $\Lambda_n(t)$ $(n = 1,2,...,\infty)$, so that $w(t,\lambda)$ is a convergent infinite product of linear term $\lambda - \Lambda_n(t)$. However, (10) shapes these products for both $w(t,\lambda)$ and $\mu^2(\lambda)$ at the same time. It cuts out the factor $\exp(g(\lambda))$, where $g(\lambda)$ is an entire function in the infinite product expansion of $w(t,\lambda)$, but fixes the relative numerical factor for $w(t,\lambda)$ and $\mu^2(\lambda)$. Next we must associate two roots $\lambda_n < \mu_n$ with every $\Lambda_n(t)$ and give another root λ_o, in addition to them, as the left-hand side of (10) is quadratic in w and linear in λ. We can choose λ_n and μ_n such that $[\lambda_n,\mu_n]$ is a range of $\Lambda_n(t)$ as it corresponds to (40) and to the consideration which attaches to it. Thus $\Lambda_n(t)$ are u.a.p. functions which move monotonically between two fixed turning points. If we take mean value of (10), we see that $\mu^2(\lambda) > 0$ for all real λ's smaller than λ_o, where λ_o is the smallest root of $\mu^2(\lambda)$. It must be negative, which follows from mean value of the Riccati equation (15) for $m(t,\lambda_o) = \frac{1}{2}w_t(t,\lambda_o)/w(t,\lambda_o)$ and from (1b). At the same time $w(t,\lambda) > 0$ for real $\lambda \leq \lambda_o$, which follows from (10) again. Hence there are two unbounded generalized Bloch functions (37) for $\lambda < \lambda_o$.

In the 'allowed' bands $(\lambda_o,\lambda_1),(\mu_1,\lambda_2),(\mu_2,\lambda_3),...$ we can also choose $w(t,\lambda)$ positive, so that there are two bounded generalized Bloch functions (37). At the endpoints of these bands $\lambda_o,\lambda_1,\mu_1,\lambda_2,\mu_2,...$ we can build up only one bounded Bloch function (39). The spectrum of (1a) consists, therefore, of the closure of these intervals.

The preceding exposition can be made clearer if we consider a finite number of intervals $[\lambda_n, \mu_n]$ $(n = 1, ..., N)$ i.e. N 'forbidden' bands. Both $\mu^2(\lambda)$ and $w(t, \lambda)$ are now polynomials

$$\mu^2(\lambda) = 4(\lambda_o - \lambda)\Pi_{n=1}^{N}(\lambda_n - \lambda)(\mu_n - \lambda), \tag{41a}$$

$$w(t, \lambda) = \Pi_{n=1}^{N}(\Lambda_n(t) - \lambda) = (-\lambda)^N + \sum_{n=0}^{N-1} w_n(t)\lambda^n \,. \tag{41b}$$

Inserting these polynomials in (10) and comparing coefficients of the same powers of λ, we get $2N + 1$ equations that $w_n(t)$ $(n = 0, ..., N - 1)$, symmetric functions of λ_n, μ_n and $\Lambda_n(t)$ respectively, and the potential $q(t)$ enter. The structure of the first N equations is such that we can calculate in a simple way $w_n(t)$ by means of $w_{n+1}(t), ..., w_{N-1}(t)$ and the other ingredients, for example,

$$\begin{aligned} 2(-1)^N w_{N-1}(t) + \lambda_o + \sum_{n=1}^{N}(\lambda_n + \mu_n) &= q(t) \\ &= \lambda_o + \sum_{n=1}^{N}[\lambda_n + \mu_n - 2\Lambda_n(t)] \end{aligned} \tag{42}$$

This means that any $w_n(t)$ $(n = 0, ..., N - 1)$ is expressible by means of $q(t)$, so that the remainig N equations must be fulfilled identically. They are, in fact, imposed conditions that the potential must satisfy in order that the number of the spectral intervals may be finite. If $N = 1$, this potential reads $q(t) = N(N + 1)p(t, g_2, g_3) + \frac{1}{3}(\lambda_o + \lambda_1 + \mu_1)$, where $p(t, g_2, g_3)$ is the Weierstrass's elliptic function, its determining parametes g_2 and g_3 being symmetric functions of the roots $\lambda_o < \lambda_1 < \mu_1$. It is worth noticing that the allowed potentials are c.p. and not u.a.p. functions of t in this case. The solution of the inverse problem consists in inserting products from (41) into (40), in solving the system of differential equations for $\Lambda_n(t)$ and substituting these $\Lambda_n(t)$ in (42).

If we have an infinite number of the forbidden bands (λ_n, μ_n) $(n = 1, ..., \infty)$, we must modify the proceding procedure. Intervals (λ_n, μ_n) cannot overlap for different n's according to their definition, so that $\lambda_n \sim \ell_n = Ln^{1+\varepsilon} \sim \mu_n$ as $n \to \infty$ where positive ε ensures the absolute convergence of the infinite products

$$\mu^2(\lambda) = 4(\lambda_o - \lambda)\Pi_{n=1}^{\infty}\frac{(\lambda_n - \lambda)(\mu_n - \lambda)}{\ell_n^2}\,, \tag{43a}$$

$$w(t, \lambda) = \Pi_{n=1}^{\infty}\frac{\Lambda_n^{(t)} - \lambda}{\ell_n} \quad (\Lambda_n(t)\varepsilon[\lambda_n, \mu_n]) \,. \tag{43b}$$

They are modified forms of the products corresponding to the Hadamard-Weierstrass's theorem for entire functions of λ with real simple zeros.

Dividing (10) by $w^2(t,\lambda)$ we expand it into the sum of partial fractions, the schematic structure of which is $c_1\lambda + c_o + c_{-1}(\lambda - \Lambda)^{-1} + c_{-2}(\lambda - \Lambda)^{-2} = 0$. Multiplying factors in (43) were chosen such that $c_1 = 0$ identically. Comparison of the constant terms gives the extended form of (42)

$$q(t) = \lambda_o + \sum_{n=1}^{\infty} (\lambda_n + \mu_n - 2\Lambda_n(t)). \tag{44}$$

Comparing the coefficients at the quadratic-pole terms one gets the system of the nonlinear differential equations of first order

$$\Lambda_{nt} = \pm 2[(\lambda_o - \Lambda_n)(\lambda_n - \Lambda_n)(\mu_n - \Lambda_n)]^{\frac{1}{2}} \\ \Pi_{k\neq n}^{\infty} \frac{[(\lambda_k - \Lambda_n)(\mu_k - \Lambda_n)]^{\frac{1}{2}}}{(\Lambda_k - \Lambda_n)} \quad (n = 1, ..., \infty), \tag{45}$$

which is the explicitly rewritten form of (40). We see that the right-hand side of (45) indeed retains its sign within the interval (λ_n, μ_n) changing it in the opposite one at the fixed turning points λ_n and μ_n only, where $\Lambda_{n,t} = 0$. Comparison of the coefficients at the simple poles gives equations for Λ_{ntt}. We do not write them down as they must follow from (45). They confirm that λ_n or μ_n are, in fact, the turning points as $\Lambda_{ntt} \neq 0$ there.

Equation (45) represents a system determining the functions $\Lambda_n(t)$ provided some initial values $\Lambda_n(t_o)\varepsilon[\lambda_n, \mu_n]$ and the initial signs $\sigma(t_o) = \pm 1 = \text{sign } \Lambda_{nt}(t_o)$ on the left-hand sides of (45) are chosen. The uniqueness of such a solution is shown by Levitan in [3]. Inserting the solutions $\Lambda_n(t)$ into (44) one reconstructs the potential $q(t)$ from the spectral data $(\lambda_o; \lambda_n, \mu_n, \Lambda_n(t_o), \text{sign } \Lambda_{nt}(t_o),\ n = 1, ..., \infty) = 0$. Although Levitan derives the system (45) primarily for periodic potentials, he proves that it also has u.a.p. solutions for some general intervals $[\lambda_n, \mu_n]$. If $\mu^2(\lambda)$ and $w(t,\lambda)$ are of the type (41), the reconstructed potential is quasi-periodic only[3].

We can fashion the formula (26) into the form of the remarkable formulas by Hochstadt. Inserting the product (43b) into (26) and making use of (40), (45) and (43b) one obtains

$$
\begin{aligned}
p(\lambda) &= \lim_{T\to\infty} -\frac{\mu(\lambda)}{4T}\int_{-T}^{T} dt \sum_{n=1}^{\infty} \frac{1}{w_\lambda(t,\Lambda_n(t))(\Lambda_n(t)-\lambda)} \\
&= \frac{1}{2c}\sum_{n=1}^{\infty}\oint_{(n)} d\Lambda \left[\frac{\lambda_o-\lambda}{\lambda_o-\Lambda}\cdot\Pi_{k=1}^{\infty}\frac{(\lambda_k-\lambda)(\mu_k-\lambda)}{(\lambda_k-\Lambda)(\mu_k-\Lambda)}\right]^{\frac{1}{2}}\frac{1}{\Lambda-\lambda},
\end{aligned}
\tag{46}
$$

where one integrates in the counterclockwise direction over a contour around every instability cut $(n) = [\lambda_n, \mu_n]$ $(n = 1, ..., \infty)$ in the complex plane of Λ and sums up the integrals. Thus $p(\lambda)$ is determined only by the spectrum again, $\lambda \notin (n)$.

We can work the third component $T_3(t,\lambda) = v(t,\lambda) = -\frac{1}{2}w_{tt}(t,\lambda) - [\lambda - q(t)]$ $w(t,\lambda)$, for which

$$
v_t(t,\lambda) = [\lambda - q(t)]w_t(t,\lambda) \tag{47}
$$

according to (9), in an analogous way to $w(t,\lambda)$. Equation (10) transforms into the equation

$$
vv_{tt} - \frac{1}{2}v_t^2 + \frac{q_t}{\lambda - q}v_t v + 2(\lambda - q)v^2 + \frac{1}{2}(\lambda - q)^2\mu^2(\lambda) = 0. \tag{48}
$$

Possible zeros $\tau_n(t)$ of $v(t,\tau)$ fill in the intervals $[\lambda_n, \mu_n]$ as $-4w(t_o,\lambda)v(t,\lambda) = \Psi_{+t}(t,t_o,\lambda)\Psi_{-t}(t,t_o,\lambda)$ due to (4), (8) and (17). This means that $\tau_t(t) - q(t)$ is always nonzero in these bands, as the equation analogous to (40)

$$
\begin{aligned}
v_\lambda(t,\tau(t))\tau_t(t) &= -v_t(t,\tau) = -[\tau(t) - q(t)]w_t(t,\tau(t)) \\
&= \perp[\tau(t) - q(t)]\mu(\tau(t))
\end{aligned}
\tag{49}
$$

just then admits the right-hand side to be zero only for $\tau(t) = \lambda_n$ or μ_n. Thus the interval [inf $q(t) < 0$, sup $q(t) > 0$] lies within an allowed band, the lowest band $[\lambda_o, \lambda_1]$ being extended to $(-\infty, \lambda_1)$ to this purpose.

The product expansion for $v(t,\lambda)$ must read

$$
v(t,\lambda) = [\tau_o(t) - \lambda]\Pi_{n=1}^{\infty}\frac{[\tau_n(t) - \lambda]}{\ell_n}, \tag{50}
$$

according to (47) and (43b). The u.a.p. function $\tau_o(t)$ moves in some interval, which must lie in the interval $(-\infty, \lambda_o)$. Dividing (48) by the squared product (50) and applying the expansion procedure we get the following equations

$$q(t) = 2\tau_o(t) - \lambda_o + \sum_{n=1}^{\infty}(2\tau_n(t) - \lambda_n - \mu_n), \tag{51a}$$

$$\sum_{n=0}^{\infty} \frac{\tau_{nt}(t)}{\tau_n - q} = 0, \tag{51b}$$

$$\tau_{ot} = \pm 2(\tau_o - q)(\lambda_o - \tau_o)^{\frac{1}{2}} \Pi_{k \neq n}^{\infty} \frac{[(\lambda_k - \tau_o)(\mu_k - \tau_o)]^{\frac{1}{2}}}{\tau_k - \tau_o} \tag{51c}$$

$$\tau_{nt} = \pm 2(\tau_n - q)\frac{[(\lambda_o - \tau_n)(\lambda_n - \tau_n)(\mu_n - \tau_n)]^{\frac{1}{2}}}{\tau_o - \tau_n} \Pi_{k \neq n}^{\infty} \frac{[(\lambda_k - \tau_n)(\mu_k - \tau_n)]^{\frac{1}{2}}}{\tau_k - \tau_n} . \tag{51d}$$

The functions $\tau_n(t)$ aperiodically oscillate between fixed λ_n and μ_n again. In virtue of (51c) and (51d), (51b) is, in fact, an implicit relation expressing $\tau_o(t)$ by means of $\tau_n(t)$ $(n = 1, ..., \infty)$. Thus $\tau_o(t)$ aperiodically oscillates, but without any fixed turning points. At the same time the spectral data D' consists of the chosen set of values $\lambda_n, \mu_n(\sim \ell_n), \tau_n(t_o)\varepsilon[\lambda_n, \mu_n]$ and of the chosen signs $\sigma_n(t_o) = \pm 1 = \text{sign}\ \tau_{nt}(t_o)$ in (51d). They determine uniquely the functions $\tau_n(t)$, which reconstruct finally the potential according to (51a). It is worth noticing that formulas (44), (45) and (51) do not depend on the auxiliary factors ℓ_n in accordance with the Hadamard-Weierstrass's theorem, in which such factors are not explicitly present.

6. The Darboux Transformation

The free choice of $\Lambda_n(t_o)$ or $\tau_n(t_o)$ and $\sigma_n(t_o)$ $(n = 1, ..., \infty)$ in the spectral data means that there is a whole variety of potentials which lead to the same common spectrum, i.e. the function $\mu^2(\lambda)$. This variety, which is 2^n-parametric for given n spectral bands, is constructed, for example, in [2] for periodic potentials by solving the classical inverse Sturm-Liouville problem. This approach cannot be applied, in principle, to the u.a.p. potentials. There exists, however, a tantamount substitute to it which uses the Darboux transformation [9,18,19].

Let us denote the energy parameter λ by σ when $\sigma \leq \lambda_o$. The functions $m_{\pm}(t, \sigma)$, which are defined by (14) and obey the Riccati equation, are u.a.p., real and well-defined functions, as $w(t, \sigma)$ are always nonzero $(w(t, \sigma) > 0)$ for any t and $\sigma \leq \lambda_o$. The new pair of the functions $w_\beta(t, \lambda)(\beta = \pm)$, which we define by the following relation

$$\begin{aligned} &4(\sigma - \lambda)w_\beta(t, \lambda)w(t, \lambda) \\ &= \mu^2(\lambda) - [w_t(t, \lambda) - 2m_\beta(t, \sigma)w(t, \lambda)]^2 , \end{aligned} \tag{52}$$

obeys (10) with the same function $\mu^2(\lambda)$ but the transformed potential

$$q_\beta(t,\sigma) = 2\sigma + 2m_\beta^2(t,\sigma) - q(t) \neq q_{-\beta}(t,\sigma) \quad (\sigma < \lambda_o) \,. \tag{53}$$

This means, however, that the functions

$$m_{\alpha,\beta}(t,\lambda) = \frac{w_{\beta t}(t,\lambda) + \alpha\mu(\lambda)}{w_\beta(t,\lambda)} \quad (\alpha = \pm,\ \beta = \pm) \tag{54}$$

satisfy the Riccati equation with the transformed potential (53). The transformation (52) is therefore an image of the Darboux transformation [9, 18] of the Schrödinger equation (1a) for the functions of the $w(t,\lambda)$ type. Mean value of the new potentials (53) is zero again, so that the Darboux transformations do not change the character of the potentials prescribed by (1b). When we differentiate (52) with respect to t, we get the linear relation

$$\begin{aligned} &w_{\beta t}(t,\lambda) + 2m_\beta(t,\sigma)w_\beta(t,\lambda) \\ &\qquad = -w_t(t,\lambda) + 2m_\beta(t,\sigma)w(t,\lambda), \end{aligned} \tag{55}$$

by means of which we can symmetrize it. The transformation (52) works also for $\lambda = \sigma$ performing the usual limit $\lambda \to \sigma$ in it.

We can now repeat the Darboux transformation substituting $w_\beta(t,\lambda)$ for $w(t,\lambda)$ and $m_{\alpha\beta}(t,\sigma')$ from (54) for $m_\alpha(t,\sigma)$. One obtains four u.a.p. solutions $w_{\alpha\beta}(t,\lambda)$ of (10) for four u.a.p. potentials $2\sigma' + 2m_{\alpha\beta}^2(t,\sigma') - q_\beta(t,\sigma) = q_{\alpha\beta}(t)$. They need not all be different; for example, there are only two different potentials for $\sigma = \sigma'$. We can continue in the same way further on, when constructing a variety of potentials which possess the same spectrum. There also exists the transformation of $v(t,\lambda)$, which is analogous to the transformation (52) and reflects the Darboux transformation of (1).

It is very likely that the Darboux transformations (52) exhibit group properties[20] analogous to those produced by the pulse potentials. In this case, the Darboux transformation removes the lowest level from the discrete part (if any) of the spectrum without touching the remaining ones. The solution of the inverse problem for u.a.p. potentials indicates, on the contrary, that there is no analogous transformation by means of which we could remove the lowest allowed band apart from the other bands.

7. Conclusion

One can solve the one-dimensional u.a.p. and c.p. Schrödinger equations on the same footing provided one makes use of the associated modified Milne equations.

One builds up their u.a.p. or c.p. solutions by means of the Weyl's solutions to the primary Schrödinger equations determining at the same time the spectral functions $\mu^2(\lambda)$, the roots of which demarcate the spectral bands. The Bloch-Weyl functions need not exist for u.a.p. potentials in the forbidden energy bands; which, however, is irrelevant from the physical point of view.

In [8], there is a nice survey of the authors who have (re)discovered the connection between the Schrödinger equation and the Milne equation. We have seen, however, that we can find the source of this connection, in fact, in the Appell paper [10], which is more than one hundred years old. Turning back to today, we cannot omit the remark that also the integro-differential operator L introduced by Calogero and Degasperis [18] can be related to the Appell equation (9). This recalls the history of the Darboux transformation which is a hundred years old, too. It was many times (re)discovered when skillfully applied in mathematics and physics.

References

1. V. A. Marchenko and I. V. Ostrovskii, Mat. Sbornik, **97**, 540 (1975).
2. H. P. McKean and P. van Moerbeke, Inventiones Math., **30**, 217 (1975).
3. B. M. Levitan, *Obratnye zadachi Sturma-Liouvilla*, Nauka, Moscow (1984).
4. M. S. P. Eastham, *The spectral theory of periodic differential equations*, Scott Acad. Phys., Edinburgh (1973).
5. E. I. Dinaburg and Ya. G. Sinai, Funk. Analiz, **9**, 8 (1975).
6. W. M. Zheng, J. Phys. A, **19**, L715 (1986).
7. W. E. Milne, Phys. Rev., **35**, 863 (1930).
8. H. J. Korsch, H. Laurent and R. Möhlenkampf, J. Phys. B, **14**, 4213 (1981).
9. E. Kamke, *Spravochnik po obyk. dif. uravneniam*, **IIL**, Moscow (1950).
10. M. Appell, Comptes Rendus Acad. Sci., Paris **XCI**, 211 (1880).
11. F. R. Gantmacher, *Teoria matric*, GITTL, Moscow (1953).
12. S. S. Schweber, *An introduction to relativistic quantum field theory Row*, Peterson C., Evanston, Elmsford (1961).
13. A. S. Besicovitch, *Almost periodic functions*, C.U.P., Cambridge (1932).
14. B. M. Levitan, *Pochti-periodicheskie funkcii*, GITTL, Moscow (1953).

15. E. C. Titchmarsh, *Eigenfunction expansions associated with second-order differential equations, Vol I + II*, Clarendon Press, Oxford (1946, 1958).

16. B. Ya. Levin, *Raspredelenie kornei celych funkcii*, GITTL, Moscow (1956).

17. H. Hochstadt, Arch. Rat. Mech. Anal., **19**, 353 (1965).

18. F. Calogero and A. Degasperis, *Spectral transform and solitons I*, North-Holland P.C., Amsterdam (1982).

19. L. Trlifaj, Czech. J. Phys., **37B**, 1209 (1987).

20. D. L. Pursey, Phys. Rev., **33D**, 2267 (1986).

A GENERALIZATION OF ROUCHÉ'S THEOREM WITH APPLICATION TO RESONANCES

Heinz Siedentop

Institut für Mathematische Physik
Carolo-Wilhelmina
Mendelssohnstraße 3
D-3300 Braunschweig
Fed. Rep. of Germany

Abstract

In this note an introduction into a method for the number of resonances (and their location) is given. Some useful definitions are collected together with a proof of the main theorem. Numerical applications to a Gauß potential are referred to as well as graphically displayed.

1. Introduction

Localizing resonance eigenvalues has been pursued by several groups during the last years. The methods used may be divided into two categories, those that exclude resonances (Moiseyev[7], Siedentop[13,15], Sölter[18]) and those that give information about the existence of resonances (Siedentop[12,15] and Engdahl and Brändas[2]). Moreover, there exist claims of two other exclusion methods (Froelich et al.[3] – for a discussion see Siedentop[15,16] – and Davidson et al.[1]).

The purpose of this note is to give an introduction into a method that - in addition to excluding resonances and thus obtaining bounds on lifetime and energy - gives information about the number of resonances in regions of the complex plane.

The organization of this note is as follows. In Section 2 we give a review of some prerequisites about formulae for determinants, trace ideals, and related things. In Section 3 we formulate the main result, a generalization of Rouché's theorem for the non-commutative case. Finally, in Section 4 we show how to apply this result to resonances, give a particular implementation for the choice of the trial operator, and show results of a numerical implementation of the procedure.

2. Trace Ideals, Determinants, and All That

Let us collect some useful definitions and facts for the following†:

1. Let $p \in [1, \infty)$, and $\mathcal{H}$ a Hilbert space that we shall always assume separable and complex. The set of all bounded operators A on $\mathcal{H}$ such that $|A|_p^p = \mathrm{tr}((A^*A)^{p/2}) = \mathrm{tr}(|A|^p)$ is finite, is called the trace ideal $\mathcal{I}_p$ of order p. $|.|_p$ is a norm on $\mathcal{I}_p$.
   The compact operators are denoted by $\mathcal{I}_\infty$, the operator norm is denoted by $|.|_\infty$.

2. The trace ideals are ideals in the bounded operators in the algebraic sense, i. e. a product of a trace ideal operator of order p with any bounded operator (left or right product) is again a trace ideal operator of order p. Furthermore the trace ideal $\mathcal{I}_p$ with the corresponding norm $|.|_p$ is a Banach space.

3. For $1 \le q \le p \le \infty$ we have $|A|_p \le |A|_q$ and thus $\mathcal{I}_q \subset \mathcal{I}_p$.

4. Let $A \in \mathcal{I}_p$ and $|z|$ be sufficiently small. Then $\exp(\mathrm{tr}\log(1 + zA))$ defines an analytic function in a neighborhood of $z = 0$. This function extends to an entire function; the extension is denoted by $\det(1 + zA)$ or $\det_1(1 + zA)$ and is called the Fredholm determinant. - We remark that the Fredholm determinant is equal to the usual determinant in the case of a finite dimensional space $\mathcal{H}$.

5. For $p \in [1, \infty)$, $A \in \mathcal{I}_p$, and n some integer not smaller than p define

$$\mathcal{R}_n(A) = (1 + A)\exp(\sum_{j=1}^{n-1} (-A)^j \supset) - 1 \tag{1}$$

   and

$$\det{}_n(1 + A) = \det(1 + \mathcal{R}_n(A)). \tag{2}$$

   $\det_n(1 + A)$ is called the regularized Fredholm determinant.

6. The regularized Fredholm determinants are entire functions and vanish, if and only if -1 is an eigenvalue of A. The Fréchet derivative of the map $\varphi : \mathcal{I}_1 \to \mathbb{C}$, $A \mapsto \det(1 + A)$ assuming -1 not in the spectrum of A is given by the formula

$$[\varphi'(A)](B) = \mathrm{tr}[(1 + A)^{-1} B \det(1 + A)]. \tag{3}$$

7. Let $A, B \in \mathcal{I}_{2p}$ for some $p = 1, 2, \ldots$ Then

$$\mathrm{tr}\,|(A - 1)B + A|^{2p} \ge \dim\ker(1 - A) \tag{4}$$

† For details see Simon[17] and Siedentop[16].

and the operator

$$B = \begin{cases} A\left((1-A)|_{(\ker(1-A))^\perp}\right)^{-1} & \text{on } (1-A)(\mathcal{H}) \\ 0 & \text{on } ((1-A)(\mathcal{H}))^\perp \end{cases} \tag{5}$$

is in I_{2p} and yields equality in (4).

- Inequality (4) thus constitutes a variational principle on the multiplicity of the eigenvalue 1 of the operator A (Müller[8] and Siedentop[11]).

- Assume A_z and B_z depending in some smooth way of a parameter z and assume for some z_0 the left hand side of inequality (4) smaller than one. Then there is a circle centered at z_0 such that it contains no eigenvalue one of A_z. Moreover, if A_z is an integral operator in some trace ideal describing resonances (see below), the diameter of such circles may estimated from below (Sölter [18]).

- The above theorem holds also for compact operators, if inequality (4) is replaced by

$$|(A-1)B + A|_\infty \geq \begin{cases} 1 & \text{if } 1 \in \sigma(A) \\ 0 & \text{if } 1 \notin \sigma(A). \end{cases} \tag{6}$$

8 Later on we shall treat resonances that are described by integral equations

$$R_k\varphi = \varphi \tag{7}$$

with R_k in some trace ideal, i. e. k^2 is a resonance energy, if and only if 1 is an eigenvalue of R_k.
Such a description of resonances is possible e. g. for zero range potentials - cum grano salis these are potentials that decay faster than any exponential, e. g. Gauß function like potentials -, for resonances in the framework of Helffer and Sjöstrand [4], and also for resonances in the dilation analytic formalism and its variants (Siedentop[10]).
Here we give only the following formulae for a three-dimensional particle. For zero range potentials we choose

$$R_k(x,y) = |V|^{1/2}(x)\frac{e^{ik|x-y|}}{-4\pi|x-y|}V^{1/2}(y). \tag{8}$$

For dilation analytic potentials we may choose

$$R_k(x,y) = |V|^{1/2}(e^\vartheta x)e^{2\vartheta}\frac{e^{ike^\vartheta|x-y|}}{-4\pi|x-y|}V^{1/2}(e^\vartheta y). \tag{9}$$

The corresponding one-dimensional kernels may be obtained from these formulae by replacing the free (dilated) Green's functions of the three-dimensional case by those of the one-dimensional one.

9 Combining (4) with the description of resonances via integral kernels yields a variational principle for the multiplicity of resonances that is suitable for excluding resonances and thus obtaining upper and lower bounds on the energy and life-time of the resonances (Siedentop[14]).
Using a partial wave analysis for spherical symmetric potentials this may again be simplified by using the (dilated) free resolvent projected onto a subspace of given angular momentum l, m.
This method has been applied to localize resonances of various potentials, e. g. δ-shell potentials (Siedentop[15]), well potentials (Krueger[5], Siedentop[10,16]), a Gauß potential (Krueger[5]), and an exponential potential multiplied by a square in the $l = m = 0$ angular momentum channel for the three-dimesional case (Engdahl[19]). This has been done by various methods for obtaining suitable trial operators B.

3. A Generalization of Rouché's Theorem for Trace Ideal Operators

We shall now formulate and prove our main result:

Theorem 1 (Siedentop[12,16]) *Let $p \in [1, \infty]$, Γ a simply connected domain in the complex plane, and γ a piecewise smooth, closed curve in Γ with index one for all enclosed points. Furthermore let $f(z) = 1 + F(z)$ and $g(z) = 1 + G(z)$ where $F(z)$ and $G(z)$ are I_p-valued functions that are analytic on γ and meromorphic on Γ such that the main part of the Laurent series of F and G about any point z_0 enclosed by γ has only finite rank coefficients. Moreover, let*

$$\max_{z \in \gamma} |f(z)g(z) - 1|_p < 1. \tag{10}$$

Then

$$\frac{1}{2\pi i} \operatorname{tr} \oint_\gamma f'(z) f(z)^{-1} dz = -\frac{1}{2\pi i} \operatorname{tr} \oint_\gamma g'(z) g(z)^{-1} dz. \tag{11}$$

We remark the following:

- In the case where $\mathcal{H} = \mathbb{C}$, i. e. the underlying Hilbert space are the complex numbers, Theorem 1 reduces to Rouché's classical theorem (see e. g. Marsden[6]).
- If $f(z)$ and $g(z)$ are matrix valued functions fulfilling the above analyticity properties for each of the matrix elements and fulfilling inequality (10) then the theorem applies.

Proof of the theorem: We regard the trace ideals imbedded in $\mathcal{B}(\mathcal{H})$. Thus the function $h(z) = 1 + H(z)$, where $H(z)$ is a function fulfilling the hypotheses of the theorem for $F(z)$ and $G(z)$, is well defined. Moreover, f and g are invertible on γ according to Fact 7 of Section 2, since

$$\begin{aligned}1 > |f(z)g(z) - 1|_p &= |(1 + F(z))G(z) + F(z)|_p \\ &= |[(-F(z)) - 1]G(z) + (-F(z))|_p\end{aligned}$$

and

$$\begin{aligned}1 > |f(z)g(z) - 1|_p &= |(g(z)^* F(z)^* + G(z)^*|_p \\ &= |[(-G(z))^* - 1]F(z)^* - G(z)^*|_p.\end{aligned}$$

The latter inequality implies the invertibility of $g(z)^*$, which, however, is invertible, if and only if g is invertible.

The invertibility of h implies that $H'(z)(1+H(z))^{-1}$ is analytic on γ, and thus integrable, i. e.

$$C_H = \frac{1}{2\pi i} \oint_\gamma H'(z)(1 + H(z))^{-1} dz$$

exists and is an operator in I_p. Now, according to the analytic Fredholm theorem of Ribaric and Vidav[9] $h(z)$ is either nowhere invertible or $h(z)^{-1}$ is a meromorphic function whose Laurent expansions have only coefficients of finite rank. The first possibility is excluded by the above remark. Thus C_H is of finite rank, implying the existence of both, the left and right hand side of equation (11).

We now choose a sequence of finite rank operators H_n that converges uniformly to H on γ.† Thus, because of equation (3)

$$\begin{aligned}m_n &= \frac{1}{2\pi i} \oint_\gamma \frac{d}{dz} \log\det(1 + H_n(z))dz \\ &= \frac{1}{2\pi i} \operatorname{tr} \oint_\gamma H_n'(z)(1 + H_n(z))^{-1} dz.\end{aligned} \tag{12}$$

The right hand side of (12) converges to $\operatorname{tr} C_H$ whereas the left hand side is an integer by the argument principle, becoming thus a constant eventually. Thus left and right hand side of equation (11) are integers.

Next we show that the left hand side of (11) may be deformed continuously into the right hand side keeping at any point of deformation the property of being an integer: We define

$$h_\kappa(z) = f(z)^{-1} + \kappa(g(z) - f(z)^{-1}).$$

h_κ is an operator $h(z)$ of the above form, since for every $A \in I_p$, there is a $B \in I_p$, such that $(1 + A)^{-1} = 1 + B$, if $-1 \notin \sigma(A)$.

Moreover, assuming the invertibility of h_κ for the moment,

$$\rho(\kappa) = \frac{1}{2\pi i} \operatorname{tr} \oint_\gamma h_\kappa'(z) h_\kappa(z)^{-1} dz$$

is an integer.

---

† Given any complete orthonormal set $\{e_1, e_2, ...\}$ define P_n to be the projection onto the space spanned by the first n of these vectors, and $H_n(z) = H(z)P_n$.

Now, $h_0(z) = f(z)^{-1}$, and therefore

$$-\operatorname{tr}\oint_\gamma h_0'(z)h_0(z)^{-1}dz = \operatorname{tr}\oint_\gamma f(z)^{-1}f'(z)f(z)^{-1}f(z)dz$$

$$= \lim_{n\to\infty}\operatorname{tr}\oint_\gamma f_n(z)^{-1}f_n'(z)dz = \lim_{n\to\infty}\oint_\gamma \operatorname{tr} f_n'(z)f_n(z)^{-1}dz$$

$$= \operatorname{tr}\oint_\gamma f'(z)f(z)^{-1}dz$$

where f_n converges uniformly to f on γ. Furthermore, $h_1(z) = g(z)$. Thus, for $\kappa = 0$

$$-\frac{1}{2\pi i}\operatorname{tr}\oint_\gamma h_\kappa'(z)h_\kappa(z)^{-1}dz$$

yields the left hand side, for $\kappa = 1$ the right hand side of (11).

Finally, we may expand the inverse of $h_\kappa(z)$

$$h_\kappa(z)^{-1} = \sum_{\nu=0}^{\infty}\left\{f(z)[g(z) - f(z)^{-1}]\right\}^\nu f(z)(-\kappa)^\nu$$

$$= \sum_{\nu=0}^{\infty}[f(z)g(z) - 1]^\nu f(z)(-\kappa)^\nu.$$

By assumption this series converges in the I_p-norm. In particular ρ is continuous which proves the theorem, since $\rho(\kappa)$ is an integer for $0 \leq \kappa \leq 1$.∎

We remark that inequality (10) may be interpreted as an approximation condition for the inverse of $f(z)$ by $g(z)$ on γ: If $g(z) = f(z)^{-1}$, then the left hand side of (11) would vanish.

4. Application to Resonances

We intend to apply the result of Section 3 to resonances. If we set $F(k) = -R_k$, $p < \infty$, the left hand side of (11) is the number of resonances N_γ enclosed by γ, if Γ is contained in a region where the description of resonances through integral equations holds (see Fact 8). This may be seen as follows by using the renormalized Fredholm determinant and the argument principle:

$$N_\gamma = \frac{1}{2\pi i}\oint_\gamma \frac{d}{dk}\log\det{}_p(1 - R_k)dk = \frac{1}{2\pi i}\oint_\gamma \frac{d}{dk}\log\det(1 + \mathcal{R}(-R_k))dk$$

where we assumed p to be integer.† Inserting the definition of $\mathcal{R}(-R_k)$ (equation (1)) yields

† If p is not an integer take $[p+1]$ instead.

$$N_\gamma = \frac{1}{2\pi i}\operatorname{tr}\oint_\gamma\{[\tfrac{d}{dk}(1+\mathcal{R}(-R_k))][1+\mathcal{R}(-R_k)]^{-1}\}dk$$

$$= \frac{1}{2\pi i}\operatorname{tr}\oint_\gamma\{\tfrac{d}{dk}[(1-R_k)\exp\left(\textstyle\sum_{j=1}^{p-1}\frac{R_k^j}{j}\right)][(1-R_k)\exp\left(\textstyle\sum_{j=1}^{p-1}\frac{R_k^j}{j}\right)]^{-1}\}dk$$

$$= \frac{1}{2\pi i}\operatorname{tr}\oint_\gamma\{-R_k'\circ\exp\left(\textstyle\sum_{j=1}^{p-1}\frac{R_k^j}{j}\right)\circ\left(\exp\left(\textstyle\sum_{j=1}^{p-1}\frac{R_k^j}{j}\right)\right)^{-1}(1-R_k)^{-1}$$

$$+(1-R_k)\tfrac{d}{dk}\left(\exp\left(\textstyle\sum_{j=1}^{p-1}\frac{R_k^j}{j}\right)\right)\exp\left(\textstyle\sum_{j=1}^{p-1}\frac{R_k^j}{j}\right)^{-1}(1-R_k)^{-1}\}dk$$

$$= -\frac{1}{2\pi i}\operatorname{tr}\oint_\gamma R_k'(1-R_k)^{-1}dk + \frac{1}{2\pi i}\operatorname{tr}\oint_\gamma(1-R_k)\tfrac{d}{dk}\left(\exp\left(\textstyle\sum_{j=1}^{p-1}\frac{R_k^j}{j}\right)\right)$$

$$\exp\left(\textstyle\sum_{j=1}^{p-1}\frac{R_k^j}{j}\right)^{-1}(1-R_k)^{-1}dk$$

$$= -\frac{1}{2\pi i}\operatorname{tr}\oint_\gamma R_k'(1-R_k)^{-1}dk.$$

Since the function g is arbitrary apart from the hypothesis of the theorem, we may choose it in such a way that the right hand side of (11) becomes calculable thus allowing for the calculation of the number of resonances in that region. One possible choice would be to set

$$g(z) = 1 + \sum_{\mu,\nu=1}^{n}\beta_{\mu,\nu}(z)|\varphi_\nu><\varphi_\mu|$$

where the coefficients $\beta_{\mu,\nu}$ are obtained by a Galerkin method, i. e. by projecting R_k onto the space generated by a set of orthonormal functions $\varphi_1,\dots,\varphi_n$ in such a way that this operator approximates the minimizing operator of the variational principle (Fact 8), i. e. choose

$$G = B = \sum_{\mu,\nu=1}^{n}(\alpha\circ(1-\alpha)^{-1})_{\mu,\nu}|\varphi_\mu><\varphi_\nu|$$

with $\alpha_{\mu,\nu} = (\varphi_\mu, R_k\varphi_\nu)$. For more details we refer to Ref. 16. – Such a procedure has been carried through numerically by Krueger[5] for a square well and a Gauß potential. One of his results is displayed in the figure. The crosses denote resonances. Their positions have been localized by the variational principle (Fact 7 of Section 2) such that the error bounds are less than the width of the crosses. Integration around each of the crosses gives the information that they represent exactly one resonance.

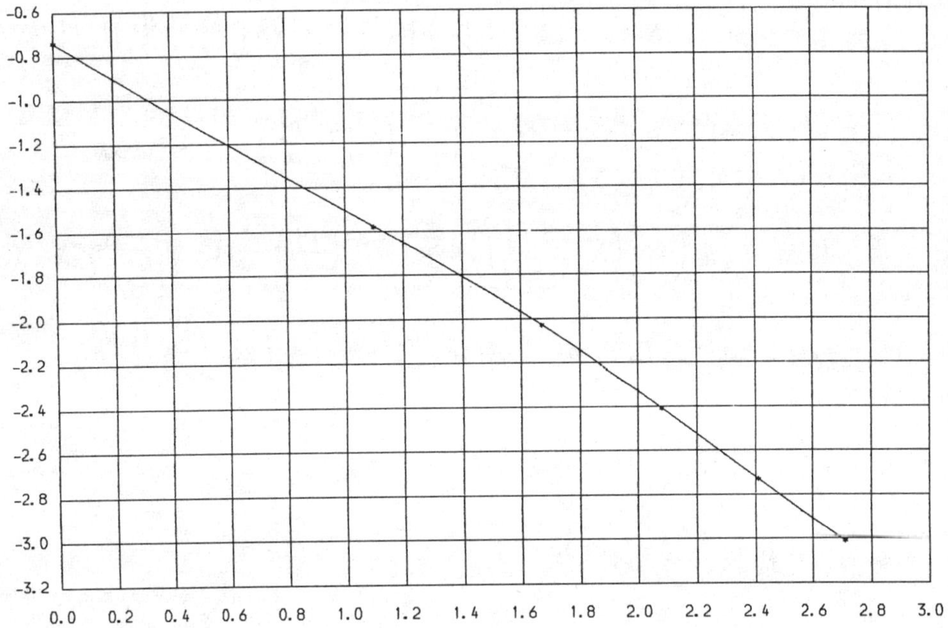

Figure 1

Resonances of the potential $-\exp(-x^2)$ in the k-plane in units where $\hbar^2/2m = 1$.

Another choice for the function g is suggested through the renormalized Fredholm series (see Simon[17]), since it yields an approximation for the inverse of $f(z)$, too.

References

1. E. R. Davidson and E. Engdahl and N. Moiseyev, Phys. Rev.,**33A**, 2436 (1986).

2. E. Engdahl and E. Brändas, Phys. Rev., **37A**, 4145 (1988).

3. P. Froelich and E. Davidson and E. Brändas, Phys. Rev., **28A**, 2641 (1983).

4. B. Helffer and J. Sjöstrand, Bull. Soc. Math. France, **114**, 1 (1986).

5. P. Krueger, *Lokalisierung von Resonanzen bei eindimensionalen Modellsystemen.* Diplomarbeit, Institut für Mathematische Physik, TU Braunschweig (1988).

6. J. E. Marsden, *Basic Complex Analysis*, W. H. Freeman., San Francisco (1973).

7. N. Moiseyev, Chem. Phys. Lett., **99**, 364, (1983).

8. A. M. K. Müller, Phys. Lett., **65A**, 273 (1978).

9. M. Ribaric and I. Vidav, Arch. Rat. Mech. Anal., **32**, 298 (1969).

10. H. K. H. Siedentop, *Dilation Analytic Methods , Models and Methods in Few-Body Physics*, (Eds. L. S. Ferreira, A. C. Fonseca and L. Streit, Springer-Verlag, Berlin (1987)) p. 630.

11. H. K. H. Siedentop, Rep. Math. Phys., **21**, 383 (1985).

12. H. K. H. Siedentop, J. Math. Anal. Appl., (1987). To be published.

13. H. K. H. Siedentop, Phys. Lett., **99A**, 65 (1983).

14. H. K. H. Siedentop, Int. J. Quant. Chem., **31**, 795 (1987).

15. H. K. H. Siedentop, Z. Phys. A, **316**, 367 (1984).

16. H. K. H. Siedentop, *Strenge Schranken für die Lebensdauer quantenmechanischer Resonanzzustände.* Habilitationsschrift. Naturwissenschaftliche Fakultät der TU Braunschweig (1986).

17. B. Simon, *Trace Ideals and their Applications*, Cambridge University Press, 1 ed., London (1979).

18. G-U. Sölter, *Generalization of Müller's Variational Principle*, These Proceedings, E. Brändas and N. Elander, Eds., Lecture Notes in Phys., Springer-Verlag, Berlin (1988).

GENERALIZATION OF MÜLLER'S VARIATIONAL PRINCIPLE

Geert-Ulrich Sölter

Institut für Mathematische Physik
Technische Universität Braunschweig
D-3300
Braunschweig,
FR Germany

Abstract

A method for getting exclusion bounds on eigenvalues of nonnormal operators by using trial (left-)inverse operators is established. This method is extended to Siedentop's version of Müller's variational principle which changes the investigation of the operator into an investigation of the affiliated Birman-Schwinger-Rollnik kernel. The method is applied to the calculation of resonances by complex scaling and yields exact exclusion bounds for the resonances. Rough exclusion bounds for resonances in the case of potentials with compact support and dilation analytic potentials are pointed out.

1. Introduction

The complex scaling method for getting resonance poles of the scattering amplitude needs the calculation of the eigenvalues of a nonnormal operator, Simon[1,2].

In contrast to selfadjoint operators, where there exist several general methods of getting upper and lower bounds for eigenvalues, there exist only a few methods for getting bounds for eigenvalues of nonnormal operators[3]. Moiseyev[4] described a method for excluding eigenvalues in certain discs of the complex plain. Froelich et al.[5] and again Davidson et al.[6] tried to get including bounds by a method similar to Weinstein's method for selfadjoint operators (Weinstein,[7] Yosida,[8] p.321) using Gerschgorin circles. Engdahl and Brändas[9] get inclusion bounds for eigenvalues by a projection operator formulation using Dunford's integral.

Other excluding methods come from Müller's variational principle (Müller,[10]

Sölter et al.,[11] Siedentop,[12] Siedentop[13]). These methods use trial operators for the resolvent of the investigated operator. For selfadjoint operators Sölter[14] got excluding and including regions for the eigenvalues by using these trial operators. In the following the excluding part will be generalized to nonnormal operators. It is even possible to apply it to the Birman-Schwinger-Rollnik kernels which appear in Siedentop's version of Müller's variational principle.

If B is the operator under investigation, λ an eigenvalue of B with (geometric) multiplicity m, f_λ the trial operator, then, with the abbreviation $a_\lambda := f_\lambda(B - \lambda) - 1$, it can be shown that $\|a_\lambda\|_2^2 \geq m$. So, if for a given $\lambda \in \mathbb{C}$ there is a trial operator f_λ, such that $\|a_\lambda\|_2 < 1$, λ cannot be an eigenvalue of B.

It is convenient to use Hilbert-Schmidt operators or other trace ideal operators for f_λ. But by the resolvent identity it is easy to show that a Hilbert-Schmidt trial operator f_λ with $\|a_\lambda\|_2 < 1$ can only exist, if B is an operator whose resolvent is Hilbert-Schmidt. This holds for every trace ideal norm: If $f_\lambda \in \mathcal{I}_n$ with $\|a_\lambda\|_n < 1$, then the resolvent of B is an element of $\mathcal{I}_n$.

For Schrödinger operators $H = T + V$, $T = -\Delta$, V the potential, Siedentop[15,16] reformulated the problem using the affiliated Birman-Schwinger-Rollnik kernel $w_\lambda := |V|^{\frac{1}{2}}(\lambda - T)^{-1}V^{\frac{1}{2}}$ with $V^{\frac{1}{2}}(x) :=$ sign$(V(x))|V(x)|^{\frac{1}{2}}$ (for a reference see Simon[17]).

In the following H should be defined as above, w_λ is the Birman-Schwinger-Rollnik kernel which belongs to H. B is used for arbitrary (densely defined) operators on a Hilbert space $\mathcal{H}$ or a Banach space X. Symbols are chosen as in the standard book of Reed and Simon[18] .

Sections 2 and 3 deal with Müller's variational principle in general, where- as in Sections 4 to 6 the application of the variational principle to the Birman-Schwinger-Rollnik kernel is investigated which leads to a method of getting exact bounds for resonances. Section 7 shows some rough but easily calculated bounds for resonances which come from the variational principle.

2. An Exclusion Theorem for Operators with Hilbert-Schmidt like Resolvents

The following theorem is the origin of the variational principle.

Theorem 1. *Let B be a densely defined operator in a Hilbert space $\mathcal{H}$. Let λ be an eigenvalue of B with (geometric) multiplicity m.*

a) *For every trial operator f_λ : If $\overline{f_\lambda(B - \lambda) - 1}$ is Hilbert-Schmidt operator, then*

$$\|f_\lambda(B - \lambda) - 1\|_2^2 \geq m.$$

b) *There exists an operator f_λ such that in (a) equality holds.*

Proof.

a) *Let $\phi_1,...,\phi_m$ be eigenfunctions for the eigenvalue λ, orthonormalized without loss of generality. Then for all i=1,..,m :*

$$(f_\lambda(B-\lambda)-1)\phi_i = -\phi_i$$

$$m = \sum_{i=1}^{m} \|(f_\lambda(B-\lambda)-1)\phi_i\|^2 \leq \left\|\overline{f_\lambda(B-\lambda)-1}\right\|_2^2$$

b) *Let K be the eigenspace of B for the eigenvalue λ, $K^\perp$ the orthogonal complement, P_K and $P_{K^\perp}$ the corresponding selfadjoint projections. Then $(B-\lambda)|_{K^\perp}$ is injective. Let*

$$C := ((B-\lambda)|_{K^\perp})^{-1}$$

on $R(B-\lambda)$. Then for every $\phi \in D(B)$:

$$(C(B-\lambda)-1)\phi = (C(B-\lambda)|_{K^\perp}P_{K^\perp}-1)\phi$$
$$= (P_{K^\perp}-1)\phi = P_K\phi$$

Theorem 1 was proven by Sölter et al.[11]. By Hölders inequality for trace ideals it can easily be generalized:

Corollary 1a. *With the assumptions of Theorem 1 and $a_\lambda \in I_p$ is*

$$tr((a_\lambda^* a_\lambda)^p) \geq tr((a_\lambda^p)^* a_\lambda^p) \geq m$$

for every p=1,2,...

But Theorem 1 is an exclusion principle only for that one point λ for which $\|a_\lambda\|_2$ is calculated. For it is in general impossible to calculate all λ, eigenvalues may be overlooked. This calamity can be removed by Theorem 2.

Theorem 2. *Let B be a densely defined operator in a Hilbert space $\mathcal{H}$. For $m \in \mathbb{N}$ and $\lambda \in \mathbb{C}$ suppose there exists a Hilbert-Schmidt operator f_λ, such that*

$$\left\|\overline{f_\lambda(B-\lambda)-1}\right\|_2^2 < m.$$

Then for every eigenvalue λ' with (geometric) multiplicity m :

$$|\lambda'-\lambda| \geq \frac{\sqrt{m}-\|a_\lambda\|_2}{\|f_\lambda\|_2}$$

That means: The set

$$\left\{\tilde{\lambda} \in \mathbb{C} \,\middle|\, |\tilde{\lambda}-\lambda| < \frac{\sqrt{m}-\|a_\lambda\|_2}{\|f_\lambda\|_2}\right\}$$

contains no eigenvalue of (geometric) multiplicity m.

Proof. *Use Theorem 1 and estimate* $\|a_{\lambda'}\|_2$ *with the choice* $f_{\lambda'} = f_\lambda$.

3. General Exclusion Theorems for Eigenvalues and other Parts of the Spectrum

It turns out that it is not necessary that the trial operator is Hilbert-Schmidt (or a trace ideal operator at all), although these norms are easier to calculate than the operator norm. But in many cases the trial operator can not be a trace ideal operator, e.g. if the investigated operator has a continuous spectrum. Nevertheless, even the case where the trial operator is only bounded can be handled.

For the next theorems we distinguish the bounded residual spectrum σ_{rb} and the unbounded residual spectrum σ_{ru} (see Appendix).

Theorem 3. *Let B be a densely defined operator in a Banach space X. Let* $\lambda \in \sigma_p(B) \cup \sigma_c(B) \cup \sigma_{ru}(B)$. *Then there exists no operator* f_λ, $f_\lambda \in \mathcal{L}(X)$, *such that* $a_\lambda^m \in \mathcal{L}(X)$ *for every* $m = 1, 2, ..., n$, *and* $\|a_\lambda^n\| < 1$. *Here n is any* $n \in \mathbb{N}$.

Proof. *Show the existence of the (bounded) operator inverse of* $(B - \lambda)$.

Theorem 4 is the analog to Theorem 2.

Theorem 4. *Let B be a densely defined operator in a Banach space X. For* $\lambda \in \mathbb{C}$ *let there exist an operator* f_λ, $f_\lambda \in \mathcal{L}(X)$, *such that* $\|a_\lambda\| < 1$. *Then for every* $\lambda' \in \sigma_p(B) \cup \sigma_c(B) \cup \sigma_{ru}(B)$:

$$|\lambda' - \lambda| \geq \frac{1 - \|a_\lambda\|}{\|f_\lambda\|}$$

That means that the set

$$\left\{ \tilde{\lambda} \in \mathbb{C} \,\middle|\, |\tilde{\lambda} - \lambda| < \frac{1 - \|a_\lambda\|}{\|f_\lambda\|} \right\}$$

contains no element of $\sigma_p(B) \cup \sigma_c(B) \cup \sigma_{ru}(B)$.

Proof. *Use Theorem 3 and estimate* $\|a_{\lambda'}\|$ *with the choice* $f_{\lambda'} = f_\lambda$.

With techniques similar to those used for Theorem 3 the following properties of the bounded residual spectrum can be demonstrated:

Theorem 5. *Let B be a densely defined operator in a Hilbert space* $\mathcal{H}$. *Then:*

a) $\sigma_{rb}(B)$ *is open.*

b) *If* λ *and* λ' *belong to the same connected component of* $\sigma_{rb}(B)$, *then*

$$\overline{R(B - \lambda)} = \overline{R(B - \lambda')}.$$

Proof. *For the proof see Reference 19.*

4. The Affiliated Birman-Schwinger-Rollnik Kernel

A potential class associated with the Birman-Schwinger-Rollnik kernel in dimension $d = 3$ and large enough for many physical applications is the Rollnik class.

Definition. *A measurable function $f(\mathbf{x})$ on $\mathbb{R}^3$ belongs to the Rollnik class R, $f \in R$, iff the Rollnik norm of f,*

$$\|f\|_R := \int\int \frac{|f(\mathbf{x})||f(\mathbf{y})|}{|\mathbf{x}-\mathbf{y}|^2}\, d^3x\, d^3y,$$

is finite.

For a proof of the next theorems see Reference 20.

Theorem 6. *(Kato[21]) Let $d = 3$. If $V \in L^{\frac{3}{2}}$, then $V \in R$.*

In fact $\|V\|_R \leq C\,\|V\|_{\frac{3}{2}}$ for a constant C.

Theorem 7. *If $p \geq \frac{3}{2}$, then $L^p + L^\infty \subset R + L^\infty$, in particular $L^2 + L^\infty \subset R + L^\infty$.*

The next theorem shows the connection between the eigenvalue (-1) of the Birman-Schwinger-Rollnik kernel and the calculation of the eigenvalues of the operator $H = T + V$.

Theorem 8. *(Simon[20]) Let $\mathcal{H} = L^2(\mathbb{R}^3)$. Let $V \in R + L^\infty$. Let λ in the canonically cut plane and w_λ as above.*

a) *If $H\psi = \lambda\psi$, then for $\phi := |V|^{\frac{1}{2}}\psi$:*

$$w_\lambda \phi - \phi$$

b) *Let $w_\lambda\phi = \phi$. Then there is a ψ with:*

(1) $\psi \in D(H)$

(2) $H\psi = \lambda\psi$

(3) $\phi = |V|^{\frac{1}{2}}\psi$

The theorem can be extended to arbitrary dimension d :

Theorem 8a. *For any V, which is T-form compact, and λ in the canonically cut plane, is*

$$(T + V)\psi = \lambda\psi$$

if and only if

$$w_\lambda \phi = \phi \qquad \text{with } \phi = V^{\frac{1}{2}}\psi.$$

Proof. *For the proof see (Simon[17], Simon[20]).*

For V T-form compact in $d = 3$ it is sufficient that $V \in R + L^\infty$, for V T-compact (and therefore T-form compact) in $d > 4$ it is sufficient that $V \in L^p$, $p > \frac{d}{2}$.

5. Estimates for Trace Ideal Norms of w_λ in Arbitrary Dimension

As a result we shall see that estimating $\|a_\lambda\|_n$ leads to the problem of estimating $\|w_\lambda\|_n$. So the next theorems will handle this, yielding an estimate for the trace ideal norms of w_λ for $\lambda \in I\!R$, $\lambda < 0$. The proof makes use of an inequality due to Seiler and Simon[22].

Lemma 1. *Let the dimension be d. If* $f \in L^p$ *and* $g \in L^p$ *with* $2 \leq p < \infty$, *then* $f(x)g(-i\nabla) \in I_p$ *and*

$$\|f(x)g(-i\nabla)\|_p \leq (2\pi)^{-\frac{d}{p}} \|f\|_p \|g\|_p \,.$$

Proof. *(Seiler and Simon[22]),(Simon[23]).*

Theorem 9. *Let the dimension be d. Let* $H = T + V$, $V = V_1 + V_2$ *such that* $w_{1\lambda} := |V_1|^{\frac{1}{2}}(\lambda - T)^{-1}V_1^{\frac{1}{2}} \in I_{2n}$ *and* $V_2 \in L^{2n}$, $n \in I\!N$. *Let* $d < 4n$, $\lambda < 0$. *Then* $w_\lambda \in I_{2n}$ *and*

$$\|w_\lambda\|_{2n} \leq \|w_{1\lambda}\|_{2n} + \left(\frac{1}{2\pi}\right)^{\frac{d}{2n}} \left[\Omega_d \left(\frac{1}{-\lambda}\right)^{2n-\frac{d}{2}} \frac{\Gamma(\frac{d}{2})\Gamma(2n-\frac{d}{2})}{2\Gamma(2n)}\right]^{\frac{1}{2n}} \|V\|_{2n}$$

$$\text{with } \Omega_d = \begin{cases} \frac{d\pi^{\frac{d}{2}}}{(\frac{d}{2})!} & \text{for } d \text{ even} \\ \frac{2^d \pi^{\frac{d-1}{2}}(\frac{d-1}{2})!}{(d-1)!} & \text{for } d \text{ odd.} \end{cases}$$

Proof. *Use Hölder's inequality for trace ideal operators and the inequality of Seiler and Simon. For the full proof see Sölter[25].*

There is a corollary in the case $d = 3$:

Corollary 9a. *Let* $d = 3$. *Let* $H = T + V$, $V = V_1 + V_2$, $V_1 \in R$, $V_2 \in L^{2n}$, $\lambda < 0$, $n \in I\!N$.
Then $w_\lambda \in I_{2n}$ *and*

$$\|w_\lambda\|_{2n} \leq \frac{1}{4\pi}\|V_1\|_R + \left(\frac{\pi}{4}\right)^{\frac{3}{4n}} \left[\left(\frac{1}{-\lambda}\right)^{2n-\frac{3}{2}} \frac{\Gamma(2n-\frac{3}{2})}{\Gamma(2n)}\right]^{\frac{1}{2n}} \|V_2\|_{2n} \,.$$

The method of calculating resonances by complex scaling leads to the complex scaled Hamiltonian $H(\theta)$ (see Reference 18).

If the unscaled Hamiltonian is $H = T + V$, then $H(\theta) = e^{-2\theta}T + V(\theta)$ with V dilation analytic in the strip $S_\alpha := \{\theta \in \mathbb{C} \,||\Im(\theta)| < \alpha\}$ and $V(\theta)$ the analytic continuation of $U_\theta V U_\theta^{-1}$. The last expression is only defined for $\theta \in \mathbb{R}$. The affiliated Birman-Schwinger-Rollnik kernel is $w_\lambda(\theta) = |V(\theta)|^{\frac{1}{2}}(\lambda - e^{-2\theta}T)^{-1}(V(\theta))^{\frac{1}{2}}$.

For real θ the operator U_θ is unitary. So for $\theta \in \mathbb{C}$ the interesting part of θ is the imaginary part. Without loss of generality let $\Im\theta > 0$ in the following.

For the estimate in the complex scaled case we need the following inequality for the free Green's functions.

Lemma 2. *Let the dimension be d, let $\lambda \in \mathbb{C}\setminus[0,\infty)$, $\kappa := \Im(\sqrt{\lambda}$ and*

$$G(\mathbf{x},\mathbf{y};\lambda) := \text{ kernel } (-\Delta - \lambda)^{-1}(\mathbf{x},\mathbf{y}).$$

Then:

$$|G(\mathbf{x},\mathbf{y};\lambda)| \le \left(\frac{\sqrt{|\lambda|}}{\kappa}\right)^{f(d)} G(\mathbf{x},\mathbf{y};-\kappa^2)$$

$$\text{with } f(d) := \begin{cases} \frac{d-2}{2} & \text{for } d \text{ even} \\ \frac{d-3}{2} & \text{for } d \text{ odd} \end{cases}$$

Proof. *Use the representation of the free Green's function in modified Bessel functions.*

Remark: By the choice of the contours it yields $\kappa > 0$ for all $\lambda \in \mathbb{C}\setminus[0,\infty)$. So without loss of generality κ can be set as $\kappa = \left|\Im\left(\sqrt{\lambda}\right)\right|$.

Now we can estimate $w_\lambda(\theta)$ in the complex rotated case:

Theorem 10. *Let the dimension be d. Let $H = T + V$, $V = V_1 + V_2$. Let the dilation value be θ, $\theta \in \mathbb{C}$, $\vartheta := \Im(\theta)$.*
Let $w_{1\lambda} := |V_1(i\vartheta)|^{\frac{1}{2}}(-\kappa^2 - T)^{-1}(V_1(i\vartheta))^{\frac{1}{2}} \in \mathcal{I}_{2n}$ with $\kappa := \left|\Im\left(\sqrt{e^{2i\vartheta}\lambda}\right)\right|$ and $V_2(i\vartheta) \in L^{2n}$
Let $\lambda \in \mathbb{C}\setminus\{\exp(-2i\theta[0,\infty)\}$ and $4n > d$.
Then $w_\lambda(\theta) \in \mathcal{I}_{2n}$ and

$$\|w_\lambda(\theta)\|_{2n} \le \left(\frac{\sqrt{|\lambda|}}{\kappa}\right)^{f(d)} \Bigg(\|w_{1\lambda}\|_{2n} + \left(\frac{1}{2\pi}\right)^{\frac{d}{2n}} \left[\Omega_d \left(\frac{1}{\kappa}\right)^{4n-d} \frac{\Gamma(\frac{d}{2})\Gamma(2n - \frac{d}{2})}{2\Gamma(2n)}\right]^{\frac{1}{2n}} \|V_2(i\vartheta)\|_{2n}\Bigg)$$

Proof. *Use Lemma 2, the inequality of Seiler and Simon and Hölder's inequality for trace ideal operators. For the full proof see Sölter*[24].

For $d = 1$, $n = 1$ and $d = 3$, $n = 1$ Theorem 6 yields the same estimate as Theorem 7 with $V_1 = 0$.

6. The Modified Variational Principle of Müller

Now Müller's variational principle is applied to the Birman-Schwinger-Rollnik kernel w_λ.

Take the abbreviations:

$$a_\lambda := f_\lambda(w_\lambda - 1) - 1, \qquad b_\lambda = f_\lambda + 1$$

Here f_λ (resp. b_λ) is the trial operator for the resolvent of $w_\lambda - 1$. The following properties of a_λ, w_λ and b_λ are easy to prove:

Proposition 1. *If $w_\lambda, b_\lambda \in \mathcal{I}_n$, then $a_\lambda \in \mathcal{I}_n$.*

Proposition 2. *If $w_\lambda, a_\lambda \in \mathcal{I}_n$ and $b_\lambda \in \mathcal{L}(\mathcal{H})$, then $b_\lambda \in \mathcal{I}_n$.*

The next theorem is the generalization of Müller's variational principle. By this theorem it is possible to get bounds for the resonances of a Hamiltonian (with a suitable potential).

Theorem 11. *(Neigbourhood Exclusion Theorem) Let the dimension be d. Let $H = T+V$ the Hamiltonian, $H(\theta) = e^{-2\theta}T+V(\theta)$ the complex scaled Hamiltonian with $\theta \in \mathbb{C}$, $\vartheta := \Im(\theta)$. Let $V(i\vartheta) \in L^{2n}$ with $4n > d$.*
Let $w_\lambda(\theta) := |V(\theta)|^{\frac{1}{2}}(\lambda - e^{-2\theta}T)^{-1}(V(\theta))^{\frac{1}{2}}$. For a λ, $\lambda \in \mathbb{C}\setminus\{e^{-2i\vartheta}[0,\infty)\}$, let there exist a trial operator b_λ, $b_\lambda \in \mathcal{I}_{2n}$, such that $\|a_\lambda\|_{2n} < 1$, where

$$a_\lambda := (b_\lambda - 1)(w_\lambda(\theta) - 1) - 1.$$

Then:
a) λ is no eigenvalue of $H(\theta)$, that means no resonance of H.
b) The set

$$\Bigg\{\lambda' \in \mathbb{C} \,\Bigg|\, |\lambda' - \lambda <$$

$$\frac{1 - \|a_\lambda\|_{2n}}{\left(\frac{1}{2\pi}\right)^{\frac{d}{2n}} \|f_\lambda\| \, \|V(i\vartheta)\|_{2n} \, \|(e^{2\vartheta}\lambda - \xi^2)^{-1}\|_{2n}^2 + (1 - \|a_\lambda\|_{2n}) \, \|(e^{2\vartheta}\lambda - \xi^2)^{-1}\|_{2n}}\Bigg\}$$

contains no eigenvalues of $H(\theta)$, that means no resonances of H.

Proof. *a) Use Theorem 1 and Theorem 8 and $\|a_\lambda\|_{2n} < 1$.*
b) Let λ' be an eigenvalue of $H(\theta)$. Then by part a): $\|a_{\lambda'}\|_{2n} \geq 1$ for all $b_{\lambda'}$. Choose $b_{\lambda'} := b_\lambda$, resp. $f_{\lambda'} := f_\lambda$. Then

$$a_{\lambda'} = f_\lambda |V(\theta)|^{\frac{1}{2}} e^{2\theta}(e^{2\theta}\lambda' - T)^{-1}(V(\theta))^{\frac{1}{2}} + f_\lambda - 1.$$

Now the resolvent identity and all those techniques used in Theorem 7 are applied to get an upper estimate for $\|a_{\lambda'}\|_{2n}$, where the first term of the series leads to $\|a_\lambda\|_{2n}$.
For the full proof see Sölter[24].

Remark: λ is on the physical sheet of $H(\theta)$, but it is on the unphysical sheet (the second sheet) of H.

A more convenient estimate can be obtained by the following corollary.

Corollary 11a. *Assume all the hypothesis of Theorem 11. Then the set*

$$\left\{\lambda' \in \mathbb{C} \,\middle|\, |\lambda' - \lambda| < \frac{1 - \|a_\lambda\|_{2n}}{\left(\frac{1}{2\pi}\right)^{\frac{d}{2n}} \|f_\lambda\| \, \|V(i\vartheta)\|_{2n} \, A(\kappa)^2 + (1 - \|a_\lambda\|_{2n}) A(\kappa)}\right\}$$

contains no eigenvalues of $H(\theta)$ and therefore no resonances of H.
Here is $\kappa := \left|\Im\left(\sqrt{e^{2\vartheta}\lambda}\right)\right|$ and

$$A(\kappa) := \left(\frac{\sqrt{|\lambda|}}{\kappa}\right)^{f(d)} \left[\Omega_d \left(\frac{1}{\kappa}\right)^{4n-d} \frac{\Gamma(\frac{d}{2})\Gamma(2n - \frac{d}{2})}{2\Gamma(2n)}\right]^{\frac{1}{2n}}.$$

Proof. *The same proof as Theorem 11 but with application of Lemma 2.*

7. Rough Estimate for Resonances

Taking a simple trial operator in general not yields a small $\|a_\lambda\|_{2n}$. But even by such methods it is possible to get a general idea of the positions of the resonances.

In the following we choose the trial operator $b_\lambda = 0$ (i.e. $f_\lambda = -1$). Then $a_\lambda = -w_\lambda$ and so λ is no eigenvalue for $\|w_\lambda\|_n < 1$. This technique can be extended to the unphysical sheet (the second sheet).

To avoid confusion because of the ambiguity of λ, it is more convenient to use k, $k := +\sqrt{\lambda}$ on the physical sheet and $k := -\sqrt{\lambda}$ on the unphysical sheet. To make the notation definite we write w_k, with $w_k = w_\lambda$ on the physical sheet and the analytic continuation of it otherwise. a_k and b_k are defined in a similar way. Further on we use the abbreviation $\kappa := \Im(k)$.

7.1 Bounds for resonances of operators with potentials with compact support in one dimension

Theorem 12. *Let $\mathcal{H} = L^2(\mathbb{R})$. Let $H = T + V$ with $V \in L^1$,* supp $(V) = [a, b]$, *$R := b - a$.*
Let k be a resonance of H. Then

$$|\Im(k)| \geq \frac{1}{R} \ln |\Re(k)| - \frac{1}{R} \ln \left(\frac{\|V\|_1}{2}\right).$$

Proof. *Choose $b_k = 0$. For k is a resonance it yields*

$$1 \le \|a_k\|_2 = \|w_k\|_2$$

and because of

$$\begin{aligned}\|w_k\|_2^2 &= \int\int |V(x)||V(y)| \frac{1}{4|k|^2} e^{-2\kappa|x-y|}\, dx\, dy \\ &\le \frac{1}{4|k|^2} e^{2|\kappa|R} \int\int |V(x)||V(y)|\, dx\, dy \\ &= \frac{1}{4|k|^2} e^{2|\kappa|R}\, \|V\|_1^2\end{aligned}$$

it yields

$$\frac{1}{4|k|^2} e^{2|\kappa|R}\, \|V\|_1^2 \ge 1$$

$$e^{2|\kappa|R} \ge \left(\frac{2}{\|V\|_1}\right)^2 \left(\Re(k)^2 + \kappa^2\right)$$

$$|\kappa| \ge \frac{1}{R}\ln\left(\Re(k)\right) - \frac{1}{R}\ln\left(\frac{\|V\|_1}{2}\right).$$

We can compare this result with the asymptotic formula for resonances of Regge for potentials with compact support in dimension d=1 (see Reference 25, p.362):

$$\begin{aligned}\Re(k_n) &= \frac{n\pi}{R} + O(1) \\ \Im(k_n) &= \frac{1+\frac{\sigma}{2}}{R}\ln n + O(1) \qquad \text{with } n = 1,2,3\ldots\end{aligned}$$

Here σ is determined by the special form of the potential and is always bigger than zero. So in the leading terms the estimate is the best possible for arbitrary form of the potential.

Using different methods Lax and Phillips[26] received a similar result for dimension $d = 3$.

7.2 Bounds for resonances of operators with dilation analytic potentials

Theorem 13. *Let the dimension be d. Let $H = T + V$, V dilation analytic with critical angle ϑ_{crit} and let $V(i\vartheta) \in L^{2n}$, $4n > d$. Let k be a resonance of H. Then for every $\vartheta < \vartheta_{crit}$ the following holds:*
a) for $d = 1$:

$$\Re(\tilde{k}) \le \left(C(\vartheta)^{\frac{1}{n}} \left(\frac{1}{\tilde{\kappa}}\right)^{\frac{2n-1}{n}} - \tilde{\kappa}^2\right)^{\frac{1}{2}}$$

b) for $d = 2, 3$:

$$\Im(\tilde{k}) \le C(\vartheta)^{\frac{1}{4n-d}}$$

c) for $d \ge 4$:

$$\Re(\tilde{k}) \le \left(\left(\frac{\tilde{\kappa}^{2nf(d)+4n-d}}{C(\vartheta)} \right)^{\frac{1}{nf(d)}} - \tilde{\kappa}^2 \right)^{\frac{1}{2}}$$

where $\tilde{k} := e^{i\vartheta}k$, $\tilde{\kappa} := \Im(\tilde{k})$ and

$$C(\vartheta) := \|V(i\vartheta)\|_{2n}^{2n} \left(\frac{1}{2\pi}\right)^d \Omega_d \frac{\Gamma(\frac{d}{2})\Gamma(2n-\frac{d}{2})}{2\Gamma(2n)}.$$

Proof. *Choose the trial operator $b_{\tilde{k}} = 0$. Then from the Theorem 10*

$$\begin{aligned} 1 &\le \left\|a_{\tilde{k}}\right\|_{2n} = \left\|w_{\tilde{k}}\right\|_{2n} \\ &\le \left(\frac{|\tilde{k}|}{\tilde{\kappa}}\right)^{f(d)} \left(\frac{1}{2\pi}\right)^{\frac{d}{2n}} \left[\Omega_d \left(\frac{1}{\tilde{\kappa}}\right)^{4n-d} \frac{\Gamma(\frac{d}{2})\Gamma(2n-\frac{d}{2})}{2\Gamma(2n)}\right]^{\frac{1}{2n}} \|V(i\vartheta)\|_{2n}. \end{aligned}$$

So, with the abbreviation for $C(\vartheta)$ from above we get

$$\left(\frac{|\tilde{k}|}{\tilde{\kappa}}\right)^{2nf(d)} \left(\frac{1}{\tilde{\kappa}}\right)^{4n-d} C(\vartheta) \ge 1$$

and

$$\left(\left(\Re(\tilde{k})\right)^2 + \tilde{\kappa}^2\right)^{nf(d)} \ge \frac{1}{C(\vartheta)} \tilde{\kappa}^{2nf(d)+4n-d}.$$

If $d = 2, 3$, then $f(d) = 0$. Otherwise

$$\Re(\tilde{k}) \le \left(\left(\frac{\tilde{\kappa}^{2nf(d)+4n-d}}{C(\vartheta)} \right)^{\frac{1}{nf(d)}} - \tilde{\kappa}^2 \right)^{\frac{1}{2}}.$$

When Theorem 13 is applied to estimate k then the curve for $\tilde{k}$ has to be rotated back with the angle $(-\vartheta)$. In Theorem 13 $\Im(\tilde{\kappa})$ is always bigger than zero, even for $\Im(k) < 0$.

In the case $d = 2, 3$ in the k-plane the boundary curves are straight lines with gradient $\tan(-\vartheta)$. The distance from the origin is given by $C(\vartheta)^{\frac{1}{4n-d}}$.

In the case $d = 1$ the curves in the k-plane are asymptotic to straight lines through the origin which have the gradient $\tan(-\vartheta)$.

7.3 Bounds for resonances of operators with Gauß potentials in one dimension

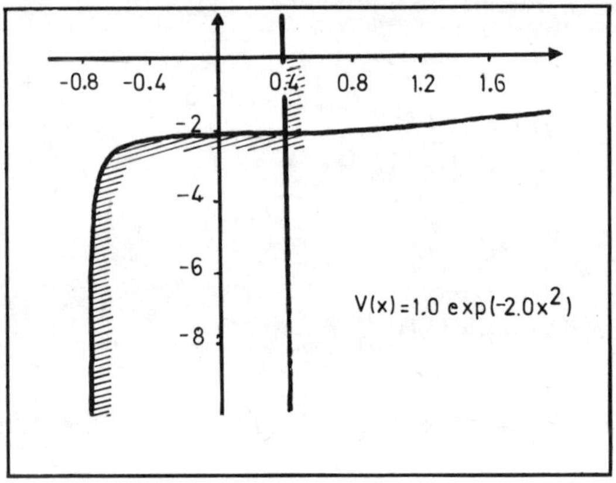

Figure 1

Excluding boundaries in the second sheet of the energy plane for $a = 1$ and $c = 2$.

In some sense in the next theorem we combine the methods of Theorem 12 and Theorem 13. We first do complex rotation and then calculate $\left\| w_{\tilde{k}} \right\|_2$ in a direct way. Finally taking one point from every ϑ-estimate we get a new estimate not depending on ϑ.

Theorem 14. *Let $d = 1$. Let $H = T + V$, $V(x) = ae^{-cx^2}$, $a \in \mathbb{C}$, $c > 0$. Then the following two estimates are valid for all resonances:*
a) In the k-plane:

$$\Im(k) \leq -\sqrt{\Re(k)^2 - \frac{\pi |a|^2}{4c}}$$

In the energy plane (second sheet):

$$\Re(\lambda) \leq \frac{\pi |a|^2}{4c}$$

b) The boundary curve is given in the k-plane by

$$k = \rho(\vartheta) \exp(i(\phi(\vartheta) + \vartheta))$$

and in the energy plane (second sheet) by

$$\lambda = \rho(\vartheta)^2 \exp(i(2\phi(\vartheta) + 2\vartheta)).$$

Here are

$$\rho(\vartheta) = \left(\frac{\pi e |a|^2}{2c \cos(2\vartheta)} \right)^{\frac{1}{2}}$$

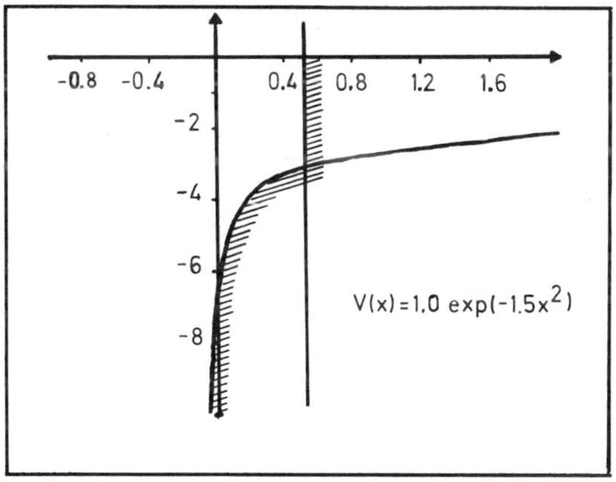

Figure 2

Excluding boundaries in the second sheet of the energy plane for $a = 1$ and $c = 1.5$.

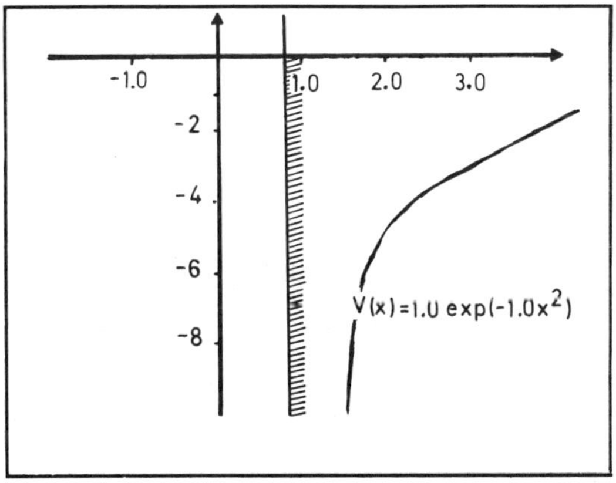

Figure 3

Excluding boundaries in the second sheet of the energy plane for $a = 1$ and $c = 1$.

and

$$\phi(\vartheta) = \arctan\left(\left(\frac{\pi e|a|^2}{(c\cos(2\vartheta))^2} - 1\right)^{-\frac{1}{2}}\right).$$

For a given argument $(\phi(\vartheta) + \vartheta)$ a resonance in the k-plane has an absolute value less or equal than $\rho(\vartheta)$. Similar in the energy plane.

Proof. *After complex rotation with angle ϑ we get for the resonance k, $\tilde{k} := e^{i\vartheta}k$,*

$$1 \le \left\|w_{\tilde{k}}\right\|_2^2 = |a|^2 \int\int e^{-c\cos(2\vartheta)(x^2+y^2)} \frac{1}{4|\tilde{k}|^2} e^{-2\tilde{\kappa}|x-y|}\, dx\, dy$$

$$= \frac{1}{4}\frac{\pi}{c\cos 2\vartheta}\left|\frac{a}{\tilde{k}}\right|^2 e^{\frac{2\tilde{\kappa}^2}{c\cos 2\vartheta}}\left(1 - \operatorname{erf}\left(\tilde{\kappa}\sqrt{\frac{2}{c\cos 2\vartheta}}\right)\right).$$

a) Taking the point with $\tilde{\kappa} = 0$ from every ϑ-estimate we get

$$\frac{\pi}{4c\cos 2\vartheta}\left(\frac{|a|}{\Re(\tilde{k})}\right)^2 \ge 1$$

and

$$\Re(\tilde{k}) \le \frac{|a|}{2}\left(\frac{\pi}{c\cos 2\vartheta}\right)^{\frac{1}{2}}.$$

Transformed back to k leads to

$$\kappa \le -\sqrt{\Re(k)^2 - \frac{\pi|a|^2}{4c}}$$

in the k-plane. In the energy plane (second sheet) we get

$$\Re(\lambda) \le \frac{|a|^2\pi}{4c}.$$

b) Estimating the error function yields

$$\left(\Re(\tilde{k})\right)^2 \le \frac{\pi|a|^2}{2c\cos 2\vartheta}\exp\left(\frac{2\tilde{\kappa}^2}{c\cos 2\vartheta}\right) - \tilde{\kappa}^2.$$

Taking the point with maximal absolut argument $\phi = \phi(\vartheta)$ from every ϑ-estimate we get for this point

$$\frac{\pi|a|^2}{2c\cos 2\vartheta}\exp\left(\frac{2\tilde{\kappa}^2}{c\cos 2\vartheta}\right) = \left(1 + (\tan\phi)^{-2}\right)\tilde{\kappa}^2$$

$$\exp\left(\frac{2\tilde{\kappa}^2}{c\cos 2\vartheta}\right) = \frac{(c\cos 2\vartheta)^2}{\pi|a|^2}\left(1 + (\tan\phi)^{-2}\right)\frac{2\tilde{\kappa}^2}{c\cos 2\vartheta},$$

and

$$e = \frac{(c\cos 2\vartheta)^2}{\pi|a|^2}\left(1 + (\tan\phi)^{-2}\right)$$

$$1 - \frac{2\tilde{\kappa}^2}{c\cos 2\vartheta}$$

From the first equation we get for the angle

$$\phi(\vartheta) = \arctan\left(\left(\frac{\pi e|a|^2}{(c\cos 2\vartheta)^2} - 1\right)^{-\frac{1}{2}}\right)$$

and from the second equation and the inequality for the absolute value of this point

$$(\rho(\vartheta))^2 = \Re(\tilde{k})^2 + \tilde{\kappa}^2 = \frac{\pi e|a|^2}{2c\cos 2\vartheta}.$$

As shown in the figures it is possible to exclude resonances with positive $\Re(\lambda)$ and large negative $\Im(\lambda)$ for parameters of the Gaussian $a = 1$ and $c > 1.5$.

8. Summary

Coming from a method to calculate bounds for eigenvalues of selfadjoint operators Müller's variational principle turns out to be a fruitful and effective method to determine bounds for resonances. First numerical calculations using Siedentop's version of the principle, but without the neigbourhood exclusion theorem (Theorem 11), have been performed yielding satisfactory results.

On the other hand the methods of rough estimating seem to be good for a first survey without doing numerical calculations.

Appendix

The Spectrum of an Operator

Let $\lambda \in \mathbb{C}$, B operator in a Banach space X.

1) Resolvent set $\rho(B)$
$\lambda \in \rho(B) \iff (B-\lambda)$ injective, $(B-\lambda)^{-1}$ bounded, $\overline{R(B-\lambda)} = X$.

2) Point spectrum $\sigma_p(B)$
$\lambda \in \sigma_p(B) \iff (B-\lambda)$ not injective

3) Continuous spectrum $\sigma_c(B)$
$\lambda \in \sigma_c(B) \iff (B-\lambda)$ injective, $\overline{R(B-\lambda)} = X$, $(B-\lambda)^{-1}$ unbounded

4) Residual spectrum $\sigma_r(B)$
$\lambda \in \sigma_r(B) \iff (B-\lambda)$ injective, $\overline{R(B-\lambda)} \neq X$

a) Bounded residual spectrum $\sigma_{rb}(B)$
$\lambda \in \sigma_{rb}(B) \iff \lambda \in \sigma_r(B)$ and $(B-\lambda)^{-1}$ bounded on $R(B-\lambda)$

b) Unbounded residual spectrum $\sigma_{ru}(B)$
$\lambda \in \sigma_{ru}(B) \iff \lambda \in \sigma_r(B)$ and $(B-\lambda)^{-1}$ unbounded on $R(B-\lambda)$

$$\sigma_r(B) = \sigma_{rb}(B) \cup \sigma_{ru}(B)$$

$$\sigma(B) = \sigma_p(B) \cup \sigma_c(B) \cup \sigma_r(B)$$

$$\rho(B) \cup \sigma(B) = \mathbb{C}$$

References

1. B. Simon, Ann. Math., **97**, 247 (1973).

2. B. Simon, Int. J. Quant. Chem., **14**, 529 (1978).

3. H. L. Cycon, R. G. Froese, W. Kirsch and B. Simon, *Schrödinger Operators*, Springer-Verlag, Berlin (1987) p. 150.

4. N. Moiseyev, Chem. Phys. Lett. **99**, 364 (1983).

5. P. Froelich, E. R. Davidson and E. Brändas, Phys. Rev., **28A**, 2641 (1983).

6. E. R. Davidson, E. Engdahl and N. Moiseyev, Phys. Rev., **33A**, 2436 (1986).

7. D. H. Weinstein, Proc. Nat. Acad. Sci. USA, **20**, 529 (1934).
8. K. Yosida, *Functional Analysis*, Springer-Verlag, Berlin (1980) p. 321.
9. E. Engdahl and E. Brändas, Phys. Rev., **37A**, 4145 (1988).
10. A. M. K. Müller, Phys. Lett., **11**, 238 (1964).
11. G-U. Sölter, A. M. K. Müller and G. Gerlich, J. Phys. A, **12**, 325 (1979).
12. H. K. H. Siedentop, Phys. Lett., **99A**, 65 (1983).
13. H. K. H. Siedentop, Z. Phys. A, **316**, 367 (1984).
14. G-U. Sölter, Diploma, Technische Universität Braunschweig (1977).
15. H. K. H. Siedentop, *Modifikation des Müllerschen Variationsprinzips*, Dissertation, TU Braunschweig (1982).
16. H. K. H. Siedentop, Rep. Math. Phys., **21**, 383 (1985).
17. B. Simon, *Functional Integration and Quantum Physics*, Academic Press, New York (1979).
18. M. Reed and B. Simon, *Methods of Modern Mathematical Physics* Vol 4, Academic Press, New York (1978).
19. G-U. Sölter, *Bounds for eigenvalues of nonnormal operators*, Unpublished, (1987).
20. B. Simon, *Quantum Mechanics for Hamiltonians Defined as Quadratic Forms*, Princeton University Press, Princeton (1971).
21. T. Kato, Math. Ann., **162**, 258 (1966).
22. E. Seiler and B. Simon, Comm. Math. Phys., **45**, 99 (1975).
23. B. Simon, *Trace Ideals and their Applications*, Cambridge University Press, Cambridge (1979).
24. G-U. Sölter, *Bounds for resonances by a modified version of Müller's variational principle*, Submitted for publication (1987).
25. R. G. Newton, *Scattering Theory of Waves and Particles*, Springer-Verlag, New York Heidelberg Berlin (1982).
26. P. D. Lax and R. S. Phillips, Arch. Rat. Mech. Anal., **40**, 268 (1971).

RESONANCE STATE EXPANSIONS IN NUCLEAR PHYSICS

Tore Berggren

Department of Mathematical Physics
Lund Institute of Technology
P.O.Box 118
S-221 00 Lund
Sweden

Abstract

It is shown how the proof[8], that the bound and scattering eigenstates in a local potential well form a complete set, can be extended so as to include also a selected set of resonance states. This requires regularization of certain divergent radial integrals, methods for which are briefly described. As an illustration, the derivation of the Breit-Wigner formula using a resonance state expansion is sketched.

1. Ancestry: Gamow, Breit and Wigner, Siegert, Humblet and Rosenfeld

Most students of scattering and reaction theory encounter already, at an elementary stage, the concept of scattering resonance as a maximum in the scattering amplitude at an energy where the phase shift is 90 degrees modulo 360 degrees. At a somewhat more advanced stage one also learns that the phase shift should rise through this value in order to produce a true physical resonance that shows up as a bump in a plot of the cross section as a function of energy. Nuclear physics textbooks[1] usually also give some derivation of the Breit-Wigner formula[2] by solving the case of scattering from a potential well, expressing the phase shift in terms of the logarithmic derivative of the wave function at the edge of the well, and then expanding the logarithmic derivative around the resonance energy. However, the first time I heard about resonance *states* was in Prof. Rosenfeld's lectures on nuclear reaction theory at the Niels Bohr Institute in the early 60's. There he exposed the formalism which he, Humblet and coworkers [3]had developed starting from Siegert's definition[4] of a resonance state as a solution of the Schrödinger equation having only outgoing waves at infinity. This idea in its turn goes back to Gamow's famous treatment[5] of alpha decay as a quantal phenomenon: an alpha particle inside a radioactive nucleus has a wave function that is non-zero also outside the confining barrier caused by the Coulomb repulsion from the residual nucleus. For physical reasons the wave function outside the barrier must behave as an outgoing wave. Gamow realized that this would imply a complex energy value for such a state, with the imaginary part of the energy inversely proportional to the life-time of the state.

1.1 The Mittag-Leffler side-line

The Humblet-Rosenfeld formalism used the fact that the Gamow-Siegert solutions are associated with poles in the scattering amplitudes taken as functions of the wave number for the incident particle. Assuming that these functions are meromorphic, we may use a theorem due to Mittag-Leffler which states that such a function can be expanded as follows:

$$S(k) = f_p(k) + \sum_{-\infty}^{\infty} \left(\frac{k}{k_n}\right)^{p+1} \frac{r_n}{k - k_n}$$

Here k is the wave number, related to the energy E according to $k = \sqrt{\frac{2\mu E}{\hbar^2}}$. The function $f_p(k)$ is an entire function of k which depends on the parameter p in the power of k in the sum. The quantities r_n are residue parameters, while k_n (with $k_{-n} \equiv -k_n^*$) is the complex wave number defining the position E_n and width Γ_n of the resonance state according to

$$W_n \equiv E_n - \frac{1}{2} i\Gamma_n = \hbar^2 k_n^2 / 2\mu$$

where W_n is the complex energy of the resonance state. If k_n is purely imaginary and thus equal to $-k_n^*$, then the state is a bound state ($ik_n > 0$) or an anti-bound state ($ik_n < 0$), in which case only one of the two equal terms is included in the sum. (The lowest bound state, if there is any, may therefore be labeled 0, of course.) This is not very controversial or problematic. The problems arise when the actual values of these parameters are to be determined in a given experimental situation; especially the interpretation of the residues r_n may be unclear in a situation where many resonance states contribute. The principal application that the authors had in mind was the parametrization of low-energy strength functions. An urgent problem at that time was the factorization of the residues into partial width amplitudes characterizing the incoming and outgoing channels, including the extraction of penetration factors depending on the charges and angular momenta in these channels. A complication in these matters was the ambiguity associated with the arbitrariness in the choice of the power p in the Mittag-Leffler formula. The formalism was some time later reformulated in such a way that no arbitrary channel radii entered explicitly, but already in the original formulation it was shown that the matching radii separating the internal and external wave functions did not affect any physical quantity such as the cross section.

Coming from the field of direct reactions, my interest in the Humblet-Rosenfeld formalism was rather mild. However, the physical meaning of resonance states was important: I studied reactions which phenomenologically seemed to lead to the 'ground state' of ^{5}He. This state may be identified with a spin 3/2 resonance with odd parity in the scattering of neutrons by alpha particles (^{4}He nuclei). How to combine the direct-reaction theory with the Humblet-Rosenfeld formalism was not clear to me. It would have been different if the resonance state could be included in a completeness relation ('a resolution of the identity') in the same way as the bound states of the final system.

2. Completeness: How To Prove It, How To Extend The Proof

The problem of incorporating the resonance states into a completeness relation and thus accomplish a resonance state expansion based on more familiar principles (to a direct-reaction theorist) than the Mittag-Leffler theorem occupied me for several years in the late 60's and early 70's. The break-through for me personally was the rediscovery of a method (already suggested[6] by Zel'dovich) to normalize the notoriously ill-behaved Gamow-Siegert resonance states: they increase exponentially with r and therefore are not absolutely square integrable. I found that if one defines the value of the normalization integral by means of a Gaussian convergence factor and chooses for normalization purposes not the absolute square of the wave function but the 'simple' square, then a finite value can be assigned to the integral and a norm can be defined for the resonance state. Moreover, the same method can be used to generalize the concept of inner product (this was also suggested[7] about simultaneously by Romo) and then show that the resonance states thus normalized form two sets which are biorthogonal in the sense that any state in one set is orthogonal to all states in the other set except one, which is the time-reverse of the state chosen in the first set. (Clearly this relationship is reciprocal.) This property opens the possibility of a completeness relation but far from proves its validity. To do this one must go to the theory of differential equations, if one is content with a model in terms of a particle in a potential (or two particles interacting via a potential). My source here was a review article[8] by Newton, who presented the potential scattering theory in a very clear way, including a proof of the completeness of the set of bound and scattering states belonging to a given local potential. I shall outline this proof briefly below, but in order to generalize the proof so as to admit resonance states, we must look at some regularization methods available for defining divergent integrals and improper limits. It was a wonderful experience to me when I learned from Hardy's book[9] on divergent series that (and how!) mathematicians did such rebellious things as summing divergent series and integrating violently oscillating functions.

2.1 Regularization methods and their justification

To regularize a divergent integral is a task related to the definition of a limit for a function which is unbounded or oscillating as the variable approaches a certain value (if the variable takes integer values this value must be $\pm\infty$). One way to define the limit is the following: Let $s(x)$ be a function whose limit as $x \to \infty$ is to be defined in such a way that if the limit exists in the ordinary sense the new definition shall yield the same value. The new definition must be linear in $s(x)$, and we may assume that it is the ordinary limit of a linear transform $t(x)$ of $s(x)$ defined by

$$t(x) = \int_0^\infty c(x,y)s(y)dy$$

This transformation is said to be *regular*, if $\lim_{x\to\infty} s(x) = s$ implies $\lim_{x\to\infty} t(x) = s$, and *totally regular*, if this is true also for the improper limit $s = \infty$. The proof can be found in Hardy's book[9] and is quoted in papers[10,11] by me and by Gyarmati and Vertse. To define an integral which is divergent on the interval $(0,\infty)$ we study the limit of the integral over $(0,r)$ as $r \to \infty$. This is equivalent to defining the integral with a convergence factor $g(x,y)$ in the integrand with the properties

$$\lim_{y\to\infty} g(x,y) = 0, \qquad \lim_{x\to\infty} g(x,y) = 1.$$

Thus,

$$\int_0^\infty s(x)dx = \lim_{x\to\infty} \int_0^\infty g(x,y)s(y)dy$$

Given a possible transformation kernel $c(x,y)$, the convergence factor is

$$g(x,y) = \int_y^\infty c(x,t)dt.$$

The convergence factor used in the 1968 paper corresponds to $g(x,y) = \exp(-y^2/x)$, which is derived from a totally regular transformation.

A slightly different regularization method is to include a factor $g(x,y)$ in the integrand, which is chosen such that the integral is an analytic function of x. This may then be continued analytically to a value of x for which $g(x,y) = 1$ for all y in the interval $(0,\infty)$. As an example we may take $g(x,y) = \exp(-xy)$ (this is e. g. Romo's choice). By this method we can, for instance, define the integral between $x = r$ and infinity of $\exp(\alpha x)$ with $\alpha > 0$ as $-\exp(\alpha r)$. This transformation is regular but *not totally* regular. A majority of mathematicians and physicists might reject it altogether, of course. However, it is very useful for obtaining an important result to which we will return later on. It is paradoxical if we try to interpret the infinite integral as the limit of a sum of positive terms. The definition in terms of a transformation ends up in a region of the complex plane which cannot be reached using the conventional method. But that method leads to an infinite result which as such is insensitive to essential properties of the integrand and the lower limit of the integral!

A third method of regularization (Hokkyo[12], García-Calderon[13]) which is just a practical consequence of any of the first two, consists in writing the normalization integral as an integral from 0 to R and then add a term that plays the role of the complementary integral from R to infinity of the wave function squared. If the actual wave function is replaced by its asymptotic form the primitive function of the integrand is sometimes known. If this function can be put $= 0$ at infinity by means of a regularizing transformation or simply by definition, then one obtains a value for the infinite integral that is independent of R, if R is such that the asymptotic form of the wave function is valid.

A formula for the primitive function of the product of two solutions to the same radial Schrödinger equation with different energies can be derived directly in terms of the wronskian of the two solutions. The corresponding formula for the square of a wave function is not much more difficult: we simply use the equation for the wronskian of the wave function and and its derivative with respect to the wave number k. This equation can be solved by quadrature, yielding a relation between the integral of the squared wave function and the values of the wronskian in question at the upper and lower limit of the integral. The value at infinity can be obtained by transformation and is zero for reasonable wave functions. Thus the correction term in this third regularization method can in principle always be calculated. In practice the Coulomb wave functions remain a problem because, to my knowledge, there are no reliable numerical methods for calculating the derivatives of these functions with respect to the Sommerfeld parameter η.

2.2 The proof of completeness

Newton proves the completeness of the set of bound states $\phi_l^{(n)}(r)$ and scattering states $\psi_l(k,r)$ generated by a local potential $V(r)$ by considering the integral

$$I(r) = \int_C k\,dk \int_0^\infty dr' h(r') \mathcal{G}_l(-k;r,r') \tag{1}$$

where $\mathcal{G}_l$ is the resolvent or complete Green's function for the radial equation of the lth partial wave. In terms of the well-known Jost solutions $\phi_l(k,r)$ (regular at $r=0$) and $f_l(k,r)$ (irregular at $r=0$) which are connected with each other via the Jost function $f_l(k)$ according to

$$\phi_l(k,r) = \frac{1}{2} ik^{-l-1}[f_l(-k)f_l(k,r) - (-1)^l f_l(k) f_l(-k,r)]$$

the scattering wave function $\psi_l(k,r)$ may be written

$$\psi_l(k,r) = [k^{l+1}/f_l(-k)]\phi_l(k,r)$$

and the resolvent

$$\mathcal{G}_l(k;r,r') = (-1)^{l+1}k^{-1}\psi_l(k,r_<)f_l(-k,r_>)$$

Here $r_<$ is the smaller, $r_>$ the larger of r and r'. The residues of the resolvent at a zero k_n of the Jost function $f_l(k)$ can be calculated using the result

$$df_l(k)/dk|_{k=k_n} = -2ck_n^{l+1}\int_0^\infty dr\{\phi_l(k_n)\}^2$$

where the constant c is defined by the equality

$$f_l(k_n, r) = c\phi_l(k_n, r)$$

valid for $k = k_n$.

The function $h(r)$ in (1) should be sufficiently regular so that the integrals exist. Newton chooses the contour C to consist of a large semicircle S of radius R in the lower half of the k-plane with center at the origin, and the segment $\mathrm{L_A} \equiv [-R, R]$ of the real k-axis closing the contour (see Figure 1). This contour will enclose those poles of the resolvent which correspond to bound states. Writing $I(r) = I_{\mathrm{S}}(r) + I_{\mathrm{L_A}}(r)$, Newton gets from the asymptotic values of $f_l(-k, r')$, $\phi_l(k, r)$ and $f_l(k)$ in the limit of large R

$$I_{\mathrm{S}}(r) = i\pi h(r).$$

Furthermore he gets

$$\begin{aligned} I_{\mathrm{L_A}}(r) &= -i\int_0^\infty dr' h(r') \\ &\times \int_{-R}^{R} dk\, k^{2l+2}\phi_l(k,r)\phi_l(k,r')/[f_l(k)f_l(-k)]. \end{aligned} \tag{2}$$

On the other hand, by Cauchy's theorem he also has from the residues at the zeroes of $f_l(k)$

$$I(r) = -2\pi i\sum_n \int_0^\infty dr' h(r')\phi_l^{(n)}(r)\phi_l^{(n)}(r')/N_n^2,$$

where

$$N_n^2 \equiv \int_0^\infty dr[\phi_l^{(n)}(r)]^2.$$

Equating the two expressions for $I(r)$ and using the symmetries in the integrand of (2) Newton obtains a result for any square integrable function $h(r)$ which may be restated as

$$\sum_b \phi_l^{(b)}(r)\phi_l^{(b)}(r')/N_b^2 + \frac{2}{\pi}\int_0^\infty dk\, \psi_l(k,r)\psi_l^*(k,r') = \delta(r - r'). \tag{3}$$

The sum goes of course over all bound state poles encircled by $\mathrm{L_A} \cup \mathrm{S}$. (In this abbreviated version of the proof we have omitted the technicality of dividing the r' integrals into two parts: one from 0 to r, the other from r to infinity due to the different forms of the resolvent in these intervals.)

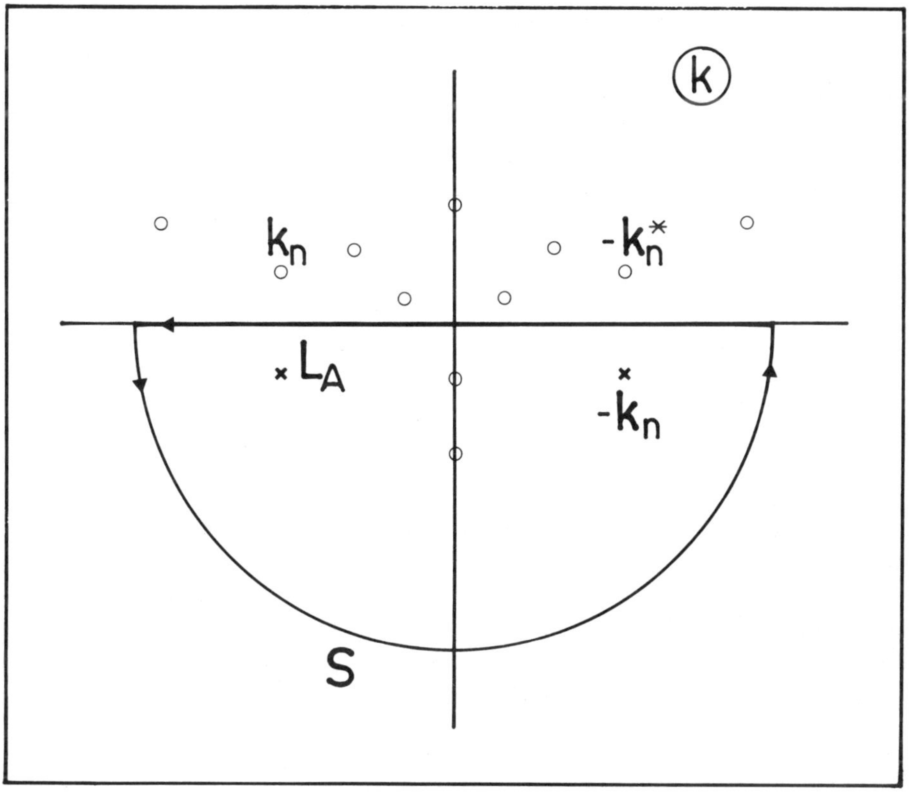

Figure 1.

Contour for proving completeness according to Newton.

2.3 The extension of the proof to resonant states

Technically, the only modifications necessary to extend Newton's proof of completeness to resonant states consist in

(i) define all r' integrals with the help of regularization;

(ii) deform the straight segment $L_A = [-R, R]$ into a curve L_B joining $k = -R$ and $k = +R$ in such a way that if k' is on L_B then so is $-k'$, and so that $L_B \cup S$ encircles the bound states and a selected number of resonance poles in the upper half of the k-plane.

The contribution from S is of course the same as before, but the finite sum now includes in addition to the bound states also the selected resonance states. Furthermore, the continuum states $\psi_l(k,r)$ will now depend on a complex k along the deformed curve L_B. Due to the general symmetry

$$f_l^*(-k^*) = f_l(k)$$

of the Jost function, the completeness relation should now be written

$$\sum_b \phi_l^{(b)}(r)\phi_l^{(b)}(r')/N_b^2 + \sum_n \phi_l^{(n)}(r)\phi_l^{(-n)*}(r')/N_n N_{-n}^* + \frac{2}{\pi}\int_{L_{B+}} dk\,\psi_l(k,r)\psi_l^*(k^*,r') = \delta(r-r'). \tag{4}$$

The integral is here taken over right-hand part of the curve L_B, denoted L_{B+}, i.e. $k \in L_{B+}$ if $\mathrm{Re}(k) \geq 0$.

3. The Off-spring: Resonance State Expansions Based On Completeness.

The main reason why a completeness relation is so desirable is that it brings the resonance states on the same footing as the bound states. We should be aware that we have to pay a price for this, namely the complex-energy scattering states associated with the new continuum curve L_B. One hopes, of course, that by making this modification one has taken the 'sting' out of the continuum, so that we can neglect it with much better confidence than before. Many people have hoped, but I fear in vain, that the continuum could be eliminated altogether if one includes all the infinitely many resonances. One should note that an essential point in the proof of the completeness relation is that the integral over the semicircle S recovers the function $h(r)$ that was to be expanded. This semicircle is drawn in the half-plane where the resolvent $\mathcal{G}_l$ is sufficiently regular so that the contribution from S can be estimated accurately. Thus we had better force the deformed path L_B back to the real k-axis for large values of $|k|$. This means that only a finite number of resonances can be included in the sum. From the physical point of view this is quite natural, and probably a blessing for the practitioner.

3.1 Some properties of bound, resonant and anti-bound states

In the first papers on resonance state expansions, only resonances with outgoing waves in the asymptotic region were included in the sum, principally because the Gaussian regularization only works if the real part of the sum of wave numbers in the asymptotic phase is numerically larger than the imaginary part. If both outgoing and incoming states are admitted, then this condition will be violated, and most particularly so if we ask for the inner product between a resonant state

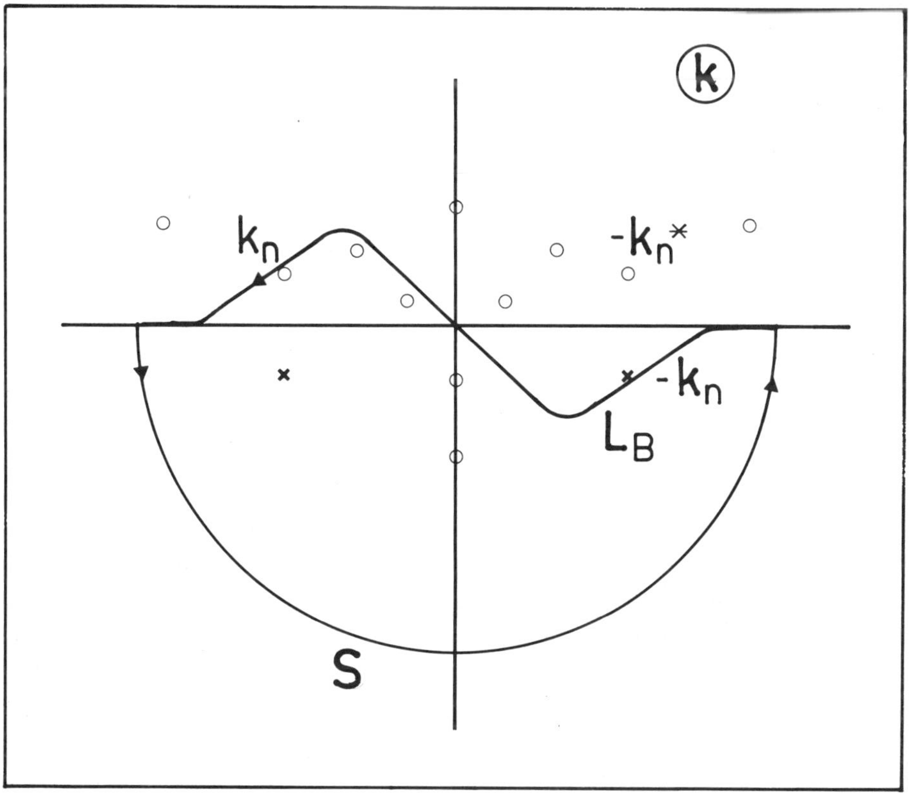

Figure 2.

Contour for inclusion of (outgoing) resonance states in the completeness relation.

and its time reverse. This is simply the absolute square integral of the wave function, and it diverges exponentially. It is out of bounds for regularization with a Gaussian convergence factor or even factors $\exp(-y^p/x)$ with $p > 1$, a requirement for totally regular transformations of this type.

But the regular, although not totally regular, convergence factor $\exp(-xy)$ with $x \to 0$ along a path in the complex plane that avoids the poles in the resulting, hopefully analytic transform, can be applied also to the absolute square integral yielding a non-infinite result. Let us introduce the notation (not necessarily restricted to positive n values)

$$\psi_l^{(n)}(r) \equiv \phi_l^{(n)}/N_n, \qquad \tilde{\psi}_l^{(n)}(r) \equiv \phi_l^{(-n)}(r)/N_{-n} = \phi_l^{(n)*}(r)/N_n^*.$$

Then we can write the completeness relation (4) as

$$\sum_b \psi_l^{(b)}(r)\psi_l^{(b)}(r') + \sum_n \psi_l^{(n)}(r)\tilde{\psi}_l^{(n)*}(r') \\ + \frac{2}{\pi}\int_{L_{B+}} dk\, \psi_l(k,r)\psi_l^*(k^*,r') = \delta(r-r') \,. \tag{5}$$

The left member of this equation should be the coordinate space representative of a projection operator P onto the space of selected bound and resonant states. If both $\psi_l^{(n)}$ and $\psi_l^{(-n)} = \psi_l^{(n)*}$ are included in the sum, then $P^2 = P$ is only consistent with

$$\langle\psi_l^{(n)}|\psi_l^{(n)}\rangle = \int_0^\infty \psi_l^{(n)*}(r)\psi_l^{(n)}(r)dr = 0. \tag{6}$$

The same condition can be derived from the radial Schrödinger equation for $\psi_l^{(n)}$ and $\psi_l^{(-n)}$. The corresponding eigenvalues are $\hbar^2k_n^2/2\mu$ and $\hbar^2(k_n^*)^2/2\mu$, respectively. The usual proof, that two eigenstates with different eigenvalues must be orthogonal, applies and yields (6). This is paradoxical if we note that the integrand in the inner product is the absolute square of the resonance wave function. If, however, we regularize the integral according to the third prescription above, we find that we get a positive contribution for a finite upper limit R of the inner integral and a negative contribution from the correction term due to increasing exponential which cancels the former contribution. (This is of course a logical inference from our result (6) but it can be checked in the case of resonance wave functions obtained numerically. The result usually agrees within the numerical precision used with the theoretical null result. It can be used as a check on the quality of the solution found.)

The theoretical importance of this result is that the absolute square integral of a resonance wave function is useless for normalization purposes: its 'natural' value is zero! One may say that instead it is the resonance wave function and its time reverse that are normalized together:

$$\langle\tilde{\psi}_l^{(n)}|\psi_l^{(n)}\rangle = \int_0^\infty [\psi_l^{(n)}(r)]^2 dr = 1.$$

The anti-bound states which, like the bound states, are their own time reverses, should most likely be normalized to -1. They are difficult to find numerically, but those who have had some experience with them tell me that such a normalization is more compatible with their behavior in perturbation calculations.

If you accept that the resonance wave functions are 'self-orthogonal' as the more generous regularization methods show, then one may admit both outgoing

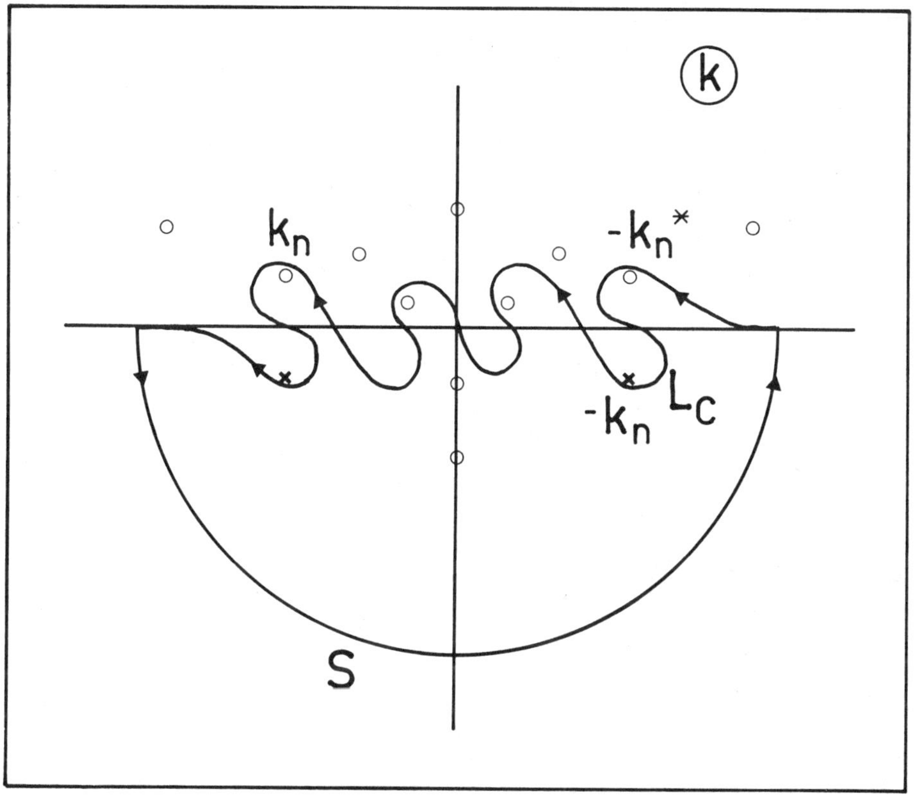

Figure 3.

Contour for inclusion of both incoming and outgoing resonance states.

and incoming resonance states in the completeness relation. The proof then needs a somewhat more complicated shape for the contour: the semicircle S is closed by means of the curve L_C in Figure 3.

3.2 The Breit-Wigner formula revisited

Now that we have a completeness relation, what can we do with it? The first application that we may think of is, of course, to derive the Breit-Wigner formula for a resonant reaction process. Take a relatively general case: a reaction

$$a + A \rightarrow b + B$$

Let $H = H_i + V_{aA} = H_f + V_{bB}$ and let $|\phi_a\rangle$ be an eigenstate of H_i with eigenvalue E_a defining the initial configuration while $|\chi_b\rangle$ is an eigenstate of H_f with eigenvalue E_b describing the final configuration. Reaction theory tells us that the transition rate ω_{if} is

$$\omega_{if} = \frac{2\pi}{\hbar}\left|\langle\chi_b|T_{fi}^{(+)}|\phi_a\rangle\right|^2 \delta(E_a - E_b)$$

where the transition operator $T_{fi}^{(+)}$ may be written

$$T_{fi}^{(+)} = V_{bB} + V_{bB}\frac{1}{E - H - i\epsilon}V_{aA}.$$

Here E is the common value of E_a and E_b, ϵ is the usual infinitesimal that defines the outgoing boundary conditions appropriate for the final state. If we now on each side of the propagator insert the completeness relation (5) involving a sum over the most important resonance states, then we will get a result that we can write

$$\omega_{if} = \frac{2\pi}{\hbar}\Big|\langle\chi_b|T^{\text{backgr}}|\phi_a\rangle + \sum_n \langle\chi_b|V_{bB}|\psi_n\rangle\frac{1}{E - W_n}\langle\tilde{\psi}_n|V_{aA}|\phi_a\rangle\Big|^2 \delta(E_a - E_b)$$

Here the first term inside brackets comes from the continuum and bound states (the single potential term usually does not contribute). The sum gives rise to a number of interference terms and terms of the form

$$\left|\langle\chi_b|V_{bB}|\psi_n\rangle\right|^2 \frac{1}{(E - E_n)^2 + \frac{1}{4}\Gamma_n^2}\left|\langle\tilde{\psi}_n|V_{aA}|\phi_a\rangle\right|^2$$

which we may identify with the Breit-Wigner term due to the resonance state $|\psi_n\rangle$. Here we get a dynamical prescription also for the numerators directly from the expansion. Their form immediately suggests a close relationship to transition rates from the initial channel to the intermediate resonance state and from the latter to the final channel, in agreement with older theories.

4. The Next Generation: What More Can We Do?

The question in the title of this chapter is a suitable point for finishing this brief and incomplete review of resonance state expansions based on completeness. The situation is, I think, that we have a lot more to do in this field. My own activity here is, at present, very low. If there is still some interest in deep hole states in nuclei when I get time for them (they are highly excited, unbound, and decay via a kind of Auger process), then I would like to play with them for a while. Many intricate problems remain to be solved. There is an abundance of possible applications of resonance state expansions. The younger generation of physicists can just pick up a problem which appeals to their taste and work happily ever after.

References

1. J. M. Blatt and V. F. Weisskopf, *Theoretical nuclear physics*, Wiley, New York (1952).

2. E. P. Wigner, Z. Phys., **83**, 253 (1933); G. Breit and E. P. Wigner, Phys. Rev., **49**, 519, 612 (1936).

3. J. Humblet and L. Rosenfeld, Nucl. Phys., **26**, 529 (1961); L. Rosenfeld, Nucl. Phys., **26**, 579 (1961); J. Humblet, Nucl. Phys., **31**, 544 (1962); **50**, 1 (1964); **57**, 386 (1964); J. P. Jeukenne, Nucl. Phys., **58**, 1 (1964); C. Mahaux, Nucl. Phys., **68**, 481 (1965); L. Rosenfeld, Nucl. Phys., **70**, 1 (1965); L. Rosenfeld, *Spectroscopic and group theoretical methods in physics*, North-Holland, Amsterdam (1968) p. 203.

4. A. F. J. Siegert, Phys. Rev., **56**, 750 (1939).

5. G. Gamow, Z. Phys., **51**, 204 (1928).

6. Ya. B. Zel'dovich, ZhETF (USSR), **39**, 776 (1960); (Translation) JETP (Sov. Phys.), **12**, 542 (1961).

7. W. J. Romo, Nucl. Phys., **116A**, 618 (1968).

8. R. G. Newton, J. Math. Phys., **1**, 319 (1960).

9. G. H. Hardy, *Divergent series*, Oxford (1949).

10. T. Berggren, Nucl. Phys., **109A**, 265 (1968).

11. B. Gyarmati and T. Vertse, Nucl. Phys., **160A**, 523 (1971).

12. N. Hokkyo, Prog. Theor. Phys., **33**, 1116 (1965).

13. G. García-Calderon, Nucl. Phys., **261A**, 130 (1976); G. García-Calderon and R. E. Peierls, Nucl. Phys., **A265**, 443 (1976).

MITTAG-LEFFLER EXPANSIONS IN NUCLEAR PHYSICS

Jens Bang

The Niels Bohr Institute
University of Copenhagen
DK-2100 Copenhagen Ø
Denmark

Abstract

Features of pole expansions of different quantities in nuclear physics are reviewed. Particular attention is payed to the expansion of the single particle Green function and its application to collective nuclear states with continuum admixtures.

The idea to use pole expansions for the S-matrix of particles scattering in a potential is very natural, since we know that the scattering amplitudes determined by this are often dominated by resonances, which could be connected with such poles. Also the fact that the bound state energies are poles of the S-matrix was recognized by Heisenberg from the early days of S-matrix theory. He actually thought they were the only ones for positive imaginary k-values.

We shall here think of all quantities as expressed by k but use $\hbar = 2m = 1$, so that k may mean the momentum, and $E = k^2$.

The pole expansions have particularly been investigated by de Alfaro and Regge and by Nussenzweig [6].

Once such a pole expansion for the S-matrix is established, it is also rather natural to use the relation of this to the Green function to get a corresponding pole expansion for this. The Green function is simply related to the scattering state wave functions, so in a sense we have automatically then also a pole expansion for them. The proof that these pole expansions exist and are convergent depend on the character of the potential.

For applications to atomic physics it is unfortunate that, for those expansions we shall discuss, we need the attractive potentials to decay faster than exponentially. So, therefore, a Gaussian potential would do; whereas the Coulomb potential is completely excluded. In nuclear physics we deal with finite range nuclear potentials, and in practice even with potentials which are 0 outside a certain range R, since we always solve the equations by matching to pure combinations of Bessel and Hankel functions (or Coulomb functions of the repulsive field).

We shall here also stick to such cut potentials, $V(r) = 0$ for $r > R$. Let us, though, start by noting that this is not what we say we are doing in nuclear physics. Already for many years it has been standard to use a Woods Saxon potential

$$V = V_o\Big(1 + \exp((r - R_A)a^{-1})\Big)^{-1} \tag{1}$$

One may, of course, argue that this is just a convenient parametrization, which should not be taken too seriously.

It is true that the quantity a determining the range of the potential seems to have little to do with large r-values; it is rather a measure for the thickness of the nuclear surface layer. If we think physically, however, we see good reasons that there actually should be an exponential tail of the potential. First, the nucleon-nucleon forces which are the basis for construction of the potential have already such a tail, given by the Yukawa expression which gives a tail with the extent determined by the mass of the Π-meson. Secondly, even if we neglect this and put the force range zero, the loosest bound nucleons will give a contribution to the potential with a tail which is determined by the square root of the binding energy, in our units

$$\exp(-\kappa r), \quad \kappa^2 = E_B.$$

Such tails may actually be considerable longer than those given by a, which is generally ~ 0.65 fm.

We shall not discuss this further here, as the description of the nuclear surface is still an unsolved problem, but below we shall look at how the exponential tail influences our pole expansions.

Let us go directly to the expansion of the Green function, from which the other expansions are easily obtained.

Let us start with the usual expression for the Green function [1] with $k^2 = E$:

$$g^+(\underline{r}, \underline{r}'; k) = \sum_{\ell m} g_\ell^+(r, r'; k) Y_{\ell m}^*(\hat{r}) Y_{\ell m}(\hat{r}') \tag{2}$$

$$g_\ell^+(r, r'; k) = \sum_{n\ bound} \frac{\varphi_n(r)\varphi_n(r')}{k^2 - k_n^2}$$

$$+ \pi^{-1} \int_{-\infty}^{\infty} k'^{-1} dk' (k^+ - k')^{-1} \Psi_\ell^{(+)*}(r; k') \Psi_\ell^{(+)}(r'; k')$$

$$k^+ = \lim_{\eta \to 0} k + i\eta \tag{3}$$

This integral can, for finite r, r' values, be calculated by Cauchy's theorem; closing the contours by large semicircles far from any pole of the integrand and chosen in such a way that they contribute but negligibly to the integral. Now the integrand has terms which go as $e^{ik(r-r')}$ and as $e^{-ik(r-r')}$ for $|k| \to \infty$. Therefore, for $r - r' > 0$, we may, in the first terms, close the contour in the upper half plane counter clockwise; in the others in the lower, clockwise. The discontinuity of Ψ^+ on the real axis plays no role. We see that we have the same poles contributing in the two cases. Our integrand is $\Psi^{+*}(r; k)\Psi^+(r'; k)$ and the complex poles occur in pairs which are symmetric about the imaginary axis, only the roles of r and r' are reversed; since the residues turn out to be symmetric in r, r', this is of no consequence. So the result is as if we were closing in the lower half plane only, neglecting the bad behavior for $|k| \to \infty$ of some terms.

It is, of course, a condition that the integration path should be sufficiently far from all poles. There are infinitely many poles but they lie more and more scattered as we go out in the k-plane. It has been proved by Romo[2] that we may really chose semicircles with larger and larger $|k|$, thus encircling more and more poles and, at the same time, get smaller and smaller integrals along these paths.

The residues of g_ℓ^+ must, obviously, be products of solutions of the Schrödinger equation for r, r' with the energy given by the pole k_n^2 satisfying

$$\left(-\frac{d^2}{dr^2} + U(r) - k_n^2\right)\varphi_n(r) = 0 \tag{4}$$

So the expansion of the Green function is

$$
\begin{aligned}
g_\ell(r,r';k) &= \sum_{i(all\ poles)} \frac{\varphi_i(r)\varphi_i(r')}{2k_i(k-sk_i)} \left(s=-sign(\Im(k_i))\right) \\
&+ \sum_{n(bound)} \frac{\varphi_n(r)\varphi_n(r')}{k^2-k_n^2} \\
&= \sum_{i(all\ poles)} \frac{\varphi_i(r)\varphi_i(r')}{2k_i(k-k_i)} \\
&= \sum_{q=-1}^{p} \frac{k^q}{q!}\frac{\partial^q}{\partial k^q} g(k) + \sum_i \left(\frac{k}{k_i}\right)^{p+1} \frac{r_i}{k-k_i} \quad ; \text{ with } \frac{\partial^{-1}}{\partial k^{-1}} g(k) \equiv 0
\end{aligned}
\tag{5}
$$

The normalization of the φ functions is, of course, now given and we see that it must be a generalization of the usual normalization of bound states.

Since the φ_i's for $r > R$ behave as $h_\ell^+(k_i r)$, they are not square integrable, but the generalization suggested by Zeldowich [3], the so-called regularization, leads to $(\ell = 0)$

$$
\begin{aligned}
\delta_{nm} &= lim_{\varepsilon\to 0} \int_0^\infty \varphi_n(r)\varphi_m(r) e^{-\varepsilon r^2} dr \\
&= \int_0^R \varphi_n\varphi_m dr + \frac{\varphi_n(R)\varphi_m(R)}{i(k_n+k_m)}
\end{aligned}
\tag{6}
$$

This latter expression is meaningful for all poles whereas the regularization is meaningful only for the bound states and the proper resonances with $|\Re(k_n)| > |\Im(k_n)|$ [3].

Note that although the regularization expression looks like an orthonormality relation with a weight function, our latter expression shows little resemblence to orthogonality.

We may use the asymptotic expression for the Green function

$$
g(r,r';k) \underset{k\to\infty}{\longrightarrow} \frac{\delta(r-r')}{k^2} + O(k^{-3}) \quad r,r' < R \tag{7}
$$

to obtain

$$
\sum \varphi_i(r)\varphi_i(r') = 2\delta(r-r') \quad r,r' < R \tag{8}
$$

$$
\sum \varphi_i(r)\varphi_i(r') k_i^{-1} = 0 \quad r,r' < R \tag{9}
$$

so we see that in the interval $(0, R)$ the φ_n-functions are far from being linearly independent. There are infinitely many (even a continuum of) relations between them.

The first relation shows that they form a complete set; but that it is even overcomplete because of the factor 2, in agreement with the linear dependence.

These properties have puzzled many. One could be tempted to ask whether the factor 2 simply means an impractical normalization. However, if we stick to $r < R$, we see that this is not so; we still have the linear dependence.

It has been shown by Regge [4] that if we go to $2R$ this linear dependence disappears, so in a way this is the answer. Another idea is that of Kravitskij [5] who says that the set is two-fold complete. This means that we can always, for two functions which are sufficiently smooth but otherwise completely independent, get two expansions (with $i = 1, 2$)

$$f_i(r) = \sum_n k_n^{i-1} a_n \varphi_n(r) \tag{10}$$

with the same a_s.

In order to throw a little more light on this, let us look at the poles.

According to Nussenzweig [6], they are asymptotically distributed as

$$k_n = R^{-1}\left(\pm n\pi - i\ell n \frac{2n\pi}{RU^{\frac{1}{2}}(R-)}\right) \tag{11}$$

and hereto come the bound states and a corresponding number of poles on the negative imaginary axis.

In parenthesis: we see here that the integration paths can be chosen far from the poles asymptotically.

If we make the potential deeper but still keep it 0 outsider R, the complex poles will move toward each other and meet at 0 or, for $\ell = 0$, at the negative imaginary axis. After that, one will move upward and become bound and the other downward to become antibound. These latter are states with a purely increasing exponential function outside R. But inside R, when the potential becomes very deep, they will be identical to the bound states. We see that in this limit the overcompleteness simply means that each state is counted twice, as bound and as antibound.

We see that, in spite of linear independence, it is not easy to use $2R$ as the interval over which we expand since it is not simple to construct an orthogonal set of functions. Nevertheless, we shall see that the pole functions can be quite useful due to the above expansion of the Green function.

Let us assume that we know the Green function for the potential V_o, and let us work with a perturbing potential V, so that

$$V = V_o + \delta V \tag{12}$$

then the corresponding Green functions are connected by the Dyson equation

$$G^+ = G(V) = G_o(V_o) + (G_o \delta V G) \tag{13}$$

Now perturbation series may be obtained by solving this equation by iteration:

$$G^0 = G_o, \quad G^1 = G_o + (G_o \delta V G_o) \;\; etc. \tag{14}$$

We may also write $\delta V = \lambda \delta V_o$ and solve by expanding everything as power series in λ [7]. Let us assume that both V_o and V are 0 outside R; then the integrals may be calculated by using the pole expansions for G and G_o.

Let us, therefore, look at the perturbation expression for k near a pole k_n. The corresponding term in the pole expansion is dominating and we may write

$$\frac{\varphi_n(r)\varphi_n(r')}{2k_n(k-k_n)} = \frac{\varphi_n^o(r)\varphi_n^o(r')}{2k_n^o(k-k_n^o)}\left(1 + \frac{\varphi_n(r')}{\varphi_n^o(r')}\frac{\int \varphi_n^o(r)\delta V \varphi_n(r) dr}{2k_n(k-k_n)}\right)$$

In the lowest order we may neglect $\varphi_n - \varphi_n^o$ and write

$$\begin{aligned} \varphi_n^o \varphi_n^o \Big(-2k_n^o(k-k_n^o) + 2k_n(k-k_n) \\ + \int dr \varphi_n^o \delta V \varphi_n^o \Big) = 0 \end{aligned} \tag{15}$$

if $k = k_n$

$$\begin{aligned} 2k_n^o(k_n - k_n^o) &= \int \varphi_n^o \delta V \varphi_n^o dr \\ \Rightarrow \; 2k_n^o \delta k_n &= \int \varphi_n^o \delta V \varphi_n^o dr \\ &= \delta(k_n^2) \end{aligned} \tag{16}$$

So, in this case, we get the same expression as in ordinary perturbation theory. This is, though, exceptional; usually the new terms will look different from the usual ones.

This expression clearly shows that something goes wrong with potentials which fall off exponentially with r. We know that for $V = 0$ outside R, the φ's go exponentially to infinity for the complex resonances. For $\ell = 0$ we simply have $\varphi_n \sim e^{ik_n r}$.

So, if we have an exponentially decaying potential $V_o \exp(-r\alpha)$, the dependence of k_n^2 on the cut off radius is,for $\ell = 0$, given by the first order expression

$$\begin{aligned}\Delta k_n^2 &= \varphi^o(R)^2 V_o \int_R^{R+\Delta R} dr\ \exp\Big((2ik_n - \alpha)(r - R)\Big) \\ &= \varphi^o(R)^2 V_o \Big(\exp\Big((2ik_n - \alpha)\Delta R\Big) - 1\Big)(2ik_n - \alpha)^{-1}\end{aligned} \tag{17}$$

So we see that even if V is very small outside R, the pole position depends crucially on R for $-\Im(k_n) \geq \alpha/2$.

It should, though, be noticed that for realistic α values there may be up to 200 poles for which $-\Im(k_n) < \alpha/2$ and among these are the good resonances, i.e. the narrow ones. That the other, and the antibound states, may depend on the cut off should probably not worry us since they play no individual role but act only as a necessary chorus in the background.

We have stressed here that the expansions in the interval $0 < r < R$ are useful in nuclear physics calculations [8]. As an example, let us take the description of giant resonances in linear response formalism.

The fundamental quantities from which everything is calculated are, here, the effective V field which arises as a response to an external field, V_o, and is given by the equation [9] (wherein i, k: channels, w: frequency of ext. field)

$$V^i(\underline{r}, w) = V_o^i(\underline{r}) + \int d\underline{r}_1, d\underline{r}_2 F^{ik}(\underline{r}, \underline{r}_1) A^k(\underline{r}_1, \underline{r}_2; w) V^k(\underline{r}_2, w) \tag{18}$$

here $F^{ik}(\underline{r};\underline{r}_1)$ is an effective particle-hole interaction which is conveniently taken to be of a separable surface form

$$\begin{aligned}F(\underline{r}, \underline{r}_1) &= \sum F^\ell [Y(\hat{r})_\ell \bullet Y_2(\hat{r}_1)_\ell]_o \\ F^m &= \kappa_m \frac{\partial U^m(r)}{\partial r} \frac{\partial U^m(r_1)}{\partial r_1}\end{aligned} \tag{19}$$

where U^m has essentially the shape of the nuclear potential.

A is the particle hole Green function

$$A(\underline{r}_1,\underline{r}_2;w) = \int G(\underline{r}_1,\underline{r}_2;\varepsilon - \frac{w}{2})G(\underline{r}_1,r_2;\varepsilon + \frac{w}{2})\frac{dw}{2\pi i} \tag{20}$$

It may be expanded in angular momentum components; then the radial function becomes (where $N_{n\ell j}$: occupation number)

$$A_L(r_1,r_2;w) = \sum_{n\ell j} N_{n\ell j} B^L_{j\ell,j'\ell'} \varphi_{n\ell j}(r_1)\varphi_{n\ell j}(r_2)$$
$$[g_{\ell' j'}(r_1,r_2;\varepsilon_{n\ell j}+w) + g_{\ell' j'}(r_1,r_2;\varepsilon_{n\ell j}-w)] \tag{21}$$
$$B^L_{j\ell,j'\ell'} = (4\pi)^{-1}\left(\hat{\ell}\hat{\ell}'\hat{j}\hat{j}' \begin{pmatrix} \ell & \ell' & L \\ 0 & 0 & 0 \end{pmatrix} \begin{Bmatrix} \ell j & \frac{1}{2} \\ j'\ell' & L \end{Bmatrix}\right)^2$$

So we see that in the integrals, due to the hole states, wave functions fall off fast and the factor F also falls off fast. We can cut our integrals at R and use a pole expansion for the g's, introducing $k^{\pm 2} = \varepsilon_{n\ell j} \pm w$ [10]. Let us note that the experiments are described in terms of the strength function

$$S_i(w) = \sum_s |<0|V^i_o|s>|^2\delta(w - w_s)$$
$$= -\frac{1}{\pi}\Im \sum_i <V^i_o A^i(w) V^i(w)> \tag{22}$$

so here we need only a calculation of V^i valid for $r < R$.

The natural possibility must be

$$V^i_{(w)} = \sum c^i_n(w)\varphi_n(\underline{r}) \tag{23}$$

With a truncated set in the expansion we will, obviously, get a set of linear equations for the determination of the c's, which for $V^i = 0$, will give an eigenvalue problem for w somehow corresponding to the usual random phase equation with bound state basis functions.

References.

1. R. G. Newton, *Scattering Theory of Waves and Particles*, Mc-Graw-Hill, New York (1966).

2. W. J. Romo, Nucl. Phys., **237A**, 275 (1975).

3. T. Berggren, Phys. Lett., **44B**, 23 (1973).

4. T. Regge, Nuovo Cimento, **9**, 491 (1958).

5. A. O. Kravitskij, Dokl. Akad. Nauk. SSSR, **170**, 1255 (1966).

6. M. M. Nussenzweig, *Causality and Dispersion relations*, Academic Press, New York (1972).

7. R. M. More and E. Gerjuoy, Phys. Rev., **7A**, 381 (1973).

8. J. Bang, F. A. Gareev, M. H. Gizzatkulov and S. A. Goncharov, Nucl. Phys., **309A**, 381 (1978).

9. F. A. Gareev, S. N. Ershov, N. I. Pyatov, S. A. Fayans and D. I. Salamov, Sov. J. Nucl. Phys., **33**, 337 (1982).

10. J. Bang and F. A. Gareev, Lett. Nuovo Cim., **32**, 420 (1981).

ON THE ANALYTICAL CONTINUATION OF THE PARTIAL WAVE S-MATRIX USING COMPLEX SCALING TECHNIQUES

Magnus Rittby[†] and Nils Elander

Manne Siegbahn Institute of Physics
Frescativägen 24
S-104 05 Stockholm
Sweden

Erkki Brändas

Quantum Chemistry Group for Research in Atomic,
Molecular and Solid State Physics
University of Uppsala
S-751 20 Uppsala
Sweden

Abstract

The analytic properties of the partial wave S-matrix and related quantities are studied and numerically investigated. The analysis is carried out by means of integration along paths in the complex k−plane. The domain for the choice of integration contours can be rigorously defined by the use of complex scaling techniques. A generalization of Levinson's theorem incorporating the poles in the lower half k−plane is proved and exemplified. An expansion theorem for the partial wave S-matrix in terms of its poles and residues is derived and analyzed. The connections between poles and associated residues and their relationships with the Breit-Wigner ansatz and the Fano line shape parameters are discussed and numerically realized. Finally, the implications of the present development in connection with the inversion problem are indicated.

1. Introduction

In recent years there has been a revived interest in the calculation of complex poles of the partial wave S-matrix and the subsequent expansion of the S-matrix using product expansions [1–8]. Such studies are not new, see for example [9–13], but with the introduction of high speed computers it has become possible to get numerical realizations of the analytic structures involved.

Although our previous studies and applications have centered largely around Weyl-Titchmarsh's m-function, the direct relationship between spectral densities and cross sections, see e.g. reference 14 and references therein, display the close connection with conventional scattering theory in terms of S-matrices and Jost functions [15]. It is well known that under suitable restrictions on the potential one can guarantee an analytic continuation of the Jost function into the lower half of the k-plane [15–17]. For a brief summary of these conditions as well as concomitant aspects referring to the possible appearance of so-called redundant poles of the S-matrix we refer to 18.

In this paper we will try to shed some more light on the connections between complex poles in the lower half of the k–plane and the scattering information obtained on the real axis. We will focus especially on the use of complex coordinate techniques to acquire such connections. The expressions 'partial wave S-matrix' and simply 'S-matrix' will be used interchangeably, but we will in this context always refer to the reduced matrix element of the S-matrix in the partial wave basis.

In section 2, we will study the analytic continuation of the Jost functions and in particular show how complex scaling allows for an analytic continuation of the S-matrix to the complex plane.

Section 3 presents the generalization of Levinson's theorem to the complex plane. We will find that the extension of the phase shift to complex k will force us to include new poles in the usual expression for the connection between the phase shift and the complex poles of S.

In section 4, we will investigate a Mittag-Leffler expansion for the S-matrix. It is not required that the S-matrix be well defined and known in the whole of the complex k–plane, but we will retain integral terms in the expansion to exemplify the possibilities of our techniques.

In section 5, the connections of the previous sections with the Breit-Wigner ansatz and the Fano parameters for the cross section line shape will be investigated and supplemented with a numerical example.

Finally, in section 6 we discuss the connection with the Marchenko integral equation for the inverse problem. Such an inversion procedure relating the analytically extended partial wave S-matrix with the scaled potential provides for a consistent relationship valid in a sector in the complex k-plane.

The derivation of the expansion of the S-matrix in terms of its poles and background that is used in sections 4 and 5 is given as an appendix.

2. Analytic Continuation of the Jost Functions and the S-Matrix

For a radial Schrödinger equation of the type

$$-\frac{1}{2}\frac{d^2\psi(r)}{dr^2} + \{(V(r) + \frac{\ell(\ell+1)}{2r^2}\}\psi(r) = E\psi(r) \tag{2.1}$$

the partial wave S-matrix can be defined by the quotient of the two Jost functions

$$S_\ell(k) = \frac{\mathcal{F}_\ell^-(k)}{\mathcal{F}_\ell^+(k)} \tag{2.2a}$$

$$S_\ell(k) = e^{\imath 2\delta(k)};\ \text{k real} \tag{2.2b}$$

where the last expression defines the phase shift $\delta(k)$.

The Jost functions fulfil the integral equation

$$\mathcal{F}_\ell^\pm = 1 + \frac{(-1)^\ell}{k}\int_0^\infty u_\ell(k,r)V(r)f_\ell^\pm(k,r)dr \tag{2.3}$$

where $u_\ell(k,r)$ are the Riccati-Bessel functions, $V(r)$ the potential, and $f_\ell^\pm(k,r)$ are the Jost solutions defined by the boundary condition

$$\lim_{r\to\infty}\{(\pm\imath)^\ell e^{\pm\imath kr} - f_\ell^\pm(k,r)\} = 0 \tag{2.4}$$

The integral representations (2.3) are useful in the study of the properties of the Jost functions as functions in the complex k-plane. As it stands (2.3) only defines $\mathcal{F}_\ell^+(k)$ ($\mathcal{F}_\ell^-(k)$) in the upper (lower) half of the complex k-plane. This leads to a well defined S-matrix on the real axis but nowhere else. For more details see, for instance, the discussion found in Newton[13], see also references 15-18.

The Jost functions may also be defined through the Wronskian relations

$$\mathcal{F}_\ell^\pm(k) = (-k)^\ell \mathcal{W}\{f_\ell^\pm(k,r), \psi_\ell(k,r)\} \tag{2.5}$$

where the Wronskian is defined by

$$\mathcal{W}\{u,v\} = u(r)\frac{dv(r)}{dr} - \frac{du(r)}{dr}v(r) \tag{2.6}$$

and $\psi_\ell(k,r)$ is the solution regular at the origin defined by

$$\lim_{r\to 0}\{(2\ell+1)!!r^{-(\ell+1)}\psi_\ell(k,r)\} = 1 \tag{2.7}$$

We will now proceed to investigate the analytic properties of the Wronskians when the potential allows for an analytic continuation to a part of the complex plane. We assume that the analytic continuation of the potential also fulfils the standard requirements of scattering theory regarding its behaviour at zero and infinity. In general, to analytically continue (2.5), we would only require the continued potential to be such that the corresponding square integrable solution at infinity is unique (Weyl's limit point case, see reference 19).

For a second order differential operator of the type below

$$H = -\frac{1}{2}\frac{d^2}{dr^2} + U(r) \tag{2.8}$$

one easily finds the standard Green's formula as given by

$$\int_0^b \{\phi^*(r)H\psi(r) - [H\phi(r)]^*\psi(r)\}dr = \mathcal{W}\{\psi,\phi^*\}|_0^b \tag{2.9}$$

Introducing now the modified scalar product i.e.

$$\langle f|g\rangle_\eta = \int_0^{b\eta} f^*(\eta^* r)g(\eta r)d(\eta r) = \int_0^b \{(\eta^*)^{\frac{1}{2}} f(\eta^* r)\}^* \eta^{\frac{1}{2}} g(\eta r)dr = \langle f(\eta^*)|g(\eta)\rangle \tag{2.10}$$

where we have defined
(note the difference between the notations $g(\eta)$ and $g(\eta r)$!)

$$g(\eta) = \eta^{\frac{1}{2}} g(\eta r) \tag{2.11}$$

one obtains a generalization of Green's formula

$$\begin{aligned} &\int_0^{b\eta} [\{\phi(\eta^* r)\}^* H(\eta)\psi(\eta r) - \{H(\eta^*)\phi(\eta^* r)\}^*\psi(\eta r)]d(\eta r) = \\ &\int_0^b \{\phi(\eta^*)\}^* H(\eta)\psi(\eta) - \{H(\eta^*)\phi(\eta^*)\}^*\psi(\eta)dr = \\ &\int_0^b \eta^{-2}\frac{d}{dr}\mathcal{W}\{\psi(\eta),\phi^*(\eta^*)\}dr = \eta^{-1}\mathcal{W}\{\psi(\eta),\phi^*(\eta^*)\}\ |_0^b \end{aligned} \tag{2.12}$$

Note that

$$H(\eta) = -\eta^2\frac{1}{2}\frac{d^2}{dr^2} + U(\eta r) \tag{2.13a}$$

where we require that the potential is analytic in some region around η, see also the discussion preceeding (2.8), such that

$$U^*(\eta^*) - U(\eta). \qquad (2.13.b)$$

This gives for $\phi^*(\eta^*) = \bar{\phi}(\eta) = \psi(\eta)$ being solutions to

$$H(\eta)\psi(\eta) = \varepsilon\psi(\eta) \qquad (2.14)$$

where ε now may be a complex resonance eigenvalue, that the Wronskian

$$\mathcal{W}\{\psi(z), \bar{\phi}(z)\} \equiv \text{constant} \qquad (2.15)$$

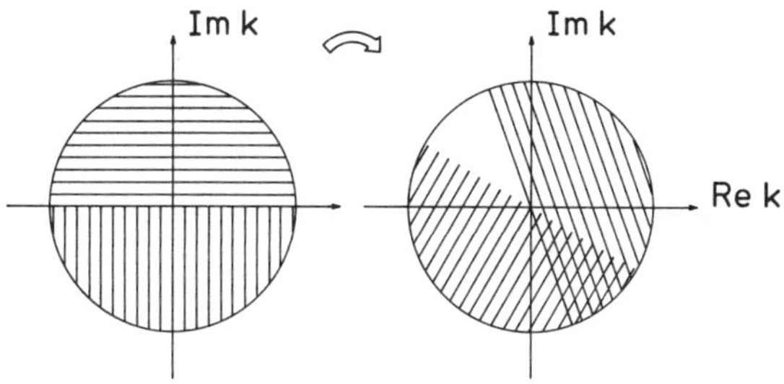

Figure 1

The picture at the left shows the areas where the respective Jost functions are defined before a complex rotation of the coordinates. Horizontal lines refer to $\mathcal{F}^+(k)$, vertical to $\mathcal{F}^-(k)$. The S-matrix is thus only well defined on the real axis. After individual complex rotations in the calcualtions of the Jost functions we find the situations at left, where the S-matrix has a meromorphic continuation in the overlapping region.

i.e. is independent of z along the path in the complex coordinate plane defined by z. This supplies us with a way to extend the Jost functions to new regions of the complex plane. Evaluating the Wronskian expression (2.5) for the Jost functions with the scaled solutions corresponding to the differential operator (2.13), the half planes, for which expression (2.5) (the same goes for (2.3) is well defined, rotate by the angle $\arg \eta^*$. This situation is displayed in figure 1. For numerical reasons we may want to scale the two Jost functions differently and thus open up a sector in the complex plane in which the expression for the S-matrix, (2.2), becomes well defined.

In the next section we will give an example of the use of this technique.

We note that the above derivation can be extended to the use of more general integration contours. The transformation defined by (2.11) is chosen for convenience only, i.e. it provides a unitary transformation for real η.

3. A Generalization of Levinson's Theorem

We will first recapitulate our knowledge on the pole structure of the partial wave S-matrix. We will, as usual, divide the poles into five different categories.

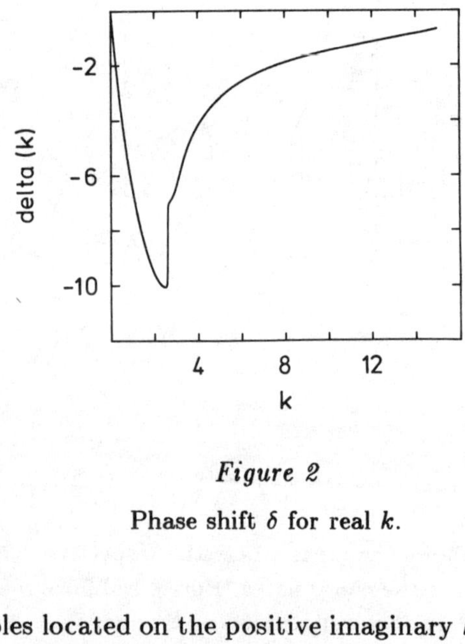

Figure 2

Phase shift δ for real k.

1. Bound state poles located on the positive imaginary k-axis (the negative real E-axis).

2. Resonance poles located in the fourth quadrant of the k-plane (the lower half plane of the second sheet of the E-plane).

3. Mirror poles located in the third quadrant of the k-plane (the upper half plane of the second sheet of the E-plane).

4. Virtual poles located on the negative imaginary k-axis (the negative real second sheet E-axis).

5. Redundant poles due to singularities in $\mathcal{F}^-(k)$.

We will here mostly be interested in the poles in categories 1-3. The effects of the poles in categories 4-5 will be dealt with indirectly.

One of the familiar results of scattering theory is the connection between the number of bound states, i.e. the number of poles in category 1, and the difference of the phase shift (2.2b) at zero and infinite energy.

$$\pi n_{\text{bound}} = \delta(0) - \delta(\infty) \tag{3.1}$$

This result is known as Levinson's theorem. We will here only treat the case of a local potential with a Jost function different from zero at the origin. For the modifications of (3.1) appropriate for these cases see [13,20].

To exemplify the method of analytic continuation presented in the previous section we will employ the model potential

$$V(r) = 7.5r^2 \exp(-r) \tag{3.2}$$

This potential fulfils the requirements of section 2, and the complex scaling can immediately be incorporated. For simplicity we will only deal with the case $\ell = 0$. Figure 2 shows the variation of the phase shift for real k. We observe that apart from a rather smooth variation with k there is a jump of nearly π near $k = 2.6$. This is a clear indication of a resonance pole near the real k-axis. The potential (3.2) does not give rise to any bound states, and so according to (3.1), the phase shift has the same limits at zero and infinite k.

We will now try to find a more general relationship between a phase shift defined for complex k and the complex poles of the S-matrix. In the next subsection we present a derivation of such a generalized Levinson's theorem, dealing with the phase shift in the complex plane and its connection with the bound, resonant and mirror poles.

3.1. Derivation of a generalized Levinson's theorem.

Here we assume that we are dealing with a local potential and that the Jost function fulfils

$$\mathcal{F}^+(0) = 0 \tag{3.3}$$

We first observe the following relations

$$\mathcal{F}^+(-k^*)^* = \mathcal{F}^+(k) = \mathcal{F}^-(-k) \tag{3.4}$$

which for real k gives

$$\mathcal{F}^{+}(k)^{*} = \mathcal{F}^{-}(k) \tag{3.5}$$

$$\mathcal{F}^{\pm}(k) = |\mathcal{F}(k)| \exp(\mp \imath\delta(k)) \tag{3.6}$$

$$S(k) = \frac{\mathcal{F}^{-}(k)}{\mathcal{F}^{+}(k)} = \exp(\imath 2\delta(k)) \tag{3.7}$$

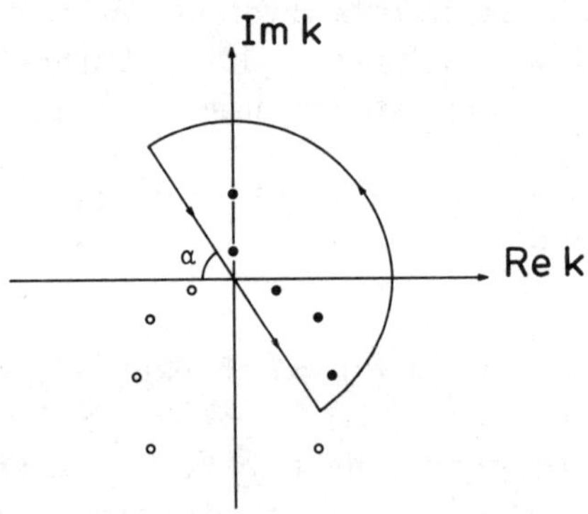

Figure 3

Integration contour C_1.

We will now derive a generalization of Levinson's theorem which connects the (generalized) phase shift for complex k and the complex zeroes of $\mathcal{F}^{+}(k)$.

We will use two different generalizations for the phase shift. The two definitions coincide for real k. First we define the phase shift as

$$\delta_I(k) = \frac{1}{2\imath} \ln S_\ell(k) \tag{3.8}$$

Let us now consider, compare e.g. the common proof of the ordinary Levinson's theorem, the integral

$$I = \int_C d\{\ln \mathcal{F}^{+}(k)\} \tag{3.9}$$

The integration contour C_1 is chosen as in figure 3. The analytic continuation of $\mathcal{F}^+(k)$ to the lower half plane fulfils the condition

$$\lim_{|k|\to\infty} \mathcal{F}^+(k) = 1 \tag{3.10}$$

This can be seen form the representation of $\mathcal{F}^+(k)$ given by (2.3) and its dilated analogue. Since the integral part is multiplied by $1/k$ it is enough to show that this integral is bounded. In the integrand $u_\ell(k,r)$ and $V(r)$ are bounded functions the product of which goes to 0 as $r \to \infty$ The analysis of the boundedness of the Jost solution $f_\ell^+(k,r)$ given by Newton[13] can furthermore be extended from the upper half plane down on the second Riemann sheet by our complex dilation scheme. In fact the Jost solution goes to 0 as $r \to \infty$ in all of the uncovered sector except for the rotated cut. Thus letting the radius of the semi-circle in figure 3 go to infinity, one obtains because of (3.10)

$$I = \lim_{R\to\infty}\Big[\int_0^{R\exp(-\imath\alpha)} d\{\ln \mathcal{F}^+(k)\} - \int_0^{-R\exp(-\imath\alpha)} d\{\ln \mathcal{F}^+(k)\}\Big] =$$

$$\begin{aligned}
&\lim_{a\to 0}\Big[\ln \mathcal{F}^+(-a\exp(-\imath\alpha)) - \ln \mathcal{F}^+(a\exp(-\imath\alpha))\Big] - \\
&\lim_{R\to\infty}\Big[\ln \mathcal{F}^+(-R\exp(-\imath\alpha)) - \ln \mathcal{F}^+(R\exp(-\imath\alpha))\Big] = \\
&\lim_{a\to 0} \ln S(-a\exp(-\imath\alpha)) - \lim_{R\to\infty} \ln S(R\exp(-\imath\alpha)) = \\
&= 2\imath\{\delta_I(0) - \delta_I(\infty\exp(-\imath\alpha)\}
\end{aligned} \tag{3.11}$$

From (3.10), (3.4) and (3.7) follows that

$$\lim_{R\to\infty} \delta_I(R\exp(-\imath\alpha)) \tag{3.12}$$

is real. On the other hand, we have by Cauchys residue theorem, that

$$I = 2\imath\pi\{n_{bound} + n_{res}\} \tag{3.13}$$

The first definition of δ thus gives

$$\pi(n_{bound} + n_{res}) = \delta_I(0) - \delta_I(\infty\exp(-\imath\alpha)) \tag{3.14}$$

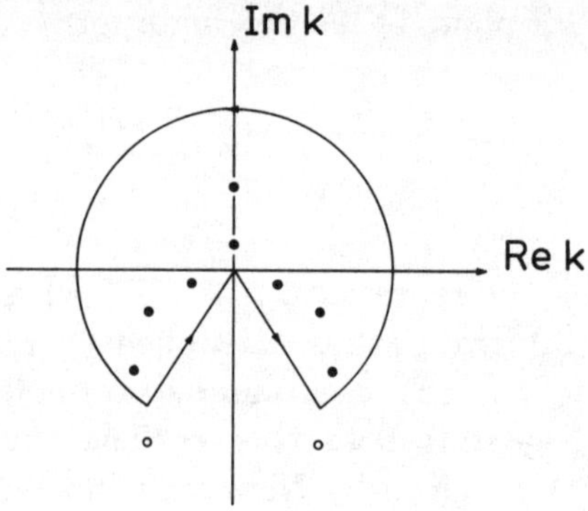

Figure 4

Integration contour C_2.

Now consider, perhaps the more natural definition

$$\delta_{II} = -\arg \mathcal{F}^+(k) \tag{3.15}$$

For the evaluation of (3.9) we choose the contour C_2 in fig. 4. This gives

$$I = \lim_{R\to\infty}[\int_0^{R\exp(-\imath\alpha)} d(\{\ln \mathcal{F}^+(k)\} - \int_0^{-R\exp(\imath\alpha)} d(\{\ln \mathcal{F}^+(k)\}] =$$

$$\begin{aligned}
&\lim_{a\to 0}[\ln \mathcal{F}^+(-a\exp(\imath\alpha)) - \ln \mathcal{F}^+(a\exp(-\imath\alpha))] - \\
&\lim_{R\to\infty}[\ln \mathcal{F}^+(-R\exp(\imath\alpha)) - \ln \mathcal{F}^+(R\exp(-\imath\alpha))] = \\
&\lim_{a\to 0}(\ln \mathcal{F}^+(-a\exp(-\imath\alpha)^*) - \ln \mathcal{F}^+(a\exp(-\imath\alpha)) - \\
&\lim_{R\to\infty}(\ln \mathcal{F}^+(R\exp(-\imath\alpha)^*) - \ln \mathcal{F}^+(R\exp(-\imath\alpha)))
\end{aligned} \tag{3.16}$$

$$= 2\imath\{\delta_{II}(0) - \delta_{II}(\infty\exp(-\imath\alpha)\}$$

Again, by Cauchys residue theorem, we have that

$$I = 2\pi\imath(n_{bound} + n_{res} + n_{mirror}) \tag{3.17}$$

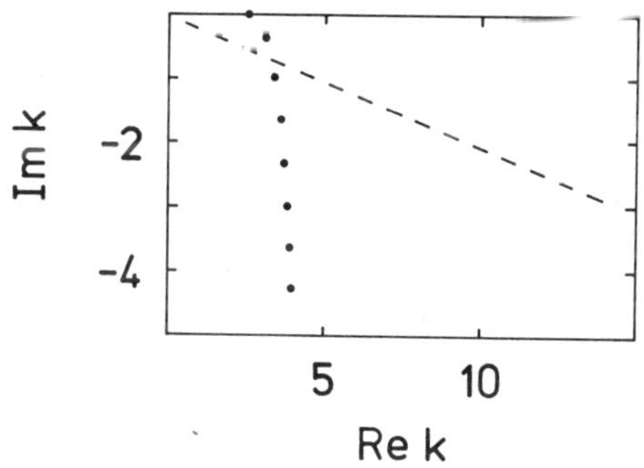

Figure 5

Pole distribution for the potential (3.2).

The second definition of δ thus gives

$$\pi(n_{bound} + n_{res} + n_{mirror}) = \delta_{II}(0) - \delta_{II}(\infty \exp(-\imath\alpha)) \qquad (3.18)$$

We have thus found two possible extensions of Levinson's theorem corresponding to two different definitions of the phase shift for complex k.

To illustrate the behaviour of the quantities introduced above we calculated the various phase shifts for some different paths in the complex k–plane. Figure 5 shows the fourth quadrant and some of the resonance poles of the potential (3.2).

We also indicate the path of calculation corresponding to figures 6 and 7. Figures 6A and 6B 7 show the variation of the real and imaginary part of δ_I along this ray. Note the rather smooth behaviour with no sign of any nearby resonance poles, and, as predicted by (3.4), the difference of 2π between δ_I at zero and infinity. Figure 7 shows a similar smooth behaviour for the real part of δ_{II} and a difference of 4π, as according to (3.6).

Figure 8A shows the variation of δ_I as calculated along a ray in the complex k-plane passing midway between the first and second resonance poles. In accordance with (3.4) we have a difference of π at zero and infinity.

In figure 8B we show the corresponding plot for δ_{II}. Note that δ_{II} shows a sharp variation of minus π near the resonance. Such a variation for the phase of the Jost

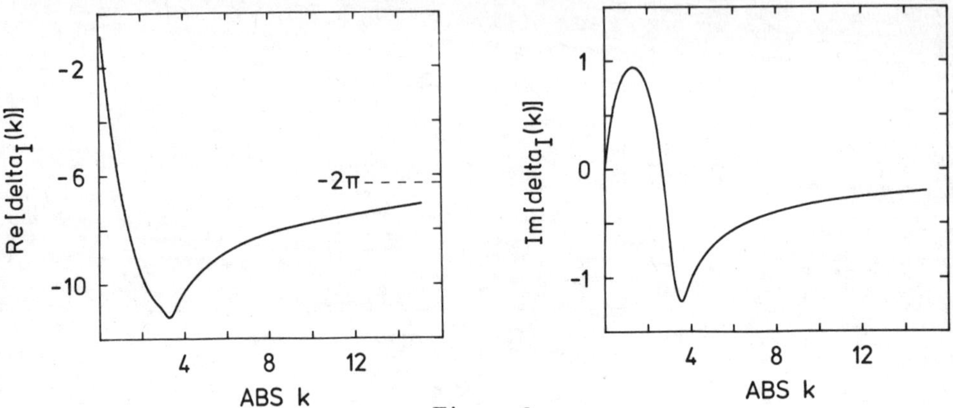

Figure 6

A) Real part of phase shift δ_I along a ray with $\arg(k) = -0.2$. **B)** Imaginary part o phase shift δ_I along a ray with $\arg(k) = -0.2$.

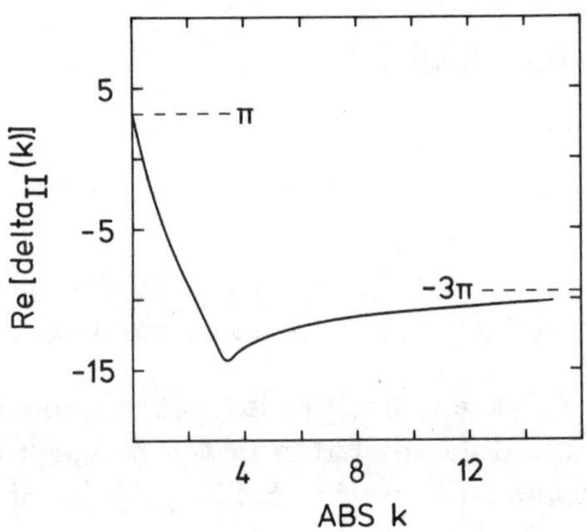

Figure 7

Real part of phase shift δ_{II} along a ray with $\arg(k) = -0.2$.

function could not occur on the real axis since there are no resonance poles in th upper half $k-$plane.

4. A Partial Wave S-Matrix Expansion

We note that the sharp variation of the phase shift, as seen from figure 2 implie a similarly sharp variation of S_ℓ and the cross section near this point.

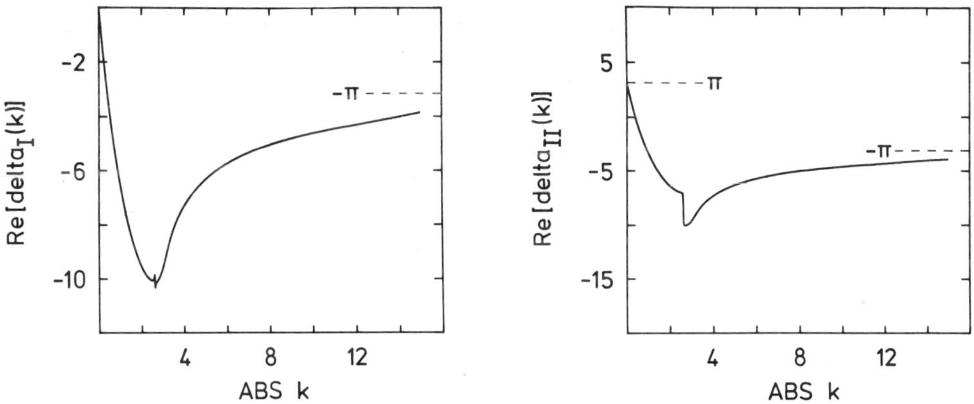

Figure 8

A)Real part of phase shift δ_I along a ray with $\arg(k) = -0.0037$. **B)** Real part of phase shift δ_{II} along a ray with $\arg(k) = -0.0037$.

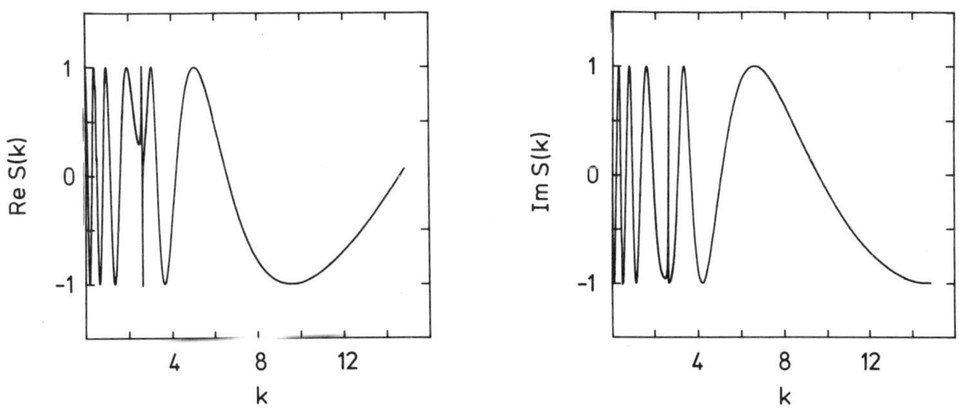

Figure 9

A) Real part of the S-matrix along the real k−axis. **B)** Imaginary part of the S-matrix along the real k−axis.

Figures 9A and 9B show the variation of the S-matrix along the real axis for the potential (3.2). In a calculation such as this, it is very easy to bypass a fast resonance-like variation if the step size of the calculation is chosen too large. From figures 6 and 7 we know, however, that we may expect a more smooth variation of the S-matrix in the complex plane, as long as resonance poles are avoided. This leads us to investigate the possibility of expanding the S-matrix in its poles and background contributions.

In the appendix we derive an expresssion for a Mittag-Leffler expansion of the partial wave S-matrix. The final form is

$$\begin{aligned} S_\ell(k) =& S_\ell(a) + \sum_{i=1}^{N} \mathcal{RES}\{S_\ell(k_i)\}\left(\frac{1}{k-k_i} + \frac{1}{k_i - a}\right) + \\ & \frac{k-a}{2\pi\imath}\int_C \frac{S_\ell(z)}{(z-k)(z-a)} dz \end{aligned} \tag{4.1}$$

The integration path can be chosen differently, as for example $C_1 - C_3$ given in figures 3,4 and 10, according to the pole information available.

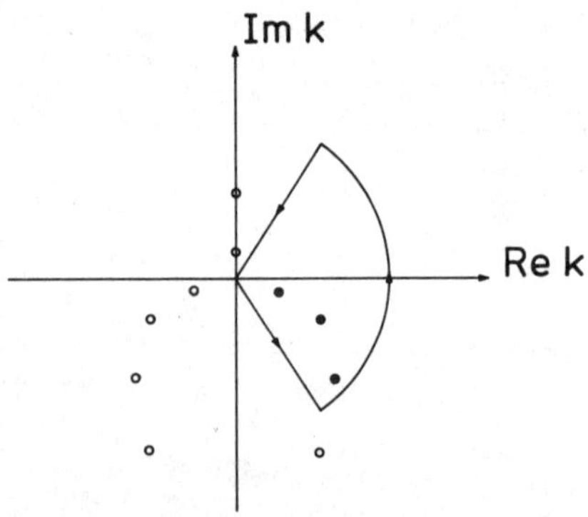

Figure 10

Integration contour C_3.

One reason for using an expression like (4.1) instead of the Weierstrass expression (A.14) is that the expansion (4.1) focuses our attention on the pole and its residue whereas the expansion (A.14) deals only with the poles. The Weierstrass expansion is of course equivalent to the Mittag-Leffler expansion when we take all the possible poles into account.

Figures 11A and 11B show the variation of the analytically continued S-matri

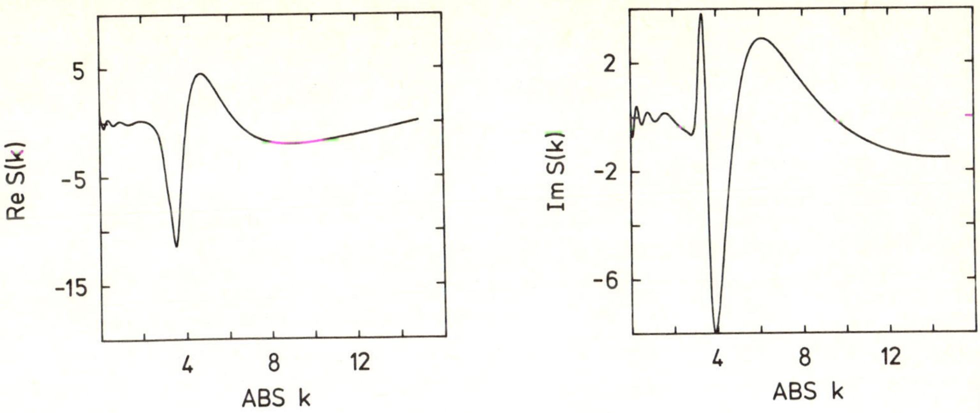

Figure 11

A) Real part of S-matrix in the complex k-plane along a ray with $\arg(k) = -0.2$.
B) Imaginary part of S-matrix in the complex $k-$plane along a ray with $\arg(k) = -0.2$

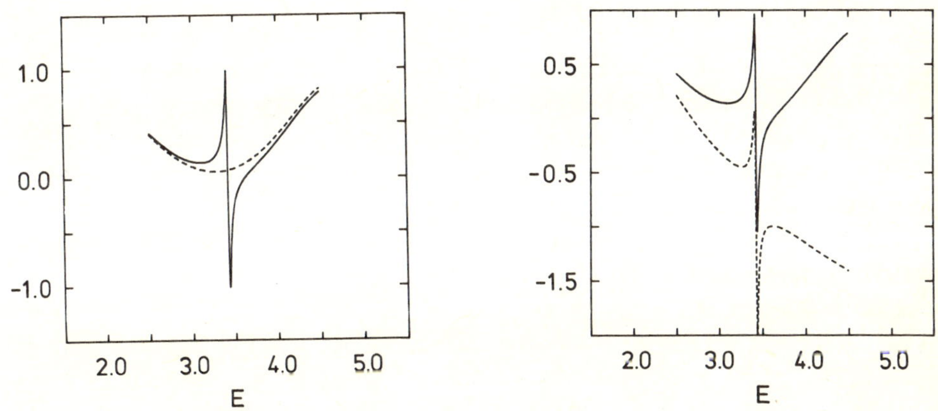

Figure 12

Comparison with exact Real part of S-matrix (solid line) and an expansion calculation with **A**)the contribution from the most narrow resonance omitted (dotted line). **B**) the contribution from the second most narrow resonance omitted (dotted line).

along the same path as for the phase shift in figures 6 and 7. Numerically, the poles can be straightforwardly evaluated by locating the zeroes of $\mathcal{F}^+(k)$, and the residues can be found by numerically evaluating

$$\lim_{k \to k_i} (k - k_i) S_\ell(k) = \frac{\mathcal{F}^-(k_i)}{\left(\frac{d\mathcal{F}^+(k)}{dk}\right)} \Big|_{k=k_i} \tag{4.2}$$

using the analytically continued Jost functions.

The present Mittag-Leffler expansion (4.1) has been numerically tested for the potential (3.2). It was found that the exact S-matrix could be reproduced to desired accuracy by using (4.1) for a finite number of poles and by incorporating the background integral contribution using, for instance, any of the paths $C_1 - C_3$ in figures 3,4 and 10. The arbitrary point a must be chosen inside the selected integration path. Such a calculation is in contrast with the ansatz of Korsch et al.[6] where the S-matrix, or rather the phase shift, is conjectured to be fully exhaustible by enumerable pole contributions. The latter procedure has several drawbacks. Firstly, the background contribution is not easily captured by a finite number of poles, see e.g. the related problem concerning the expansion of the Green's function[21]. Secondly, a pole ansatz must, in addition to bound state and resonance (and mirror) poles, also account for virtual and redundant poles as well as other analytic structures. In the present background integral, see (4.1), these effects are automatically included.

Figures 12 and 13 show, some results of S-matrix calculations based on (4.1). The phase shift, $\delta_I(k)$, calculated as in figures 6A and 6B, was used for the evaluation of the background integral. Since the phase shift is more smoothly varying than S itself, it is advantageous to use an interpolation between calculated phase shifts rather than between S-matrix points. Figure 12A shows the real part of the exact S-matrix compared with an expansion calculation where the contribution from the most narrow resonance pole has been omitted. Note that such a resonance indeed has a very local influence on the variation of S. This indicates that a Breit-Wigner ansatz for the S-matrix is justifiable. We will deal with such an ansatz in the following section. Figure 12B shows a similar comparison where the second most narrow resonance pole contribution has been omitted. Here the contribution from the pole is spread over a large interval. We also note that the omission of one pole in the expansion destroys the unitarity of S. This is in contrast with the case of the Wierstrass-type expansion.

5. Connections with the Breit-Wigner Ansatz and the Fano Parameters

The Breit-Wigner parameterization of the S-matrix[22,23] at the proximity of a resonance pole is

$$S_\ell(E) = \{1 + \frac{\imath\Gamma}{E_0 - E - \imath\Gamma/2}\} \exp(\imath 2\gamma) \tag{5.1}$$

where γ is the background phase shift, assumed to be constant in a small interval near E_0.

Assuming that such a parameterization is valid, we find that the residue of S at the pole $E_j = E_0 - \imath\Gamma/2$, can be written

$$\mathcal{RES}_E\{S_\ell(E_j)\} = \lim_{E\to E_j}(E - E_j)S_\ell(E) = -\imath\Gamma\exp(\imath 2\gamma) \tag{5.2}$$

and from the expansion (4.1) we see that after a change of variables we have

$$S_\ell(E) = \left\{G_\ell(E)\exp(-\imath 2\gamma) + \frac{\frac{\imath\Gamma}{\sqrt{2E_j}}[\sqrt{(2E\,)} + \sqrt{(2E_j)}]}{2(E_0 - E - \imath\Gamma/2)}\right\}\exp(\imath 2\gamma) \tag{5.3}$$

where we have defined

$$\begin{aligned} G_\ell(E) = g_\ell(\sqrt{2E}) = S_\ell(a) &+ \frac{\mathcal{RES}_k\{S_\ell(k_j)\}}{k_j - a} \\ &+ \sum_{i=1;i\neq j}^{N} \mathcal{RES}_k\{S_\ell(k_i)\}\Big(\frac{1}{k_i - k} + \frac{1}{k_i - a}\Big) \\ &+ \frac{k-a}{2\pi\imath}\int_{C_R}\frac{S_\ell(z)}{(z-k)(z-a)}dz \end{aligned} \tag{5.4}$$

If we now further assume that the approximation

$$\Gamma << E_0 \tag{5.5}$$

is valid, we find that

$$S_\ell(E) = \left\{G_\ell(E)\exp(-\imath 2\gamma) + \frac{\imath\Gamma}{E_0 - E - \imath\Gamma/2}\right\}\exp(\imath 2\gamma) \tag{5.6}$$

Thus, in the case of validity of the approximations (5.1) and (5.5), we may express the background phase shift either by the pole and the residue of S_ℓ

$$\exp(\imath 2\gamma(E_j)) = \imath\frac{\mathcal{RES}_E\{S_\ell\}(E_j)}{\Gamma} \tag{5.7}$$

or by the background term $G_\ell(E)$

$$\exp\Big(\imath 2\gamma(E_j)\Big) = G_\ell(E_j) \tag{5.8}$$

The Fano parameters, see Fano [24,25], are defined by

$$\left\{\begin{aligned} q &= \cot\gamma \\ \varepsilon &= \frac{2(E_0 - E)}{\Gamma} = \cot\delta^{(R)}(E) \end{aligned}\right. \tag{5.9}$$

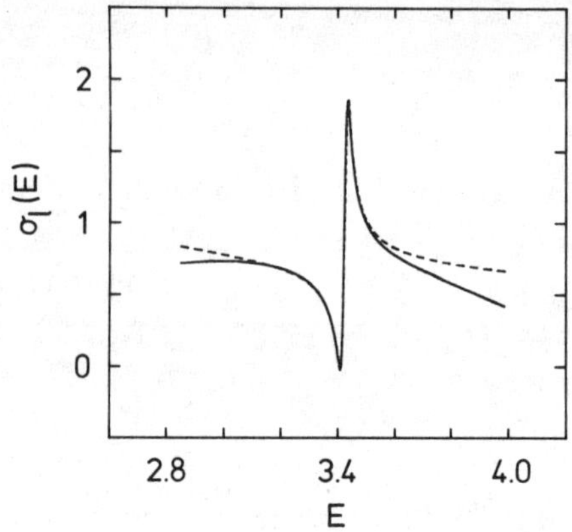

Figure 13

Breit-Wigner form with background term calculated from the residue for the most narrow resonance of the potential as compares with the cross section from the full S-matrix.

and give a parameterization of the partial cross section line shape

$$\sigma_\ell(E) = \frac{4\pi}{2E}(2\ell+1)\sin^2\gamma\frac{(q+\varepsilon)^2}{(\varepsilon^2+1)} = \sigma_\ell^{backg}(E)\frac{(q+\varepsilon)^2}{(\varepsilon^2+1)} \tag{5.10}$$

Using the potential (3.2), a pole and residue calculation was performed for the most narrow l=0 resonance. Since the relation (5.7) is not exact, γ was obtained by evaluating the average $\gamma-$value obtained from the real and imaginary parts of (5.7). Using (5.9) and (5.10) the cross section was calculated and compared with the true cross section, see figure 13. We thus find that the knowledge of the resonance pole and the residue of S gives a rather good description of the partial wave cross section in a large interval.

6. Connections with the Inverse Problem and Conclusions

It is interesting to note that the complex information of sections 3 and 4 is precisely the input required for an analytic extension of the Marchenko inversion procedure [25]. For a related analysis in connection with the Gelfand-Levitan[25] integral equation see[21].

The Marchenko inversion procedure based on the integral equation

$$A(r,t) - \int_r^\infty A(r,s)A_0(s+t)ds = A_0(r+t) \qquad ; \qquad t > r \tag{6.1}$$

with

$$A_0(t) = \frac{1}{2\pi\imath}\int_C (S_\ell(k)-1)\exp(\imath kt)dk + \sum_j s_j \exp(\imath k_j t) \tag{6.2}$$

The solution to the integral equation (6.1) can be expressed as

$$A(r,t) = \frac{1}{2\pi\imath}\int_C (S_\ell(k)-1)f(k,r)\exp(\imath kt)dk + \sum_j s_j f(k_j,r)\exp(\imath k_j t) \tag{6.3}$$

where

$$s_j = \mathcal{RES}_k\{S_\ell(k_j)\} \tag{6.4}$$

and $f(k,r)$ is the Jost solution.

The potential can be obtained via

$$U(r) = -2\,\frac{dA(r,r)}{dr} \tag{6.5}$$

The contour C appearing in (6.2) and (6.3) is the interval $(-\infty,+\infty)$ for real k. For the extension to the complex plane, any of the paths in figures 3,4 and 13, for instance, can be chosen for C. The relations above can then be identified as generalized functions, see e.g.[26], and the contributions to the sum in (6.2) and (6.3) must be taken to be those complex poles inside the chosen integration contour. For instance, using complex scaling the choice of contour given by C_1, C_2 or C_3 yields a version of the Marchenko procedure where the coordinates r, s, t are replaced by $\eta r, \eta s, \eta t$. The inversion relation (6.5) then gives the scaled potential $U(\eta r)$.

A proper analytic continuation of the S-matrix an the phase shift is directly linked with the appropriate continuation of the potential, and conversely as presently investigated. Here we have discussed the use of the complex scaling technique to find the analytic continuation of the reduced S-matrix element in the partial wave basis. Although information about narrow resonance states can be obtained by various means, the formulation above enables us to study scattering quantities in large domains of the complex plane.

In closing, we would like to stress the possibility of a generalization of the techniques above to the multi-channel case. The complex pole and the residue S-matrix would then provide us with information about the energy and width as well as the partial widths of the resonance. Work along these lines is in progress.

Appendix

Derivation of expansions for the S-matrix.

Let us study the meromorphic function

$$D_\ell(k) = S_\ell(k) - S_\ell(a) \tag{A.1}$$

Cauchy's integral theorem immediately gives

$$\begin{aligned} S_\ell(k) =& S_\ell(a) + \sum_{i=1}^{N} \mathcal{RES}\{S_\ell(k_i)\}\left(\frac{1}{k-k_i} + \frac{1}{k_i - a}\right) \\ &+ \frac{k-a}{2\pi\imath}\int_C \frac{S_\ell(z)}{(z-k)(z-a)}dz \end{aligned} \tag{A.2}$$

where C_R is some closed contour including the poles k_i with residues $\mathcal{RES}\{S_\ell(k_i)\}$, see figures 3,4 and 13. If the function furthermore fulfils

$$D_\ell(k) < Const \quad ; \quad |k| > R \tag{A.3}$$

then the radius of the contour C_R can be extended to infinity, and the contribution from the circle segment will vanish.

Imagine now that we extend the integration contour to include the whole k-plane and thus all the poles. If $(A.3)$ still holds we obtain

$$\begin{aligned} S_\ell(k) =& S_\ell(a) \sum_{i=1}^{\infty} \mathcal{RES}\{S_\ell(k_i)\}\left(\frac{1}{k-k_i} + \frac{1}{k_i - a}\right) + \\ & \frac{k-a}{2\pi\imath}\int_C \frac{S_\ell(z)}{(z-k)(z-a)}dz \end{aligned} \tag{A.4}$$

We will now show how this Mittag-Leffler expansion gives rise to the classical expansion formula for the S-matrix in terms of its poles. Assuming that the poles are due to simple zeroes of $\mathcal{F}^+(k)$ at the points k_i, we construct the function

$$h(k) = \frac{d\mathcal{F}^+(k)}{dk} / \mathcal{F}^+(k) \tag{A.5}$$

which has poles at k_i with residues equal to 1. According to the above expansion, with $a = 0$, we obtain

$$h(k) = h(0) + \sum_{i=1}^{\infty}\left(\frac{1}{k-k_i} + \frac{1}{k_i}\right) \tag{A.6}$$

Integrating from 0 to z', putting $z' = k$, one obtains

$$\ln \mathcal{F}^+(k) = \ln \mathcal{F}^+(0) + h(0)k + \sum_{i=1}^{\infty}\left(\ln\{1 - \frac{k}{k_i}\} + \frac{k}{k_i}\right) \tag{A.7}$$

so that

$$\mathcal{F}^+(k) = \mathcal{F}^+(0)\exp\{h(0)k\}\prod_{i=1}^{\infty}\{\left(1 - \frac{k}{k_i}\right)\exp(k/k_i)\} \tag{A.8}$$

Remembering the Schwartz reflection principle, a similar expansion of $\mathcal{F}^-(k)$ gives

$$\mathcal{F}^-(k) = \mathcal{F}^+(0)^*\exp\{h(0)^*k\}\prod_{i=1}^{\infty}\{\left(1 - \frac{k}{k_i^*}\right)\exp(k/k_i^*)\} \tag{A.9}$$

Putting

$$\mathcal{F}^+(0) = r\exp(\imath\alpha) \quad ; \quad h(0) = s\exp(\imath\beta) \tag{A.10}$$

we obtain for the partial wave S-matrix

$$S_\ell(k) = \frac{\mathcal{F}^-(k)}{\mathcal{F}^+(k)}$$

$$= \exp(-2\imath\alpha)\exp[(-2\imath s\sin\beta)k]\prod_{i=1}^{\infty}\frac{\left(1 - \frac{k}{k_i^*}\right)}{\left(1 - \frac{k}{k_i}\right)}\exp[k(1/k_i^* - 1/k_i)] \tag{A.11}$$

From (3.6) we find that

$$\alpha = -\delta(0) \tag{A.12}$$

We may define

$$\lim_{k\to\infty}\delta(k) = 0 \tag{A.13}$$

and Levinson's theorem (5.1) leads to the final canonical expression

$$S_\ell(k) = \exp(-2\imath bk)\prod_{i=1}^{\infty}\frac{\left(1 - \frac{k}{k_i^*}\right)}{\left(1 - \frac{k}{k_i}\right)}\exp(2\imath k\sigma\Im[\frac{1}{k_i}]) \tag{A.14}$$

with

$$b = s\sin\beta \tag{A.15a}$$

$$\sigma = \begin{cases} 0 & ; \text{if } \Im k_i = 0 \\ 1 & ; \text{if } \Im k_i \neq 0 \end{cases} \tag{A.15}$$

† Present Adress : Quantum Theory Project
Williamsom Hall
University of Florida
Gainesville, Florida 32611
U.S.A.

References.

1. M. Rittby, N. Elander and E. Brändas, Phys. Rev. **A24**, 1636 (1981).
2. M. Rittby, N. Elander and E. Brändas, Mol. Phys. **45**, 553 (1982).
3. H. J. Korsch and R. Möhlenkamp, J. Phys. **B15**, L559 (1982).
4. O. Atabek, R. Lefebvre and M. Jacon, J. Phys. **B15**, 2689 (1982).
5. H. J. Korsch, R. Möhlenkamp, J. Phys. **B15**, L559 (1982).
6. H. J. Korsch, R. Möhlenkamp and H.-D. Meyer, J. Phys. **B17**, 2955 (1984).
7. H.-D. Meyer, J. Phys. **B16**, 2265 (1983).
8. H.-D. Meyer, J. Phys. **B16**, 2785 (1983).
9. P. L. Kapur and R. E. Peierls, Proc. Roy. Soc. **A166**, 277 (1938).
10. R. E. Peierls, Proc. Roy. Soc. **A253**, 16 (1959).
11. H. M. Nussenzveig 'Causality and Dispersion Relations', Academic Press, New York (1972).
12. R. E. Peierls, in 'The Physics of Electronic and Atomic Collisions' IX ICPEAC (24-30 July 1975) ed. by Risley J . and Geballe R., Univ. of Wash. Press, Seattle London (1976).
13. R. G. Newton, 'Scattering Theory of Waves and Particles', Springer Verlag, New York Heidelberg Berlin (1982).
14. E. Brändas, M. Rittby and N. Elander, J. Math. Phys. **26**, 2648 (1985).
15. R. Jost, Helv. Phys. Acta **20**, 256 (1947).
16. V. Bargmann, Rev. Mod. Phys. **21**, 48 (1949).
17. K. Kodaira, Am. J. Math. **71**, 921 (1949).
18. P. Krylstedt, N. Elander and E. Brändas, Int. J. Quant. Chem. **31**, 755 (1987).

19. E. Engdahl, E. Brändas, M. Rittby and N. Elander, J. Math. Phys. **27**, 2629 (1986).

20. P. Swan, Proc. Roy. Soc. **A228**, 10 (1955).

21. E. Brändas, E. Engdahl, M. Rittby and N. Elander, Int. J. Quant. Chem. **19**, 457 (1986).

22. G. Breit and E. Wigner , Phys Rev. **49**, 519 (1936).

23. A. Böhm, 'Quantum Mechanics', Springer Verlag, New York (1979).

24. U. Fano, Phys. Rev. **124**, 1866 (1961).

25. K. Chadan and P.C. Sabatier, "Inverse Problems in Scattering Theory" (Springer-Verlag, New York, Heidelberg, Berlin, 1977).

26. I.M. Gel'fand and N. Ya. Vilenkin , "Generalized Functions", volume 4 (Academic Press, New York, 1964).

CALCULATION OF RESONANT WAVE FUNCTIONS IN NUCLEAR PHYSICS

Borbála Gyarmati

Institute of Nuclear Research of the
Hungarian Academy of Sciences,
Debrecen
H-4001 P.O. Box 51, Hungary

Abstract

In this review the single-particle resonance is identified with the Gamow function, the solution of the stationary Schrödinger equation with purely outgoing asymptotics. Its properties and the working rules related to it are presented. Two numerical methods and the corresponding codes developed in nuclear physics are reviewed : the direct numerical integration and an efficient and flexible approximation based on the separable expansion of the potential. Examples are given for calculating neutron and proton resonances in spherical and deformed potentials.

1. Introduction

The title is much too restrictive and much too general at the same time.

It is restrictive because the methods used for the calculation of resonances in nuclear physics surely have very much in common with those used in other branches of quantum mechanics-based physics irrespective of the fact that the "inventors" were or were not aware that similar or even identical methods are in existence in other fields. Sometimes the disappointment due to the rediscovery was counterbalanced by the satisfaction of clearing up the interrelations between the various methods.

On the other hand, it is too general because the overview of all methods developed in nuclear physics would certainly be a highly nonrealistic aim in the framework of a lecture.

I am going to give a short outline of some methods for the description of resonances involving one degree of freedom called one-particle or shape resonances. I do so for two main reasons. Firstly, they are qualitatively different from the many-body resonances in the sense that they cannot be looked upon as bound states embedded in and coupled to the continuum. In the many-body case the close resemblance between the bound states and resonance states is manifest, the coupling of a bound state to the continuum causes only some shift and spreading(s) which can be calculated in the region of validity of the orthodox quantum mechanics [1]. The beautiful and flexible projection operator technique [2] provides us in principle with the position and width, as well as the wave function of a resonance involving any number of freedoms up to the compound resonances, without leaving the safe ground. The extension of the shell model to the continuum [3] does the same on the same safe ground with approximately the same results if the coupling of continuum states to continuum states does not play an important role. In the frame of the orthodox quantum mechanics the shape resonances can be approached only as a phenomenon of the scattering. From the energy dependence of the phase shift, the position and width of a resonance may be defined but it requires some amount of arbitrariness to assign a function to the resonance as a "resonant wave function." The use of the energy dependent scattering wave function completed by an integration over an energy range containing the resonance is a more convincing practice. To regard the resonance as an extension of the bound state leads off the realm of the orthodox quantum mechanics. Secondly, and less importantly, it is the field to which the Debrecen group contributed most. For shorter or longer periods J. Cseh, B. Gazdy, B. Gyarmati, F. Krisztinkovics, A. T. Kruppa, R. G. Lovas, Z. Papp, K. F. Pál, T. Vertse and G. Wolf worked as students, fellows or staff members on topics related to Gamow states. (In this lecture I use freely their published works and theses.) That is why perhaps unproportionally large emphasis will be put on the Gamow functions as the tools we have found to be most elegant and best suited for the description of a resonance.

2. One-Particle Resonances in Nuclear Physics

Before dwelling on how to compute them we should ask why compute them. Two distinct questions may occur in the case of any artifacts of theoretical physics: is it necessary somewhere, is it useful somewhere? Certainly if there are many affirmative answers to the first question the artifact will be subjected to thorough scrutiny and, possibly after some detour, it will reach a highly developed phase. If this phase is reached, more often than not, the answers to the second question will also be in the affirmative, i.e. the application of the developed technique starts growing. In nuclear physics, letting alone the old alpha-decay problem, the pressing necessity to describe one particle resonances was brought about by the nuclear reactions leading to unstable but long-lived final states which manifested themselves in the spectra of the emitted particles as peaks belonging to positive energies of the product nuclei. Let us quote here one of the early observations in the

^{24}Mg(d,n)^{25}Al* reaction [4]. Many others followed. Since the angular distribution of the ejected particle suggested the stripping mechanism, the natural theoretical framework to describe the reaction was the Distorted Wave Born Approximation which - here it is enough to know this much about it - requires the (one-particle) eigenfunction of the captured particle in the target nucleus; this function occurs in a matrix element containing otherwise long-ranged and/or oscillatory functions. Here we are!

Since the particle is for a long time, but all the same temporarily captured, a decaying state should be described.

Positive-energy states may be scattering states belonging to continuous energy, or ones belonging to discrete complex energies, the real parts of which are positive. The latter complex energies are eigensolutions of the one-particle Schrödinger equation ($\Im E < 0$, Gamow or Siegert solution) or of the Sturm equation ($\Im E = 0$, Weinberg solution) belonging to purely outgoing asymptotics. In the Weinberg case the potential is determined so as to give $\Re E$ a prescribed value and make $\Im E$ disappear. This results in a complex (eigen) potential the imaginary part of which acts as a source to make up for the purely outgoing flux. In all cases the wave functions oscillate up to infinity at least with constant amplitudes; in the discrete cases their imaginary parts cannot be removed by the application of a phase factor. This poses difficulties for the computation of the DWBA matrix element which is at least very slowly converging; in the Gamow case even diverging. To remedy the problem three methods were invoked: the application of a convergence factor [5,6,7], the continuation of the integrand onto the complex r-plane and the use of a complex contour [8,9], the taking into account of the exact boundary conditions [10,11]. The often attempted simple idea of using just a weakly bound wave function did not lead to reliable results [10,8].

As the consistent application of the scattering wave function described in the introduction is a very time-consuming procedure, it has become commonplace to choose a particular scattering wave function as the resonant wave function (the wave function belonging to the $\delta = \pi/2$ phase shift, to the energy at which the ratio of the inner and the first outer maxima of the amplitude is maximal, etc.) and to prescribe the norm in various ways (e.g. the function is normalized to 1 up to the first outer node with sharp or soft cut-off, a proton function is normalized up to the classical turning point). The Weinberg and Gamow functions were also often used without their imaginary parts normalized in a number of more or less sensible ways. There were so many variants that it would be senseless to pick out quotations haphazardly. It was a relief when the papers by T. Berggren [12] and W. Romo [13] appeared on the Gamow functions.

3. The Gamow Functions

The Gamow function was invented back in 1928 [14] for the description of the alpha-decay. It is the solution $|\Psi\rangle$ of the stationary Schrödinger equation with purely outgoing boundary condition. It belongs to complex energy $E = E^{res} - i\Gamma/2$ or wave number $k = \sqrt{2mE/\hbar^2} = \kappa - i\gamma$, $\kappa > \gamma > 0$. This means that it describes states with purely exponential decay law

$$L(t) \equiv |\langle\psi(\vec{r},0)|\psi(\vec{r},t)\rangle|^2 = \exp[-\Gamma t/\hbar], \tag{1}$$

that in coordinate space oscillate with exponentially growing amplitudes:

$$\lim_{t\to\infty} \psi(\vec{r},t) = \exp[i\kappa r - iE^{res}t/\hbar]exp[\gamma r - \Gamma t/2\hbar]. \tag{2}$$

It is known that there is a deviation from the exponential decay law at small t values because the quantum mechanical system remembers how it was prepared, but it loses its memory soon, as it is supposed to whenever the Breit-Wigner formula or Niels Bohr's compound nucleus concept is used. Another limit [15] on the validity of Equation (1) comes from the natural requirement that

$$\lim_{t\to\infty} L(t) = 0$$

If the initial energy distribution is a Breit-Wigner distribution it imposes the limit on the time interval where the exponential decay law is valid

$$(\Gamma t/\hbar)exp[-\Gamma t/\hbar] >> (\Gamma/E^{res})^2. \tag{3}$$

Examples were given [16] showing that (3) is not more restrictive than the natural requirement that allows us to call a formation resonance only if its lifetime $\tau = \hbar/\Gamma$ is longer than the time T that takes the particle to cross the volume of the potential/nucleus.

As for the asymptotic behavior, it proved to be the long-lived, obstinate objection against the use of Gamow states [17]. We feel that Berggrens's prescriptions opened the way to their consequent and fruitful application.

The idea is to generalize the concept of the quantum mechanical inner-product. The *sine qua non* of the generalization is to put in bra position the time-reverse of the Gamow state. With this the matrix element can be regularized in the Abel sense [18] with a suitable convergence factor and a limiting procedure.

The resonant states and their time-reverse satisfy the equations

$$H|\Psi_n\rangle = E_n|\Psi_n\rangle \tag{4a}$$

and

$$\langle\tilde{\Psi}_n|H = E_n\langle\tilde{\Psi}_n| \tag{4b}$$

with the boundary conditions

$$u_{\ell,n}(0) = \tilde{u}_{\ell,n}(0) = 0 \tag{5a}$$

$$\lim_{r\to\infty} u_{\ell,n}(r) = O_\ell(k_n, r) \tag{5b}$$

$$\lim_{r\to\infty} \tilde{u}_{\ell,n}(r) = I_\ell(k_n^*, r). \tag{5c}$$

The functions $u_{\ell,n}(r), \tilde{u}_{\ell,n}(r)$ are defined by

$$\langle\vec{r}|\Psi_n\rangle \equiv \Psi_n(\vec{r}) = \frac{u_{\ell,n}(r)}{r} Y_{\ell,m}(\hat{r}). \tag{6}$$

$$\langle\vec{r}|\tilde{\Psi}_n\rangle \equiv \tilde{\Psi}_n(\vec{r}) = \frac{\tilde{u}_{\ell,n}(r)}{r} Y_{\ell,m}(\hat{r})$$

The functions $O_\ell(k_n r)$ and $I(k_n^* r)$ are outgoing and incoming spherical waves, respectively. If $H^+ = H$, $\tilde{\Psi}_n(\vec{r}) = (u_{\ell,n}^*(r)/r)Y_{\ell,m}(\hat{r})$. Let us fix here the definition of the continuum states belonging to an arbitrary $\vec{k}$

$$\langle\vec{r}|\Psi_{\vec{k}}\rangle = \frac{1}{kr}\sum_{\ell,m} i^\ell u_\ell(k,r) Y_{\ell,m}(\hat{r}) Y_{\ell,m}^*(\hat{k}). \tag{7}$$

$$\langle\vec{r}|\tilde{\Psi}_{\vec{k}}\rangle = \frac{1}{k^* r}\sum_{\ell,m} i^\ell \tilde{u}_\ell(k^*,r) Y_{\ell,m}(\hat{r}) Y_{\ell,m}^*(\hat{k})$$

Berggren showed [12] that

$$\lim_{\varepsilon\to 0}\int_0^\infty exp[-\varepsilon r^2]\tilde{u}_{\ell,n}^*(r) u_{\ell',n'}(r) dr = \delta_{\ell n,\ell' n'} \tag{8}$$

i.e. the Gamow functions are orthonormalized in the generalized sense. The scattering functions obey the relation

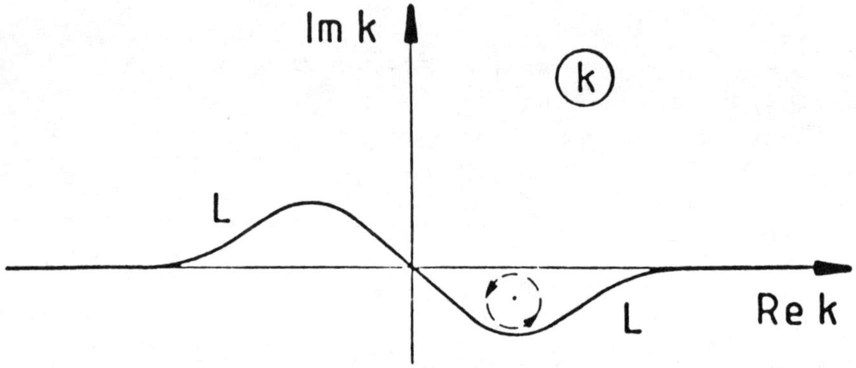

Figure 1

The curve L in the complex k-plane the $\Re k > 0$ part of which is L^+.

$$\int_0^\infty \tilde{u}_\ell^*(k^*, r) u_\ell(k', r) dr = \delta(k - k'). \tag{9}$$

The starting point of the derivation of the generalized inner product and Fourier transformation is a completeness relation. Berggren showed [12] that any function $h(r)$ that contains asymptotically the functions $exp[ikr]$ with k values taken from above the curve L of Fig. 1, or goes to zero faster than any exponential if $r \to \infty$ can be cast in the form

$$h(r)\exp[-\varepsilon r^2] = \int_0^\infty dr' h(r') \exp[-\varepsilon r'^2] \left\{ \sum_n u_{\ell,n}(r) \tilde{u}_{\ell,n}(r')^* \right.$$
$$\left. + \int_{L^+} u_\ell(k, r) \tilde{u}_\ell(k^*, r')^* dk. \right. \tag{10}$$

Here $u_{\ell,n}(r)$ is a normalized Gamow or bound state, $u_\ell(k,r)$ is a normalized scattering function with k taken from L^+ which is the $\Re k > 0$ part of the curve L. The curve L is characterized by the following properties: if k is on L so is $-k$, $|\Re k| > |\Im k|$, for large $|\Re k|$ L coincides with the real axis.

(It is worth noting that there exist more general convergence factors [19] and other forms of the completeness relations [20] which require different forms of the curve L.) In the following it is enough to use a special form of (10) given for a free particle [21]

$$h(r)exp[-\varepsilon r^2] = (2/\pi)\int_{L^+} dkJ_\ell(kr)\int_0^\infty J_\ell(kr')h(r')exp[-\varepsilon r'^2]dr' \tag{11}$$

$$= (2/\pi)\int_0^\infty [h(r')exp[-\varepsilon r'^2]\int_{L^+} J_\ell(kr)J_\ell(kr')dk]dr'$$

which shows that

$$(2/\pi)\int_{L^+} J_\ell(kr)J_\ell(kr')dk = \delta(r-r'). \tag{12a}$$

Here $J_\ell(kr) = krj_\ell(kr)$ is the spherical Bessel function. Using the (12a) form of the $\delta(r-r')$ function in the generalized (Abel) integral and changing the order of the integration we obtain a new representation of the function $h(r)$

$$h(r) = \lim_{\varepsilon\to 0}\int_0^\infty exp[-\varepsilon r'^2]h(r')\delta(r-r')dr' \tag{13a}$$

$$= \lim_{\varepsilon\to 0}\int_0^\infty exp[-\varepsilon r'^2]h(r')(2/\pi)\int_{L^+} J_\ell(kr)J_\ell(kr')dk]dr'$$

$$= (2/\pi)^{1/2}\int_{L^+} [J_\ell(kr)\lim_{\varepsilon\to 0}(2/\pi)^{1/2}\int_0^\infty exp[-\varepsilon r'^2]h(r')J_\ell(kr')dr']dk$$

$$= (2/\pi)^{1/2}\int_{L^+} J_\ell(kr)\hat{h}(k)dk.$$

It seems natural to regard the expression

$$\hat{h}(k) = \lim_{\varepsilon \to 0} (2/\pi)^{1/2} \int_0^\infty exp[-\varepsilon r^2] h(r) J_\ell(kr) dr \tag{13b}$$

and its inverse (13a) as a transformation between the coordinate and momentum representation. The counterpart of (12a) can also be derived by substituting (13b) in (13a) and changing the order of integration

$$\lim_{\varepsilon \to 0} (2/\pi) \int_0^\infty exp[-\varepsilon r^2] J_\ell(k'r) J_\ell(kr) dr = \delta(k - k'), \quad k, k' \in L^+. \tag{12b}$$

Now, the generalized inner product for the functions $h(r)$ and $g(r)$ satsifying the conditions given above is,

$$\langle g|h \rangle = \lim_{\varepsilon \to 0} \int_0^\infty exp[-\varepsilon r^2] g(r) h(r) dr. \tag{14a}$$

By using the formulas (13a-b) and changing the order of integration, the inner product in the momentum representation emerges

$$\langle g|h \rangle = \int_{L^+} \hat{g}(k) \hat{h}(k) dk. \tag{14b}$$

The generalized physical norm N of the function $h(r)$ is the square root of

$$N^2 = \langle h|h \rangle = \lim_{\varepsilon \to 0} \int_0^\infty exp[-\varepsilon r^2] h(r) h(r) dr \tag{15a}$$

$$= \int_{L^+} \hat{h}(k) \hat{h}(k) dk. \tag{15b}$$

These are the formulas we relied upon in numerical work.

One should notice that all of the above formulas go back to the corresponding orthodox quantum mechanical formulas when $h(r)$ is a bound state wave function. This is due to the regular nature of the regularization procedure, namely if the integral exists in the Riemann sense it exists in the generalized sense too, and the values coincide.

A remark is in order at this place. Instead of regularization, sometimes analytic continuation is preferred as in References [13,22], but the results are equivalent with those obtained with the above rules [19,23].

The Gamow function together with the listed instructions is a good generalization of the bound state and to take it as a definition of the single particle resonance emphasizes the similarities between the bound and resonant states. It is remarkable that even if one tends to despise the Gamow state because of its "unphysical tail behavior" one has to admit that its use in the generalized matrix element is completely satisfactory. Not only because the generalized matrix element converges (e.g. the DWBA matrix element we referred to in Section 2) but also because it virtually dispenses with the "unphysical" region. Similarly to the δ-function the usefulness and physical significance of a bare Gamow function may be doubted but (generalized) matrix elements containing Gamow function(s) should be accepted.

There is one difficulty, namely the matrix elements, including the squared norm, are complex. Obviously, this has happened because we have left the Hilbert space. So we have to answer the question: what is the meaning of a complex matrix element? As far as the diagonal matrix elements, in other words the quantum mechanical mean values, are concerned the complex eigenenergy itself suggests the definition

$$\langle A \rangle \equiv \Re \langle \tilde{\Psi}_{\ell,m,n} | A \Psi_{\ell,m,n} \rangle . \tag{16}$$

The application of (16) to the energy operator yields $\langle H \rangle = E^{res}_{\ell,n}$. The customary definition

$$(\Delta A)^2 \equiv |\langle A^2 - \langle A \rangle^2 \rangle| \tag{17}$$

for the standard deviation yields in the case of the energy $(\Delta H)^2 = \Gamma^2/4$ suggesting that the fluctuation around the mean value has something to do with the imaginary part of the matrix element.

To check how dangerous it would be to draw general conclusions, we made another step forward and calculated the matrix element $R \equiv \langle \tilde{u} | r^2 | u \rangle$ in a square-well + square barrier model potential [16]. We chose four different values for the barrier height U_0 $A > B > C > D$. In Fig. 2-3 we show the mean values $\langle r^2 \rangle$ and deviations $(\Delta r^2)^2$ along pole trajectories generated by the change of the well-depth. The trajectories from deeply bound states are followed up well beyond the limits imposed by the validity of the exponential decay law, Equation (3). Omitting the close neighborhood of the essential singularity at $E = 0$ we see a nice correspondence: mean values in Gamow state can be considered as an

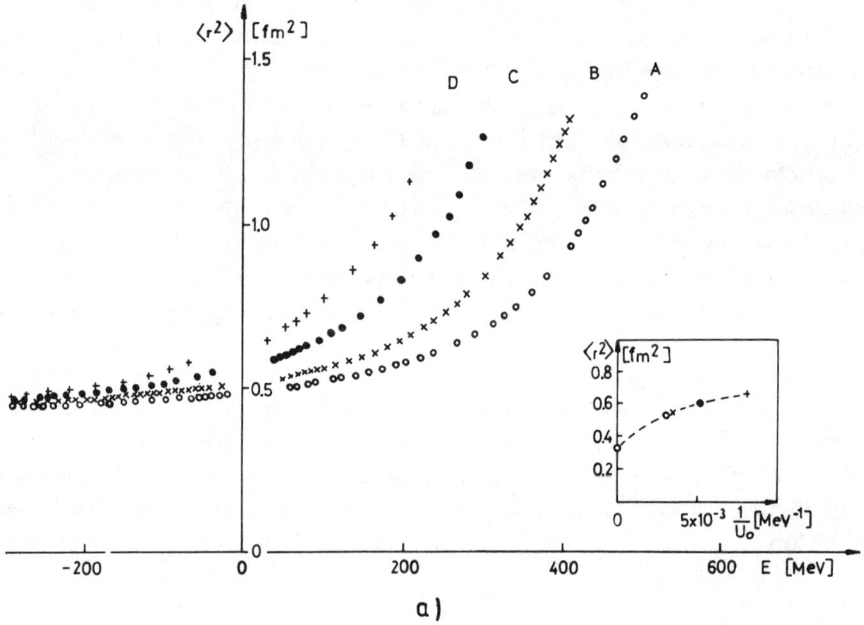

Figure 2

The real part of $R \equiv \langle \tilde{u}|r^2|u\rangle$ matrix element in a square well + square-barrier model potential, $A > B > C > D$ represent different heights of the barrier. The well-depth is varying along the trajectories.

extrapolation of the bound state values. The inserts show that also with $U_0 \to \infty$ the bound state value (particle in a box) is smoothly reached.

It is worth noting that the deviation from the mean value has two distinct sources in Gamow states. One source is the fact that $|u\rangle$ is not an eigenvector of r^2, the other is finite energy width of the Gamow state. Let us identify the latter with $(\Im R)^2$. The figure shows that in narrow resonances this contribution is small, and that the points $(\Delta r^2)^2 - (\Im R)^2$ nicely correspond to the bound state values.

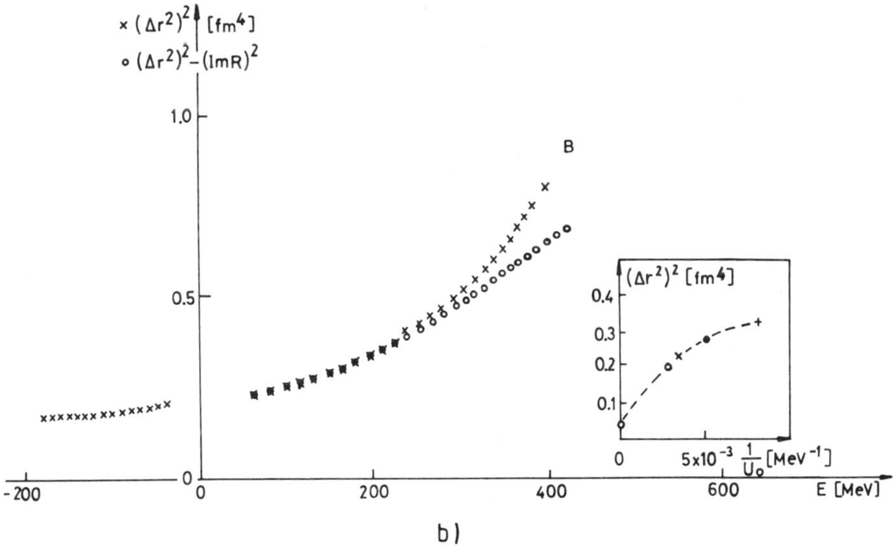

Figure 3

The full deviation from the mean-value (x) and the deviation diminished by the quantity $(\Im R)^2$ taken as a measure of the uncertainty coming from the energy uncertainty (o).

This example encourages us to interpret, in general, the real part of a matrix element as the actual value of the physical quantity. The value of the imaginary part tells us how seriously it must be taken. It is, of course, another question what piece of information, if any, is hidden in the imaginary part beyond that.

4. The Computation of Gamow Functions

From now on we stick to the formulas given in the preceding section because we find them the most complete system of rules given so far for Gamow states. They enable us to present the normalized Gamow state both in coordinate and impulse representations and also provide a mode of switching from one representation to the other. Some people find the limiting procedure in coordinate representation inconvenient and to be a major disadvantage of this sort of regularization, especially when one resorts to numerical integration of the Schrödinger equation. We have found a way out of this difficulty [19]. Below the direct numerical integration (DNI) of the Schrödinger equation and an approximation method especially well suited for the Gamow function will be presented.

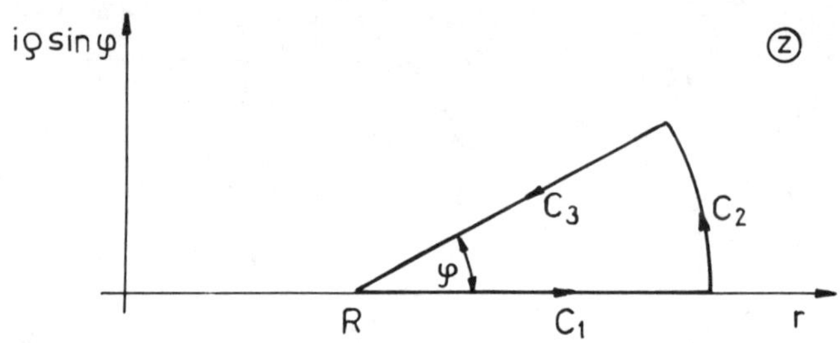

Figure 4

The contour C of integration

4.1 The direct numerical integration: the program GAMOW [24]

The numerical integration of the Schrödinger equation is a well elaborated technique for bound state. Normally it consists of an integration from zero outward to a certain matching point R_m and of an inward integration from infinity to the same matching point. The energy is determined from the condition that at the matching point the two pieces of the function match smoothly. Then the function is renormalized with the norm. It is obvious that this can be achieved for Gamow states, too, with the aid of the generalized formulas; however, to perform numerically, the limiting procedure is awkward. To circumvent this difficulty it is enough to remember that the regularization is regular, i.e. if the integrand is integrable in the Riemann sense we can simply omit the convergence factor and the limit. So any transformation that makes the integrand integrable and leaves valid those properties of the convergence factor that ensure the regularity of the regularization, may be helpful. For the case of short-ranged potentials or short-ranged+Coulomb potentials we proposed [19] the extension of the integral into the complex r-plane and the integration over a C contour determined by

$$Z = \begin{cases} r & R \leq r < \infty \quad on\ C_1 \\ |Z| \exp[i\alpha] |Z| \to \infty & 0 \leq \alpha \leq \varphi \quad on\ C_2 \\ R + \varrho \exp[i\varphi] \ \tan^{-1}(\gamma/\kappa) < \varphi < \pi/4 & 0 \leq \varrho < \infty \quad on\ C_3 \end{cases} \tag{18}$$

(see Figure 4).

R is outside the range of the short-ranged potential. By virtue of the Cauchy theorem

$$\lim_{\varepsilon \to 0} \int_R^\infty exp[-\varepsilon r^2] f(r) dr = \lim_{\varepsilon \to 0} \int_{C_3} exp[-\varepsilon Z^2] f(Z) dZ \tag{19}$$

for any function *f(Z)* that is analytic inside the contour and vanishes on C_2. The square of both the Hankel and Coulomb functions are like this. It was shown in References [19,25] that the square of the Hankel function or Coulomb function is integrable in the Riemann sense and that $exp[-\varepsilon Z^2]$ is a convergence factor on C_3 so

$$\lim_{\varepsilon \to 0} \int_R^\infty exp[-\varepsilon r^2] f(r) dr = \int_{C_3} f(Z) dZ. \tag{20}$$

Based on (20) the $[R, \infty]$ contribution to the norm can be given in closed form for any short-ranged potential. For the short-ranged+Coulomb case the inward integration (of the transformed differential equation) is performed along the C_3 path from a point where the asymptotic forms of the Coulomb functions are valid and also the $[R, \infty]$ contribution to the norm is calculated numerically on C_3. In this way also for the Coulomb case the normalized Gamow function in the interval $0 \leq r \leq R$ is yielded by the code without any numerical limiting procedure. All results to be presented in the next subsection, numbers and curves denoted by DNI were generated with the code Gamow.

4.2 An approximation method: The Potential Separable Expansion Method: the code PSEUDO [26]

Since the Gamow states can be looked upon as a generalization of the bound state, most of the approximation methods can be generalized so as to cope with resonant states. The most conspicuous property of a resonant state is its violent tail behavior. So only the bound-state methods that handle the asymptotics correctly stand a fair chance for working efficiently in a generalized form. The Potential Separable Expansion method is such a method by construction. The Hamiltonian H is split into two parts H_0 and $H_S \equiv V$ so that H_0 contains the terms affecting the infinitely far region, H_S comprises the terms disappearing at infinity. A method is searched for that does not truncate H_0, approximates only V. The solution to

$$(H_0 + V)|\Psi\rangle = E|\Psi\rangle \tag{21}$$

is searched for as the $N \to \infty$ limit of the intermediate problems

$$(H_0 + V_N)|\Psi_N\rangle = E_N|\Psi_N\rangle, \quad N = 0, 1, ... \tag{22}$$

here V_N is the projection of V to a subspace of *N+1* dimension of the Hilbert space. It was shown [27] that with

$$V_N = \sum_{i,j=0}^{N} |i\rangle V_{ij} \langle j|, \quad V_{ij} = \langle i|V|j\rangle \tag{23}$$

the procedure is convergent for bound states. Here $|i\rangle$ is an element of a basis $\{|i\rangle\}$. Substituting (23) into (22) and arranging the equation into the form of a homogeneous Lippmann-Schwinger equation

$$|\Psi_N\rangle = (E_N - H_0)^{-1} V_N |\Psi_N\rangle \equiv G_o^+(E_N) \sum_{i,j=0}^{N} |i\rangle V_{ij} \langle j|\Psi_N\rangle$$

or denoting $\langle j|\Psi_N\rangle = C_j^N$

$$|\Psi_N\rangle = \sum_{i,j=0}^{N} C_j^N V_{ij} G_o^+(E_N)|i\rangle, \tag{24}$$

by multiplying with $\langle k|$ a system of homogeneous algebraic equations follows for C_j^N

$$\sum_{j=0}^{N} \left\{ \delta_{kj} - \sum_{i=0}^{N} \langle k|G_o^+(E_N)|i\rangle V_{ij} \right\} C_j^N = 0, \quad k = 0, 1, ...N. \tag{25}$$

The equation

$$D(E) \equiv det \left\{ \delta_{kj} - \sum_{i=0}^{N} \langle k|G_o^+(E_N)|i\rangle V_{ij} \right\} = 0 \tag{26}$$

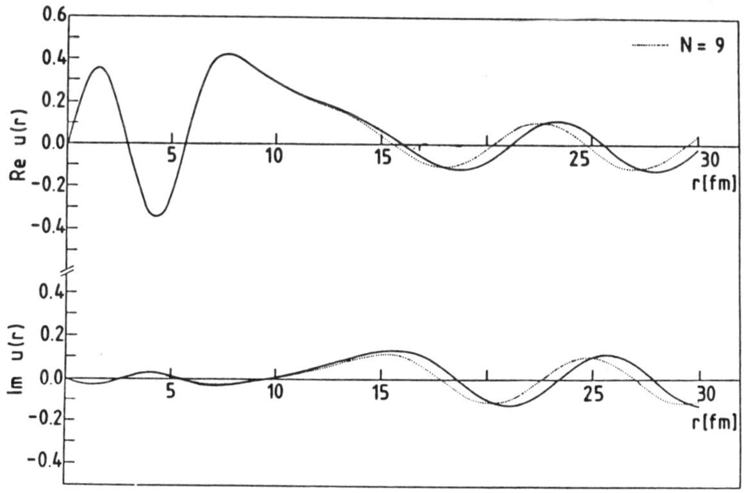

Figure 5

The wave function of the (9.849-i 0.1628) MeV $\ell = 0$ resonance developed in (27) with V=40 MeV, $R_o = R_c = 7.06$ fm, $a = 0.75$ fm, $Z_1 = 82$, $Z_2 = 1$, $R_s = 14.12$ fm, $\sigma = 3.53$. The heavy line is the result of DNI, the dotted is the PSE result with N = 9, for N > 9 the PSE result cannot be distinguished from the DNI line.

ensuring the solvability of (25) is a transcendent eigenvalue equation for the approximate energy eigenvalue $E_{N,\alpha}$. The solution of (25) at $E_{N,\alpha}$ yields the coefficients $C^N_{j,\alpha}$ for (24) which, in turn, gives the approximate eigenvectors $|\Psi_{N,\alpha}\rangle$. From Equations (24-26) the ingredients and region of applicability of the approximation can be read. First, we need a basis to expand the potential. Still it is not only the potential that imposes conditions on the basis. We should find a set of functions with which the matrix elements $\langle k|G_o^+|i\rangle$ can be exactly calculated. For short-ranged potentials $H_o \equiv T$, the kinetic energy, we can invoke the

Table I

The energy ($E^{res} - i\Gamma/2$)(MeV) of an $\ell = 0$ resonance in the potential (27)^a at the indicated basis size.

N	Eigenvalue
9	9.881 - i 0.1419
14	9.848 - i 0.1637
19	9.849 - i 0.1628
24	9.849 - i 0.1628
29	9.849 - i 0.1628
DNI	9.858 - i 0.1641

$^{a)}$ The values of the parameters are given in the caption to Fig. 5

well-manageable harmonic oscillator wave functions (HOWF) because the matrix elements of the free Green operator can exactly be given by a recursion [26]. Second, the potential matrix elements are needed. In general, they are to be computed numerically. If the potential is spherical it mixes only the HOWFs belonging to different main quantum numbers, if it depends on the angles (deformed potential) or contains e.g. tensor forces it may mix also the functions belonging to different values of the orbital angular momentum, total angular momentum and magnetic quantum numbers. The roots of *D(E)* of Equation (26) are to be determined once G_{ki} and V_{ij} are at our disposal. The real roots yield the bound states, the complex ones with $k_{N,\alpha} = \sqrt{2mE_{N,\alpha}/\hbar^2} = \kappa_{N,\alpha} - i\gamma_{N,\alpha}$; $\kappa_{N,\alpha} > \gamma_{N,\alpha} > 0$, yield the Gamow states [28]. It is needless to say that the generalized matrix elements are to be used throughout, the bound states being automatically included.

Out of the numerous examples three are presented here for the short- ranged potential case. The energy and wave function of an $\ell = 0$ resonance developed in a Saxon-Woods + screened Coulomb potential [23]

$$V(r) = -V_o(1 + exp[(r - R_o)/a])^{-1} + V_{coul}$$

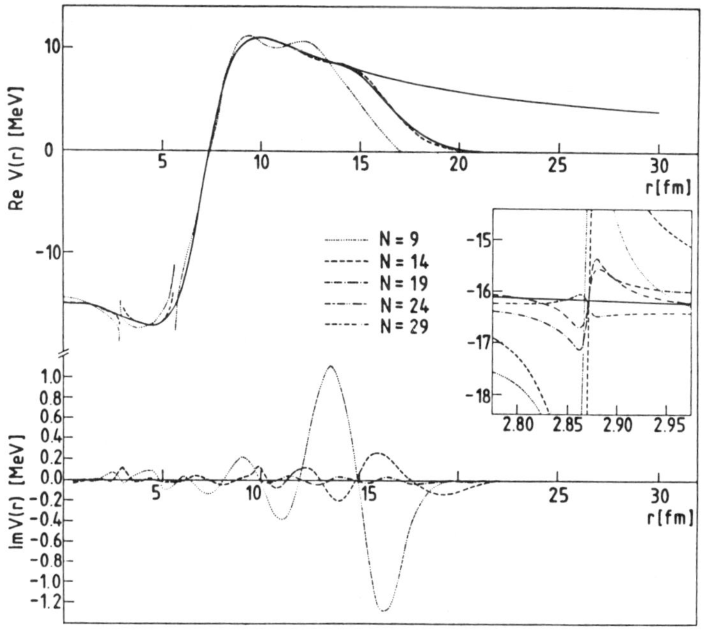

Figure 6

The potentials (28) at the indicated basis size. The heavy continuous line represents the Saxon-Woods + Coulomb potential, the light continuous line shows potential (27). The insert is a blown up part of $\Re V(r)$. Note the difference between the scales!

$$V_{coul} = \begin{cases} (Z_1 Z_2 e^2/2R_c)(3 - r^2/R_c^2) & r < R_c \\ Z_1 Z_2 e^2/r & \mathrm{R}_c < r < R_s \\ (Z_1 Z_2 e^2/r)\ \exp[-(r - R_s)^2/\sigma^2] & \mathrm{R}_s < r \end{cases} \tag{27}$$

are displayed in Table I and Figure 5, respectively.

Figure 6 shows the result of a consistency check. The convergence of the series

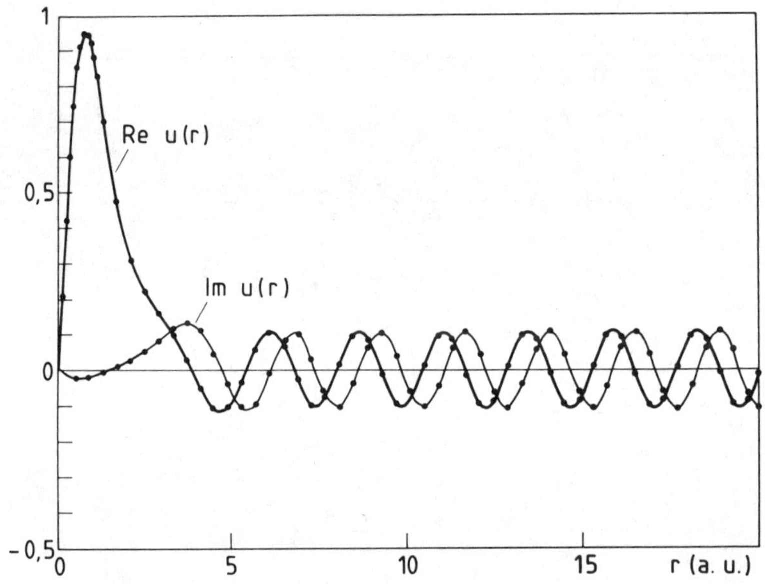

Figure 7

The real (imaginary) parts of the radial wave function. The heavy (light) lines represent the results of the direct numerical integration, the dots are the PSE values.

of the trivially equivalent local potentials

$$V_N^{TEL}(r) = \frac{\langle r|E_N - T|\Psi_N\rangle}{\langle r|\Psi_N\rangle} \tag{28}$$

is presented. It is seen that the local potential recovers, even the pseudo-singularity heals as it is shown by the insert. The next example is the well-known $S_{1/2}$ resonance in the $7.5r^2e^{-r}$ potential. Table II presents the convergence to the exact value 3.42639-i 0.0127745, Figure 7 shows the wave function.

Till now the PSE results have been compared to those of the DNI, which means that the examples were feasible also for the DNI. Let us see an example which is

Table II

The energy F of the $S_{1/2}$ resonance in the potential $7.5r^2e^{-r}$ as a function of the basis size.

N	Eigenvalue	N	Eigenvalue
12	3.42834 - i 0.0126291	13	3.42741 - i 0.0131535
14	3.42701 - i 0.0126980	15	3.42674 - i 0.0127830
16	3.42674 - i 0.0128572	17	3.42667 - i 0.0127485
18	3.42663 - i 0.0128073	19	3.42652 - i 0.0127838
20	3.42652 - i 0.0127681	21	3.42648 - i 0.0127948
22	3.42645 - i 0.0127711	23	3.42642 - i 0.0127786
24	3.42642 - i 0.0127794	25	3.42642 - i 0.0127721
26	3.42641 - i 0.0127799	27	3.42640 - i 0.0127737
28	3.42640 - i 0.0127753	29	3.42640 - i 0.0127761

non-trivial even for bound states and to our knowledge has never been considered for the Gamow case. It is the case of a deformed Saxon-Woods potential

$$V(\vec{r}) = V_o\, f(\vec{r}) \qquad f = f_o + f_D$$

$$f_o = f(r) + (V_{so}/V_o)(\hbar^2/m_pC)^2(1/r)(df(r)/dr)(\vec{\ell}\vec{\sigma}) \tag{29}$$

$$f_D = R_o(d\, f(r)/dr)(\beta_2 Y_{20}(\vartheta) + \beta_4 Y_{40}(\vartheta))$$

with

$$f(r) = (1 + exp[(r - R_o)/a])^{-1}.$$

here $R_o = r_o A^{1/3}, \beta_2$ and β_4 measure the quadrupole and hexadecapole deformations. In Table III the energy of an $\Omega^\pi = 3/2^+$ neutron resonance is shown as a

Table III

The $(E^{res} - i\Gamma/2)$(MeV) energy of an $\Omega^{\pi} = 3/2^{+}$ neutron resonance in the potential (29)[a] at the indicated basis size.

N	L = 2	L = 4	L = 6	L = 8	L = 10
2	1.88-i0.360	1.84-i0.342	1.82-i0.355	1.82-i0.336	1.82-i0.336
3	2.03-i0.429	1.99-i0.401	1.98-i0.394	1.98-i0.394	1.98-i0.394
4	2.13-i0.438	2.09-i0.419	2.07-i0.412	2.07-i0.411	2.07-i0.411
5	2.21-i0.443	2.17-i0.424	2.15-i0.416	2.15-i0.416	2.15-i0.415
6	2.23-i0.441	2.19-i0.422	2.17-i0.415	2.17-i0.415	2.17-i0.415
7	2.23-i0.439	2.19-i0.420	2.17-i0.413	2.17-i0.413	2.17-i0.413
8	2.24-i0.437	2.20-i0.419	2.18-i0.412	2.18-i0.411	2.18-i0.411
9	2.24-i0.437	2.20-i0.418	2.18-i0.411	2.19-i0.411	2.18-i0.411

[a]) The actual values of the paramters in eq(29) are
$V = 40$ MeV, $V_{so} = 7.2$ MeV, $r_0 = 1.233$ fm, $A = 25$, $a = 0.6$ fm, $\beta_2 = 0.3$ and $\beta_4 = 0.2$

function of the basis size. The potential mixes the HOWFs with different n (up to N) and ℓ (up to L).

Instead of giving the many components of the wave function, we present Table IV to show how our result for some single-particle resonances of ^{239}U compares with the numbers given in [29] where the positions and widths of the resonances were calculated with the tan $\delta = \infty$ condition (quasistationary states).

Unfortunately, in the case of the short-range + Coulomb potential the procedure is not automatic. The Coulomb potential does influence the asymptotic region so, according to the philosophy of the PSE method, its place would be in H_o. But the matrix element of the Green operator belonging to $H_o = T + V_c$ cannot be computed exactly on HOWF basis. A direct way out is to find a suitable new basis for the expansion. This has been done in [30]. Here we present a methodologically more interesting HOWF-based approximation [31]. We keep $H_o = T$ and try to expand the whole potential including the Coulomb term on HOWF basis. In bound state the PSE is convergent for any bounded potential, and sure enough,

Table IV

The positions and widths of three single-neutron resonances of ^{239}U with quasi stationary (Q) and Gamow (G) prescription

State	$\frac{9^+}{2}$ [844]		$\frac{13^+}{2}$ [817]		$\frac{13^+}{2}$ [804]	
	E_r	Γ	E_r	Γ	E_r	Γ
Q	10.477	0.508	6.547	0.950	12.377	0.301
G	10.067	0.559	7.065	1.367	12.868	0.380

our code produces good results for charged-particle bound states. It fails for resonances. Since the same expansion of the potential is involved in both cases one guesses that perhaps the different tail behavior of the wave functions may matter. Therefore, we changed the problem to be solved to one offering square integrable wave functions; in other words, we invoked complex scaling. Instead of solving $H|\Psi\rangle = E|\Psi\rangle$ with purely outgoing boundary condition we solved

$$H(\Theta)|\Psi_\Theta\rangle = E(\Theta)|\Psi_\Theta\rangle$$

$$|\Psi\rangle = U(\Theta)^{-1}|\Psi_\Theta\rangle. \tag{30}$$

$U(\Theta)$ is the operator of the complex scaling defined by

$$(U(\Theta)f)(\vec{r}) = exp[3\Theta/2]f(\vec{r}\ exp\ \Theta) \equiv f_\Theta(\vec{r}),$$

Θ is complex, $\Im\Theta > \tan^{-1}(\gamma/\kappa)$; $H(\Theta) = U(\Theta)^{-1}HU(\Theta)$. It is known [32] that under certain conditions on the potential $E(\Theta) \equiv E$ and $\langle\vec{r}|\Psi_\Theta\rangle$ is square integrable for discrete eigenvalues. We approximated (30) with our HOWF-based PSE. It turned out that for analytically well-behaved potentials this procedure is equivalent with approximating the unscaled equation with complex HOWF based PSE. Complex HOWF means a HOWF with complex size parameter. So without any

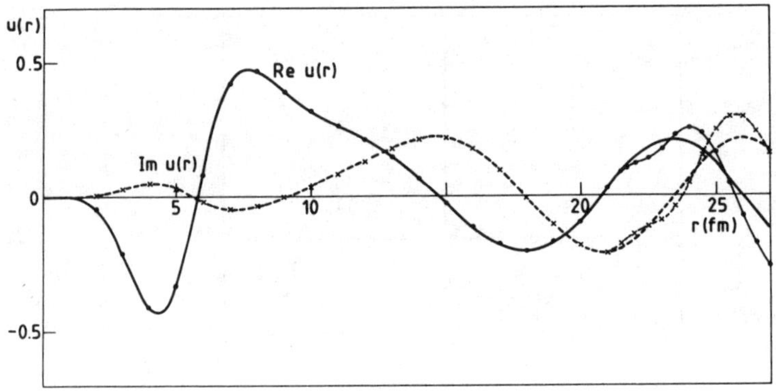

Figure 8

The wave function of the resonance. Heavy continuous and dashed lines are the DNI values, the dots and crosses are from the PSE calculation.

change in the code we attempted to compute resonances. The last example shows an $\ell = 4$ j $= 9/2$ proton resonance in a Saxon-Woods + hard-sphere Coulomb potential mocking a ^{208}Pb nucleus. The DNI value of energy of the resonance is (13.5701-i 0.3394) MeV. Table V presents the PSE values to different $\varphi = \Im\Theta$ and basis size N. Figure 8 displays the wave function transformed back from $\varphi = 0.12$, N = 29.

It is to be noted that the agreement between the back-transformed and DNI wave functions decreases with increasing φ. This is a general drawback of the complex scaling technique. This limits its use to narrow resonances if the wave function is needed. As far as obtaining the position and width of the resonance is concerned it works excellently even for broad resonances.

5. Summary

In this lecture we explain the historical background of, as well as the reasons for, using Gamow wave functions together with Berggren's rules for the description of single-particle resonances. Then the ways and means we have developed for

Table V

The energy of the $\ell = 4, j = 9/2$ proton resonance in a Saxon-Woods + Coulomb potential[a] as a function of the rotation angle and basis size.

N	$\varphi = 0.04$	$\varphi = 0.08$	$\varphi = 0.12$
18	13.606-i0.326	13.587-i0.337	13.579-i0.340
21	13.564-i0.314	13.570-i0.329	13.572-i0.336
24	13.549-i0.344	13.564-i0.340	13.570-i0.339
27	13.575-i0.357	13.571-i0.345	13.571-i0.340
30	13.584-i0.338	13.575-i0.340	13.572-i0.340
	$\varphi = 0.16$	$\varphi = 0.20$	$\varphi = 0.24$
18	13.575-i0.339	13.547-i0.333	13.558-i0.299
21	13.572-i0.339	13.573-i0.341	13.588-i0.352
24	13.571-i0.339	13.571-i0.339	13.563-i0.337
27	13.571-i0.340	13.572-i0.340	13.576-i0.340
30	13.571-i0.340	13.571-i0.340	13.569-i0.340

[a] The actual parameters of the potential are :
$V_0 = 50.9$ MeV, $V_{so} = 5.8$ MeV, $r_0 = 1.19$ fm, $a = 0.75$ fm, $A = 208$, $Z_1 = 82, Z_2 = 1, R_c = r_0 A^{\frac{1}{3}}$.

the computation of the Gamow wave functions are presented. The performance of the direct numerical integration and the approximation based on a separable expansion of the potential is illustrated with a couple of trivial and less trivial examples. Some applications of the single-particle resonance concept to more complicated nuclear systems will be shown in other lectures. An application to the orthogonality condition model, which is a very successful approximation to the cluster model, can be found in [33].

Finally, let us answer a natural question. What is the relation between the description of a single-particle resonance with the well-established Gamow functions and using any of the recipes mentioned in Section 2? In an early paper [34] we compared, in a square well + square barrier model, the pole solutions (Gamow and Weinberg) to some scattering solutions normalized to 1 up to the first outer node. We found that for narrow resonances ($\Gamma/E < 0.1$) the real parts of the pole solutions and the scattering solutions cannot be distinguished up to 50 fm. The essential difference is that the pole solutions have imaginary parts that vanish inside the potential but are of the same magnitude as the real parts outside the potential. Therefore, the difference may manifest itself at most in the description of phenomena that are sensitive to the region just outside the nucleus. It is an interesting and partly open question how large this difference may be and what kind of information it may contain.

References.

1. P. A. M. Dirac, *The principles of quantum mechanics*, Clarendon Press, Oxford (1956).

2. H. Feshbach, Ann. Phys. N. Y., **5**, 357 (1958); **19** 287 (1962); H. Feshbach, A. K. Kerman and R. H. Lemmer, Ann. Phys., N. Y., **41**, 41 (1967).

3. C. Mahaux and H. A. Weidenmüller, *Shell model approach to nuclear reactions*, North Holland, Amsterdam, (1969).

4. E. Goldberg, Phys. Rev., **89**, 760 (1953).

5. R. Huby and J. R. Mines, Proc. of the Conf. on Direct Interactions and Nuclear Reaction Mechanisms, E. Clemental and C. Villi, Eds., Gordon and Breach, New York-London, (1963) p. 530.

6. J. L. Alty, L. L. Green, R. Huby, G. D. Jones, J. R. Mines and J. F. Sharpey-Shafer, Phys. Lett., **20**, 664 (1966); J. L. Alty, L. L. Green, R. Huby, G. D. Jones, J. R. Mines and J. F. Sharpey-Shafer, Nucl. Phys., **97A**, 541 (1967).

7. J. Bang and J. Zimányi, Nucl. Phys., **139A**, 534 (1969).

8. H. T. Fortune and C. M. Vincent, Phys. Rev., **185**, 1401 (1969).

9. C. M. Vincent and H. T. Fortune, Phys. Rev., **2C**, 782 (1970).

10. V. E. Bunakov, Nucl. Phys., **140A**, 241 (1970).

11. V. E. Bunakov, K. A. Gridnev and L. V. Krasnov, Phys. Lett., **32B**,587 (1970).

12. T. Berggren, Nucl. Phys., **109A**, 265 (1968).

13. W. Romo, Nucl. Phys., **116A**, 618 (1968).

14. G. Gamow, Z. Phys., **51**, 204 (1928).

15. L. A. Khalfin, Sov. Phys. JETP, **6**, 1053 (1958).

16. B. Gyarmati, F. Krisztinkovics and T. Vertse, Phys. Lett., **41B**, 110 (1972).

17. C. Mahaux and A. M. Saruis, Nucl. Phys., **177A**, 103 (1971).

18. G. H. Hardy, *Divergent Series*, Clarendon Press, Oxford (1949).

19. B. Gyarmati and T. Vertse, Nucl. Phys., **160A**, 523 (1971).

20. W. J. Romo, Nucl. Phys., **398A**, 525 (1983).

21. T. Berggren, Phys. Lett., **38B**, 61 (1972).

22. E. Hernández and A. Mondragón, Phys. Rev., **29C** 722 (1984).

23. B. Gyarmati, A. T. Kruppa and Z. Papp, Phys. Rev., **31C**, 2317 (1985).

24. T. Vertse, K. F. Pál and Z. Balogh, Comp. Phys. Comm., **27**, 309 (1982).

25. Z. Balogh, Diploma thesis, ATOMKI, Debrecen Hungary, (1977).

26. A. T. Kruppa and Z. Papp, Comp. Phys. Comm., **36**, 59 (1985).

27. B. Gyarmati, A. T. Kruppa and J. Revai, Nucl. Phys., **326A**, 119 (1979).

28. B. Gyarmati, A. T. Kruppa, Z. Papp and G. Wolf, Nucl. Phys., **417A**, 393 (1984).

29. J. Bang, F. A. Gareev, I. V. Puzynin and R. M. Jamalejev, Nucl. Phys., **261A**, 59 (1975).

30. Z. Papp, J. Phys. A, **20**, 153 (1987).

31. B. Gyarmati and A. T. Kruppa, Phys. Rev., **34C**, 95 (1986).

32. M. Reed and B. Simon, *Methods of modern mathematical physics*, Academic Press, New York (1978).

33. K. F. Pál, J. Phys. A, **18**, 1665 (1985).

34. B. Gyarmati, R. G. Lovas and J. Zimányi, Phys. Lett., **35B**, 549 (1971).

THE USE OF GAMOW FUNCTIONS IN NUCLEAR PROBLEMS

Tamas Vertse†, Patricia Curutchet‡, and Roberto J. Liotta

Research Institute of Physics
Frescativägen 24
S-10405 Stockholm
SWEDEN

Abstract

The inclusion of single particle Gamow functions as building blocks in nuclear calculations is presented here through three different examples. We show that it is a quick method which enables us to account for the most important features of the continuum effect.

1. Introduction

Resonances are characteristic features of nuclear reactions and therefore reaction theories should be able to account for their properties. Theories describing resonances by the aid of Gamow states, i.e. eigenstates of the time-independent Schrödinger equation with complex eigenvalues[1,2] have proved to be superior in several aspects respect to the others, for example the parameters of the resonances do not depend on the channel radii as in the R-matrix theory[3] or on the bombarding energy like in the theory of Kapur and Peierls[4]. Gamow first used these states, almost half a century ago, in the description of the spontaneous alpha decay[5]. Somewhat later Siegert[2] gave a dispersion formula in which the resonant terms were represented by "radioactive" states of the system having asymptotically outgoing wave behaviour. Therefore these states are called Gamow states or Gamow-Siegert states and the corresponding wave functions are called Gamow functions.

A detailed study of the position of the poles in a square well potential has been carried out by Nussenzveig[6] but the use of Gamow states was hindered by the lack of regularization procedure for calculating their norm. An important step was taken when Zel'dovich[7] suggested the use of a Gaussian convergence factor for this purpose. Hokkyo[8] introduced another type of recipe for the calculation of the norm and used the adjoint pair of the Gamow function in order to form a bi-orthonormal set of wavefunctions. It was Berggren[9] who gave a completeness relation in which bound and Gamow states were treated in an unified way and he formulated a theory including resonances by generalizing the concept of scalar product. The new scalar product was based on the bi-orthogonality of the states and used the regularization procedure of Zel'dovich. Romo[10] also introduced a prescription for calculating the norm of Gamow functions and matrix elements using the technique of analytical continuation and he showed that in perturbation methods Gamow functions can also be used. Soon another type of regularization technique using contour integration method on the complex r-plane was introduced by Gyarmati and Vertse [11] and the existence of the norm of the Gamow function for the case of charged particles was proved. The method used became known later as "exterior complex scaling"[12]. It was also shown in the same work[11] that a wider class of convergence factors can be used for regularization purposes and they all lead to identical results which for the chargeless case coincide with that of the analytical continuation of Romo[10]. We shall see below that for the potentials used in nuclear physics the exterior complex scaling can be generally applied and it is better suited for numerical calculations than the other methods.

The works mentioned before opened the possibility of applying Gamow functions in nuclear physics calculations. They were applied with success indeed in several works[13–18] being too many to list them here.

In this lecture we want to show only a few examples in which Gamow states have been used with success in nuclear physics. These examples were selected quite arbitrarily. The common characteristic of them is that whenever regularization procedures were needed the exterior complex scaling was used. For the sake of simplicity we will restrict ourselves to potential scattering by a spherically symmetric potential and use r-representation only. In this case the Gamow functions can be generated easily by solving the radial Schrödinger equation using the very powerful technique of direct numerical integration[19]. Nevertheless other methods[20–25] applicable to more general potentials have been developed recently and some of them treat the problem in momentum representation[21–23].

The building blocks in our calculations are the "ill-behaving" complex single particle Gamow functions which can not be handled so simply as the bound state wave functions but they can still be applied in nuclear calculations if we use adequate regularization procedures. Perhaps we should emphasize that our aim is always to take into account the most apparent effect of the continuum in an economic way and give some insight into the physical process. With the aid of Gamow wave functions we can only take the resonant part of the cross section into account but not the smooth background part. Therefore our aim is not to calculate the effect of the continuum exactly (this can be achieved by other more time consuming methods) but to present a quick method which is able to reproduce the most characteristic features of the process. The accuracy of our approximation can be judged by comparing our results to that of the "exact" methods. Here we are going to present three examples where this sort of comparison has been carried out.

They are the following ones:

i. the almost trivial case of one channel potential scattering,

ii. a two channel case which describes isobaric analogue resonances,

iii. an RPA description of giant multipole resonances where the escape of the particles is taken into account by complementing the basis with Gamow resonances.

The first two examples help us to understand the third one which represents a real life use of Gamow functions in nuclear structure calculations. In chapter 2 we summarize the basic definitions and notations. In chapter 3 we compare some of the well known regularization procedures for calculating the norm of the Gamow functions and comment on their range of applicability. Finally in chapter 4 we present the three applications mentioned before.

2. Basic concepts

Let us review briefly the basic concepts concerning Gamow functions for the sake of simplicity in the case of potential scattering. A spinless particle is scattered in a spherically symmetric potential $V(r)$, which is the sum of a short-range nuclear and an infinite range Coulomb term: $V(r)=V_N(r)+V_C(r)$ which satisfies that $V_N(r) = 0$ if $r > a$ and $V_C(r) = Z_1Z_2e^2/r$ for $r \geq a$. The solution of the Schrödinger equation $H\Psi = E\Psi$ can be expanded into partial waves and u_ℓ will be a solution of the radial equation

$$u_\ell''(r,k) + [k^2 - \frac{l(l+1)}{r^2} - U(r)]u_\ell(r,k) = 0, \tag{1}$$

where $c = 2\mu/\hbar^2, U(r) = cV(r)$ and $k^2 = cE$.

The solution must be regular at the origin

$$u_\ell(0,k) = 0 \tag{2}$$

and for $r \geq a$ it can be expressed by the two linearly independent Jost solutions or the O_ℓ and I_ℓ outgoing and incoming Coulomb waves and the S-matrix as follows:

$$u_\ell(r,k) = x_\ell(k)O_\ell(kr) + y_\ell(k)I_\ell(kr) \tag{3}$$

$$u_\ell(r,k) = y_\ell(k)(I_\ell(kr) - S_\ell(k)O_\ell(kr)). \tag{4}$$

Imposing the Siegert boundary condition of having only outgoing waves at infinity, i.e. $y_\ell(k) = 0$ which is satisfied for only those points k_n of the complex k-plane where S_ℓ has a pole, three types of solutions are obtained corresponding to bound states ($\mathrm{k}_n = i\gamma_n$), antibound states ($\mathrm{k}_n = -i\gamma_n$) and resonances ($\mathrm{k}_n = \pm\kappa_n - i\gamma_n$). Since the far asymptotic behaviour is governed by $O_\ell(kr) \to \exp(ikr)$, it is clear that only the bound states are square integrable since they behave as $exp(-\gamma_n r)$, while the antibound states behave like $\exp(\gamma_n r)$. The resonances which are lying on the lower half of the k-plane in pairs: $k_n = \kappa_n - i\gamma_n$ and $\tilde{k}_n = -\kappa_n - i\gamma_n = -k_n^*$ are also divergent for large value of r. The complex energy corresponding to k_n is E_n=E_r - i $\Gamma/2$, where E_r is the position of the resonance and Γ is its width so that the decay in the wave function is exponential with a relaxation time $\tau = 2\hbar/\Gamma$.

The function $\tilde{u}_\ell(r, \tilde{k}_n)$ which satisfies the same Schrödinger equation and the same boundary condition is usually called "mirror state" and fulfills the property that $\tilde{k} = -k^*$ and $\tilde{u}_\ell(r, \tilde{k}_n) = u_\ell^*(r, k_n)$.

Since the complex energy-eigenvalues can have only discrete values, resonance states rather resemble bound states than scattering states having continuous spectra. Resonances can be considered as generalizations of the bound states since they satisfy the same equation and the same boundary conditions as the bound states, namely they are regular at r=0 and beyond the range of the nuclear potential they behave as outgoing spherical waves. Besides a generally small imaginary component, Gamow functions are very similar to bound state wave functions inside the nuclear potential however at large distances they oscillate with exponentially growing amplitude, therefore they are not square integrable. Using the traditional definition of the scalar product Gamow functions can be normalized only on a finite radial distance and the value of the norm will depend on this distance. This is an inconvenient feature and several methods have been proposed to overcome this difficulty.

3. Normalization procedures

With the standard prescription $\int_0^\infty |u_\ell(r, k_n)|^2 dr$ only the bound states can be normalized in an infinite interval. Therefore we have to reconsider the way we normalized wavefunctions before. We have seen in Berggren's lecture that we have to use a biorthogonal basis which is a generalization of that containing bound states only. Therefore the state in bra position should always be replaced by its mirror image ($\tilde{u}$) so that the integrand of the norm becomes

$$\tilde{u}_\ell^*(r, \tilde{k}_n) u_\ell(r, k_n) = u_\ell^2(r, k_n). \tag{5}$$

The internal contribution to the norm

$$N_i^2 = \int_0^{\mathbf{a}} u_\ell^2(r, k_n) dr \tag{6}$$

which depends on $V_N(r)$, is always finite and in most of the cases can only be calculated numerically. The real problem is how to deal with the tail region in which the integrand is $\exp(2\gamma_n r)\exp(2i\kappa_n r)$ for resonances and $\exp(2\gamma_n r)$ for antibound states.

Since the magnitude of the integrand of (6) is growing exponentially, the norm will depend on the range of the finite interval a that is used.

A method which allows to get rid of the dependence on the parameter a was developed by Hokkyo et al.[8] and used later by Peierls and his coworkers[18]. The normalization is made in a finite interval but an extra term is added in order to cancel the dependence on a:

$$N^2 = N_i^2 - C_\ell(\mathrm{a}, k_n). \tag{7}$$

Then to have $dN^2/d\mathrm{a} = 0$, the condition

$$\frac{dC_\ell(r,k)}{dr} = u_\ell^2(r,k) \tag{8}$$

has to be fulfilled. Hokkyo actually used a simple form for C_ℓ which is valid if the Coulomb potential is zero and for $\ell = 0$

$$C_0(\mathrm{a},k) = \frac{A^2 \exp(2ik\mathrm{a})}{2ik}. \tag{9}$$

In the no Coulomb case for an arbitrary value of ℓ, $u_\ell(r,k_n)$ asymptotically behaves as

$$u_\ell(r,k) = AH_\ell(kr) \tag{10}$$

where the Riccati-Hankel function $H_\ell(k_n r)$ is defined as:

$$\begin{aligned} H_0(x) &= \exp(ix) \\ H_1(x) &= (\frac{1}{x} - i)\exp(ix) \\ H_{\ell+1}(x) &= ((2\ell+1)/x)H_\ell(x) - H_{\ell-1}(x)) . \end{aligned} \tag{11}$$

Gyarmati and Vertse[11] gave a closed form for the C_ℓ term in this case:

$$\begin{aligned} C_\ell(\mathrm{a},k) = A^2\frac{\mathrm{a}}{2}[&H_\ell(k\mathrm{a})^2 + H_{\ell+1}(k\mathrm{a})^2 \\ &- ((2\ell+1)/k\mathrm{a})H_\ell(k\mathrm{a})H_{\ell+1}(k\mathrm{a})]. \end{aligned} \tag{12}$$

Unfortunately for the case when Coulomb interaction is present the form of C_ℓ is unknown. This is a big disadvantage of the Hokkyo's method, which is otherwise very general and can formally be used even for normalizing antibound states.

Another prescription was proposed by Romo[10]. He suggested to calculate integrals (in our case the norm integral) on the upper half of the k-plane and to continue the result analytically to the pole k_n in the lower half k-plane. This procedure for the chargeless case gives:

$$N^2 = N_i^2 + N_e^2 \tag{13}$$

where

$$\begin{aligned} N_e^2 &= \lim_{k\to k_n} \int_{\mathbf{a}}^{\infty} u_\ell^2(r,k)dr = \lim_{k\to k_n} [C_\ell(r,k)]_{\mathbf{a}}^{\infty} \\ &= - \lim_{k\to k_n} C_\ell(\mathbf{a},k) = -C_\ell(\mathbf{a},k_n). \end{aligned} \tag{14}$$

So we have proved that Romo's procedure in this case gives the same norm as eq. (7), i.e. the prescription of Hokkyo and it has also the drawback that can not be applied to the charged case since the form of C_ℓ is not known.

A different type of normalization procedure[9] is based on the inclusion in the integrand of a convergence factor $e^{-\epsilon r^2}$ as first suggested by Zel'dovich[7]

$$N_e^2 = \lim_{\epsilon\to 0} \int_{\mathbf{a}}^{\infty} e^{-\epsilon r^2} u_\ell^2(r,k)dr. \tag{15}$$

The limit of the integral exists if the resonances considered fulfill the condition $|\kappa_n| > \gamma_n > 0$.

Even a more general convergence factor $e^{-\epsilon r^p}$ can be used[11] if we replace the former condition by $1 < p < \pi/(2\mathrm{arctg}(\gamma_n/\kappa_n))$. It has been shown[11] that within the range of applicability the result does not depend on p and this procedure leads to the same norm as calculated with Romo's method. The proof has been carried out by using the Cauchy theorem on a contour $C = C_1 + C_2 + C_3$ (see Fig.1) in the complex r-plane :

$$r = \begin{cases} r & \text{for } a \le r < \infty \\ |r| \exp{(i\alpha)} & \text{for } 0 \le \alpha \le \pi \\ a + |r-a| \exp{(i\varphi)} & \text{for } arctg(\frac{\gamma_n}{\kappa_n}) < \varphi < (\frac{\pi}{2p}). \end{cases} \tag{16}$$

It was shown[11] that

$$\lim_{\epsilon\to 0} \int_{\mathbf{a}}^{\infty} e^{-\epsilon r^p} u_\ell^2(r,k_n)dr = - \int_{C_3} u_\ell^2(r,k_n)dr = - \int_{C_3} A^2 O_\ell^2(k_n r)dr. \tag{17}$$

This transformation of the coordinate r makes $O_\ell^2(rk)$ integrable on C_3 since $\arg(k_n)+\arg(r) > 0$. Presently this procedure is called "exterior complex scaling". It is a most powerful regularization method well suited for numerical calculation. Moreover, it has been shown[11] that for the chargeless case this procedure gives the same results as the above mentioned methods, namely:

$$\lim_{\epsilon\to 0} \int_{\mathbf{a}}^{\infty} e^{-\epsilon r^p} A^2 H_\ell(kr)dr = -C_\ell(\mathbf{a},k). \tag{18}$$

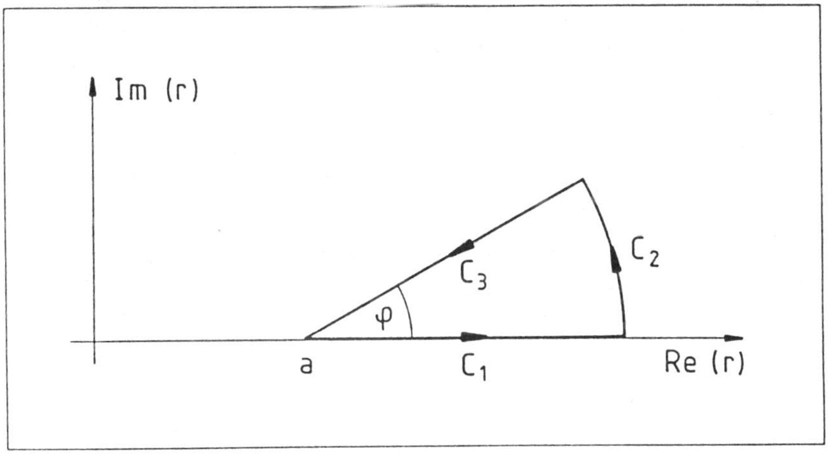

Figure 1

Contour of integration in the complex r-plane used in ref. 11 for calculating the integral of eq. (17)

In fact the exterior complex scaling can be extended to φ angles larger than the limit $\pi/2p < \pi/2$ posed by the convergence factor. If we use a rotational angle $\varphi > \pi/2$ we can normalize even antibound states which can not be normalized with a convergence factor. The price we pay for it is that at this rotation angle bound states become non normalizable. There is no rotation angle where all three pole solutions are square integrable in an infinite interval.

Nevertheless complex scaling offers a way to assign a finite value to the norm of antibound states at $k_n = -i\gamma_n$ and this norm for the chargeless case is given by

$$N^2 = N_i^2 = \quad C_\ell(\mathbf{a}, -i\gamma_n). \tag{19}$$

The result coincides with that obtained from Hokkyo's procedure and also with the results of analytical continuation. The code GAMOW[19] uses this procedure (i.e. exterior complex scaling) to assign norm to antibound states with the rotational angle $\varphi = \pi$ if Coulomb potential is present.

4. Application of the resonance states

4.1 One channel case: potential scattering

The simplest nuclear problem in which it is possible to use single-particle resonant states for describing resonance phenomena is the motion of a nucleon in a potential well. Our aim is to compare the energy dependence of the S-matrix for a given partial wave S_ℓ calculated:

1) "exactly", using an optical model code for real potential, and

2) approximately, calculating the complex energy eigenvalue and using the one pole approximation for S_ℓ.

The formalism and equations have been already described in section 2. In case 1 ("exact" treatment) the radial equation

$$[\frac{d^2}{dr^2} + k^2 - \frac{l(l+1)}{r^2} - cV(r)]\chi_\ell(r,k) = 0. \tag{20}$$

is solved by integrating numerically in an interval $[0, R_{max}]$ with the usual boundary conditions, i.e.

$$\chi_\ell(0,k) = 0 \tag{21}$$

$$\chi_\ell(r,k) \rightarrow_{r\rightarrow\infty} I_\ell(kr) - S_\ell O_\ell(kr) \tag{22}$$

where $k^2 = cE$ and $c = 2\mu/\hbar^2$. From the matching of the solutions at each E it is possible to obtain the function S_ℓ.

In case 2 the radial equation is considered on the complex E-plane and we search for the Gamow solution for which $S_\ell \rightarrow \infty$. The radial wave function $u_\ell(r,k_n)$ behaves asymptotically as an outgoing wave O_ℓ and the wave number is complex

$$k_n^2 = (\kappa_n - i\gamma_n)^2 = c(E_R - i\frac{\Gamma}{2}) = cW_n. \tag{23}$$

With the pole energy W_n thus calculated it is possible to evaluate the S-matrix in the one pole approximation,

$$S_\ell(E) = \frac{E - W_n^*}{E - W_n} = 1 - i\frac{\Gamma}{E - W_n}. \tag{24}$$

Both methods were applied for the $^{48}Ca + p$ case[14] and the results for $\ell = 1$, $j = 3/2$ are shown in Fig.2. One can see that a very good agreement has been obtained.

4.2. Two channel case: Isobaric Analogue Resonance in the $^{208}Pb(p,p)^{208}Pb$ reaction

When a proton collides with an A nucleon system narrow resonances appear at energies corresponding to bound states of the A+1 particle system. The Hamiltonian describing the system is

$$H = H(\xi) + K + (1/2 - t_3)V_C(r) + V_o(r) + \vec{t}.\vec{T}V_1(r) \tag{25}$$

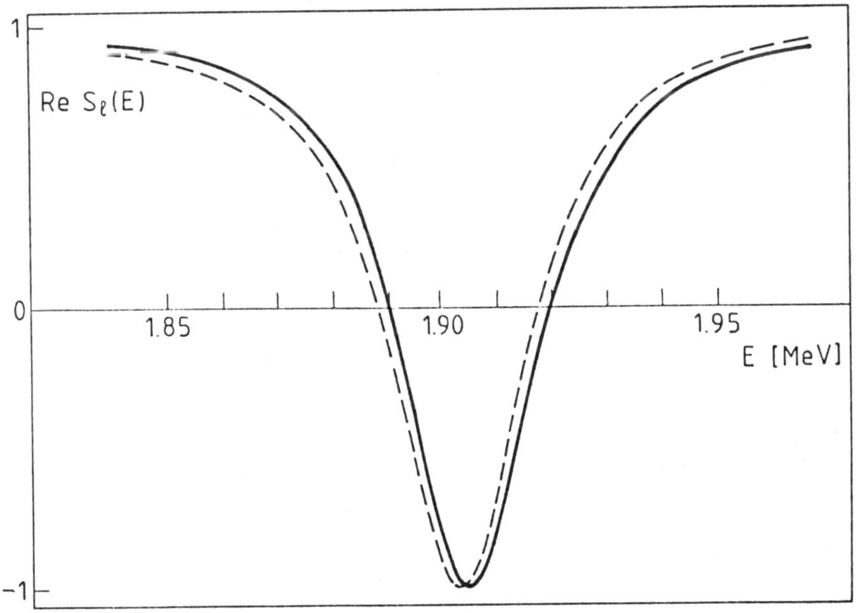

Figure 2

The full line is the result of the optical model calculation and the dashed line is the one pole approximation of eq. (24) for a $2p_{\frac{3}{2}}$ resonance in the $^{48}Ca + p$ system taken from ref. (14).

where $H(\xi)$ stands for the target, K is the kinetic energy of the nucleon, $V_o(r)$ is the isoscalar part of the nuclear potential, $V_C(r)$ is the Coulomb potential, $V_1(r)$ is the symmetry potential and $\vec{T}$ and $\vec{t}$ are the isospin operator of the target and the nucleon respectively. The last term of eq. (25) couples the two channels: channel 1 is the target+proton partition, while in channel 2 we have the analog of the target+the neutron. For their coupling we use the shorthand notation: $V_{ii'}(r)$, and the diagonal potential in channel i is denoted by $V_i(r)$. The "exact solution" of this Hamiltonian is provided by the numerical integration of the coupled Lane equations. We assume that the solution is of the form:

$$\Psi_{CL} = \chi_p(r)\psi_{T_0T_0} + \chi_n(r)\psi_{T_0T_0-1} \tag{26}$$

where the first term corresponds to channel 1 and the second term to channel 2. The functions $\chi_p(r)$ and $\chi_n(r)$ are the wave function for the proton and the neutron respectively and ψ_{T_0,T_3} stands for the core. Taking into account that $V_{ii'}$ does not couple different angular momenta $\ell, \jmath$, we have for a given $\ell, \jmath$:

$$[K - \Delta_C + V_i(r) - E_i\chi_i(r)] + V_{ii'}\chi_{i'}(r) = 0 \tag{27}$$

where $i, i' = 1, 2$. The asymptotic behaviour is the same as in eq.(21) and (22) if k is real but with

$$O_\ell(i\kappa r) \sim \exp(-\kappa r) \tag{28}$$

for the closed channel, i.e. $E_2 = E - \Delta_C = -\kappa^2/c$, where Δ_C is the Coulomb shift.

In our approach using Gamow states we are diagonalizing the Hamiltonian in a basis of the solutions of the decoupled equations for both the proton and the neutron. The Hamiltonian is splitted into two parts :

$$H = H_o + V \tag{29}$$

The functions Ψ^i satisfying

$$H_o\Psi^i = E^i\Psi^i \tag{30}$$

are factorized in the following way

$$\Psi_p^i = \varphi_p^i(r)\psi_{T_o,T_o} \tag{31}$$

$$\Psi_n^i = \varphi_n^i(r)\psi_{T_o,T_{o-1}} \tag{32}$$

$$\Psi^i = \Psi_p^i + \Psi_n^i \tag{33}$$

Assuming that the completeness relation of Berggren[9] holds for the case when V=V$_N$+V$_C$ we can expand $\varphi^i(r)$ in terms of a basis which includes bound states, resonant states and an integral along an L contour in the k-plane:

$$\varphi_i(r) = \sum_n a_n^i u_n(r, k_n) + \int_{L^+} a(k)u(r,k)dk \tag{34}$$

The simplest description of the IAR would include one proton resonance at the position of the IAR and one neutron bound state, which will form the "basis" and satisfy

$$[K + V_i(r) - \varepsilon_i]u_i(r, k_i) = 0 \tag{35}$$

with

$$\varepsilon_1 = E_R^p - i\frac{\Gamma^p}{2} = k_1^2, \qquad for \quad u_1(r) \to O_\ell(k_1 r) \tag{36}$$

$$\varepsilon_2 = E_n = -\kappa_2^2, \qquad for \quad u_2(r) \to O_\ell(i\kappa_2 r) \tag{37}$$

If we diagonalize H on this basis we end up with a second order equation for the (complex) energy eigenvalues. The eigenvalue having the larger real part corresponds to the IAR:

$$\mathcal{E}_{IAR} = \frac{1}{2}[(\varepsilon_1 + \varepsilon_2 + \Delta_C)+$$
$$+\sqrt{\varepsilon_1 - (\varepsilon_2 + \Delta_C)^2 + 4 < \tilde{u}_1|V|u_2 >< \tilde{u}_2|V|u_1 >}] \tag{38a}$$
$$= E_{IAR} - i\frac{\Gamma^{IAR}}{2}. \tag{38b}$$

For the coefficients a_n and a_p of the wave function of the IAR we get the simple relation

$$\frac{a_n}{a_p} = \frac{\mathcal{E}_{IAR} - \varepsilon_1}{< \tilde{u}_1|V|u_2 >}, \tag{39}$$

which with the normalization condition

$$a_n^2 + a_p^2 = 1 \tag{40}$$

determine them uniquely. Naturally in bra position in the matrix elements the mirror image states should be used. No special regularization method is needed to calculate these matrix elements because the fall off of the formfactor of the potential V makes the integrand converge to large distances. But the normalization of the Gamow state can be conveniently done with exterior complex scaling. And again as in the former case the energy dependence of the S-matrix can be calculated in the one pole approximation

$$S_\ell(E) = e^{2i\delta_\ell^0}\left(1 - i\frac{\Gamma^{IAR}}{E - E_{IAR}}\right) \tag{41}$$

It was found in reference[15] that a very good agreement with the exact solution can be achieved if the real part of the proton energy is pushed to the E_{IAR} through the adjustement of the potentials in H_o as it can be seen in fig. 3 for the $\ell, j = g_{\frac{7}{2}}$.

Other cases gave similar results, that is to say that for other partial waves the resonance energy was also reproduced giving a width which in all the cases is smaller than the one obtained with the "exact" method. This, in fact is a promising feature of the method which may mean that an enlargement of the basis would result in a desired convergence.

4.3. Giant resonances

Giant resonances in nuclear physics are collective particle-hole states which are immersed deep in the continuum. They are usually rather broad. The width of these resonances consists of two major parts. The spreading width is produced

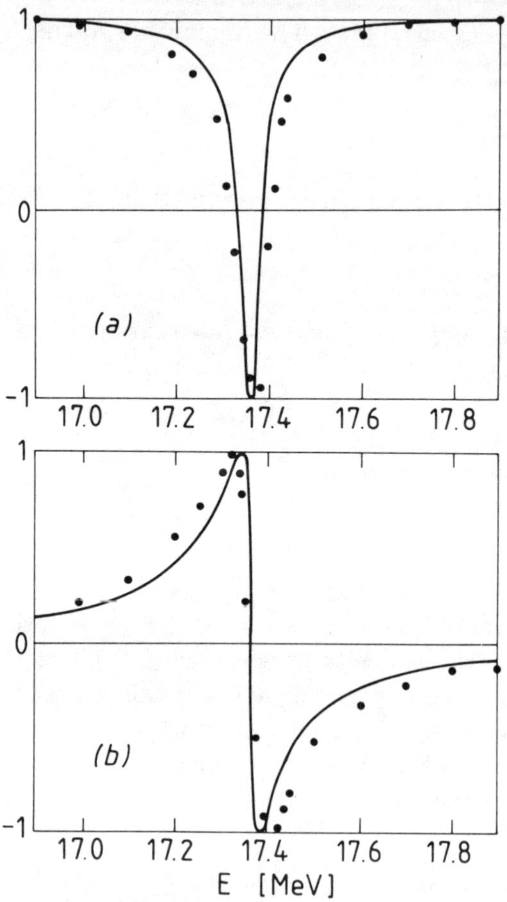

Figure 9

Solid line: (a) real and (b) imaginary part of the S of eq.(41) for the $g_{\frac{7}{2}}$ isobaric analogue resonance in the $^{208}Pb + p$ system (from ref. (15)). The dotted line is the solution of the coupled Lane equations (27).

by mixing with more complicated configurations than pure particle-hole. The escape width is produced by particle emission. The calculation of the energy and the spreading width of the particle-hole giant resonances is usually performed by using a large harmonic oscillator representation[26]. The calculation of the escape width requires a proper treatment of the continuum. This might be a very difficult task because to treat the continuum numerically one has to discretize it and one is then left with an infinite set of coupled equations which is difficult to solve[27,28]. There are methods available which take the effect of the continuum into account exactly as for example the TDHF or the method of coupled-channels

(CC)[27] or approximately as the continuum RPA[26,29]. The "exact" calculation that we use for checking our results[30] calculated in "resonant" RPA approach is the CC calculations of Buck and Hill[27].

The crucial point of our resonant RPA approach to the calculation of the giant resonances is to use bound states and Gamow functions as single-particle representation. In this way we are using the completeness relation of Berggren and again we neglect the integral in the equation (34). By including the resonances lying close to the real E-axis into the basis we are taking a major part of the effect of the continuum into account.Thus, the escape width is automatically included in the calculation, with a reasonable accuracy. One expects that the narrow Gamow states in the energy range of interest play an important role while the others should have little effect on the results. As basis vectors we choose only bound single-particle states and those Gamow states where the imaginary part of the eigenenergy is not larger (in absolute value) than the real part.

Giant resonances are strongly correlated particle-hole states which can be described within the RPA formalism. The form of our RPA equations is similar to those obtained within a bound (real) representation[31,32]

$$\begin{pmatrix} A & B \\ -B & -A \end{pmatrix} \begin{pmatrix} X_n \\ Y_n \end{pmatrix} = W_n \begin{pmatrix} X_n \\ Y_n \end{pmatrix} \tag{42}$$

but the matrix elements of matrix A and B should be calculated with the biorthogonal basis mentioned before. The normalization of the RPA solution is now

$$X_n^T X_n - Y_n^T Y_n = 1, \tag{43}$$

and *not* $X_n^{T*} X_n - Y_n^{T*} Y_n = 1$ as in the real representation. In the biorthogonal basis the spin and angular part of the matrices A and B are the same as in the bound single-particle representation[32] and the difference appears only in the radial integrals which have the following form:

$$I_{pqrs} = Reg \int_0^\infty r_1^2 dr_1 \int_0^\infty r_2^2 dr_2 R_p(r_1) R_q(r_2) V(r_1, r_2) R_r(r_1) R_s(r_2) \tag{44}$$

where the radial part of the single particle wavefunction is

$$R_i(r) = \frac{u_i(r)}{r} \tag{45}$$

and *Reg* denotes the regularization of the integral which is otherwise divergent. We have used two kinds of interaction in the RPA calculations, namely: a) the Landau-Migdal and b) the separable multipole-multipole force.

The form of the Landau-Migdal force is

$$V(r_1, r_2) = \delta(r_1 - r_2)[f(\rho) + f'(\rho)\tau_1.\tau_2 + g\sigma_1.\sigma_2 + g'(\sigma_1.\sigma_2)(\tau_1.\tau_2)] \tag{46}$$

where

$$f(\rho) = f_o^e + (f_o^i - f_o^e)\rho(r) \tag{47}$$

$$\rho(r) = \frac{1}{1+e^{\frac{r-R}{a}}} \tag{48}$$

This leads to the radial integral to be calculated

$$I_{pqrs} = Reg \int_0^\infty r^2 dr R_p(r) R_q(r) v(r) R_r(r) R_s(r) \tag{49}$$

$$v(r) = \begin{cases} 1 \\ \rho(r) \end{cases} \tag{50}$$

The regularization procedure we use here is the exterior complex scaling. The angle φ of the complex rotation should be chosen so that all kinds of radial wavefunctions (bound states and resonances) go to zero along the complex path of eq. (16). This condition is satisfied if we use

$$\varphi = \frac{1}{2} \max|\arg(k_n)| + \frac{\pi}{4}. \tag{51}$$

The form of the separable multipole-multipole force is

$$V_\lambda(r_1, r_2) = -\kappa_\lambda Q_\lambda(r_1) Q_\lambda(r_2), \tag{52}$$

where

$$Q_\lambda(r) = \begin{cases} r\frac{\partial V(r)}{\partial r} \\ r^\lambda \end{cases} \tag{53}$$

The radial integral in this case is

$$I_{pqrs} = M_{pr} M_{qs} \tag{54}$$

where

$$M_{pr} = Reg \int_0^\infty r^2 dr R_p(r) Q_\lambda(r) R_r(r). \tag{55}$$

With the separable force eq.(42) reduces to the well known dispersion relation[32]

$$F_\lambda(W) = \sum_{ki} \frac{2(\epsilon_k - \epsilon_i) < k|Q(\lambda)|i >^2}{W^2 - (\epsilon_k - \epsilon_i)^2}; \tag{56a}$$

$$F_\lambda(W_n) = \frac{1}{\kappa_\lambda} \tag{56b}$$

where κ_λ is the strength of the multipole-multipole interaction, $k(i)$ stands for singel particle (hole) and W_n is the energy of the state n. In the real case eq.(56) can be conveniently solved graphically as the intersection of the curve $F_\lambda(W)$ with the horizontal line $1/\kappa_\lambda$. In our case, however, all quantities in eq.(56) are complex and those curves become surfaces in the complex space.

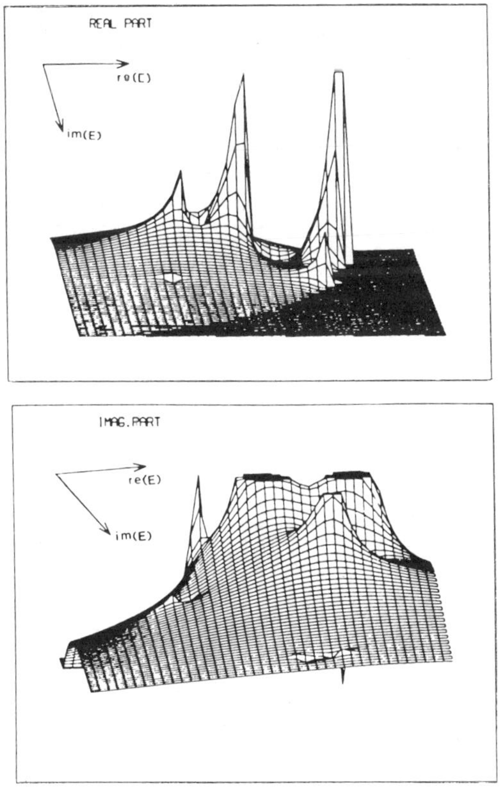

Figure 4.

a) Real and b) imaginary part of the dispersion relation $F_\lambda(E)$ of eq. (56) as functions of the complex energy E=W for the isoscalar 2^+ giant resonance in ^{208}Pb calculated in ref. 30.

We can generalize the graphical solution of the dispersion relation to the complex case as follows. The real and the imaginary parts of F_λ are surfaces with singularities at the unperturbed particle-hole energies: $W = \pm(\varepsilon_k - \varepsilon_i)$. An example is shown in Fig. 4.

The strength of the separable interaction κ_λ is also complex in general and eq.(56) should hold both for the real and the imaginary part

$$Re(F_\lambda(W)) = Re(1/\kappa_\lambda) \tag{57a}$$
$$Im(F_\lambda(W)) = Im(1/\kappa_\lambda) \tag{57b}$$

The W values satisfying eq.(57a) lie on the intersection of the surface $Re(F_\lambda(W))$ and the plane $Re(1/\kappa_\lambda)$. Let us denote this curve as C_R. Then C_I is defined analogously by eq.(57b). The curves C_R and C_I are shown as dotted and full lines respectively in Fig.5. The roots of eq.(56) are the points lying both on C_R and C_I: $W_n \in C_R, W_n \in C_I$.

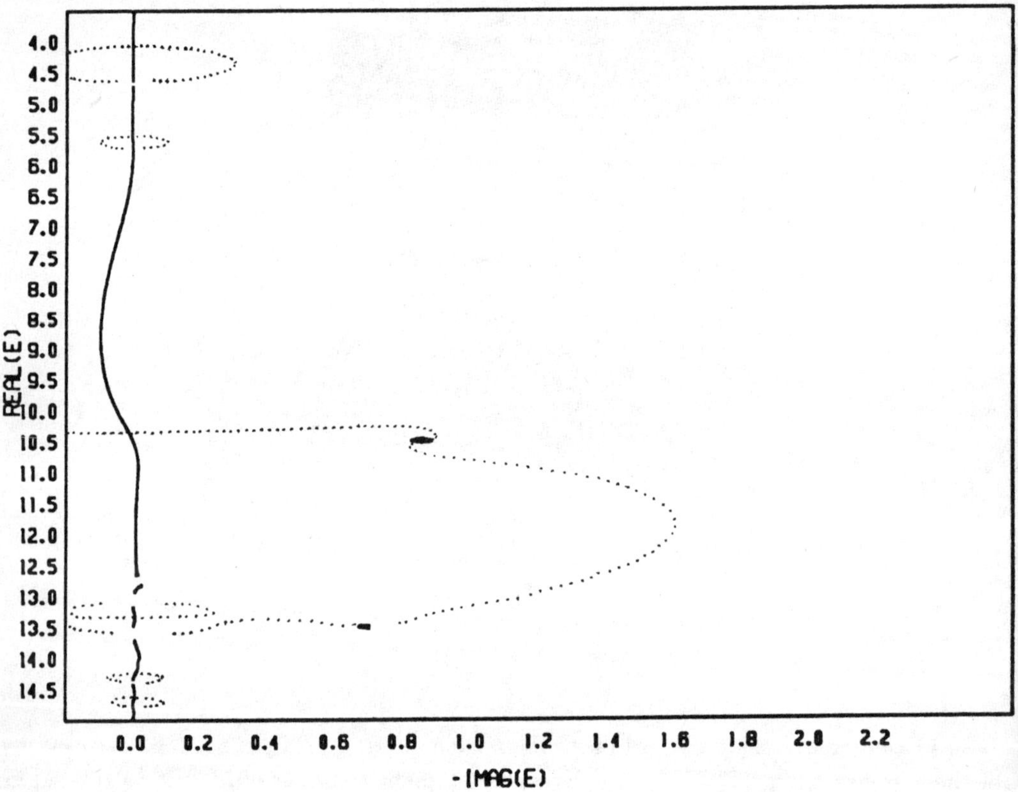

Figure 5

The curves C_R and C_I as defined by the eqs. (57a) and (57b) respectively. Their intersections represent the solution of eq. (56).

The graphical solution offers an insight into the building up of the collectivity as the strength of the interaction increases. With weak interaction the planes cut the surfaces at large distances where C_R and C_I are (unclosed) circles around the unperturbed energies and their intersections remain also close to these particle-hole energies. Increasing the strength the planes corresponding to $1/\kappa_\lambda$ move closer to the zero plane. The features of the perturbed state depend on the geography of the dispersion relation. If the plane cuts hills standing alone, the root remains close to the particle-hole energies and the solution is non-collective. On the other hand we can observe that a large mountain is formed from many of the individual peaks and if we cut this mountain at its foot with the plane, the intersection goes far from the poles. The root corresponding to this situation has a collective feature since many particle-hole states contribute to its formation. Since the collective state lies on the slope of the mountain its position strongly depends on the strength of the interaction while the positions of the non-collective rooots do not.

The complicated structure of the dispersion relation explains why even the most sophisticated complex root finding methods failed in finding all the roots of the dispersion relation. Therefore we turned to another method, namely we diagonalized the corresponding RPA matrix making use of complex diagonalization routines of the NAG library[33].

The calculated RPA energies are complex. The real part corresponds to the position of the resonance while the imaginary part is half the escape width. Our strength function is also complex. It is not completely clear what meaning should be assigned to the imaginary part of the strength function. A reasonable interpretation of this quantity is that it is related to the uncertainty of the interference between the resonance and the background of the process being studied[34].

We applied the "resonant RPA" method described above to calculate giant multipole resonances in ^{16}O and ^{208}Pb. First we tested the reliability of our method by comparing with the classical calculation for ^{16}O of ref.[27], where the continuum is treated exactly. For this we chose the potential parameters as in ref.[27] (case without absorption so that only processes that contribute to the escape width are included). The results thus obtained agree with those in ref.[27] within 100 keV. Moreover, we also compared with the continuum RPA calculation of ref.[35] (here we used the particle-hole Migdal force of ref.[35]) and the results of both calculations agree within 150 keV. We then calculated the giant dipole resonance in ^{16}O using a separable interaction. The strength was adjusted to obtain a zero-energy solution (the spurious 1^- state). The corresponding giant resonance was obtained at an energy of (23.69-i0.28) MeV, exhausting 94%of the EWSR.

Finally, we calculated particle-hole states in the heavy nucleus ^{208}Pb. The single-particle basis was obtained solving the one-particle Schrödinger equation with a Wood-Saxon potential, including the spin-orbit term using the computer code GAMOW[19]. Together with the calculated states we give in table 1 the experimental values which are uncertain for high lying states. When comparing the experimental and calculated values in table 1 one has to keep in mind that the experimental values do not refer to pure single particle states[36]. In fact, the

Table I

Single-particle states used in the calculations for ^{208}Pb. The proton (neutron) energies E_p (E_n) are in MeV. The experimental data are from ref.[37]. Note that all imaginary values are negative, as it should be for outgoing (decaying) resonant states. The parameters in the Wood-Saxon potential are a=0.75 (0.70) fm, r_o =1.19 (1.27) fm, V_o =66.0 (44.4) MeV and V_{so} =9.5 (8.25) MeV for protons (neutrons).

No	State	E_p	Exp	E_n	Exp
1	$0f_{7/2}$	-22.67	-	-	-
2	$0f_{5/2}$	-20.17	-	-	-
3	$1p_{3/2}$	-18.32	-	-	-
4	$1p_{1/2}$	-17.33	-	-	-
5	$0g_{9/2}$	-16.23	-	-20.99	-
6	$0g_{7/2}$	-12.37	-11.48	-18.06	-
7	$1d_{5/2}$	-11.04	-9.68	-17.06	-
8	$0h_{11/2}$	-9.26	-9.35	-14.96	-
9	$1d_{3/2}$	-9.10	-8.36	-15.51	-
10	$2s_{1/2}$	-8.71	-8.01	-15.30	-
11	$0h_{9/2}$	-3.78	-3.80	-10.69	-10.78
12	$1f_{7/2}$	-3.54	-2.90	-10.49	-9.71
13	$0i_{13/2}$	-1.84	-2.19	-8.57	-9.00
14	$2p_{3/2}$	-0.69	-0.68	-8.35	-8.27
15	$1f_{5/2}$	-0.52	-0.98	-8.08	-7.94
16	$2p_{1/2}$	0.49	-0.17	-7.41	-7.37
17	$1g_{9/2}$	4.03-i0.00	-	-3.93	-3.94
18	$0i_{11/2}$	5.43-i0.00	-	-2.80	-3.16
19	$0j_{15/2}$	5.96-i0.00	-	-1.88	-2.51
20	$2d_{5/2}$	6.75-i0.00	-	-2.07	-2.37
21	$3s_{1/2}$	7.84-i0.04	-	-1.44	-1.90
22	$1g_{7/2}$	8.09-i0.00	-	-0.77	-1.45
23	$2d_{3/2}$	8.53-i0.03	-	-0.78	-1.40
24	$2f_{7/2}$	-	-	2.10-i0.87	-
25	$1h_{11/2}$	-	-	2.25-i0.03	-
26	$2f_{5/2}$	-	-	2.70-i2.32	-
27	$0k_{17/2}$	-	-	5.03-i0.00	-
28	$1h_{9/2}$	-	-	5.40-i0.73	-
29	$0j_{13/2}$	-	-	5.41-i0.01	-
30	$1i_{13/2}$	-	-	7.66-i1.04	-

calculated bound single particle energies in table 1 are approximately the same as those obtained by previous calculations[37]. From the point of view of this work, the interesting feature is the set of unbound states. Although the proton resonances are narrow, the width of the neutron resonances varies widely.

With the single-particle states of table 1 we formed the RPA matrix of eq.(42) and also the functions used in the dispersion relation eq.(56). The irregular behaviour of these functions can be seen in Fig. 5.

Table 2

Correlated particle-hole energy E_c up to 13MeV and the corresponding contribution to the isoscalar energy weighted sum rule S_c for the operator $r\partial V/\partial r$ in ^{208}Pb. The energies are in MeV and the sum rule is given in arbitrary units

E_c	S_c
4.09-i0.00	67-i5
5.44-i0.00	10-i3
5.51-i0.00	5-i2
5.81-i0.00	0+i0
10.13-i2.33	2-i2
10.18-i0.87	0+i0
10.44-i0.06	349-i23
10.49-i0.86	10-i10
10.78-i2.33	0+i0
11.06-i2.33	0+i0
12.59-i0.88	0+i0
12.78-i0.03	0+i0
12.79-i0.87	0+i0
12.94-i0.03	0+i0

The diagonalization of the RPA matrix provided us with the correlated states, the first few of which are shown in table 2. There are some striking features in this table worthwhile to be commented. In the correlated spectrum of table 2 it appears a very collective state (it exhausts 76% of the EWSR) at 10.44 MeV with a width of only 0.12 MeV just engulfed by non collective but very wide states. This collective state is the Isoscalar Quadrupole Giant Resonance(SQGR). In the experiment the SQGR is at 10.6MeV, it exhausts 70% of the EWSR and its total width is about 2 MeV[38]. The escape width is not known but can be estimated if we assume that it amounts to 15% of the total width as in the case of GDR. This estimate gives 0.3MeV for the escape width which is in agreement with our results. Although already in itself this is a surprising result, it is also remarkable that the giant resonance is not the lowest state among the $\Delta N = 2$ excited states as it would be within a real (bound) basis. To analyse this in more detail we also calculated the quadrupole states using the computer code RPAPH[39] within a harmonic oscillator representation with standard parameters[32]. We found that bound states are about the same in both calculations, but the

SQGR is indeed the first $\Delta N = 2$ excited state in the harmonic oscillator case. It lies at 8.66 MeV of excitation energy. This value is very low but agrees with the one given in ref.[40]. In our case, the first $\Delta N = 2$ excited state is the very broad one at 10.13 MeV, which is only weakly excited by the external field. This indicates that our method differs considerably from the usual treatment of resonant states. In general, the main features of the SQGR mentioned above appear for the isovector case as well. Just opposite to what one would have predicted from the uncorrelated spectrum, our calculated giant resonances turn out to be narrow. The main reason for this is that the most important components of these states are built upon high spin single-particle states, all of which are either bound or narrow. Although the calculated position of the SQGR agrees well with the corresponding experimental value, the calculated width seems to be too small. However, one has to consider that the escape width extracted from the experimental spectrum is rather uncertain. Moreover, only a small mixing of the giant resonance with the near lying wide states would enlarge it considerably. In our case, however, the EWSR has small imaginary components. We also calculated the sum rule corresponding to an electromagnetic field to see if the long range components of this field would be unrealistically enhanced by the divergent tail of the Gamow function but we found that the structure of the corresponding strength function is very similar to the $r\partial V/\partial r$ case analysed above.To see the effect of the size of the basis upon the results we increased the number of states in table 1 by 20% but the complex energies of the giant resonances remained practically the same while the total EWSR increased by only 3%. We also verified that a known property of the RPA energy weighted sum rule is satisfied , namely the total values of the uncorrelated and correlated EWSR exactly coincide[41].

† On leave from: Institute of Nuclear Research of the Hungarian Academy of Sciences, H-4001 Debrecen, Pf. 51, Hungary

‡ Fellow of the CONICET, Buenos Aires, Argentina

References

1. J. Humblet and L. Rosenfeld, Nucl. Phys., **26**, 529 (1961).

2. A. J. F. Siegert, Phys. Rev., **36**, 750 (1939).

3. A. M. Lane and R. G. Thomas, Rev. Mod. Phys., **30**, 257 (1958).

4. P. L. Kapur and R. E. Peierls, Proc. Roy. Soc., **166A**, 277 (1938).

5. G. Gamow, Z. Phys., **51**, 204 (1928).

6. H. M. Nussenzveig, Nucl. Phys., **11**, 499 (1959).

7. Y. B. Zel'dovich, Zh. Eksp. i Theor. Fiz., **39**, 776 (1960); Sov. Phys. JETP, **12**, 542 (1961).

8. N. Hokkyo, Prog. Theor. Phys., **33**, 1116 (1965).

9. T. Berggren, Nucl. Phys., **109A**, 265 (1968).

10. W. J. Romo, Nucl. Phys., **116A**, 617 (1968).

11. B. Gyarmati and T. Vertse, Nucl. Phys., **160A**, 523 (1971).

12. B. Simon, Phys. Lett., **71A**, 211 (1979).

13. J. Bang and J. Zimányi, Nucl. Phys., **139A**, 534 (1969).

14. B. Gyarmati, T. Vertse, J. Zimányi and M. Zimányi, Phys. Rev., **1C**, 1 (1970).

15. B. Gyarmati, T. Vertse, Nucl. Phys., **182A**, 315 (1972).

16. W. R. Coker, Phys. Rev., **9C**, 784 (1974); **7C** 2426 (1973).

17. H. S. Bradlow, W. D. M. Rae, P. S. Fisher, N. S. Godwin, G. Proudfoot and D. Sinclair, Nucl. Phys., **314A**, 171 (1979).

18. G. Garcia-Calderon and R. Peierls, Nucl. Phys., **265A**, 443 (1976).

19. T. Vertse, K. F. Pál and Z. Balogh, Comp. Phys. Comm., **27**, 309 (1982).

20. M. Baldo, L. S. Ferreira, L. Streit, Nucl. Phys., **467A**, (1987).

21. B. Gyarmati, A. T. Kruppa, Z. Papp and G. Wolf, Nucl. Phys., **417A**, 393 (1984).

22. Z. Papp, J. Phys. A, **20**, 153 (1987).

23. B. Gyarmati and A. T. Kruppa, Phys. Rev., **C34**, 95 (1986).

24. K. F. Pál, J. Phys. A, **18**, 1665 (1985).

25. B. G. Giraud, M. V. Mihailovic, R. G. Lovas and M. A. Nagarajan, Ann. Phys. (N.Y.), **140**, 29 (1982).

26. G. F. Bertsch, P. F. Bortignon and R. A. Broglia, Rev. Mod. Phys., **55**, 287 (1983).

27. B. Buck and A. D. Hill, Nucl. Phys., **A95**, 271 (1967).

28. C. Mahaux and H. A. Weidenmüller, *Shell model approach to nuclear reactions*, North Holland, Amsterdam (1984) p. 393.

29. S. Shlomo and G. Bertsch, Nucl. Phys., **243A**, 507 (1975).

30. T. Vertse, P. Curutchet, O. Civitarese, L. S. Ferreira and R. J. Liotta, Phys. Rev., **37C**, 876 (1988).

31. A. Bohr and B. Mottelson, *Nuclear Structure*, Vol 2, Benjamin, New York, (1975).

32. D. R. Bes, R. A. Broglia and B. S. Nilsson, Phys. Rep., **16C**, 1 (1975).

33. NAG library manual, Oxford, (1982).

34. T. Berggren, Phys. Lett., **73B**, 389 (1978).

35. G. Co' and S. Krewald, Nucl. Phys., **333A**, 392 (1985).

36. G. F. Bertsch and I. Hamamoto, Phys. Rev., **26C**, 1323 (1982).

37. L. Rydström and J. Blomqvist, Annual Report Research Institute of Physics, Stockholm (1980) p. 86.

38. F. E. Bertrand, E. E. Gross, D. J. Horen, R. O. Sayer, T. P. Sjoreen, D. K. Mc Daniels, J. Lisanti, J. R. Tinsley, L. W. Swenson, J. B. Mc Clelland, T. A. Carey, K. Jones and S. J. Seestrom-Morris, Phys. Rev., **34C**, 45 (1986).

39. O. Civitarese, computer code RPAPH, available at the Niels Bohr Institute, Copenhagen.

40. D. R. Bes, P. Curutchet, S. L. Reich, N. N. Scoccola and H. M. Sofia, Nucl. Phys., **452A**, 531 (1986).

41. S. G. Nilsson, C. F. Tsang, A. Sobiczewski, Z. Szymanski, S. Wyceck, G. Gustafson, I. L. Lamm, P. Moller and B. Nilsson, Nucl. Phys., **131A**, 1 (1969).

RESONANCES
AS
AN EIGENVALUE PROBLEM

Lidia S. Ferreira*

The Niels Bohr Institute
Blegdamsvej 17
DK-2100 Copenhagen Ø
Denmark

Abstract

The theory of resonances as developed primarily in the field of nuclear physics is briefly reviewed. In particular the resonance problem is posed as general eigenvalue problem. The connection with scattering theory as well as a numerical application to the deuteron wavefunction are displayed in some detail.

1. Introduction

It is common in collisions of low and high energy particles that two particles might stay together for a while without forming a bound state, but instead an excited state of positive energy called a resonance. The implication of this physical situation to scattering theory is the existence of unstable states of the dynamical system composed by the scattering center and scattered particle. These states can be associated with a complex energy whose imaginary part gives the width of the resonance thus measuring the time the two particles stay together.

Resonances were originally associated with acoustical phenomena and a theory was later developed for resonances of electromagnetic waves in cavities. They have been observed in low energy nuclear physics, where the cross sections display various narrow peaks; as well as in elementary particle collisions, where they generally have a substantial width typically of order of 100 Mev. The data is well fitted by a Breit-Wigner parametrization of the scattering matrix in terms of the energy and width.

The criteria [1] adopted to identify a resonance is given by the behaviour of the scattering amplitude as a function of energy in the Argand plot. If there is a resonance, the scattering amplitude will trace a resonance circle in the left-handed sense. The phase shift will go through $\pi/2$ and the angular velocity of a point describing the resonance circle will be $v = \frac{\Gamma/2}{\left[(E - E_r)^2 + \Gamma^2/4\right]}$, which is maximum at the resonance energy E_r, and slows down as the energy E increases. The quantity Γ is the width. For this type of situation the scattering matrix has complex poles at the resonance energy.

The theory of resonances has been developed by many authors [2], where expansions for the S-matrix were obtained in terms of the complex energies associated with these states. However in such expansions, since the interaction was considered to be of finite range, there are terms associated with poles which do not fulfill the requirements of a physical resonance and are of mathematical origin. If no distinction is made between these *physical* and *unphysical* resonances they can be gathered under the generic name of Gamow state vectors. Their mathematical properties, as generalized eigenfunctions in scattering theory were studied by Berggren [3], and have been extensively used in the literature [4] in a wide context.

Gamow vectors can be obtained by solving the radial Schrödinger equation with purely outgoing wave boundary conditions. In practice, the potential is approximated by a potential of finite range and the Gamow vector is matched beyond this range to a pure outgoing free wave with complex momentum k ($\Im\, k < 0$).

Because of the boundary condition imposed on these states, they are not solutions of an ordinary eigenvalue problem. Exception is made in the case of bound states. Since they have an exponential growth it is impossible to expand them in a basis set of functions; and the Schrödinger equation cannot be reduced to the characteristic matrix form of an eigenvalue problem.

In this lecture the possibility of reducing this problem to an eigenvalue one is considered. In Section 1 a brief review of scattering theory notions connected with the general framework of Gamow states is made. The method is explained and illustrated in Section 2. Applications are made in Section 3. We conclude in Section 4 with a brief discussion of the criteria adopted.

2. Gamow States In The General Context Of Resonances

Gamow state vectors are defined as solutions for complex energies of the Schrödinger equation, with purely out-going wave boundary conditions. They correspond to complex energies and grow exponentially for large radii. For negative energies they coincide with the corresponding bound states and decay exponentially.

They can be understood, in a very simple way, if one considers the scattering by a potential of compact support, namely a potential that goes to zero beyond a range R. Outside the range of the potential, the radial wave function for $l = 0$, can be written as $\phi = I sin(kr)/k + Se^{ikr}$ where I and S are the incident and scattered wave amplitudes, respectively. At $r = R$, ϕ and $\frac{d\phi}{dr}$ give the general boundary conditions,

$$Ie^{-ikR} = \left(\frac{d\phi}{dr}\right)_R - ik\phi(R) . \tag{1}$$

In order to be a resonance, the incoming wave can at most be a perturbation; so I should be equal to zero, leading to the boundary condition $\left(\frac{d\phi}{dr}\right)_R = ik\phi(R)$. This last condition together with continuity at the origin described by $\phi(0) = 0$, form a set of homogeneous boundary conditions defined at the ends of a finite interval. In terms of the Sturm-Liouville theory of differential equations, the energy spectrum must be discrete and have a complex quantity since the boundary condition has a complex form.

Then let k_n be the discrete momenta and ϕ_n the corresponding solution of the Schrödinger equation. The orthogonality for this functions is defined in terms of a real scalar product.

In order to prove that, we consider the equations for two different solutions. Subtracting them and integrating over the range of the interaction we obtain

$$(k_n - k_m) \int_0^R \phi_n \phi_m dr + \left[\phi_m \frac{d\phi_n}{dr} - \phi_n \frac{d\phi_m}{dr}\right] = 0 \tag{2}$$

The second term in (2) vanishes as a consequence of the boundary condition, therefore $\int_0^R \phi_n \phi_m dr = 0$ for $k_n \neq k_m$. Note that the integrand for $n = m$ is not the modulus of the wave function as in the usual scalar product, but the square of the function. In a more mathematical point of view, Gamow vectors can be obtained from the zeros of the Jost function and correspond to the poles of the S-matrix if the potential is of compact support. Still considering, for simplicity, the s-wave scattering example and a potential less singular at $r = 0$ then r^{-2}, two types of solutions can be defined for the radial Schrödinger equation. The ones which vanish at $r = 0$, called the regular solution $\phi(k,r)$, with $\phi'(k,0) = 1$; and the ones which diverge at $r = 0$. The latter, called the irregular solutions $f_\pm$, are defined by the boundary condition dependent on momentum, $\lim_{r\to\infty} e^{\mp ikr} f_\pm(k,r) = 1$. The functions f_+ (f_-) are well defined only for $\Im k \geq 0$ (≤ 0). If the potential is of compact support, $f_\pm$ are entire analytic functions of the momentum k, and are connected by analytic continuation such that $f_-(k,r) = f_+^*(k^*,r)$.

Since they are two independent solutions of the Schrödinger equation the regular solution can be expressed in terms of them by $\phi = af_+ + bf_-$, where the quantities a and b are defined in terms of the Wronskians $\mathcal{L}_\pm(k) = W(f_\pm, \phi)$ by $a = \mathcal{L}_-(k)/2ik$ and $b = -\mathcal{L}_+(k)/2ik$. The Wronskian of any two solutions of a differential equation is independent of r. From the boundary condition for the regular solution at $r = 0$ the Jost function $\mathcal{F}(k) = W(f_+, \phi)$, is equal to $f_+(k, 0)$, therefore an entire analytic function of k. Both scattering states and S-matrix can be expressed in terms of $\mathcal{L}_\pm(k)$. For s-wave scattering, the S-matrix is simply $S(k) = \dfrac{\mathcal{L}_-(k)}{\mathcal{F}(k)}$.

If the potential is of compact support, the poles of the S-matrix are the zeros of the Jost function, or simply the zeros of the Wronskian $W(f_+, \phi)$. They correspond then to resonances since the coefficient b in the expression for the regular solution, will be zero, describing a situation with no incoming wave. If these zeros are defined by $k_n = \alpha_n + i\gamma_n$ (by taking complex conjugation in $W(f_+, \phi)$), it can be seen from the properties of $f_\pm$ and ϕ that $-k_n^* = -\alpha_n - i\gamma_n$ is also a solution. The poles come in pairs and without loss of generality α can be set as a positive quantity. It can be proved that $\gamma_n > 0$ if $\alpha \neq 0$. They are then distributed symmetrically with respect to the imaginary axis in order of increasing modulus and lie in the lower half of the complex k plane.

The asymptotic behaviour of these poles was studied by Regge[5] which proved that for n large enough, they behave like

$$\begin{aligned} \alpha_n &\rightarrow \frac{\pi}{2R} n + \mathcal{O}(1) \\ \gamma_n &\rightarrow \frac{\sigma + 2}{2R} \ln n + \mathcal{O}(1) \end{aligned} , \tag{3}$$

where σ is a constant.

The Gamow vectors and momenta reflect the fact that the potential is of finite range. They are then strongly dependent on the range R. However, if the imaginary part is small, they can describe physical resonances in the sense that the phase shift will go through $\pi/2$ at the energy corresponding to the real part and satisfy the requisites discussed in the introduction. In this situation they will not depend on the cutoff radius, as it should be expected from physical grounds. The other ones are purely mathematical resonances.

It is interesting to consider the transition from an infinite range potential to a finite one and look at the motion of the poles in the complex plane. A potential with exponential decay $V(r) = r^\beta \, e^{-\mu r}$ is known to have a cut on the imaginary axis starting from $-1/\mu$. If we consider the potential $V(r) = r^\beta \, e^{-\mu r^\alpha}$ with $\alpha > 1$ which decays faster than the exponential one, it will produce poles symmetrically distributed in the k plane. Varying α and making it approach one[6], the poles ge closer to each other and the trajectories rotate until they finally collapse in the cut for $\alpha = 1$.

3. Eigenvalue Problem For Gamow Vectors

In this section we follow closely the derivation presented in ref. 7. In the case of s-wave potential scattering, the Gamow state vectors can be obtained exactly from the radial Schrödinger equation by solving an eigenvalue problem.

Let $u_0(r)$ be the radial wave function for $l = 0$ obtained from a potential $V(r)$ of support R. Making the transformation

$$u_0(r) = \begin{cases} e^{ikr}\phi_0(r) & r \leq R \\ e^{ikR}\phi_0(R) & r > R \end{cases} \tag{4}$$

in the original radial equation, the function $\phi(r)$ will satisfy the equation

$$\phi_0''(r) + 2ik\phi_0'(r) - V(r)\phi_0(r) = 0 \tag{5}$$

with the boundary condition $\phi_0(0) = 0$ and $\phi_0'(R) = 0$ which guarantees that $u_0'(R)/u_0(R) = ik$. Expanding $\phi_0(r)$ in a complete set of functions $P_m(r) = Nrj_0(k_m r)$ where $j_0(x)$ are the spherical Bessel functions, equation (5) can be recast in the form of an eigenvalue equation for a non-hermitean matrix

$$\sum_m \left[A_{nm} - ikB_{nm}\right] a_m = 0. \tag{6}$$

The components of the matrices A and B correspond to the matrix elements of the Schrödinger and first derivative operators, taken between the basis functions $P_m(k_m r)$. For details see reference 7. The quantities k_m are fixed, by the boundary condition, as $k_m = (2m+1)\pi/2R$. The solutions of (6) correspond to all the Gamow vectors for the given potential.

The method was applied to potentials with physical resonances giving the correct results. The two cases considered were a model potential with an attractive spherical rectangular well and a barrier able to produce a resonance, and the ${}^1S_0$ channel of the Nucleon-Nucleon interaction described by the Argonne potential[8]. The latter has a virtual state at $\Im(k) = -0.053 fm^{-1}$ which was correctly reproduced with 30 basis functions in the expansion.

For higher partial waves the method is not exact. The transformation of (7) has to be generalized to

$$u_l(r) = \begin{cases} rh_l^{(+)}(kr)\phi_l(r) & r \leq R \\ rh_l^{(+)}(kr)\phi_l(R) & r > R \end{cases} \tag{7}$$

where $h_l^{(+)}$ are the spherical Hankel functions of the first kind. The function $\phi_l(r)$ satisfies the same boundary condition as in the $l = 0$ situation. This condition is not ik, but a rational polynomial of kr. This complicated dependence on k, introduced by the boundary condition, prevents a reduction to an eigenvalue problem if one proceeds as before and expand $\phi_l(r)$ in a suitable basis. However, if (7) is considered as

$$u_l(r) = \begin{cases} e^{ikr}\psi_l(r) & r \leq R \\ e^{ikr}\psi_l(R) & r > R \end{cases} \tag{8}$$

the expansion of $\psi_l(r)$ in the complete orthogonal set of functions $Q_m^l = Nrj_l(k_m r)$ immediately gives an eigenvalue form for the radial Schrödinger equation.

If R is large enough, one is only approximating the Hankel function by an exponential which is asymptotically correct. Otherwise one is cutting off the centrifugal potential for $r > R$. This procedure, the centrifugal cutoff approximation CCA, will mainly affect the imaginary part of the poles. Actually, it is well known from shell model calculations that cutting the potential at the top of the barrier produces practically the correct real energies for the single particle levels.

The N-N interaction did not have any resonance or virtual state in the single channels for $l \neq 0$. The method was then tested with a model potential with a Woods-Saxon type of well and a barrier with parameters adjusted to produce sharp physical resonances in the $l = 1$ and $l = 2$ channels. It was possible to reproduce the exact resonance energy and width with just a few functions (see reference 7).

The most interesting case in the N-N system is certainly the ${}^3S_1 - {}^3D_1$ coupled channel, the deuteron. The method was generalized in a straightforward way to the coupled channel problem by imposing the transformation of (7) for each channel and then expanding in a suitable basis of functions, as in the single channel case. The application to the deuteron problem calculated with the Reid soft core[9] potential with a cutoff radius of 6.5 fm, produced the momentum $k = i\,0.2315$ with 30 functions to be compared with the value given by Reid as $k = i\,0.2316$. The results for the wave function were also very precise up to the cutoff radius. Outside this range the wave function is assumed to decrease as an exponential and deviations appeared in the D component. If the cutoff radius is increased better accuracy is reached, but the size of the basis has to be enlarged.

Since the deviations are small, the method can be improved with perturbation theory for eigenvalue equations. This is described in detail in reference 7, therefore we only quote the results in Table 1.

Table 1

Deuteron wave function for $x = 0.7\ r$. The CCA approximation is displayed in the first column for the S and D components. The second column shows the results corrected by perturbation theory. Comparison is made with the exact results from Reference 9.

x	CCA		CCA+pert		exact	
	u	w	u	w	u	w
0.0725	0.0001	0.0001	0.0001	0.0001	0.0000	0.0000
0.1975	0.0132	0.0051	0.0132	0.0051	0.0127	0.0042
0.3225	0.0747	0.0306	0.0747	0.0306	0.0754	0.0308
0.4475	0.1898	0.0789	0.1898	0.0789	0.1899	0.0789
0.5725	0.3136	0.1292	0.3136	0.1292	0.3139	0.1293
0.6975	0.4124	0.1650	0.4124	0.1650	0.4132	0.1653
0.8850	0.4991	0.1869	0.4991	0.1869	0.4995	0.1871
1.135	0.5313	0.1793	0.5313	0.1793	0.5317	0.1795
1.385	0.5187	0.1573	0.5187	0.1573	0.5193	0.1574
1.760	0.4747	0.1232	0.4746	0.1232	0.4750	0.1231
2.510	0.3785	0.0742	0.3785	0.0742	0.3786	0.0739
3.510	0.2739	0.0402	0.2739	0.0402	0.2740	0.0391
4.510	0.1971	0.0246	0.1971	0.0246	0.1972	0.0220
5.010	0.1671	0.0205	0.1671	0.0201	0.1672	0.0169
6.010	0.1201	0.0147	0.1201	0.0121	0.1201	0.0102

Up to now we were only concerned with Gamow vectors which described physical resonances. From the discussion of Section 2, finite range potentials display other resonances of mathematical origin. All the model and realistic potentials for the N-N interaction mentioned before were approximated by finite range forms, therefore they give this type of solutions for the Schrödinger equation.

Considering for example the case of the $^1P_1$ channel of the N-N interaction described by the Argonne potential, all the poles of the S-matrix are of mathematical origin. The calculation made according to the CCA approximation, is compared in Table 2 with an exact calculation where the Schrödinger equation is solved numerically with outgoing wave boundary conditions; a cutoff radius of 6.5 fm was used in both cases. The results for the real part are quite close to the exact values but the imaginary is quite affected by the CCA. Since the discrepancy is large, perturbation theory is unable to completely correct the results. For example, the first order corrections to the first two poles of Table 2 give $k = 0.649, -i\,0.400$ and $k = 1.178, -i\,0.500$, respectively.

Table 2

First few Gamow momenta in fm^{-1} for the $^1P_1$ channel obtained with a cutoff radius $R = 6.5$ fm for the CCA and exact calculations.

exact		CCA	
Real (k)	Im (k)	Real (k)	Im (k)
0.657	-0.578	0.644	-0.300
1.172	-0.662	1.182	-0.411
1.703	-0.731	1.728	-0.470
2.273	-0.777	2.263	-0.508
2.767	-0.809	2.790	-0.535

Instead of the perturbation theory, one can use an iterative process to correct for the transformation of (8) and introduce the correct boundary condition. This was developed[10] using a spline expansion for the single channel problem and proved to be very accurate after a sufficient number of iterations.

In the coupled channel problem the same situation occurs. Different poles with increasing modulus are obtained and displayed in Table 3 for the $^3S_1 - ^3D_1$ channel calculation with the Argonne potential. An exact calculation could not be made, but we expect that the deviations for the imaginary part of the unphysical poles should be comparable to the ones found in the single channel and the corrections from perturbation theory should slightly improve the results.

4. Gamow Vectors As A Basic Set Of Functions

The Gamow vectors constitute a set of functions that, in spite of their exponential growth, can be normalized. The norm is not defined as the integral of the square of the modulus but as the direct product of two functions as in the case of (2), where the integral is restricted to a range up to the cutoff radius. This can be proved[11] by rotating this state through a complex dilation in order to make it a bound state. The integral is then evaluated with standard methods to handle diverging integrals[12].

Table 3

Same as in Table 2 for the $^3S_1 - ^3D_1$ channel of the Argonne V_{14} interaction in the CCA approach.

Real (k)	Im (k)
0.0000	0.2316
0.4269	-0.7513
0.8804	-0.2727
1.0242	-0.7903
1.4231	-0.3355
1.5390	-0.8246
1.9388	-0.3759

A natural application for the set of Gamow vectors is the construction of form factors to represent a separable interaction[13]. Separable representations of the N-N interaction have been the subject of research for many years since they simplify greatly the few body and nuclear matter equations.

The N-N interaction should in principle be non local. It is however normally described by a local potential model of finite range which can then be approximated up to the required accuracy by a non local separable one.

Using the general method of Adhikari and Sloan[14] a rank N separable representation V_N of an original potential V was obtained which has the form

$$V_N = \sum_{n,m=1}^{N} V \,|\chi_n> D_{nm} < \chi_m|V$$
$$\left(D^{-1}\right)_{nm} = <\chi_n|V|\chi_m> \qquad (9)$$

where $|\chi_n>$ are the first N Gamow vectors obtained from the original potential V.

This procedure was applied to the Reid soft core and Argonne V_{14} potentials which are realistic N-N interactions in the sense that they explain phenomenologically the available data. Separable representations were made for the different partial wave single and coupled channels[15] with the Gamow vectors obtained from the CCA approximation.

The phase shifts and the off-shell Kowalski-Noyes function are presented in Figure 1 for the $^1P_1$ and $^3D_1$ channels of the rank $N = 4$ approximation to the

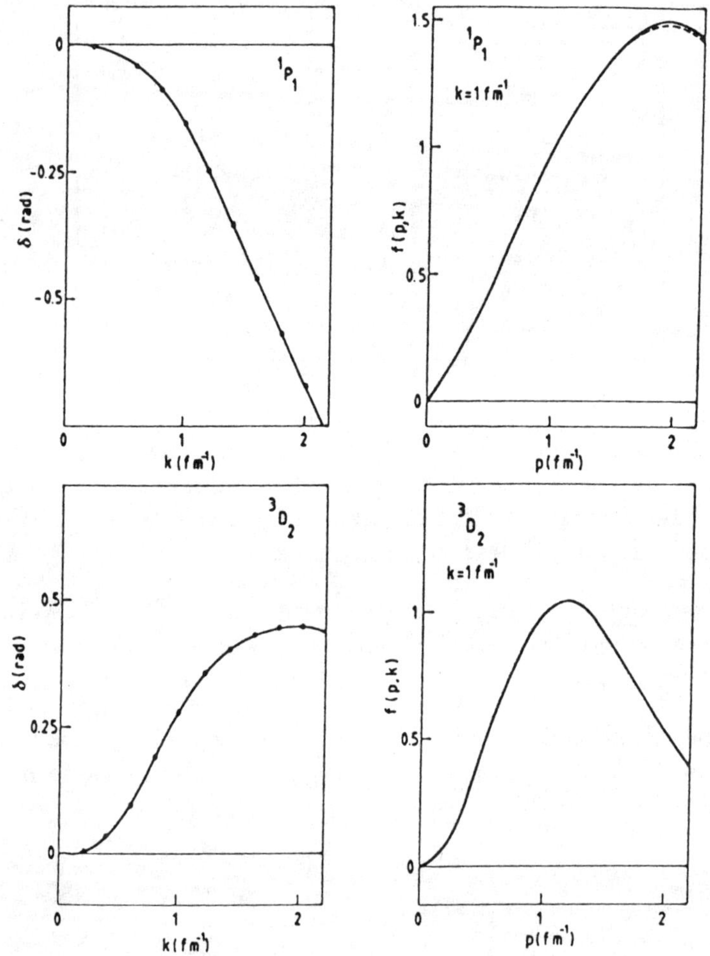

Figure 1

Phase shifts and Kowalski-Noyes functions for the Reid soft core potential in the $^1P_1$ and $^3D_1$ partial wave channels. The solid line represents the exact phases and the circles the rank $N = 4$ approximation. For the Kowalski-Noyes function the rank $N = 4$ approximation is represented by the dashed line. Inside graphical accuracy it cannot be practically distinguished from the exact calculation represented by the full line. The cutoff radius was taken as $R = 6.5$ fm.

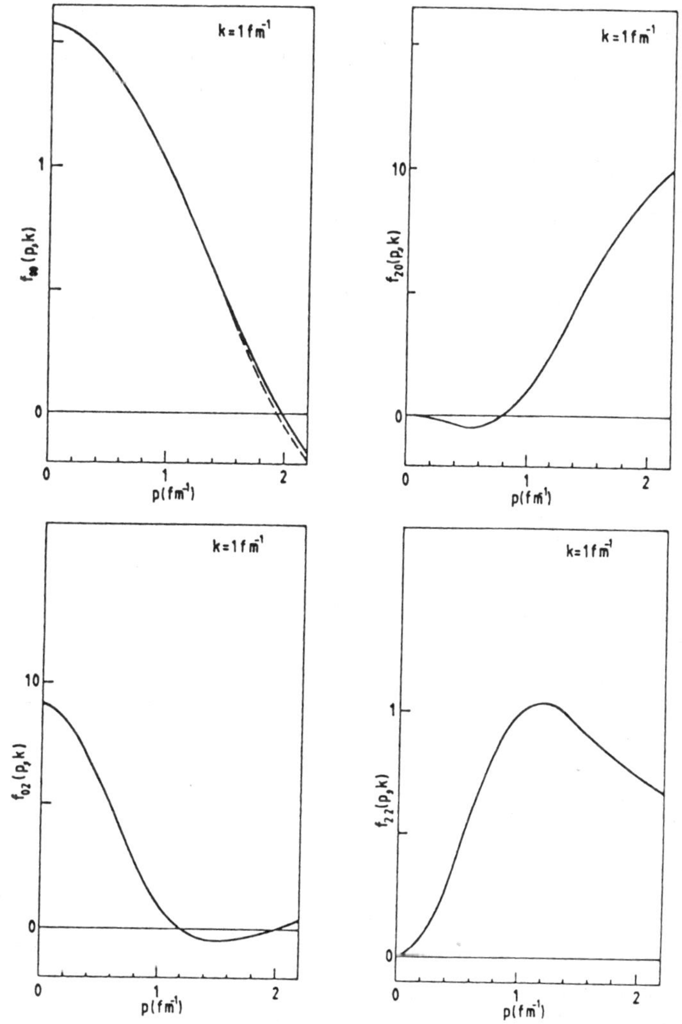

Figure 2

The Kowalski-Noyes function for the four K-matrix elements of the $^3S_1 -^3 D_1$ channel of the Argonne V_{14} interaction. The full and dashed lines correspond to the exact and rant $N = 9$ calculations.

Argonne potential, showing excellent agreement with the exact results.

The coupled $^3S_1 - {}^3D_1$ channel is presented in Table 4 and Figure 2 for the same interaction. The phases are calculated using the Stapp parametrization for the

Table 4

Phase parameters in the $^3S_1 - ^3D_1$ channel of the Argonne V_{14} potential for the rank $N = 9$ and exact calculations. The quantities δ_0, δ_2 and $\rho = \sin(2\epsilon)$ are the $^3S_1$, $^3D_1$ phase shifts and mixing parameter respectively.

k (fm^{-1})		N=9	exact
0.4	δ_0	1.6643	1.6659
	ρ	0.0640	0.0610
	δ_2	-0.0168	-0.0172
1.0	δ_0	0.8304	0.8340
	ρ	0.0872	0.0845
	δ_2	-0.1873	-0.1874
1.6	δ_0	0.3198	0.3277
	ρ	0.1428	0.1351
	δ_2	-0.3453	-0.3498
2.0	δ_0	0.0564	0.0666
	ρ	0.1861	0.1885
	δ_2	-0.3936	-0.4245

S-matrix. The agreement between the rank $N = 9$ calculation is perfect on and off the energy shell.

In spite of the deviations introduced by the CCA in the Gamow momenta and vectors calculation, the Gamow separable expansion seems to be essentially insensitive to these differences.

5. Conclusions

The method discussed above treats bound states and resonances in a similar way, as an eigenvalue problem for a real non symmetric matrix, consequently it can be used to calculate *physical* and *mathematical* resonances for the single and coupled channel problem. In the case of *unphysical* resonances, the CCA procedure introduces deviations which might be improved by the perturbation theory. The *physical* poles are not affected by the approximations involved.

The application made to the construction of a separable representation of an interaction seemed to be insensitive to the shifts introduced by the CCA in the location of the poles.

I would like to thank M. Baldo for reading the manuscript and E. Maglione for the $H_I{}^G H$ – $T_E X$ esthetics. A grant from the Danish Research Council is also acknowledged.

* Permanent address:
Instituto Superior Tecnico and
Centro de Física da Matéria Condensada
Avenida Gama Pinto 2, 1699 Lisboa
Portugal

References

1. R. H. Dalitz, *Resonances Models and Phenomena*, Lecture Notes in Phys., **211**, Springer-Verlag, Berlin (1984) p. 1.

2. J. Humblet and L. Rosenfeld, Nucl. Phys., **26A**, 529 (1961); C. Bloch and V. Gillet, Phys. Lett., **16**, 62 (1965); **18**, 58 (1965); H. A. Weidenmüller, Nucl. Phys., **65**, 189 (1966); P. L. Kapur and R. E. Peierls, Proc. Royal Soc., **166A**, 277 (1938).

3. T. Berggren, Nucl. Phys., **109A**, 265 (1968).

4. B. Gyarmati, These proceedings, E. Brändas and N. Elander, Eds., Lecture Notes in Phys., Springer-Verlag, Berlin, (1988); *Resonances Models and Phenomena*, Lecture Notes in Phys., **211**, S. Albeverio, L. S. Ferreira, L. Streit, Eds., Springer-Verlag, New York (1985).

5. T. Regge, Nuovo Cimento, **8**, 671 (1958).

6. E. Balslev, Private communication.

7. M. Baldo, L. S. Ferreira and L. Streit, Nucl.Phys., **467A**, 44 (1987).

8. R. B. Wiringa, R. A. Smith and T. L. Ainsworth, Phys. Rev., **29C**, 1207 (1984).

9. R. V. Reid, Ann. Phys. (N. Y.), **50**, 411 (1968).

10. T. Pena and A. Fonseca, Centro de Fisica Nuclear preprint, IFM 5/87, Lisboa (1987).

11. B. Simon, Ann. Math., **97**, 247 (1973).

12. Ya. B. Zel'dovich, JETP (Sov. Phys.), **12**, 542 (1961).

13. M. Baldo, L. S. Ferreira and L. Streit, Phys. Rev., **32C**, 685 (1985).

14. S. D. Adhikari, I. H. Sloan, Phys. Rev., **11C**, 113 (1975).

15. M. Baldo, L. S. Ferreira and L. Streit, Phys. Rev., **C36**, 1743 (1987); M. Baldo and L. S. Ferreira, Phys. Lett., In press.

A MODEL FOR STUDYING TIME DEPENDENT QUANTUM MECHANICAL PROCESSES AND ITS APPLICATION FOR QUASI-STATIONARY STATES

János Révai

Central Research Institute for Physics
H-1525 Budapest
P.O.B. 49
Hungary

Abstract

A particle moving in an external field, consisting of a stationary attractive and a time dependent repulsive part, is considered. Due to the use of separable interactions the time evolution dynamics can be solved exactly. The model is applied for studying time evolution properties of quasi stationary states.

1. Introduction

Thumbing through the textbooks on quantum mechanics, it is surprising how poor our methods are for solving the time dependent Schrödinger equation, compared to other fields. The approximate methods are designed for certain extreme situations (e.g. small, time dependent perturbations; adiabatic, sudden or periodic time dependence) and it is not easy to extend them beyond their natural limits or to control their accuracy. Direct numerical calculations are restricted to the simplest cases, mainly due to computational complexity arising from the mixing of time and space variables, which leads inevitably to multidimensional problems. There are very few cases where exact solution of "non-extreme" time dependent problems can be attempted, they are mainly connected with transitions in the discrete spectrum.

The aim of the present work is to construct a model, which, on one hand, is flexible enough to imitate certain physical properties of real systems and, on the other hand, allows exact solution of its time dependent dynamics. We propose to achieve this double goal by the use of separable interactions, which reduce the level of difficulty of solving the time dependent Schrödinger equation to that of integrating one–variable Volterra–type integral equations. As a first application, we shall discuss certain questions connected with time evolution of quasi stationary states. Section 2 sets up our model; Section 3 contains the formal solution of the time evolution problem; Section 4 discusses the calculation of the physical transition amplitudes; and Section 5 contains the results of numerical calculations.

2. Formulation of the Model

We consider a particle moving in an external potential field, which consists of a stationary and a time dependent part. The Hamiltonian of the system reads

$$\hat{H}(t) = \hat{H}_0 + \hat{V}_1 + \hat{V}_2(t) = \hat{H}_1 + \hat{V}_2(t) \tag{1}$$

where $\hat{H}_0 = \hat{\mathbf{p}}^2$ denotes the kinetic energy operator (for the sake of simplicity we use the units, $\hbar = 1,\ 2m = 1$) and the potential operators $\hat{V}_i$ are taken in separable form:

$$\hat{V}_1 = |\beta_1\rangle\lambda_1\langle\beta_1| \tag{2a}$$

$$\hat{V}_2(t) = |\beta_2\rangle\lambda_2(t)\langle\beta_2| \tag{2b}$$

The use of separable potentials in certain quantum mechanical calculations has been widely discussed in the literature, mainly in connection with practical solution of few–body integral equations. Here it will be appropriate to quote only two important points concerning separable interactions:

a.) their use essentially simplifies the dynamical (equation–solving) calculations;

b.) there exist a variety of powerful methods for obtaining separable approximations to almost any given interaction operator.

While point a.) will be demonstrated again in the following considerations, point b.) indicates, that the choice (2) is not hopelessly unrealistic and by increasing the number of separable terms in (2) the model Hamiltonian (1) can approach a more realistic one.

The most frequently used concepts and formulas, that occur in treating one- and two–term separable potential stationary problems and upon which the treatment of the time dependent case is based, are collected in the Appendix.

Our time dependent physical problem can be formulated as follows: the particle, which at $t = 0$ is in the bound state $|\phi_B\rangle$ of the potential $\hat{V}_1$ is subject to a time dependent repulsive potential $\hat{V}_2(t)$, which depends on time via its coupling constant $\lambda_2(t)$. Two conditions are imposed on $\lambda_2(t)$:

$$\lambda_2(0) = 0, \qquad \lambda_2(t) \underset{t\to\infty}{\longrightarrow} 0,$$

otherwise its functional dependence is arbitrary. (As it will be seen later, the condition of existence of the integrals, defining the physical transition amplitudes requires $(t\lambda_2(t)) \underset{t\to\infty}{\longrightarrow} 0$.)

We are interested in the probability of finding the particle for $t \to \infty$ either in the original bound state $|\phi_B\rangle$ or leaving the potential region in a free state $|\mathbf{k}\rangle$. By definition, these probabilities are given by

$$w_B = |\lim_{t\to\infty} \langle\phi_B(t)|\Psi(t)\rangle|^2 \tag{3a}$$

$$w(\mathbf{k}) = |\lim_{t\to\infty} \langle\mathbf{k}(t)|\Psi(t)\rangle|^2 \tag{3b}$$

where

$$|\phi_B(t)\rangle = |\phi_B\rangle e^{-iE_B t} \tag{4a}$$

$$|\mathbf{k}(t)\rangle = |\mathbf{k}\rangle e^{-i\mathbf{k}^2 t} \tag{4b}$$

and $|\Psi(t)\rangle$ is the solution of the Schrödinger equation

$$i\frac{\partial}{\partial t}|\Psi(t)\rangle = \hat{H}(t)|\Psi(t)\rangle \tag{5}$$

with the boundary condition

$$|\Psi(t=0)\rangle = |\phi_B\rangle\ . \tag{6}$$

3. Solution of the Time Evolution Problem

The formal solution of (5) satisfying boundary condition (6) can be written as

$$|\Psi(t)\rangle = \hat{\Gamma}(t,0)|\Psi(0)\rangle = \hat{\Gamma}(t,0)|\phi_B\rangle \tag{7}$$

where $\hat{\Gamma}(t,t')$ is the time evolution operator associated with the Hamiltonian (1). It satisfies the differential equations

$$i\frac{\partial}{\partial t}\hat{\Gamma}(t,t') = \hat{H}(t)\hat{\Gamma}(t,t'), \qquad i\frac{\partial}{\partial t'}\hat{\Gamma}(t,t') = -\hat{\Gamma}(t,t')\hat{H}(t') \tag{8}$$

which together with the boundary condition

$$\hat{\Gamma}(t,t) = \hat{I}$$

may be combined into the integral equations

$$\hat{\Gamma}(t,t') = \hat{I} - i\int_{t'}^{t} \hat{H}(t'')\hat{\Gamma}(t'',t')dt'' = \hat{I} - i\int_{t'}^{t} \hat{\Gamma}(t,t'')\hat{H}(t'')dt'' \tag{9}$$

The formal solution of (9) is usually expressed as

$$\hat{\Gamma}(t,t') = \hat{T}\ e^{-i\int_{t'}^{t}\hat{H}(\tau)d\tau} \tag{10}$$

$\hat{T}$ being the time-ordering operator. For time-independent Hamiltonians, $\hat{T}$ can be omitted and (10) yields the familiar expression

$$\hat{\Gamma}(t,t') = e^{-i(t-t')\hat{H}} \tag{11}$$

Apart from $\hat{\Gamma}(t,t')$, one can introduce the operators $\hat{\Gamma}_0(t,t')$ and $\hat{\Gamma}_1(t,t')$, which describe time evolution governed by the sub–Hamiltonians $\hat{H}_0$ and $\hat{H}_1$, respectively. According to (11), they can be written as

$$\hat{\Gamma}_0(t,t') = e^{-i(t-t')\hat{H}_0}, \qquad \hat{\Gamma}_1(t,t') = e^{-i(t-t')\hat{H}_1} \tag{12}$$

The integral equations

$$\begin{aligned}\hat{\Gamma}(t,t') &= \hat{\Gamma}_1(t,t') - i\int_{t'}^{t} \hat{\Gamma}_1(t,t'')\hat{V}_2(t'')\hat{\Gamma}(t'',t')dt'' \\ &= \hat{\Gamma}_1(t,t') - i\int_{t'}^{t} \hat{\Gamma}(t,t'')\hat{V}_2(t'')\hat{\Gamma}(t'',t')dt''\end{aligned} \tag{13}$$

connect the operators $\hat{\Gamma}$ and $\hat{\Gamma}_1$.

Using (7) and (13), we can obtain the following integral equation for $|\Psi(t)\rangle$:

$$|\Psi(t)\rangle = |\phi_B(t)\rangle - i\int_0^t \hat{\Gamma}_1(t,t')\hat{V}_2(t')|\Psi(t')\rangle dt' \tag{14}$$

Substituting the separable form (2b) of $\hat{V}_2(t)$ we get

$$|\Psi(t)\rangle = |\phi_B(t)\rangle - i\int_0^t \hat{\Gamma}_1(t,t')|\beta_2\rangle X(t')dt', \tag{15}$$

where the function $X(t)$ is defined as

$$X(t) = \lambda_2(t)\langle\beta_2|\Psi(t)\rangle\ . \tag{16}$$

From (15) and (16) we obtain, for $X(t)$, the integral equation

$$X(t) = \lambda_2(t)\langle\beta_2|\phi_B(t)\rangle - i\lambda_2(t)\int_0^t \langle\beta_2|\hat{\Gamma}_1(t,t')|\beta_2\rangle X(t')dt' \tag{17}$$

In principle, (17) can serve as a starting point for determination of $X(t)$ but in practice the calculation of its kernel

$$\langle\beta_2|\hat{\Gamma}_1(t,t')|\beta_2\rangle = \langle\beta_2|e^{-i(t-t')\hat{H}_1}|\beta_2\rangle$$

is not too easy, even if all quantities associated with $\hat{H}_1$ (spectrum, eigenfunctions, Green operator, etc.) are known. Therefore, using the fact, that $\hat{\Gamma}_1(t,t')$ is connected with $\hat{\Gamma}_0(t,t')$ via an integral equation similar to (13), we can rewrite (15) as

$$|\Psi(t)\rangle = |\phi_B(t)\rangle - i\int_0^t \hat{\Gamma}_0(t,t')|\beta_2\rangle X(t')dt' - i\int_0^t \hat{\Gamma}_0(t,t')|\beta_1\rangle Y(t')dt' \tag{18}$$

and in this case for $X(t)$ and $Y(t)$ we get

$$X(t) = \lambda_2(t)\langle\beta_2|\phi_B(t)\rangle + \lambda_2(t)\int_0^t [K_{22}(t,t')X(t') + K_{21}(t,t')Y(t')]\,dt'$$
$$Y(t) = \lambda_1\int_0^t [K_{12}(t,t')X(t') + K_{11}(t,t')Y(t')]\,dt' \tag{19}$$

where the kernels are defined as

$$K_{ij}(t,t') = -i\langle\beta_i|\hat{\Gamma}_0(t,t')|\beta_j\rangle = -i\langle\beta_i|e^{-i(t-t')\hat{H}_0}|\beta_j\rangle \tag{20}$$

As it can be seen, at the price of introducing an extra unknown function $Y(t)$, the kernel of the corresponding integral equation became much simpler: it contains only the operator $\hat{H}_0 = \hat{\mathbf{p}}^2$. For suitably chosen form factors $\langle\mathbf{k}|\beta_i\rangle$ the calculation of all matrix elements needed to set up the integral equation (19) can be performed analytically in momentum representation.

The structure of (18)–(20) shows the great advantage of using separable potentials: solving a single–variable Volterra–type integral equation for the c-number functions $X(t), Y(t)$, one is able to calculate the wave function $|\Psi(t)\rangle$ for any time t, while for ordinary potentials the same goal can be achieved only by solving integral (or differential) equations with mixed space and time variables.

4. Calculation of Transition Amplitudes

According to definition (3), the probabilities of the physical processes can be calculated from the amplitudes

$$a_B = \lim_{t\to\infty}\langle\phi_B(t)|\Psi(t)\rangle \tag{21a}$$

$$a(\mathbf{k}) = \lim_{t\to\infty}\langle\mathbf{k}(t)|\Psi(t)\rangle \tag{21b}$$

An expression for the bound state amplitude a_B can be obtained from (15):

$$a_B = 1 - i\langle\phi_B|\beta_2\rangle\int_0^\infty e^{iE_Bt}X(t)dt \tag{22}$$

To calculate the emission amplitude $a(\mathbf{k})$ using (15), we have to use the complete set of eigenstates of $\hat{H}_1$ $\{|\phi_B\rangle, |\phi^{\pm}_{\mathbf{k}'}\rangle\}$ and performing the integration over $\mathbf{k}'$ we have to remember,that the t–limit should be taken in combination with the usual representation of the non–square integrable state $|\mathbf{k}(t)\rangle$ as a limit of wave packets of decreasing widths. This leads us to the expression

$$a(\mathbf{k}) = \lim_{t\to\infty}\langle\phi_{\mathbf{k}}^{-}(t)|\Psi(t)\rangle = -i\langle\phi_{\mathbf{k}}^{-}|\beta_2\rangle\int_0^\infty e^{i\mathbf{k}^2 t}X(t)dt, \tag{23}$$

which demonstrates the fact, that $|\phi_{\mathbf{k}}^{-}(t)\rangle$ is just *that* scattering eigenstate of $\hat{H}_1$, which for $t\to\infty$ approaches a pure plane wave $|\mathbf{k}(t)\rangle$, and thus in the $t\to\infty$ limit $|\mathbf{k}(t)\rangle$ can be replaced by $|\phi_{\mathbf{k}}^{-}(t)\rangle$.

Thus we see from (22) and (23), that the function $X(t)$ contains all the physical information about the system. Since $X(t)$ is proportional to $\lambda_2(t)$ (see (16)), if $\lambda_2(t)$ vanishes for $t > T$, the integrals in (22) and (23) extend only up to T and we need to solve the Volterra equation (19) only in the finite interval $[0, T]$. On the other hand, if $\lambda_2(t)$ for $t\to\infty$ is a slowly decreasing function, the $X(t)$ and $Y(t)$ functions might be required for very large t–values, and their determination can be rather time–consuming.

Therefore it is desirable to design a procedure, which would enable us to calculate the amplitudes (22) and (23) using only finite t–values of $|\Psi(t)\rangle$. For this aim we define two more solutions of the Schrödinger equation (5) with the boundary conditions

$$\begin{aligned} |\Psi_B^{-}(t)\rangle &\underset{t\to\infty}{\longrightarrow} |\phi_B(t)\rangle \\ |\Psi_{\mathbf{k}}^{-}(t)\rangle &\underset{t\to\infty}{\longrightarrow} |\phi_{\mathbf{k}}^{-}(t)\rangle \end{aligned} \tag{24}$$

Similarly to (14), we can derive for them the following integral equations:

$$|\Psi_B^{-}(t)\rangle = |\phi_B(t)\rangle + i\int_t^\infty \hat{\Gamma}_1(t,t')\hat{V}_2(t')|\Psi_B^{-}(t')\rangle dt' \tag{25}$$

$$|\Psi_{\mathbf{k}}^{-}(t)\rangle = |\phi_{\mathbf{k}}^{-}(t)\rangle + i\int_t^\infty \hat{\Gamma}_1(t,t')\hat{V}_2(t')|\Psi_k^{-}(t')\rangle dt' \tag{26}$$

In terms of these states, the amplitudes (21) can be rewritten as

$$\begin{aligned} a_B &= \langle\Psi_B^{-}(t)|\Psi(t)\rangle \\ a(\mathbf{k}) &= \langle\Psi_{\mathbf{k}}^{-}(t)|\Psi(t)\rangle \end{aligned} \tag{27}$$

and the expressions (27) are valid for *any* time $t > 0$, since the overlap of two solutions of the *same* Schrödinger equation is independent of time.

The states $|\Psi_B^-(t)\rangle$ and $|\Psi_{\mathbf{k}}^-(t)\rangle$ can be viewed as developping backward in time: they are pure states at $t \to \infty$ and, due to the backward switching on of $\hat{V}_2(t)$, they become more and more complicated for smaller times. Our aim is to derive their large–t asymptotic form from (25) and (26) and then the forward determination of the physical state $|\Psi(t)\rangle$ need not follow the slow decrease of $\lambda_2(t)$: the amplitudes can be calculated as overlaps of $|\Psi(t)\rangle$ and $|\Psi_B^-(t)\rangle_{as}$, $|\Psi_{\mathbf{k}}^-(t)\rangle_{as}$ for large but finite t values, for which the asymptotic expressions $|\Psi_B^-(t)\rangle_{as}$ and $|\Psi_{\mathbf{k}}^-(t)\rangle_{as}$ approximate the true ones to the desired accuracy.

Since the phases of the states $|\Psi_B^-(t)\rangle$ and $|\Psi_{\mathbf{k}}^-(t)\rangle$ are unimportant, for the derivation of asymptotic expressions, it will be convenient to intrduce

$$|\Psi_B^-(t)\rangle = e^{-iE_B t} e^{i\int_t^\infty \triangle(t')dt'} |\tilde{\Psi}_B^-(t)\rangle \tag{28}$$

$$|\Psi_{\mathbf{k}}^-(t)\rangle = e^{-ik^2 t} |\tilde{\Psi}_{\mathbf{k}}^-(t)\rangle \tag{29}$$

where $\triangle(t) = E_B(t) - E_B$ and $E_B(t)$ is the bound state eigenvalue of $\hat{H}(t)$[1]. The phase factor in (28) takes into account the phase change due to the gradual change of the energy eigenvalue.

The states $|\tilde{\Psi}_B^-(t)\rangle$ and $|\tilde{\Psi}_{\mathbf{k}}^-(t)\rangle$ satisfy the differential equations

$$i\frac{\partial}{\partial t}|\tilde{\Psi}_B^-(t)\rangle = (\hat{H}(t) - E_B(t))|\tilde{\Psi}_B^-(t)\rangle \tag{30}$$

$$i\frac{\partial}{\partial t}|\tilde{\Psi}_{\mathbf{k}}^-(t)\rangle = (\hat{H}(t) - \mathbf{k}^2)|\tilde{\Psi}_{\mathbf{k}}^-(t)\rangle \tag{31}$$

and the integral equations

$$|\tilde{\Psi}_B^-(t)\rangle = |\phi_B\rangle + i\int_t^\infty e^{-i(\hat{H}_1 - E_B)(t-t')} \left(\hat{V}_2(t') - \triangle(t')\right) |\tilde{\Psi}_B^-(t')\rangle dt' \tag{32}$$

[1] In considering the asymptotics of $|\Psi_B^-(t)\rangle$ we suppose, that t is large enough to ensure the existence of the bound state of $\hat{H}(t)$

$$|\tilde{\Psi}_{\mathbf{k}}^{-}(t)\rangle = |\phi_{\mathbf{k}}^{-}\rangle + i\int_{t}^{\infty} e^{-i(\hat{H}_1-\mathbf{k}^2)(t-t')}\hat{V}_2(t')|\tilde{\Psi}_{B}^{-}(t')\rangle dt' \tag{33}$$

The simplest way to derive asymptotic expressions for the states $|\tilde{\Psi}_{B}^{-}(t)\rangle$ and $|\tilde{\Psi}_{\mathbf{k}}^{-}(t)\rangle$ is to iterate (32) and (33), obtaining this way a (converging) series representation in powers of $\lambda_2(t)$. The simplest case is to consider the first iteration of (32) and (33) as asymptotic representation of the states $|\tilde{\Psi}_{B}^{-}(t)\rangle$ and $|\tilde{\Psi}_{\mathbf{k}}^{-}(t)\rangle$:

$$|\tilde{\Psi}_{B}^{-}(t)\rangle_{as} = |\phi_B\rangle + i\int_{t}^{\infty} e^{-i(\hat{H}_1-E_B)(t-t')}\left(\hat{V}_2(t') - \triangle(t')\right)|\phi_B\rangle dt' \tag{34}$$

$$|\tilde{\Psi}_{\mathbf{k}}^{-}(t)\rangle_{as} = |\phi_{\mathbf{k}}^{-}\rangle + i\int_{t}^{\infty} e^{-i(\hat{H}_1-\mathbf{k}^2)(t-t')}\hat{V}_2(t')|\phi_{\mathbf{k}}\rangle dt' \tag{35}$$

In the case of perturbative expansion of a bound state, it is usually advantageous to separate from the Hilbert space the unperturbed state. Introducing the projection operators

$$\hat{P} = |\phi_B\rangle\langle\phi_B|, \qquad \hat{Q} = \hat{I} - \hat{P}$$

we can rewrite (34) as

$$|\tilde{\Psi}_{B}^{-}(t)\rangle_{as} = |\phi_B\rangle + i\int_{t}^{\infty} e^{-i(\hat{H}_1-E_B)(t-t')}\hat{Q}\hat{V}_2(t')|\phi_B\rangle dt' \tag{34a}$$

where we used the fact that up to first order in $\hat{V}_2(t)$, which is the order of accuracy of (34),

$$\triangle(t) = \langle\phi_B|\hat{V}_2(t)|\phi_B\rangle$$

The substitution of the form (2b) of $\hat{V}_2(t)$ into (34a) and (35) leads to the following integrals

$$\begin{aligned} &i\int_{t}^{\infty} e^{-i(\hat{H}_1-E_B)(t-t')}\lambda_2(t')dt' \\ &i\int_{t}^{\infty} e^{-i(\hat{H}_1-\mathbf{k}^2)(t-t')}\lambda_2(t')dt' \end{aligned} \tag{36}$$

The simplest way to evaluate these integrals is to approximate $\lambda_2(t)$ for $t \geq t'$ by a sum of decaying exponentials:

$$\lambda_2(t') \approx \sum_{\nu} c_\nu e^{-b_\nu(t'-t)}, \qquad t' \geq t \tag{37}$$

and to fix the parameters c_ν, b_ν by requiring the matching of a certain number of derivatives of the exact and approximate expressions at t. The simplest case is to take one term in (37), which yields

$$\lambda_2(t') \approx \lambda_2(t) e^{-\gamma(t)(t-t')}, \qquad t' \geq t \tag{38}$$

where

$$\gamma(t) = \frac{\lambda_2'(t)}{\lambda_2(t)} < 0$$

and the matching of $\lambda_2(t)$ and $\lambda_2'(t)$ is ensured. Thus the integrations (36) can be performed to give, finally

$$|\tilde{\Psi}_B^-(t)\rangle_{as} = |\phi_B\rangle + \hat{G}_1\left(E_B + i\gamma(t)\right)\hat{Q}|\beta_2\rangle\lambda_2(t)\langle\beta_2|\phi_B\rangle \tag{39}$$

$$|\tilde{\Psi}_{\mathbf{k}}^-(t)\rangle_{as} = |\phi_{\mathbf{k}}^-\rangle + \hat{G}_1\left(\mathbf{k}^2 + i\gamma(t)\right)|\beta_2\rangle\lambda_2(t)\langle\beta_2|\phi_B\rangle \tag{40}$$

where $\hat{G}_1(z)$ denotes the resolvent (Green) operator of $\hat{H}_1$: $\hat{G}_1(z) = (z - \hat{H}_1)^{-1}$. Eq. (39) shows why was it necessary to introduce the projection operators $\hat{P}$ and $\hat{Q}$ for the bound state case: the operator $\hat{G}_1(z)$ is singular for $z = E_B$ while $\hat{G}_1(z)\hat{Q}$ is not.

Using (39) and (40) we finally get, for the transition amplitudes,

$$\begin{aligned} a_B = e^{-iE_B t}&\left\{1 - i\langle\phi_B|\beta_2\rangle\left[1 - i\langle\phi_B|\beta_2\rangle^2\frac{\lambda_2(t)}{\gamma(t)}\right]\int_0^t e^{iE_B t'}X(t')dt'\right\} \\ &- i\langle\phi_B|\beta_2\rangle\lambda_2(t)\int_0^t\langle\beta_2|\hat{G}_1\left(E_B - i\gamma(t)\right)\hat{\Gamma}_1(t,t')|\beta_2\rangle X(t')dt' \end{aligned} \tag{41}$$

$$a(\mathbf{k}) = -ie^{-i\mathbf{k}^2 t}\langle\phi_{\mathbf{k}}^-|\beta_2\rangle \int_0^t e^{i\mathbf{k}^2 t'} X(t')dt' + e^{-iE_B t}\frac{\langle\phi_{\mathbf{k}}^-|\beta_2\rangle\lambda_2(t)\langle\beta_2|\phi_B\rangle}{\mathbf{k}^2 - E_B - i\gamma(t)}$$
$$- i\langle\phi_{\mathbf{k}}^-|\beta_2\rangle\lambda_2(t)\int_0^t \langle\beta_2|\hat{G}_1\left(\mathbf{k}^2 - i\gamma(t)\right)\hat{\Gamma}_1(t,t')|\beta_2\rangle X(t')dt' \tag{42}$$

The amplitudes (41) and (42) are calculated at a certain large t value. Their independence of t is a check of the accuracy of asymptotic expressions (39) and (40). If the accuracy is not sufficient it can be increased either by taking a larger t value or by considering more terms in the asymptotic expansion of the states (25) and (26).

5. Numerical Results

As an example of applicability of the described model, we studied some questions connected with time evolution of quasi stationary states. For this sake the time dependence of $\lambda_2(t)$ was chosen in the following form:

$$\lambda_2(t) = \begin{cases} \lambda_2(1 - e^{-t/\tau}), & \text{if } t \le T; \\ \lambda_2(1 - e^{-T/\tau})e^{(T-t)/\tau}, & \text{if } t > T. \end{cases} \tag{43}$$

i.e. $\lambda_2(t)$ exponentially (with characteristic time τ) reaches a certain value λ_2 and preserves it until T, and then again decreases exponentially (for simplicity with the same time constant τ). The value of λ_2 was chosen in such a way, that the Hamiltonian

$$\hat{H} = \hat{H}_0 + \hat{V}_1 + |\beta_2\rangle\lambda_2\langle\beta_2| \tag{44}$$

should have a resonant state at $E_r = E_0 - i\Gamma/2$ and no bound state.

The calculations were performed with s-wave Gaussian form factors

$$\langle\mathbf{k}|\beta_i\rangle = \left(\frac{\beta_i}{\sqrt{\pi}}\right)^{\frac{3}{2}} \exp\left\{-\frac{1}{2}\mathbf{k}^2\beta_i^2\right\}, \qquad i = 1,2 \tag{45}$$

which made possible the analytic (up to complex error function) evaluation of all the necessary matrix elements.

The range β_1 of the stationary potential $\hat{V}_1$ was chosen as the unit of distance ($\beta_1 = 1$), while its strength λ_1 was fixed to give $E_B = -0.49$. For the repulsive potential parameters λ_2, β_2 we took three different sets, which give three different resonance types:

(i) a "broad" one ($\beta_2 = 2$, $\lambda_2 = 1.6$) with $E_0 = 0.84$, $\Gamma = 1.19$,

(ii) a "normal" one ($\beta_2 = 1.8$ $\lambda_2 = 1.6$) with $E_0 = 0.8$, $\Gamma = 0.12$,

(iii) a "narrow" one ($\beta_2 = 1.6$ $\lambda_2 = 1.12$) with $E_0 = 0.63$, $\Gamma = 0.011$.

The chosen time dependence of $\lambda_2(t)$ is characterized by two parameters: the "life–time" T and the "switch–time" τ. We were interested how these parameters effect the physical observables w_B and $w(\mathbf{k})$ in the three cases (i)–(iii).

For a given set of parameters λ_i, β_i our system has two characteristic times. The first one (t_1) is of the order of the inverse of the typical energies of the system $(1/E_B, 1/E_0)$ and it determines the speed of oscillatory changes in the phase of $|\Psi(t)\rangle$. For achieving certain accuracy in numerical integration of (19), the step–length h has to be adjusted to t_1. They are roughly the same for the considered three cases: $t_1 \approx 1$, $h = 0.1$.

The second characteristic time t_2 is connected with the imaginary part of the resonance pole position Γ ($t_2 = 1/\Gamma$) and it characterizes the large T behavior of the system: both w_B and $w(\mathbf{k})$ contain parts which "die out" as $\exp\{-T/t_2\}$.

We performed two sets of calculations for each resonance type:

(a) for fixed τ we gradually increased T;

(b) for fixed T we gradually increased τ.

The (a)-type calculations were intended to demonstrate how the system goes over to the often treated case of large time behavior of an unstable state. In this limit we expect an exponentially decreasing bound state probability (exponential decay) and a Breit–Wigner spectrum of emitted particles with the parameters E_0 and Γ. To see this T-asymptotic regime, we must have $T \gg t_2$ and, therefore, in the "normal" and "narrow" cases we ought to follow the time evolution of the system for quite large times (compared to t_1). To avoid lengthy numerical integrations, for these two cases we used in the $T \to \infty$ studies the "potential jump" limit, which we get from (43) by taking $\tau \to 0$. For this case, our model problem allows an exact solution:

$$|\Psi(t)\rangle = \int d\mathbf{k} |\chi_{\mathbf{k}}^{\pm}\rangle e^{-i\mathbf{k}^2 t} \langle \chi_{\mathbf{k}}^{\pm} | \phi_B \rangle \quad 0 < t \leq T \tag{46}$$

$$a_B = \langle \phi_B | \Psi(T) \rangle, \qquad a(\mathbf{k}) = \langle \phi_{\mathbf{k}}^{-} | \Psi(T) \rangle \tag{47}$$

where the $|\chi_{\mathbf{k}}^{\pm}\rangle$ are the scattering eigenstates of the Hamiltonian (44) and, in this case, form a complete set. The explicit analytic form of the $|\chi_{\mathbf{k}}^{\pm}\rangle$ (see Appendix) enables us to perform the integration involved in evaluating (46) and to calculate the amplitudes (47). We checked the validity of the potential jump approximation for $\tau \neq 0$, and found, that if $T \gg \tau$ is fulfilled, the potential jump and exact amplitudes are practically identical.

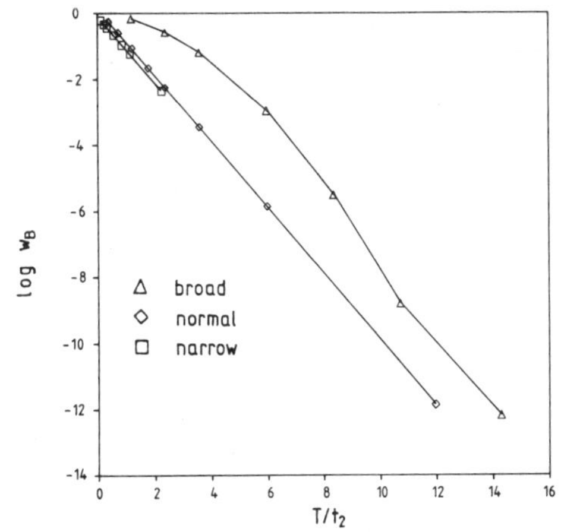

Figure 1.

Bound state probabilities w_B as a function of the relative lifetime T/τ_2 for different resonance types.

The results of the (a)-type calculations are presented in Figures 1–4. Figure 1 shows the logarithm of the bound state probabilities versus T/t_2 for the three resonance types. It can be seen, that in the "broad" case there are deviations from the exponential law.

The spectra are displayed in Figures 2–4. All cases show a certain "shrinking" of the main spectrum peak with growing T. The pure Breit-Wigner limit with the natural line widths Γ is characteristic only for the "normal" and "narrow" cases. It is intersting to note that in the "narrow" case for $T \ll t_2$, one can observe certain oscillations in the spectrum. These can probably be interpreted as effects of the interference between asymptotically stable and decaying, but "still-not-died-out," contributions. Indeed, a closer observation of the potential jump formula for the spectrum $w(\mathbf{k})$ - which is valid in this case - reveals a term in it proportional to

$$\exp(-\Gamma T) \cos \left[(k^2 - E_0)T + \delta\right]$$

which perfectly describes these oscillations.

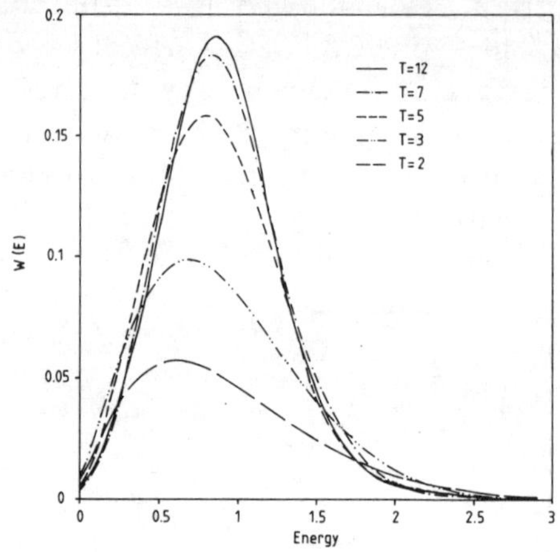

Figure 2.

Particle emission spectra from the "broad" resonance for different life times T.

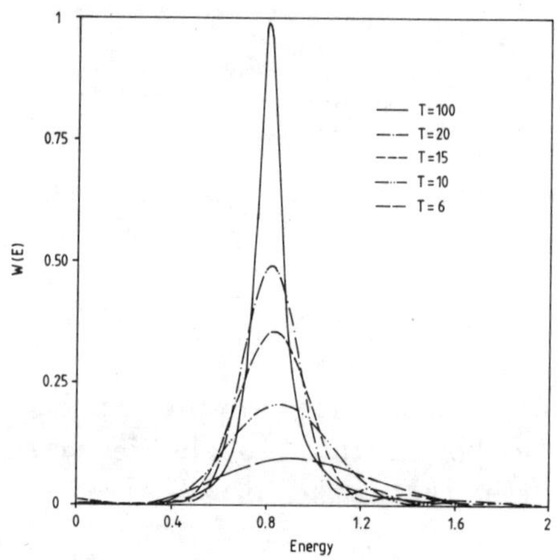

Figure 3.

Particle emission spectra from the "normal" resonance for different life times T.

In the (b)-type calculations we fixed T ($T = 10$) and varied τ ($0 \leq \tau \leq 4$) changing the time dependence of $\lambda_2(t)$ from abrupt to quite smooth. The results for the three cases are seen in Figures 5a,b and c. In all cases the spectra for smoother time dependence were displaced toward smaller energies. This observation might be connected with a kind of Fourier transformation effect: abrupt changes in time correspond to high-energy components in the spectrum

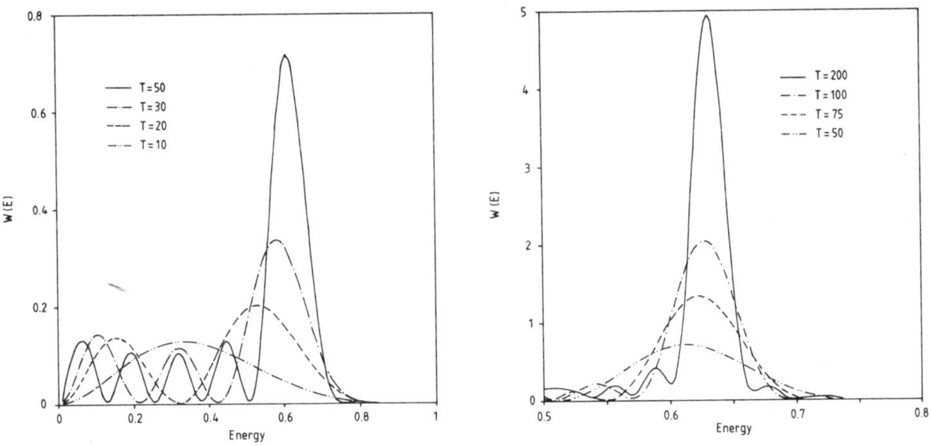

Figure 4.

Particle emission spectra from the "broad" resonance for different life times T. **a)** For $T >> t_2$ **b)** For $T \geq t_2$.

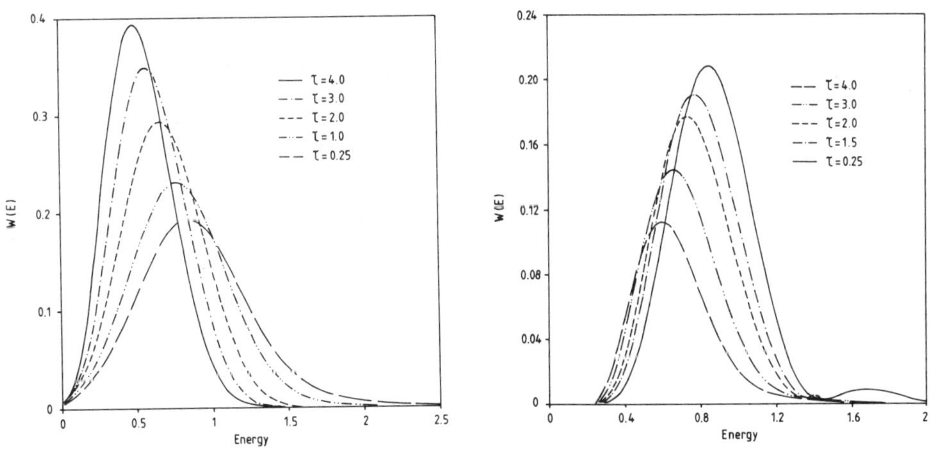

Figure 5 a,b.

Dependence of the emission spectrum from the resonance on the switch time τ for a fixed life time $T = 10$. **a)** For the "broad" resonance. **b)** For the "normal" resonance.

and smoothing in time reduces the weight of this high energy part.

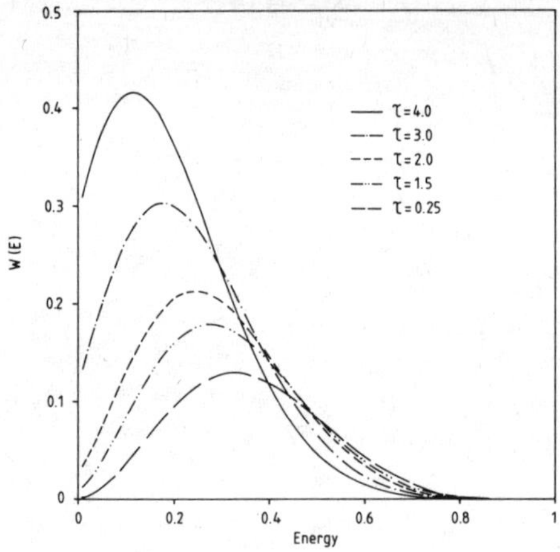

Figure 5c.

Dependence of the emission spectrum from the resonance on the switch time τ for a fixed time $T = 10$ For the "narrow" resonance.

Appendix
Formulas for Stationary Problems With One- and Two–term Separable Potentials

One–term case

$$\hat{H}_1 = \hat{H}_0 + |\beta_1\rangle\lambda_1\langle\beta_1|$$

Bound state:

$$|\phi_B\rangle = \hat{G}_0(E_B)|\beta_1\rangle N, \quad \hat{G}_0(z) = (z - \hat{H}_0)^{-1}, \quad N = \sqrt{\langle\beta_1|\hat{G}_0^2(E_B)|\beta_1\rangle}$$

Eigenvalue equation:

$$D_1(E_B) = 1 - \lambda_1\langle\beta_1|\hat{G}_0(E_B)|\beta_1\rangle = 0, \qquad E_B < 0$$

Scattering states:

$$|\phi_{\mathbf{k}}^{\pm}\rangle = |\mathbf{k}\rangle + \hat{G}_0(\mathbf{k}^2 \pm i\epsilon)|\beta_1\rangle \frac{\lambda_1\langle\beta_1|\mathbf{k}\rangle}{D_1(\mathbf{k}^2 \pm i\epsilon)}$$

Resolvent operator:

$$\hat{G}_1(z) = (z - \hat{H}_1)^{-1} = \hat{G}_0(z) + \frac{\hat{G}_0(z)|\beta_1\rangle\lambda_1\langle\beta_1|\hat{G}_0(z)}{D_1(z)}$$

Completeness and orthogonality:

$$|\phi_B\rangle\langle\phi_B| + \int d\mathbf{k}|\phi_{\mathbf{k}}^{\pm}\rangle\langle\phi_{\mathbf{k}}^{\pm}| = \hat{I}$$

$$\langle\phi_{\mathbf{k}}^{\pm}|\phi_{\mathbf{k}'}^{\pm}\rangle = \delta(\mathbf{k} - \mathbf{k}'), \qquad \langle\phi_B|\phi_{\mathbf{k}}^{\pm}\rangle = 0$$

The zeros of the analytic continuation of $D_1(\mathbf{k}^2)$ into the lower half of the complex k–plane correspond to resonances.

Two–term case

$$\hat{H} = \hat{H}_0 + |\beta_1\rangle\lambda_1\langle\beta_1| + |\beta_2\rangle\lambda_2\langle\beta_2| = \hat{H}_1 + \hat{V}_2 = \hat{H}_2 + \hat{V}_1$$

Let $\lambda_1 < 0$ and$\lambda_2 > 0$, then again only one bound state is possible:

$$\begin{aligned}|\chi_B\rangle &= \hat{G}_0(E_B^{(2)})|\beta_1\rangle N_1 + \hat{G}_0(E_B^{(2)})|\beta_2\rangle N_2 \\ &- \hat{G}_1(E_B^{(2)})|\beta_2\rangle N_2 = \hat{G}_2(E_B^{(2)})|\beta_1\rangle N_1\end{aligned}$$

Eigenvalue equation:

$$D(E_B^{(2)}) = \begin{vmatrix} M_{11}(E_B^{(2)}) & M_{12}(E_B^{(2)}) \\ M_{21}(E_B^{(2)}) & M_{22}(E_B^{(2)}) \end{vmatrix} = 0$$

$$M_{ij}(z) = \delta_{ij} - \lambda_i\langle\beta_i|\hat{G}_0(z)|\beta_j\rangle, \qquad i,j = 1,2$$

$$N_1 = \sqrt{\frac{M_{22}(E_B^{(2)})\lambda_1}{D'(E_B^{(2)})}}, \qquad N_2 = -\sqrt{\frac{M_{11}(E_B^{(2)})\lambda_2}{D'(E_B^{(2)})}}$$

Again, the resonances are given by the zeros of $D(\mathbf{k}^2)$ in the lower–half of the k–plane.

Scattering states:

$$|\chi_{\mathbf{k}}^{\pm}\rangle = |\mathbf{k}\rangle + \hat{G}_0(\mathbf{k}^2 \pm i\epsilon) \sum_{i,j=1}^{2} |\beta_i\rangle \left(M^{-1}(\mathbf{k}^2 \pm i\epsilon)\right)_{ij} \lambda_j \langle\beta_j|\mathbf{k}\rangle$$

$$= |\phi_{\mathbf{k}}^{\pm}\rangle + \hat{G}_1(\mathbf{k}^2 \pm i\epsilon)|\beta_2\rangle \frac{\lambda_2\langle\beta_2|\phi_{\mathbf{k}}^{\pm}\rangle}{1 - \lambda_2\langle\beta_2|\hat{G}_1(\mathbf{k}^2 \pm i\epsilon)|\beta_2\rangle}$$

The resolvent operator:

$$\hat{G}(z) = \hat{G}_0(z) + \hat{G}_0(z) \sum_{i,j=1}^{2} |\beta_i\rangle \left(M^{-1}(z)\right)_{ij} \langle\beta_j|\hat{G}_0(z)$$

$$= \hat{G}_1(z) + \hat{G}_1(z) \frac{|\beta_2\rangle\lambda_2\langle\beta_2|}{1 - \lambda_2\langle\beta_2|\hat{G}_1(z)|\beta_2\rangle} \hat{G}_1(z)$$

TIME-DEPENDENT DYNAMICS APPLIED TO ELECTRON TRANSFER

Erik Deumens and Yngve Öhrn

The Quantum Theory Project
Departments of Chemistry and Physics
University of Florida
Gainesville FL 32611
U.S.A.

Abstract

A time-dependent approach to molecular dynamics involving both nuclei and electrons is presented. Some of our findings from an application of the theory to a two-level model, with the nuclei treated classically, are discussed in some detail as an illustration of the importance of choice of appropriate basis states and of how nuclear-electron coupling is accounted for. The need for an appropriate time-averaging scheme for the fast electron and nuclear modes is demonstrated and such a procedure is outlined.

1. Introduction

Molecular processes as photodissociation [1], non-radiative transitions [2], and gas-surface scattering [3] have recently been studied in terms of the time evolution of Gaussian wave packets [4]. We present an approach to the study of processes involving both electrons and nuclei that employ generalized wave packets. A preliminary study of the time evolution for electron transfer processes applies the theory to a simple two-state model.

For systems of even modest complexity it is not possible to propagate an initial state exactly in time, and one therefore invokes variational principles as those proposed by Dirac [5], Frenkel [6], and McLachlan [7] leading to tractable equations for the evolution of the wave packet parameters. We prefer to start from a quantum action principle known as the time-dependent variational principle (TDVP) [8], which has been preferred in nuclear physics. Both approaches lead to formally equivalent equations for the time evolution of the wave packet parameters [9], but the mathematical structure of the TDVP has been further developed into the form of Hamilton's equations for a classical dynamics on an abstract phase space [8,10].

The applications to molecular systems have primarily dealt with wave packet dynamics on potential energy surfaces. In contrast to this, we have been interested [11] in a time-dependent approach to molecular dynamics involving both electrons and nuclei and formulated to study the interaction between electronic structure and nuclear geometry during various molecular processes.

In this paper we summarize some of our findings [11,12] from an application of the theory to a two-level model, in which the nuclei are treated classically. The model is employed to describe electron transfer and the dynamics is analyzed in terms of 'wave packets' for the electronic wave function. In terms of an 'adiabatic' wave packet the transfer process is described very much like molecular dynamics on a potential energy surface, while a description in terms of a 'diabatic' wave packet contains the interaction of the nuclear motion with the electronic degrees of freedom. When the Born-Oppenheimer approximation (BOA) or the adiabatic approximation (AA) are valid, both descriptions are in complete agreement. When it fails, the simpler dynamics using the 'diabatic' wave packet gives the more accurate description.

In the next section we give a brief overview of the theory, and a time- averaged dynamics is then presented, which is designed to handle the problem of disparate dynamical clocks in the system. The two-state model is then presented with a sample of results illustrating the effects of electron-nuclear coupling.

2.Theory

As stated in the introduction, our approach can be characterized as a generalized wave packet dynamics involving both electrons and nuclei and can best be introduced from the TDVP [8] in terms of a quantum mechanical action functional:

$$F(\alpha^*, \alpha) = \int_{t_1}^{t_2} \langle \Psi(\alpha) | i\partial/\partial t - H | \Psi(\alpha) \rangle dt. \tag{1}$$

In (1) $\Psi(\alpha)$ is a model wave function labeled by a family of (complex) parameters α. The variation of F with respect to α and α^* yields the equations ($\dot{\alpha}$ denotes time derivative) [8,11]:

$$i\dot{\alpha} = \{\alpha, E\}, \textit{ and } i\dot{\alpha}^* = \{\alpha^*, E\} \tag{2}$$

with

$$E(\alpha,\alpha) = \langle\Psi(\alpha)H|\Psi(\alpha)\rangle/\langle\Psi(\alpha)|\Psi(\alpha)\rangle, \tag{3}$$

and with the generalized Poisson bracket

$$\{u,v\} = (\partial u/\partial\alpha, \partial u/\partial\alpha^*)\begin{pmatrix} 0 & -(C^T)^{-1} \\ C^{-1} & 0 \end{pmatrix}\begin{pmatrix} \partial v/\partial\alpha \\ \partial v/\partial\alpha^* \end{pmatrix} \tag{4}$$

In (4) the symplectic form is defined in terms of the hermitian matrix C with elements

$$C_{kl} = \partial^2 ln\langle\Psi(\alpha)|\Psi(\alpha)\rangle/\partial\alpha_k\partial\alpha_1^*. \tag{5}$$

Our approximate wavefunction $\Psi(\alpha)$ can of course be chosen in a variety of ways. Both electrons and nuclei can be treated quantum mechanically. In fact some of the nuclei could be treated classically and some quantum mechanically. The classical treatment of the nuclei can be viewed as the limit of narrow (Gaussian) wave packets [11]. The electrons are treated quantum mechanically with the wave function $|g\rangle$ labeled by the complex parameters g. In our model studies $|g\rangle$ will have a very simple representation, but in principle it can be an SCF state or a correlated wave function [10].

The dynamical equations for the case of classical nuclei are:

$$\begin{aligned} i\sum_\beta C_{\alpha\beta}\dot{g}_\beta^* &= \partial E/\partial g_\alpha \\ -i\sum_\beta C_{\alpha\beta}^*\dot{g}_\beta &= \partial E/\partial g_\alpha^* \\ \dot{P}_{kj} &= -\partial E/\partial R_{kj} \\ \dot{R}_{kj} &= \partial E/\partial P_{kj}\ , \end{aligned} \tag{6}$$

with

$$\begin{aligned} E &= \sum_j \underline{\mathrm{P}}_j^2/2m_j + \sum_{i\neq j} Z_iZ_j/|\underline{\mathrm{R}}_i - \underline{\mathrm{R}}_j| + U(g,g^*;R) \\ C_{\alpha\beta} &= \partial^2 lnS(g,g^*)/\partial g_\alpha\partial g_\beta,\ and\ S(g,g^*) = \langle g|g\rangle\ . \end{aligned} \tag{7}$$

Here the dot denotes derivative with respect to time; $\underline{\mathrm{P}}_j$ is the momentum, $\underline{\mathrm{R}}_j$ the coordinate, and Z_j the charge of nucleus j; U is the electronic energy with respect to a wave function $|g\rangle$ depending on the, in principle complex, parameters g. The parameters in our Hamilton equations are now g, P_{kj}, and R_{kj}

The symplectic structure [8] of our Hamilton's equations (6) is not important here and we express them schematically as

$$\dot{x} = f(x, t), \tag{8}$$

where f is the function giving the time derivative of (the parameters) x at every point x. The dynamical evolution $x(t; t_0, x_0)$ which is the solution of (8) with initial condition $x(t_0) = x_0$, describes the sequence of states the molecule goes through.

We will occasionally, in the following, refer to the parameters describing the electronic wave function as 'electronic dynamic coordinates' (EDC); to the parameters describing the nuclear wave function as 'nuclear dynamic coordinates' (NDC); and to the total set x as 'molecular dynamic coordinates' (MDC).

We have implemented the numerical integration of the first-order coupled equations (1)[25] and some preliminary work on real systems. The model calculations to be discussed have shown that in order to maintain accuracy and stability, i.e. a constant total energy E to within a precision of, say, a microhartree or better, very short time steps (0.001 a.u.) are required and the computing time becomes impractically long even to trace a diatomic potential energy curve (half a vibrational period). This problem obviously derives from the very short time scale of the detailed electronic dynamics and arises as soon as one tries to describe the electronic part of the reaction at a level beyond the Born-Oppenheimer molecular potential surface. These preliminary results show that the TDVP yields a possible, albeit inefficient, description of the interaction of nuclear and electronic degrees of freedom.

Because we believe that a more accurate description of the electronic part of the dynamics would considerably improve our understanding of many molecular processes, as e.g. electron transfer, we want to address the problem of speeding up the numerical integration during most of the process when nothing interesting is happening, without giving up the possibility of having a full description during the short periods, say, where the electronic dynamics strongly influences the process. Our aim is to sufficiently increase the efficiency of the description of the full molecular dynamics to make it practical.

This characteristic that the electronic variables change on a time scale that is much shorter than the time scale for the nuclear variables lies at the basis of molecular dynamics and of the BOA and the AA. It is more efficient to determine the molecular potential energy surfaces, e.g. by optimizing the electronic parameters for every nuclear configuration, and then to follow the dynamics of only the nuclei which occur on a slower time scale, than to follow the full dynamics at the electronic time scale. Such treatment of the electronic component, however, excludes a detailed description of the dynamical interaction between electrons and nuclei, which may be essential at certain stages of a microscopic process.

We want to apply the BO principle to speed up the determination of the dynamics in regions where the BOA and AA are valid, but we want to avoid the determination of the molecular potential energy surface, because that locks in the approximation in a way that is very difficult to undo. We propose to consider the dynamics of both the electronic and nuclear wave functions in a wave-packet formulation, but whenever the electronic state changes adiabatically, i.e. much faster than the nuclei, we only follow a time-averaged state of the system. When the time scales approach each other, we can reduce the averaging time and follow the dynamics in full detail.

3. Average Dynamics

Let τ be a time that is large compared to the electronic time scale, i.e. large compared to electronic vibration periods; but small compared to the times at which the nuclear configuration changes appreciably, i.e. nuclear vibration- rotation periods and molecular reaction times. We define the 'average' or 'approximate dynamics' of the molecule described by (8) as the evolution $X(t;t_0,X_0)$ obtained by solving the equation

$$\dot{X} = F(X,t) = \frac{1}{\tau} \int_{t-\tau/2}^{t+\tau/2} f\left(x(s;t,X)\ ,t\right) ds \ . \tag{9}$$

The right hand side of this equation gives the average force in the average state at time t. It is constructed as follows: For given average molecular dynamical coordinates X at time t the solution $x(s;t,X)$ of (8) with initial condition $x(s = t) = X$ is constructed for times s ranging from $t-\tau/2$ to $t+\tau/2$. Then the forces $f(x,t)$ to which the molecule is subject during this evolution $x(s;t,X)$ are averaged to obtain the force acting on the average coordinate X in (9). Note that when τ goes to 0, (9) becomes (8), which means that in that limit the instantaneous dynamics are recovered from the averaged one.

When we choose τ to be larger than the largest electronic period, but still smaller than any nuclear time period, (9) allows us to skip over the instantaneous changes in the electronic state, taking into account their average effect, but without neglecting their dynamical presence as is done in the BOA and AA. This choice of τ will be possible in reactions, or parts of reactions, where the BOA applies. This is the first assumption for speeding up the solution of (8).

As it stands, (9) still requires the calculation of the fast evolution of all MDC. The next step in our implementation of the BO principle is to introduce a linear approximation to the evolution in (8) during the averaging time τ, i.e. we assume the evolution to be essentially harmonic. Again this will be valid very often when the BOA is. It will, however, pose restrictions on the size of τ

Within the linear approximation, the instantaneous evolution $x(s;t,X)$ is obtained in the following way: We expand the right hand side of (4) in x to first order around X

$$\begin{aligned} \dot{x} &= f(x,t) = f(X,t) + f'(X,t)(x-X) \\ &= a + b(x-X) , \end{aligned} \tag{10}$$

where $a = f(X,t)$ and $b = f'(X,t)$ is the derivative of $f(x,t)$ with respect to x at X. Introducing the variable $y = x - c$, with c a constant vector to be determined,

$$\dot{y} = a + b(y + c - X), \tag{11}$$

if we choose c such that $a + b(c - X) = 0$, i.e. $c = X - b^{-1}a$, this is a linear equation

$$\dot{y} = by , \tag{12}$$

with initial condition $y(s = t) = x(s = t) - c = X - c = b^{-1}a$. The linear system of coupled differential equations (8) can be solved in closed form with an exponential

$$y(s) = b^{-1}a \; e^{b(s-t)} . \tag{13}$$

The exponential involving the matrix b can be defined e.g. by diagonalizing b and exponentiating the eigenvalues. We conclude that, within the harmonic approximation the solution $x(s;t,X)$ needed to take the average is given by

$$x(s;t,X) = b^{-1}a \; (e^{b(s-t)} - 1) + X . \tag{14}$$

With the explicit instantaneous solution (13) it is straightforward to obtain the average force from (9)

$$\begin{aligned} \dot{X} = F(X,t) &= \frac{1}{\tau} \int_{t-\tau/2}^{t+\tau/2} \frac{d}{ds} x(s;t,X) ds \\ &= b^{-1}a \left(e^{b\tau/2} - e^{-b\tau/2} \right) \tau \end{aligned} \tag{15}$$

$$= f'(X,t)^{-1} f(X,t) \left(e^{f'(X,t)\tau/2} - e^{-f'(X,t)\tau/2}\right)/\tau \tag{16}$$

Now we are in a position to solve a differential equation that describes the nuclear as well as the electronic motion; that assumes the time scale of the nuclear motion wherever the BOA approximation holds; and that requires no greater computational effort than the potential-energy-surface-plus-dynamics approach, viz. electronic energies, gradients $f(x,t)$ and second derivatives $f'(x,t)$. Note that, even within the harmonic approximation, we indeed recover (8) when we take the limit $\tau \to 0$

The determination of the optimum τ in each case is still an open question. Several possible approaches are presently under testing.

The harmonic approximation introduced above to accelerate the solution of (9) clarifies the meaning of the averaging procedure. Consider the case where the electronic wave function is a HF state. Then the harmonic approximation, or the linear approximation on the level of differential equations, corresponds as far as the electrons are concerned to linearized time dependent Hartree-Fock or the Random Phase Approximation (RPA) [16]. The instantaneous motion consists of all electronic 'vibrations' the HF state executes by mixing in all excited RPA states. This motion is well understood and not very interesting but because of the often high frequencies involved it slows down numerical integration of (8) tremendously. It is this motion that is averaged out in (9). Note that because of the coupling between nuclei and electrons, the nuclei will also oscillate at these frequencies, albeit with a very small amplitude because of their heavy masses. (Corrections for this fast motion can be built in to the ordinary wave function optimization approach by the Generator Coordinate Approximation [17] to go one step beyond the AA.)

4. Model

The model we employ has the advantages of intuitive simplicity and clarity and is at the same time rich enough to simulate physically interesting situations. Similar models have been introduced in the literature, sometimes in completely different contexts [18], and sometimes with suitable embellishments for studies of certain aspects of electron transfer [19,20].

The Hamiltonian is given as

$$H = T + H_e \tag{17}$$

$$T = \sum_{i=1}^{N}\sum_{k=1}^{d} P_{ki}^2/2m_i \tag{18}$$

$$H_e = \begin{pmatrix} V_\alpha(R) & \varepsilon \\ \varepsilon & V_\beta(R) \end{pmatrix} = V(R)\sigma_0 + W(R)\sigma_z + \varepsilon\sigma_x \tag{19}$$

where R and P are the nuclear coordinates and momenta, respectively, in d dimensional space, $V(R) = (V_\alpha(R)+V_\beta(R))/2$, $W(R) = (V_\alpha(R)-V_\beta(R))/2, \sigma_i(i = x, y, z)$ are the Pauli spin matrices and σ_0 is the unit matrix. The electronic surfaces $V_\lambda(R)$ for $\lambda = \alpha, \beta$ are given the sum of two-body forces to guarantee translational and rotational invariance:

$$V_\lambda(R) = c_\lambda + \sum_{i<j}^{N} \left(k_{\lambda ij}/81^2_{\lambda ij}\right)\left(|\vec{R}_i - \vec{R}_j|^2 - 1^2_{\lambda ij}\right)^2 \tag{20}$$

where $k_{\lambda ij}$ and $l_{\lambda ij}$ are 'effective' force constants and bond lengths, respectively, (derived from a given equilibrium geometry) and c_λ are the equilibrium electronic energies. The choice of quartic potentials will simplify the calculations of matrix elements with nuclear wave packets and also introduce some anharmonicity, but we can of course equally well choose other potentials, e.g., that can account for dissociation processes.

The electronic structure in this model is described as a two-level system. The two basis states $|\alpha\rangle$ and $|\beta\rangle$ correspond to the electronic energies $V_\alpha(R)$ and $V_\beta(R)$, which might cross. Thus $|\alpha\rangle$ and $|\beta\rangle$ behave as diabatic states in this respect. For the purpose of the following illustration we assume $|\alpha\rangle$ and $|\beta\rangle$ as well as ε to be independent of R. This restriction is not very fundamental and can readily be removed should this prove necessary or interesting.

The adiabatic states are obtained by diagonalizing H_e, yielding the eigenvalues and eigenvectors:

$$V_\pm(R) = V(R) \pm [W(R)^2 + \varepsilon^2]^{1/2} \tag{21}$$

$$\begin{aligned} |+\rangle &= \cos\phi|\alpha\rangle + \sin\phi|\beta\rangle \\ |-\rangle &= -\sin\phi|\alpha\rangle + \cos\phi|\beta\rangle \end{aligned} \tag{22}$$

where

$$\tan\phi = \varepsilon/\left(W(R) + [W(R)^2 + \varepsilon^2]^{\frac{1}{2}}\right) \tag{23}$$

We note that the adiabatic states $|+\rangle$ and $|-\rangle$ do depend on R. This gives the model sufficient richness to describe avoided crossings in diatomics and conical intersections in in polyatomic systems.

This model can be viewed as a simple but realistic approximation to *ab-initio* Hamiltonians in the BOA or the AA. If one restricts the *ab-initio* molecular Hamiltonian within two BOA electronic states it has the form of (17) with the functions $V_\lambda(R)$ and $\varepsilon(R)$given by the electronic structure calculation. The quartic surfaces in (20) are qualitative approximations to such ab-initio potential energy surface, at least in the parts of space that excludes dissociative processes. Because each quartic surface has only one stable configuration, they have more the appearance of diabatic surfaces than adiabatic ones. Within the two-level approximation to the ab-initio Hamiltonian, the electronic state is restricted to mixtures of the two basis states and the dynamics of the interaction between the nuclear and electronic degrees of freedom can be expected to be essentially the same as that of the model Hamiltonian in (17).

In order to apply the TDVP we choose the 'wave packet' for the electronic degrees of freedom to be the wave function:

$$|g\rangle \equiv |g_1 g_2\rangle = g_1|1\rangle + g_2|2\rangle \; . \tag{24}$$

For $|1\rangle = |\alpha\rangle$, and $|2\rangle = |\beta\rangle$ we call $|g\rangle$ the diabatic wave packet, and for $|1\rangle = |-\rangle$, and $|2\rangle = |+\rangle$ we denote $|g\rangle$ the adiabatic wave packet. The coefficients g_1 and g_2 are assumed to be complex. The Hamiltonian function for the electronic TDVP equations is:

$$\begin{aligned} U(g,R) = \Big[& H_{11}(R)|g_1|^2 + H_{22}(R)|g_2|^2 \\ & + H_{12}\,(g_1^* g_2 + g_1 g_2^*) / \left(|g_1|^2 + |g_2|^2\right) \end{aligned} \tag{25}$$

with $H_{kl} = \langle k|H_e|1\rangle$, where unlike (7) the nuclear-nuclear repulsion terms are included in U.

The equations for the electron dynamics are:

$$-i\sum_l C_{kl}\dot{g}_1 = \partial U / \partial g_k^* \tag{26}$$

with

$$C_{kl} = \partial^2 ln \left\{ |g_1|^2 + |g_2|^2 \right\} / \partial g_k^* \partial g_1 \; . \tag{27}$$

Explicitly we can write:

$$\begin{pmatrix} g_2^* g_2 & -g_2^* g_1 \\ -g_1^* g_2 & g_1^* g_1 \end{pmatrix} = \begin{pmatrix} \dot{g}_1 \\ \dot{g}_2 \end{pmatrix}$$

$$= i\left[(H_{11}(R) - H_{22}(R))\, g_1 g_2 + H_{12}(R)(g_2^2 - g_1^2)\right] \begin{pmatrix} g_2^* \\ -g_1^* \end{pmatrix} . \tag{28}$$

This Hamiltonian system reduces to the nonlinear ordinary differential equation

$$\dot{x} = i\left[(H_{11}(R) - H_{22}(R))\, x + H_{12}(R)(1 - x^2)\right] \tag{29}$$

for the complex variable $x = g_1/g_2$. This equation is equivalent to the Schrödinger equation for this two-level system because our choice of wave packets cover the entire state space. (The ratio x suffices to describe the dynamics since the norm of the wave packet is irrelevant for the time-evolution.)

An advantage of the TDVP is the ease with which classical and quantum mechanical descriptions can be combined. As a first approximation the nuclei can be treated classically. The molecular dynamics is then generated by the Hamiltonian function $E = T + U$ resulting in the equations of motion (28) for the electronic 'coordinates' g_1, and g_2 coupled to the equations

$$\dot{R}_{ki} = \partial E/\partial P_{ki} , \tag{30}$$

$$\dot{P}_{ki} = -\partial E/\partial R_{ki} , \tag{31}$$

for the nuclear position and momentum coordinates R_{ki} and P_{ki}. In reference 11 we discuss the quantum mechanical treatment of the nuclei within the TDVP framework and we derive the classical dynamics as a limit.

The coupled equations (28), (30), and (31) describe the dynamics of the full molecular system within the TDVP approximation. For the simple model studied here the electronic wave packet ansatz can describe the exact electronic evolution. This is coincidental and will not be the case for a real molecular system, but the TDVP approach applies equally well to an approximate electronic wave function. When, say, a Hartree-Fock ansatz is used for the electronic wave packet within an *ab-initio* or semi-empirical treatment, the TDVP leads to a similar set of coupled differential equations. The difference being a greater number of variables corresponding to the more elaborate description of the system.

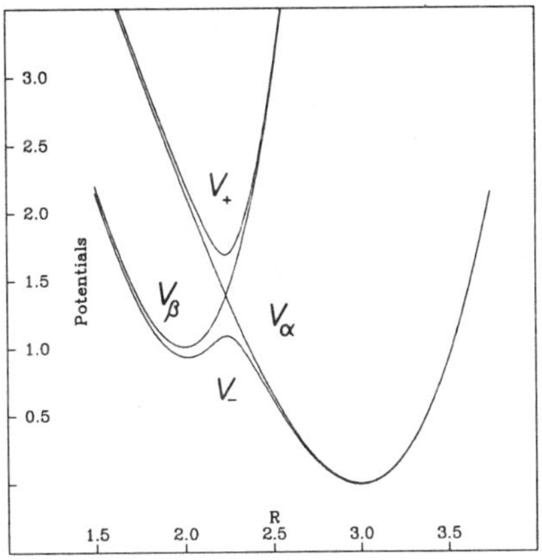

Figure 1

Normal electron transfer in the two-moiety two-level model. The case with large gap 2ε. Diabatic and adiabatic potentials.

5. Transition Probabilities

For illustrative purposes we apply our simple model to a situation of electron transfer, where we consider the (diabatic) states $|\alpha\rangle$ and $|\beta\rangle$ to have different charge densities. We let state $|\alpha\rangle$ correspond to, say, a donor acceptor complex DA and state $|\beta\rangle$ to a charge separated D^+A^- complex. Then the probability to find the system in state $|\beta\rangle$, i.e. the probability of the electron to transfer from D to A, is given by

$$p(t) = |\langle g(t)|\beta\rangle|^2 \tag{32}$$

This probability can be used to follow the electron transfer during the 'reaction.' Using the diabatic wave packet we obtain:

$$p(t) = |g_2|^2 / \left(|g_1|^2 + |g_2|^2\right) = 1/(1 + |x|^2) \tag{33}$$

When p varies from zero to one the electronic state changes from $|1\rangle = |\alpha\rangle$ to $|2\rangle = |\beta\rangle$. For the adiabatic wave packet, however, transfer will have occurred if the electronic state remains $|1\rangle = |-\rangle$ throughout. (See Figure 1 for the potential energy curves V_α, V_β, V_+, and V_-.)

We consider the two moieties D and A interacting through the quartic potential in (24). In Figure 2a we show the time evolution of the electronic energy $U(g, R)$

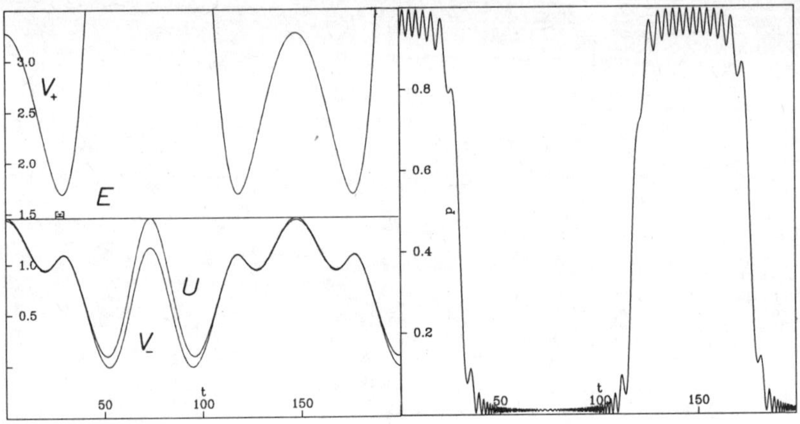

Figure 2

a)Time evolution of total molecular energy E and electronic energy h using the "adiabatic" wave packet. Also shown are the adiabatic energy levels for the nuclear separation $R(t)$. **b)**Time evolution of the probability that the electronic state $|z\rangle$ is state $|\beta\rangle$ for the evolution in **a**).

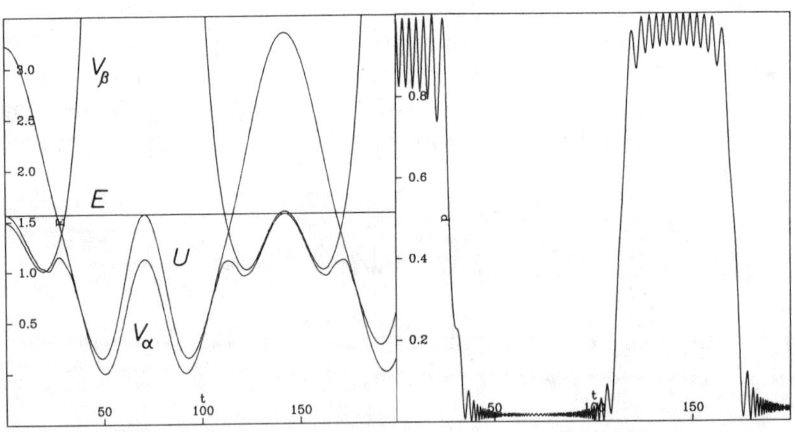

Figure 3

a)Time evolution fo energies as in Fig. 2a but using the"diabatic" wave packet. **b)**Time evolution probability that the electronic state $|z\rangle$ is state $|\beta\rangle$ for the evolution in **a**)

for the adiabatic wave packet. In the same graph we display, for reference, the

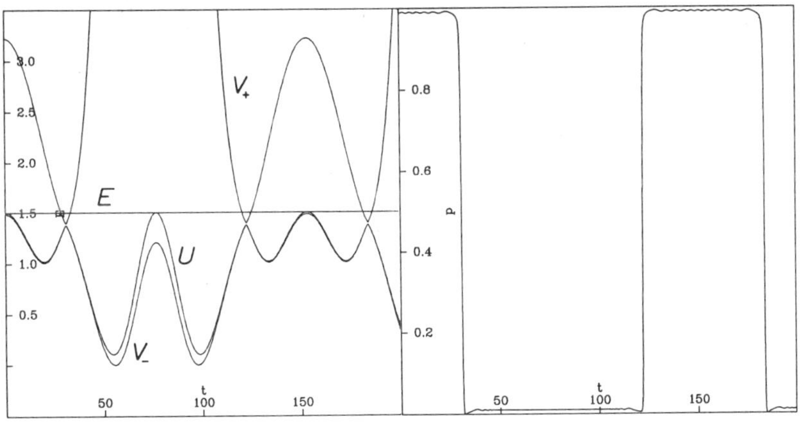

Figure 4

a)Normal electron transfer in the two-moiety two-level model. The case with large gap 2ε. Time evolution of total molecular energy E and electronic energy h using the "diabatic" wave packet. Also shown are the adiabatic energy levels for the nuclear separation $R(t)$. **b)**Time evolution of the probability that the electronic state $|z\rangle$ is state $|\beta\rangle$ for the evolution in **a)**

total energy, which, of course, is constant, and the adiabatic potential energies V_+ and V_- with a substantial splitting 2ε at the avoided crossing. The initial condition has the intermoiety separation 1.7 a.u. and 99% of the adiabatic wave packet in $|-\rangle$. Figure 3a shows the same situation for the diabatic wave packet and, for reference, the total energy and the potential energies V_α and V_β. We note that the time evolution of $U(g, R)$ is practically identical in the two cases. The corresponding transfer probabilities p(t) are displayed in Figures 2b and 3b. They come out almost identical for the diabatic and the adiabatic wave packets, showing the transfer back and forth as expected for a system with one nuclear degree of freedom.

In Figures 4 and 5 we show the same information as in Figures 2 and 3, but now with ε so small that $1/\varepsilon$ corresponds to typical nuclear periods. With the adiabatic wave packet, the probability flips back and forth between 0 and 1 within a very short time whenever the system reaches the avoided crossing. When the diabatic wave packet is used for the same situation no transfer takes place and p(t) stays constant.

This example shows the essence of the BOA. When the adiabatic wave packet is used the effect of the coupling between electronic states is fully taken into account by the basis no matter how slow or fast the nuclear configuration changes. In other

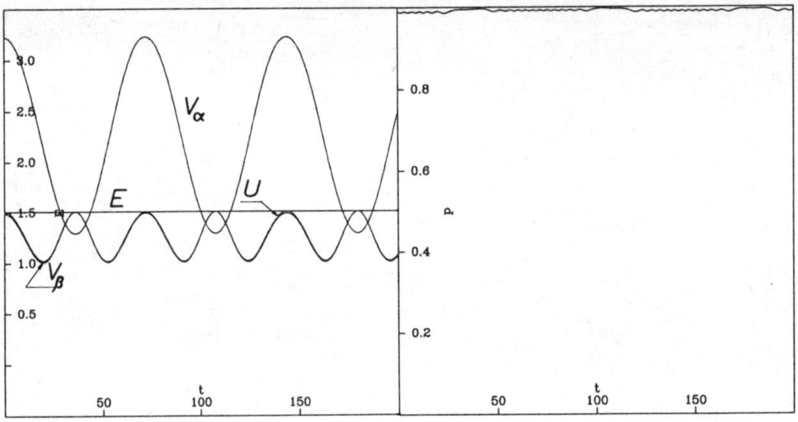

Figure 5

a)Time evolution of energies as in Fig. 4a but using the "diabatic" wave packet **b)**Time evolution of the probability that the electronic state $|z\rangle$ is state $|\beta\rangle$ for the evolution in **a)**.

words, the coupling is fixed at the outset by the assumption that the electronic degrees of freedom change much faster than the nuclear ones, and the dynamical equations from the TDVP have no control over the coupling between electronic states. The resulting dynamics becomes molecular dynamics on one potential surface. This might be an average surface if the initial electronic state contains both adiabatic states. The dynamics is then very much like the 'mean trajectory approximation' [21].

On the other hand, when the diabatic wave packet is used, the coupling between the electronic states appears explicitly in the TDVP equations. The resulting dynamics will therefore reflect the interaction between the electronic and nuclear degrees of freedom. When ε is large, the molecular dynamics using the adiabatic wave packet agrees with the explicitly coupled dynamics using the diabatic wave packet, and the BOA is valid. But, when ε is small, i.e. $1/\varepsilon$ is comparable to nuclear periods, then the adiabatic wave packet forces the transfer of the electron, while (the correct) explicitly coupled dynamics shows that the nuclear motion and the coupling at the crossing are such that no transfer occurs. Thus, we have a breakdown of the BOA.

The use of the diabatic wave packet is to be preferred, since it permits the detailed description of the interaction between nuclear and electron dynamics, say, during an electron transfer process. Calculations with an adiabatic wave packet use fewer time steps as a consequence of the fact that the electronic coupling, responsible for the short electronic cycles when the system is away from the crossing, is absent when the adiabatic basis is used. The extra time steps with the diabatic wave packet are interesting close to the crossing, but they only reconstruct the adiabatic basis when the system is far from the crossing. Which approach is then computationally more efficient depends on the efficiency of constructing the adiabatic basis, since optimization of, say, SCF orbitals requires extra work making each step in the time integration slower. This is where the time averaged dynamics discussed above becomes important in allowing the system, in a manner of speaking, to adjust the basis to its own dynamics.

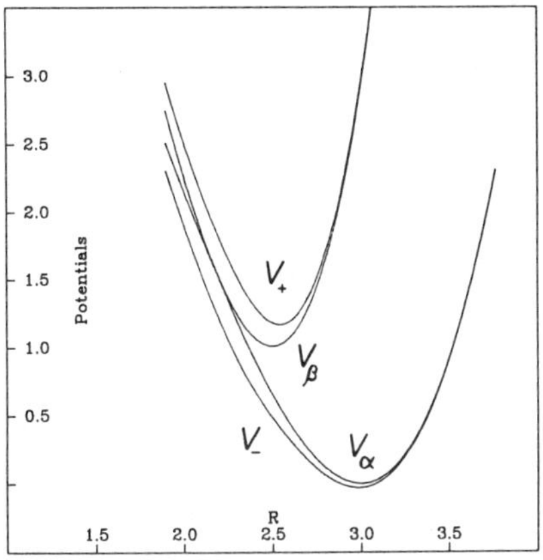

Figure 6

Abnormal electron transfer in the two-moiety two-level model. The case with intermediate gap 2ϵ. Diabatic and adiabatic potentials.

6. Nonadiabatic Effects

In the previous paragraphs, we considered what is known [22] as the 'normal' case for electron transfer. It is generally recognized that in the 'abnormal' case, the transfer is made possible by the existence of nonadiabatic effects. (See Figure 6 for the potentials V_α, V_β, V_- and V_+.) Since our calculations so far have treated the nuclei classically, we cannot describe this case properly. The quantum mechanical description of nuclei in molecular reactions has found a very flexible and powerful tool in wave packet dynamics [23]. The TDVP, as mentioned above, is of course perfectly suited when using wave packets for nuclei. Computationally, Gaussian wave packets are most appropriate.

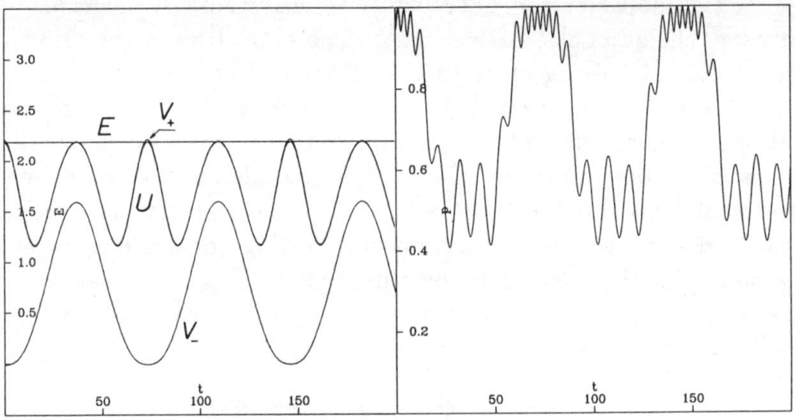

Figure 7

a)Time evolution of total molecular energy E and electronic energy h using the "adiabatic" wave packet and without nonadiabatic corrections. Also shown are the adiabatic energy levels for the nuclear separation $R(t)$. **b)**Time evolution of the probability p that the electronic state $|z\rangle$ is state $|\beta\rangle$ for the evolution in **a**).

We indicate in this section how nonadiabatic couplings are fully included in the TDVP approach. It has been pointed out [24] that the proper treatment of the nuclear motion in the neighborhood of a conical intersection, shows a phenomenon like the Aharanov-Bohm effect. We can therefore expect the Hamiltonian equations from the TDVP to exhibit a vector potential. This can be readily verified by a short calculation using the so-called frozen-wave-packet approximation [23] and using the classical limit for the nuclear wave packet [11]. Then we find in the adiabatic representation some extra terms in the Hamiltonian function E, which can now be written as

$$E = \sum_{i=1}^{N} \sum_{k=1}^{d} \frac{1}{2m_i} (g_1^* g_2^*) \left(P_{ki} \sigma_0 + \frac{\partial \phi}{\partial R_{ki}} \sigma_y \right)^2 \begin{pmatrix} g_1 \\ g_2 \end{pmatrix} + U(g, R) \tag{34}$$

with U and ϕ as given in (18), (25) and (23). The dynamical equations follow from (26) with the Hamiltonian function E from (34) instead of U.

In Figures 7, 8 and 9 we show the results of calculations with an intermediate ε. The initial conditions put the intermoiety distance at 2.9 a.u. and the electronic state is 99% $|+\rangle$, which is mostly $|\beta\rangle$ at that separation. In Figure 7a the time evolution without nonadiabatic corrections is displayed and in Figure 7b we show the corresponding transfer probability. We see that the electron is transferred

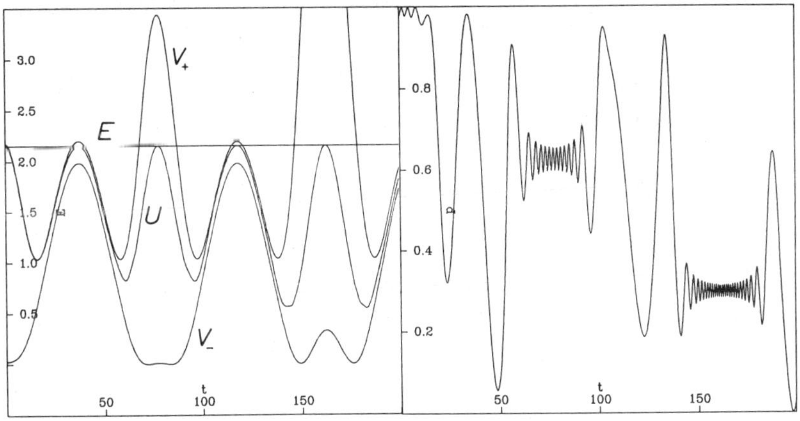

Figure 8

a) Time evolution of energies as in Fig. 7a, but using the "adiabatic" wave packet and with nonadibatic corrections. Also shown are the adiabatic energy levels for the nuclear separation $R(t)$. **b)** Time evolution of the probability p that the electronic state $|z\rangle$ is state $|\beta\rangle$ for the evolution in **a**).

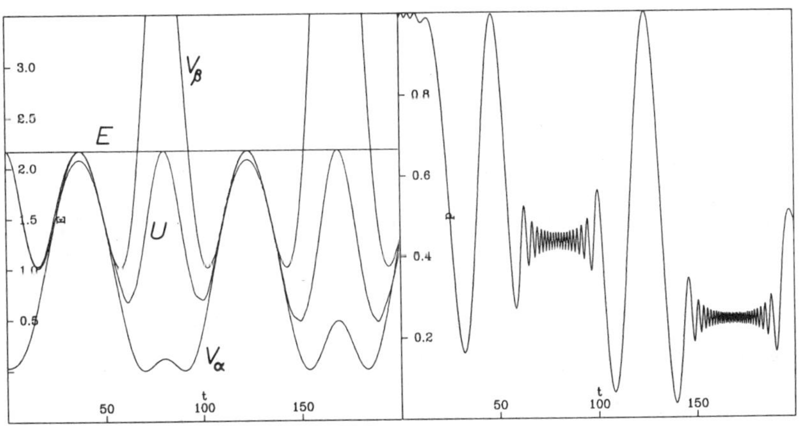

Figure 9

a) Time evolution of energies as in Fig. 7a but using the "diabatic" wave packet. **b)** Time evolution of the probability p that the electronic state $|z\rangle$ is state $|\beta\rangle$ for the evolution in **a**) half way to the $|\alpha\rangle$ state in the region where the two states are close, but no full transfer occurs.

Figure 8a and 8b give the evolution of the energy and the transfer probability, respectively, of the adiabatic wave packet with the nonadiabatic coupling as given by (34). We see that a transfer is now occurring. While the system is in the crossing region, the electronic state changes as slowly as the nuclear configuration from purely $|\beta\rangle$ to purely $|\alpha\rangle$.

In Figure 9a and 9b, we show the same functions, but now with the diabatic wave packet and the Hamiltonian function E = T + U *without the nonadiabatic corrections*, because these do not exist when the electronic Hamiltonian is represented in the diabatic basis. We find the same results as with the adiabatic wave packet *with the nonadiabatic corrections*. This shows that if the TDVP equations are allowed to determine both the electronic and nuclear motion they produce the correct dynamics.

We followed this evolution for a long time, 3000 a.u., (Figure 9b is for 200 a.u.) and found that the electron transfers $|\alpha\rangle$ and then moves back to $|\beta\rangle$ in several successive essentially random jumps. This confirms in our simple model the common notion that the electron transfer does not occur in one step, but in general happens by many passages through the crossing region and finally get trapped on one side.

References

1. S. Y. Lee and E. J. Heller, J. Chem. Phys., **76**, 3035 (1982).

2. H. Köppel and D. H. Meyer, J. Chem. Phys., **81**, 2605 (1984).

3. G. Drolshagen and E. J. Heller, J. Chem. Phys., **71**, 2072 (1983).

4. *Dynamics of Wave Packets in Molecular and Nuclear Physics*, Proceedings, Priorij Corsendonck, Belgium 1985, Eds. J. Broeckhove, L. Latouwers, and P. van Leuven, Lecture Notes in Phys., Springer-Verlag.

5. P. A. M. Dirac, Proc. Cam. Phil. Soc., **26**, 376 (1930).

6. J. Frenkel, *Wave Mechanics*, Oxford University Press, Oxford (1934).

7. A. D. McLachlan, Mol. Phys., **8**, 39 (1964).

8. P. Kramer and M. Saraceno, *Geometry of the Time-Dependent Variational Principle in Quantum Mechanics*, Springer-Verlag, (1981).

9. F. Arickx, J. Broeckhove, E. Kesteloot, L. Lathouwers and P. van Leuven, Chem. Phys. Lett., **128**, 310 (1986).

10. E. Deumens, B. Weiner and Y. Öhrn, Nucl. Phys., **466A**, 85 (1987).

11. E. Deumens and Y. Öhrn, J. Phys. Chem., To appear.

12. E. Deumens, L. Lathouwers and Y. Öhrn, Int. J. Quant. Chem., **S21**, 321 (1987).

13. L. F. Shampine and M. K. Gordon, *Computer solution of ordinary differential equations: the initial value problem*, Freeman, (1975).

14. A. C. Hindmarsh, *GEAR.. Ordinary differential equation system solver*, UCID-30001, Rev 3, Lawrence Livermore Laboratory, (1974); A. C. Hindmarsh, GEARB.. Solution of ordinary differential equations having a banded Jacobian, UCID-30059, Rev 1, Lawrence Livermore Laboratory, (1975).

15. W. H. Press, B. P. Flannery, S. A. Teutolsky and W. T. Vetterling, *Numerical Recipes: The art of scientific computing*, Cambridge, (1986), Section 15.4.

16. J. Linderberg and Y. Öhrn, *Propagators in Quantum Chemistry*, Academic Press, London (1973).

17. L. Lathouwers and P. van Leuven, Adv. Chem. Phys., **49**, 115 (1982).

18. B. De Raedt and H. De Raedt, Phys. Rev., **29B**, 5325 (1984); P. G. Wolynes, J. Chem. Phys., **86**, 1957 (1987); R. E. Cline and P. G. Wolynes, it ibid., **86**, 3836 (1987).

19. A. Garg, J. N. Onuchic and V. Ambegaokar, J. Chem. Phys., **83**, 4491 (1985).

20. R. E. Cline, Jr. and P. G. Wolynes, J. Chem. Phys., **86**, 3836 (1987).

21. S. Sawada, A. Nitzan and H. Metiu, Phys. Rev., **32B**, 851 (1985); A. E. DePristo, Surf. Sci., **137**, 130 (1984).

22. M. D. Newton and N. Sutin, Ann. Rev. Phys. Chem., **35**, 437 (1984).

23. E. J. Heller, J. Chem. Phys., **62**, 1544 (1975); Chem. Phys. Lett., **34**, 321 (1975); J. Chem. Phys., **64**, 63 (1976); J. Chem. Phys., **65**, 4979 (1976); J. Chem. Phys., **68**, 2066 (1978); J. Chem. Phys., **68**, 3891 (1978); J. Chem. Phys., **75**, 2923 (1981).

24. C. A. Mead and D. G. Truhlar, J. Chem. Phys., **70**, 2284 (1979).

25. We use several programs to integrate the coupled system of first order differential equations. The first one uses a modified divided difference form of the Adams-Peace formulas and local extrapolation [13]. The second uses Adams methods as programmed by Hindmarsh [14]. The third is the method of Bulirsch and Stoer [15]. We found that all these methods give essentially the same

results, but the first one seems to be more efficient in minimizing the number of function calls. This is particularly relevant because every function call requires a molecular energy and gradient calculation.

SEMICLASSICAL DESCRIPTION OF RESONANCES

H. Jürgen Korsch

Fachbereich Physik,
Universität Kaiserslautern,
D-6750, Kaiserslautern,
Germany

Abstract

The semiclassical quantization of complex energy resonance poles of the S-matrix is reviewed. The method leads to closed form results, which are valid also for broad shape resonances in the vicinity or above potential barriers. The theory is extended to various Feshbach-type resonances in curve crossing systems. The results are in very good agreement wiht exact quantum computations when available.

1. Introduction

During the last few years a semiclassical method for complex energy quantization has been developed [1−12], which can be viewed as an extension of the well known WKB quantization of bound states to resonances.

There seems to be some confusion concerning the meaning of a "resonance": in the present article the term will be used for states satisfying the Siegert boundary condition, i.e. states with wavefunctions which vanish at the origin and which are purely outgoing at infinity [13]. This condition is only satisfied for discrete complex values of the energy $E_n = E_R - \Gamma/2$. For the one-dimensional single channel case these energies have been proven to be identical with the complex energy poles of the S-matrix.

Due to limited space, we cannot derive the basic semiclassical results in this article and we cannot discuss the delicate subtleties of some semiclassical results, which are, of course, very fascinating for the experts in this field. Instead, we will simply present some of the basic semiclassical formulas in the following section and try to illustrate their applicability to various problems in quantum mechanics, where resonance effects occur.

We further assume that the reader is familiar with the basic idea of semiclassical mechanics, namely the construction of a limit of quantum mechanics in the region where $\hbar$ can be considered as a small quantity. This semiclassical mechanics has developed into a very rich and beautiful theory during the last decades and the interested reader is referred to the literature (for an introduction see the review articles [14–16]).

2. Semiclassical Background

The semiclassical treatment of complex energy resonance states makes extensive use of the diagrammatic method developed by Child [17–18]. The essential idea behind this approach is the construction of semiclassical approximations for complicated situations from the knowledge of simple substructures. Here we can describe a whole variety of resonance problems from only four of the subunits: free propagation, reflection at a turning point, penetration through a barrier and transmission through a crossing of two potential curves. In the following, we will describe these elementary building blocks (see References 8, 11, and 12).

2.1 Free propagation

The well known WKB-wavefunction

$$\Psi(r) = \frac{A'}{\sqrt{p(r)}} e^{\frac{i}{\hbar}\int_a^r p(r)dr} + \frac{A''}{\sqrt{p(r)}} e^{-\frac{i}{\hbar}\int_a^r p(r)dr} \tag{1}$$

with classical momentum

$$p(r) = \sqrt{2m(E - V(r))} \tag{2}$$

is the general solution of the Schrödinger equation in the semiclassical limit $\hbar \to 0$. Equation (1) is valid in a region where no classical turning points r_t with $p(r_t) = 0$ are encountered. The coefficients A' and A'' in (1) are defined with respect to a reference point r = a. Changing the reference point to r = b gives

$$\Psi(r) = \frac{B'}{\sqrt{p(r)}} e^{\frac{i}{\hbar}\int_b^r p(r)dr} + \frac{B''}{\sqrt{p(r)}} e^{-\frac{i}{\hbar}\int_b^r p(r)dr} . \tag{3}$$

The coefficients A', A'' and B', B'' are related by

$$\begin{pmatrix} B' \\ B'' \end{pmatrix} = \begin{pmatrix} e^{i\sigma} & 0 \\ O & e^{-i\sigma} \end{pmatrix} \begin{pmatrix} A' \\ A'' \end{pmatrix} \tag{4}$$

with

$$\sigma = \frac{1}{\hbar} \int_a^b p(r) dr. \tag{5}$$

2.2 Reflection from a single turning point

At a classical turning point the classical momentum (2) vanishes and the WKB wavefunction (1) diverges. Actually, the turning points are branch points of the semiclassical solution (1). It can be shown that different representations of the form (1) are valid in different sectors separated by the so-called Stokes lines originating from the turning point r_t. Along a Stokes line the action integral

$$S(r) = \int_{r_t}^{r} p(r) dr \tag{6}$$

is purely imaginary. Typically there are three Stokes lines emerging from a turning point in complex r-space and therefore three sectors with WKB wavefunctions of type (1). (See Figure 1; note that the turning points r_t are complex for complex E.)

The continuation of a WKB wavefunction from one sector to the other is governed by the principle of exponential dominance [19,20]. The final result is a factor of $-i$ caused by reflection on a turning point:

$$A' = -iA''. \tag{7}$$

The same result can be obtained by matching the problem under consideration onto the standard form of a linear potential, whose solution can be expressed in terms of Airy functions.

2.3 Reflection from a potential barrier

The problem of reflection from a potential barrier is governed by two classical turning points b and c. The semiclassical result in this case depends on the pattern of Stokes lines emerging from these turning points. This pattern is usually structurally stable if the (complex) energy E is slightly changed. The pattern can, however, undergo abrupt structural changes if the energy crosses certain discontinuity lines D_{bc} in the complex E-plane. The line D_{bc} is defined by the

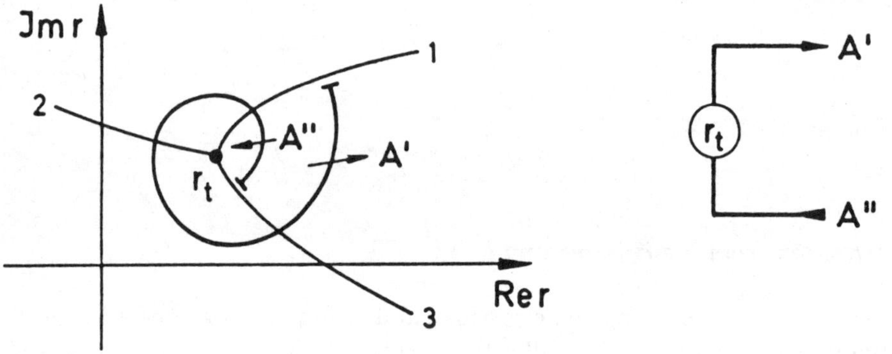

Figure 1.

Reflection from a turning point r_t. The incoming flux A'' can be continued across the Stokes lines 1,2,3 to the outgoing flux A' .

complex energies for which the turning points b and c are linked by a common Stokes line (i.e.$\Re \int_b^c p(r)dr = 0$). For real valued potentials with a single maximum this is the case for real energies. Figure 2 shows the evolution of the pattern of Stokes lines if the energy is varied along a path encircling the energy of the barrier maximum.

By mapping the potential under consideration onto a parabolic barrier (the solutions are parabolic cylinder functions) a connection formula can be derived which connects the semiclassical amplitudes of the incoming and outgoing flux on both sides of the barrier [1-6,8,18,21-23] (here and in the following the primes denote flux to the right ($'$) or to the left ($''$)):

$$\begin{pmatrix} C' \\ C'' \end{pmatrix} = \begin{pmatrix} A^- & -ie^{-\pi\varepsilon} \\ ie^{-\pi\varepsilon} & A^+ \end{pmatrix} \begin{pmatrix} B' \\ B'' \end{pmatrix} \tag{8}$$

with

$$A^{\pm} = \frac{\sqrt{2\pi}}{\Gamma\left(\frac{1}{2} \mp i\varepsilon\right)} \exp\left(-\frac{\pi}{2}\varepsilon \pm i[\varepsilon - \varepsilon\ln(-\varepsilon)]\right), \tag{9}$$

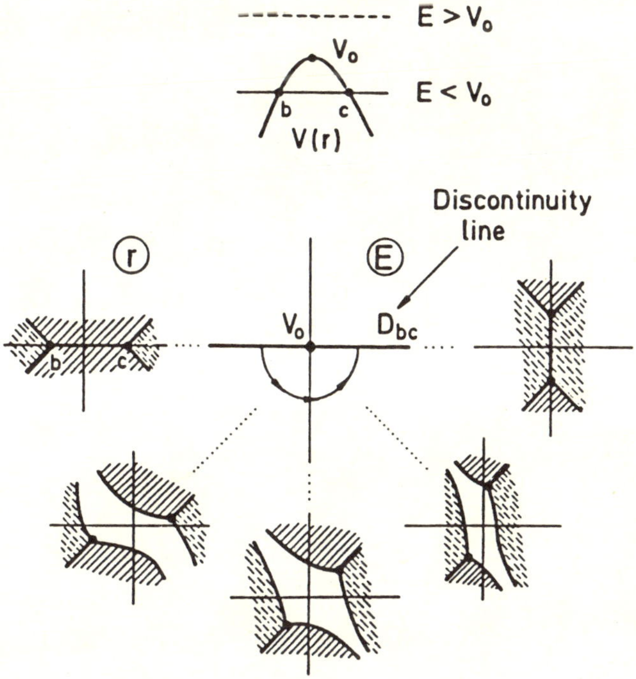

Figure 2.

Evolution of the Stokes line pattern if the energy is varied along a contour around the maximum of the potential.

where ε is the action integral between the turning points:

$$\varepsilon = -\frac{i}{\pi\hbar}\int_b^c p(r)dr. \tag{10}$$

It is important to note that the result in (8) is uniformly valid for all energies E. In particular, Equation (8) shows no discontinuity if the energy crosses a discontinuity line or if it passes through the barrier maximum.

In the limit of an infinitely thick barrier ε goes to $-\infty$ and (8) reduces to

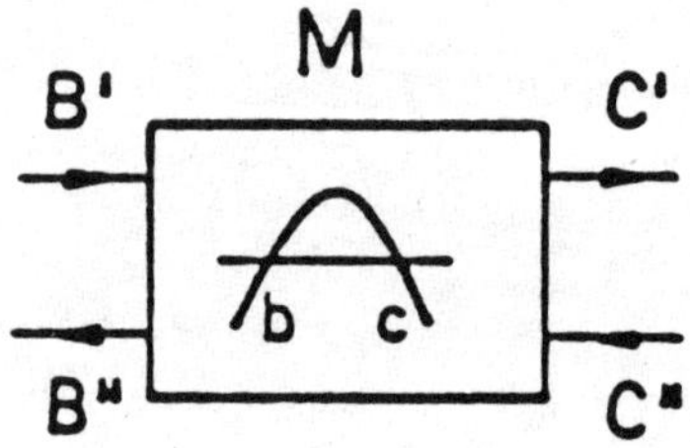

Figure 3.

Connection diagram for barrier reflection.

$$\begin{pmatrix} C' \\ C'' \end{pmatrix} = \exp(-\pi\varepsilon) \begin{pmatrix} 1 & -i \\ i & 1 \end{pmatrix} \begin{pmatrix} B' \\ B'' \end{pmatrix} \tag{11}$$

from which we easily deduce pure reflection on both sides of the barrier as well as the familiar phasefactor $-i$ (see Section 2.2).

2.4 Passage through a curve crossing

For the two-state curve crossing system we are concerned with the solution of a set of two coupled Schrödinger equations

$$\begin{aligned} \left\{ \frac{\hbar^2}{2m} \frac{d^2}{dr^2} + E - V_1(r) \right\} \Psi_1(r) &= V_{12}(r) \Psi_2(r) \\ \left\{ \frac{\hbar^2}{2m} \frac{d^2}{dr^2} + E - V_2(r) \right\} \Psi_2(r) &= V_{21}(r) \Psi_1(r) \end{aligned} \tag{12}$$

with radial coupling $V_{12}(r) = V_{21}(r)$ and diabatic potentials $V_1(r)$ and $V_2(r)$ which cross at a certain crossing point.

Diagonalizing the potential matrix, the adiabatic potentials

$$V^{\pm}(r) = \frac{1}{2} \left[V_1 + V_2 \pm \sqrt{(V_1 - V_2)^2 + 4V_{12}^2} \right] \tag{13}$$

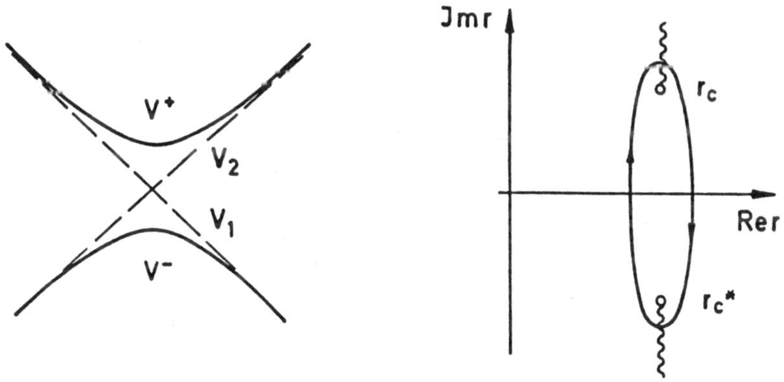

Figure 4.

Diabatic (V_1, V_2) and adiabatic (V^+, V^-) potentials for a two-state curve-crossing system. V^+ and V^- intersect at complex r_c, r_c^* .

are obtained. These potentials show a typical avoided crossing with separation $V^+ - V^- = 2V_{12}$

at the point where the diabatic potentials cross. For the semiclassical description it is essential to observe that the adiabatic potentials possess an actual crossing at complex values r_c and r_c^* , where the square-root in Eq. (13) vanishes.

The amplitudes of the semiclassical flux for passing through a crossing region (see Figure 5) are related by [8,18,24–28].

$$\begin{pmatrix} W'_+ \\ W'_- \end{pmatrix} = \begin{pmatrix} B^- & -e^{-\pi\nu} \\ e^{-\pi\nu} & B^+ \end{pmatrix} \begin{pmatrix} U'_+ \\ U'_- \end{pmatrix} \tag{14}$$

Here the + or - subscripts denote flux on the upper (+) or lower (-) adiabatic potential.

$$B^{\pm} = \frac{\sqrt{-2\pi i\nu}}{\Gamma(1 \pm i\nu)} \exp(-\frac{1}{2}\pi\nu \mp i(\nu - \nu ln\nu)), \tag{15}$$

ν is given by a contour integral

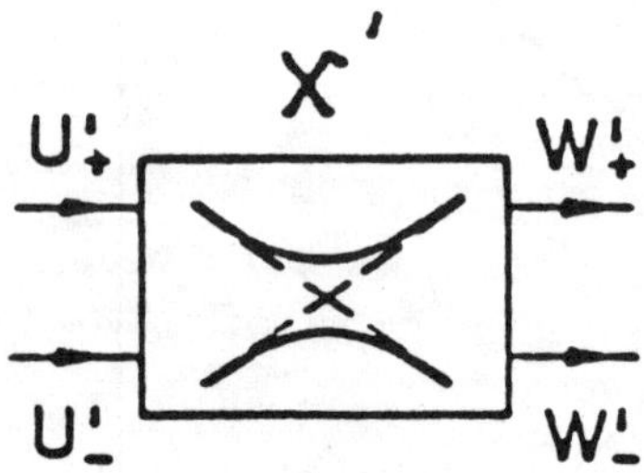

Figure 5.

Connection diagram for curve-crossing.

$$\nu = -\frac{i}{2\pi\hbar} \oint p(r)dr \tag{16}$$

where the contour encircles the complex intersection points r_c and r_c^*. p(r) is the classical momentum on the adiabatic potential surface (13). (It is important to note that the two adiabatic potentials V^+ and V^- are two sheets of a common Riemann surface which are connected at r_c and r_c^*!).

The fluxes from right to left are related by an equation similar to (14):

$$\begin{pmatrix} U''_+ \\ U''_- \end{pmatrix} = \begin{pmatrix} B^- & e^{-\pi\nu} \\ -e^{-\pi\nu} & B^+ \end{pmatrix} \begin{pmatrix} W''_+ \\ W''_- \end{pmatrix}, \tag{17}$$

where again U_+ and U_- denote the amplitudes on the left and W_+ and W_- on the right side of the crossing.

The various elementary substructures ('diagrams') can now be used in combination to describe more complicated situations. Examples will be discussed in the following sections.

3. Shape- And Orbiting-Resonances

One of the most popular resonance systems is the case of a potential with an attractive well and a potential barrier. A typical potential of this type is shown in Figure (6). The resonance arises from a coupling of the oscillatory motion in the well with the tunneling through the barrier. They are termed 'shape resonances' (mainly in spectroscopy) or 'orbiting resonances' (mainly in scattering).

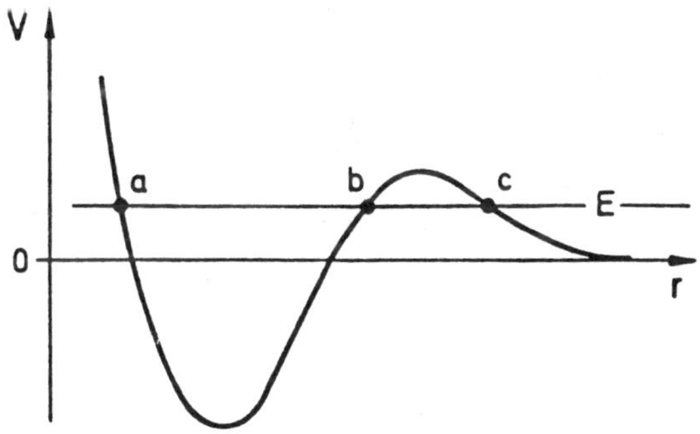

Figure 6.

Typical resonance potential with a well and a barrier, showing three turning points a,b,c.

States with negative values of the energy are bound and the usual WKB bound-state quantization formula applies, i.e. the action integral

$$\alpha = \frac{1}{\hbar}\int_a^b p(r)dr \tag{18}$$

is quantized:

$$N_{WKB}\ (E) = \frac{\alpha}{\pi} = n + 1/2. \tag{19}$$

For positive energies the system shows three turning points a,b,c and the connection of fluxes can be schematically described by the diagram shown in Figure (7). Imposing now purely outgoing (Siegert) boundary conditions, we set $C'' = 0$, i.e.

$$0 = C'' = i\exp(-\pi\varepsilon)B' + A^{+}(\varepsilon)B''. \tag{20}$$

From reflection at the turning point a and free motion between a and b (executed two times!) we set (compare Sections 2.1 and 2.2)

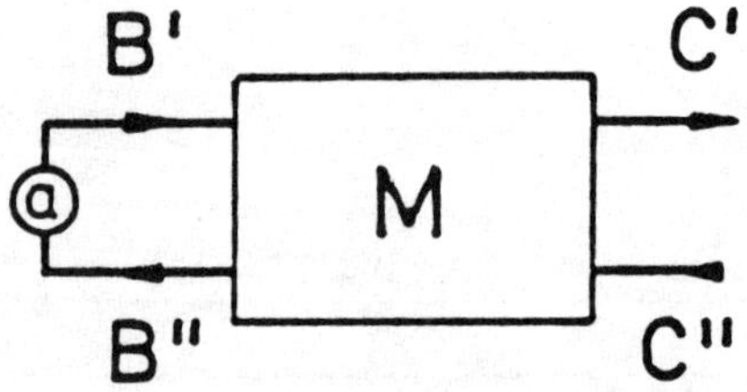

Figure 7.

Connection diagram for potentials shown in Figure 6.

$$B' = -i\exp(2i\alpha)B'', \tag{21}$$

where α is given in (18), and hence

$$O = \Big(\exp(-\pi\varepsilon + 2i\alpha) + A^+(\varepsilon)\Big)\, B'' \tag{22a}$$

or - because B'' is nonvanishing for a nontrivial solution -

$$\exp(-2i\alpha + \pi\varepsilon)A^+(\varepsilon) = -1 \tag{22b}$$

or

$$N_o(E) = \frac{\alpha}{\pi} - \frac{1}{2\pi i}\Big(\pi\varepsilon + lnA^+(\varepsilon)\Big) \tag{23}$$

$$= n + 1/2, \quad n = 0, 1, ...$$

The subscript 'o' of N_o stands for 'orbiting.' This equation determines the complex energy resonances E_n. A few remarks are, however, necessary concerning the validity of (23). The derivation is based on the fact that the system can be considered as a three turning point problem. For complex energy the turning points get complex as well. In addition, there can be more complex turning points. Of all these turning points only the 'active' ones need to be considered, i.e. those which can be joined by a path extending from the origin to infinity without crossing a Stokes line [8,20]. If we now change the complex energy arbitrarily, the turning points move and along with them, their Stokes lines. It may then happen that two turning points, e.g. a and b, are connected by a common Stokes line. In this case the real part of the action integral between a and b vanishes:

$$\Re\, \alpha = \Re \frac{1}{\hbar} \int_a^b p(r) dr = 0 \tag{24}$$

The condition (24) defines a line in complex E-space, the so-called discontinuity line D_{ab}. If the energy crosses this discontinuiy line the Stokes line pattern changes in structure, the turning point b becomes inactive and the complex energy quantization condition (23) reduces to the simpler two-turning point version

$$N_o(E) = \frac{1}{\pi} \int_a^c p(r) dr = n + 1/2. \tag{25}$$

As an example, Figure (8) shows the complex energy resonance poles for the effective potential ($m = \hbar = 1$)

$$V(r) = (\frac{1}{2}r^2 - J) \exp(-\lambda r^2) + \ell(\ell+1)/2r^2 \tag{26}$$

for $\ell = 0$(the semiclassical approximation uses the usual Langer modification $\ell(\ell+1) \to (\ell+1/2)^2$). The exact quantum results computed by the complex rotated Milne method [2,3] are compared with the semiclassical approximation. Good agreement is observed and the results are indistinguishable on the scale of drawing in Figure (8). Also shown in Figure (8) are the two discontinuity lines D_{bc} and D_{ab}, as well as the different pattern of Stokes lines in different regions. For energies between the discontinuity line D_{ab} and the real axis the three-turning point formula (23) has been used, for higher resonances the two point formula (25). Similar agreement has been found for other potentials.

4. Resonances For Purely Repulsive Potentials

Resonances are frequently associated with potentials of the type shown in Figure (6). It should be pointed out, however, that resonance poles are also found for purely repulsive potentials, as for instance the exponential potential

$$V(r) = A \exp(-\alpha r) \ , \ A > 0, \tag{27}$$

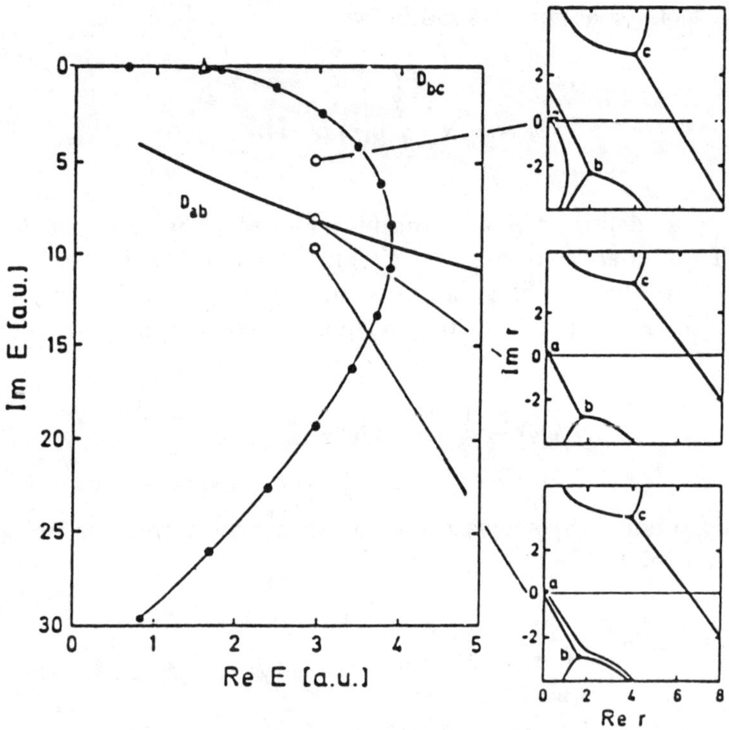

Figure 8.

Complex energy resonances for potential (26) (full circles). The triangle (△) marks the potential barrier. Also shown are the discontinuity lines D_{ab} and D_{bc} and the Stokes line patterns in the different regions.

where the resonance poles for s-wave scattering are given by the zeros of Bessel-functions with imaginary argument and complex index [29]. In addition to these 'true' poles, so called 'false' or 'redundant' poles appear for the case of the exponential potential (27) at

$$k_p = ip\alpha/2, \quad p = 1, 2, 3, \ldots \tag{28}$$

Table I.

The first eight true poles for $2A^{\frac{1}{2}}/\alpha = 3, \alpha = 2/3$. Exact computations [29] are compared with the semiclassical results [10].

Exact		Semiclassical	
$\Re(k)$	$\Im(k)$	$\Re(k)$	$\Im(k)$
0.54011743	-0.7488785	0.5421793	-0.7444103
-0.54011743	-0.7488785	-0.5421793	-0.7444103
0.16914553	-1.2860294	0.1696130	-1.2858070
-0.16914553	-1.2860294	-0.1696130	-1.2858070
0.0	-1.6811980	0.0	-1.6813268
0.0	-1.9989372	0.0	-1.9989236
0.0	-2.3333841	0.0	-2.3333846
0.0	-2.6666648	0.0	-2.6666645

The potential (27) possesses an infinite number of classical turning points - due to the periodicity of the exponential function with period $2\pi i/\alpha$. Semiclassically the contributions from all of these turning points must be summed. This has been achieved recently [10] and a closed form result has been obtained in very good agreement with exact computations. The 'redundant' poles are reproduced exactly by the semiclassical method and for some of the poles exact quantum results [29] are compared with the semiclassical approximation in Table I.

These resonances are, of course, not visible in scattering cross sections because of their large width: resonances lead to observable effects only if their width is smaller than their distance.

5. Curve-Crossing: Feshbach Resonances And Predissociation

Here we will discuss the case of resonances in a typical curve-crossing system (compare Section 2.4), where a repulsive diabatic potential V_1 crosses a potential V_2, which possesses a minimum. Due to coupling between the potential curves an incoming particle can be temporarily trapped in the other channel, which leads to resonance effects. Schematically, this is shown in Figure (9). We have one turning point a^- on the lower potential curve V^- and two turning points a^+ and b^+ on the upper adiabatic potential V^+. The interrelation between the fluxes is shown in the diagram.

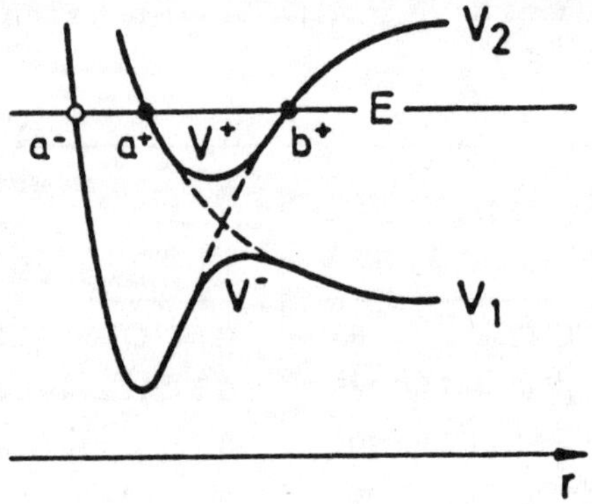

Figure 9.

Potential curves for a predissociating two-state system.

In an application of the connection formulas (14) and (16) one must use a phase-reference point $\tilde{r}$ in the middle of the crossing region. $\tilde{r}$ is conveniently defined by the phase-matching condition [11]

$$-\frac{i}{2\pi\hbar}\int_{r_c^*}^{\tilde{r}}(p_- - p_+)dr = -\frac{i}{2\pi\hbar}\int_{\tilde{r}}^{r_c}(p_- - p_+)dr = \nu/2 \tag{29}$$

where p_+ and p_- are the classical momenta on the upper (+) or lower (-) adiabatic potential.

A derivation of the quantization condition for curve crossing predissociation has been presented in [12] . Here we give a pictorially simpler derivation. Starting with the Siegert boundary condition $W_-'' = 0$ and the arbitrary value of $W_+'' = 1$ we obtain, behind the first passage through the crossing, the amplitudes

$$u = B^-, \qquad v = -e^{-\pi\nu} \tag{30}$$

(compare Figure (10) and Secion 2.4). Reflection at the turning points gives the additional factors

$$d^\pm = -ie^{2i\alpha^\pm} \quad , \quad \alpha^\pm = \frac{1}{\hbar}\int_{a^\pm}^{\tilde{r}} p_\pm(r)dr, \tag{31}$$

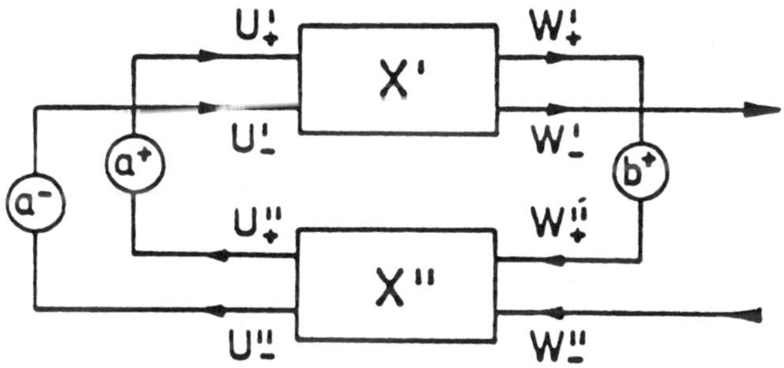

Figure 10.

Connection diagram for the predissociating two-state system.

$$f = -ie^{2i\beta^+} \quad , \quad \beta^+ = \frac{1}{\hbar}\int_{\tilde{r}}^{b^+} p_+(r)dr, \tag{32}$$

and the second passage through the crossing gives

$$zf = (B^- ud^+ - e^{-\pi\nu} vd^-)f = 1, \tag{33}$$

where the last equality is given by the requirement of consistency with the initial choice $W''_+ = 1$.

Equation (33) can be rewritten as

$$\left[(B^-)^2 e^{2i\alpha^+} + e^{-2\pi\nu+2i\alpha^-}\right] e^{2i\beta^+} = -1, \tag{34}$$

or in terms of the quantum number function

$$N_x(E) = \frac{1}{\pi}\left[\beta^+ + \frac{1}{2i} ln[(B^-)^2 e^{2i\alpha^+} + e^{-2\pi\nu+2i\alpha^-}]\right] \tag{35}$$

$$N_x(E_n) = n + 1/2 \tag{36}$$

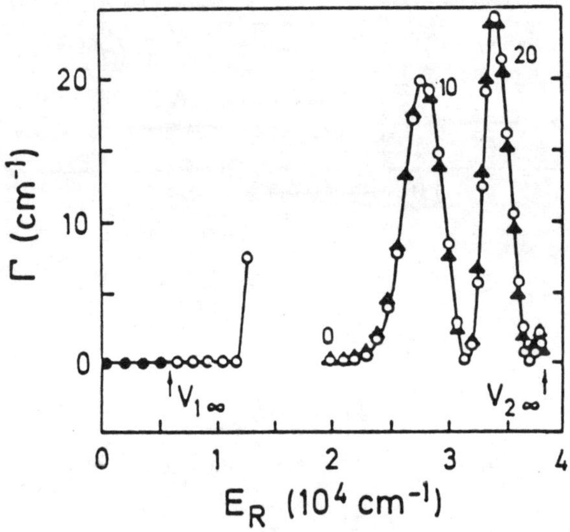

Figure 11.

Resonances for the predissociating MgH model system [12]. Quantum computations for bound (•) and resonance (○) states are compared with semiclassical results (△). The numbers give the adiabatic quantum number.

with integer values of n.

In the diabatic limit $V_{12} \to 0$ we have $\nu \to 0$ and hence

$$N_x(E) \to \frac{1}{\pi}(\alpha^- + \beta^+) = \frac{1}{\pi\hbar}\int_{a^-}^{b^+} p_2(r)dr, \tag{37}$$

i.e. WKB quantization of the diabatic potential V_2. In the adiabatic limit of large V_{12} we have $e^{-2\nu} \to 0, B^- \to 0$ and

$$N_x(E) \to \frac{1}{\pi}(\alpha^+ + \beta^+) = \frac{1}{\pi\hbar}\int_{a^+}^{b^+} p_+(r)dr, \tag{38}$$

i.e. WKB quantization of the upper adiabatic potential V^+.

As an example, Figure (11) shows a comparison of quantum computations [30,31] and semiclassical results obtained from equation (36) for a potential modeling the MgH system [12]. Here semiclassical resonances have only been computed for states above the minimum of the upper adiabatic potential V^+. The results are shown in the (E_R, Γ)-plane $(E = E_R - i\Gamma/2)$, they show the familiar oscillatory behavior of the width [17,18,32–36]. The agreement is good, with the notable exception of the semiclassical resonance n = 7 which is missing in the quantum case.

Because of the ease of numerical computation of semiclassical resonances, one could carry out quite extensive numerical calculations. For example, we have traced all the 36 predissociating states shown in Figure (11) when the interaction strength A_{int} is varied $(V_{12} \sim A_{int})$, i.e. we have calculated so-called correlation diagrams. Let us recall the basic findings (for details see Reference 12):

(i) In the limit $A_{int} \to 0$ the system decouples and we have bound states in the diabatic potential V_2.

(ii) In the limit of large A_{int} the system decouples as well and we have bound states and resonances in the adiabatic potentials V_+ and V_-.

(iii) With increasing A_{int} all the 36 predissociating semiclassical resonances approach the bound states of the upper adiabatic potential V_+ (in particular the width goes to zero). The resonances could hence be classified in terms of the adiabatic quantum numbers $n_a = 0, 1, ...35$.

(iv) In the limit $A_{int} \to 0$ most of the semiclassical resonances approach the bound states of the diabatic potential V_2 and their width goes to zero. There is, however, an unexpexcted exception: The resonance with quantum number $n_a = 18$ behaves differently. For $A_{int} \to 0$ the real part of the energy of resonances with adiabatic quantum numbers $n_a = 17$ and $n_a = 18$ both approach the diabatic bound state energy with diabatic quantum number $n_d = 30$. Because of the nondegeneracy of states, one can conjecture that at most one of these states can converge to zero width, i.e. actually correlate to a diabatic bound state. And indeed the width of state $n_a = 18$ goes to infinity for $A_{int} \to 0$. This behavior is illustrated in Figure (12). Similar observations have been made for other systems. The origin of this behavior is not clear up to now. It has not even been shown that the exact quantum resonances show the same kind of phenomenon.

(v) Some of the resonance states are very unsensitive to the varying interaction strength: The real parts of the energy change very smoothly and the width always remains very small, i.e. these states are long-lived for all interaction strengths. There are other states, however, which are very sensitive to A_{int}: The real parts show a step-like variation and the widths are large, i.e. these states will easily decay. This behavior is clearly related to the oscillatory

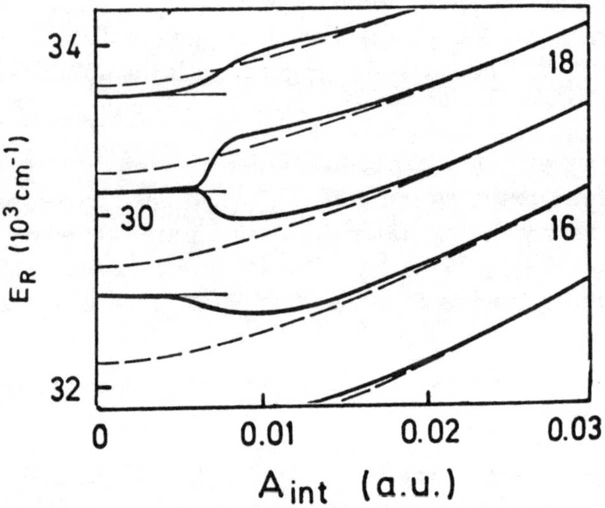

Figure 12a.

Correlation diagram: semiclassical resonance position E_R for varying interaction strength. Shown are states with adiabatic quantum numbers $n_a = 15 - 19$, diabatic bound states of potential V_2 are shown as thin lines for $n_d = 29 - 31$. Bound states of the potential V^+ are shown as dashed lines.

variation of the resonance width with energy shown in Figure (11), but its precise origin is still not completely understood.

(vi) Finally it should be noted that a critical coupling A^c_{int} can be defined: All the states show a transition from diabatic to adiabatic limits in the vicinity of this value of the coupling.

Closing this section, we would like to mention that more complicated two state curve crossing systems have also been studied, as for example, potentials with a potential barrier inside or outside the crossing region (termed internal- or external-orbiting). The semiclassical treatment is based on a combination of barrier penetration diagrams (Section 2.3) and curve crossing diagrams. Applications to the $(NH)^{3+}$ system showed successful agreement with existing quantum computations. Details can be found in Reference 11 .

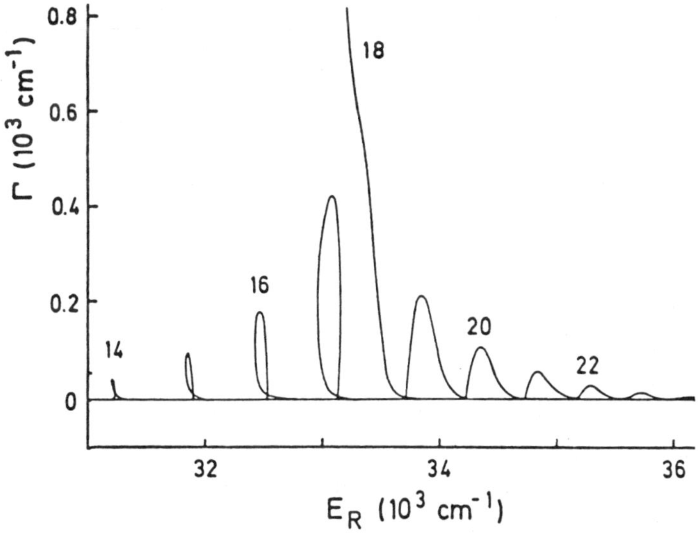

Figure 12b.

Trajectories of the states with adiabatic quantum number $n_d = 14, .., 23$ for varying coupling strength A_{int}.

6. Resonances In Multiple Curve-crossings

The semiclassical treatment of the two-state predissociation resonances outlined in the preceeding section can be generalized to the multiple curve-crossing case. Let us start with a consideration of the three-state case, where an attractive diabatic potential V^- curve is crossed by two repulsive potentials V_1^+ and V_2^+. This case is schematically shown in Figure (13), where also an obvious notation is introduced. The flux-diagram for this case is given in Figure (14), again with an obvious notation: The two crossings are denoted by 1 and 2, respectively, and so on. We introduce the action integrals

$$\alpha^- = \frac{1}{\hbar} \int_{a^-}^{\tilde{r}_1} p_-(r) dr \tag{39}$$

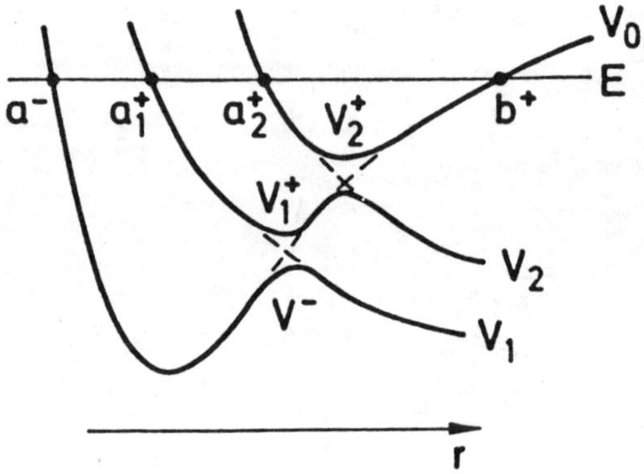

Figure 13.

Three-state curve-crossing system.

$$\alpha_k^+ = \frac{1}{\hbar} \int_{a_k^+}^{\tilde{r}_k} p_{+k}(r)dr, \quad k = 1, 2 \tag{40}$$

$$\beta^+ = \frac{1}{\hbar} \int_{\tilde{r}_2}^{b^+} p_{+2}(r)dr \tag{41}$$

$$p_{+k}(r) = \sqrt{2m(E - V_k^+(r)}, \quad k = 1, 2 \tag{42}$$

$$p_-(r) = \sqrt{2m(E - V^-(r))}, \tag{43}$$

as well as the phase integral ν_1 and ν_2 for contours encircling the crossings 1 or 2 and define the associated quantities $B_1^\pm$ and $B_2^\pm$ (compare Section 2.4). Simplifying the diagram 14 for vanishing incoming flux (Siegert condition) we find the flux-diagram shown in Figure (15).

Now we repeat the procedure outlined for the two-state case in Section 5: We assign the arbitrary value of 1 to W_{2+}'', compute the amplitudes

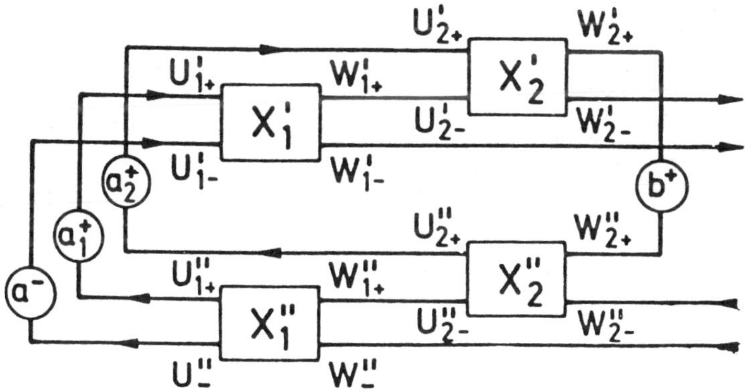

Figure 14.

Connection diagram of the fluxes for a three-state curve-crossing system.

$$u = B_2^-, \tag{44}$$

$$v = -e^{-\pi\nu_2 + i\sigma} \tag{45}$$

as well as

$$d^- = -ie^{2i\alpha^-}, \quad f = -ie^{2i\beta^+} \tag{46}$$

$$d_k^+ = -ie^{2i\alpha_k^+}, \quad k = 1, 2. \tag{47}$$

For the fluxes to the left we obtain

$$w = B_1^-, \tag{48}$$

$$x = -e^{-\pi\nu_1} \tag{49}$$

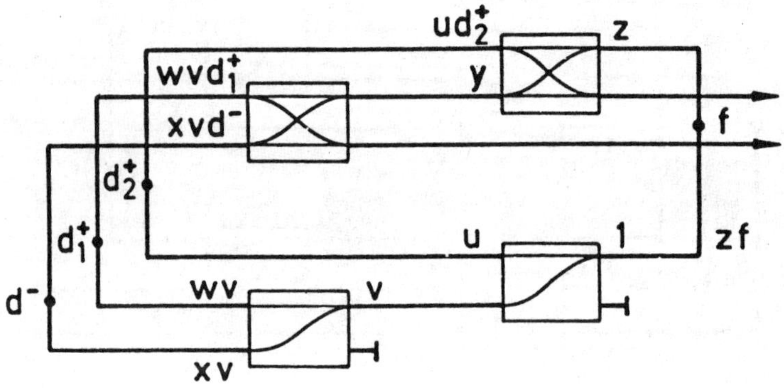

Figure 15.

Simplified diagram 14 for Siegert boundary conditions (no incoming flux).

and for the second passage of the crossings (fluxes to the right) we have

$$y = e^{i\sigma} W'_{1+} \tag{50}$$

$$= e^{i\sigma}(B_1^- w d_1^+ \; - e^{-\pi\nu_1} x d^-) v,$$

with

$$\sigma = \frac{1}{\hbar} \int_{\tilde{r}_1}^{\tilde{r}_2} p_{+1}(r) dr. \tag{51}$$

The factor $e^{i\sigma}$ stems from free propagation on the adiabatic potential V_1^+ between the crossings.

Transmission through the second crossing gives

$$z = B_2^- u d_2^+ - e^{-\pi\nu_2} y \tag{52}$$

and finally we obtain by comparison with the initial assumption $W''_{2+} = 1$ the consistency equation:

$$1 = fz \tag{53}$$

$$= -if\left\{(B_2^-)^2 d_2^+ + e^{-2\pi\nu_2+2i\sigma}\left[(B_1^-)^2 d_1^+ + e^{-2\pi\nu_1} d^-\right]\right\},$$

which can be rewritten as

$$\left[(B_2^-)^2 e^{2i\alpha_2^+} + e^{-2\pi\nu_2+2i\sigma}\left[(B_1^-)^2 e^{2i\alpha_1^+} + e^{-2\pi\nu_1+2i\alpha^-}\right]\right] e^{2i\beta^+} = -1 \tag{54}$$

or - again in terms of a quantum number function -

$$N_{xx}(E_n) = n + 1/2 \tag{55}$$

with

$$N_{xx}(E) = \frac{1}{\pi}\left\{\beta^+ + \frac{1}{2i}\ln\left[(B_2^-)^2 e^{2i\alpha_2^+} + e^{-2\pi\nu_2+2i\sigma}[(B_1^-)^2 e^{2i\alpha_1^+} + e^{-2\pi\nu_1+2i\alpha^-}]\right.\right. . \tag{56}$$

Switching off one of the crossings Equation (56) reduces to the two-state result (35).

This derivation can be easily extended to the case where the attractive diabatic state V_o crosses N repulsive states $V_1, ..., V_N$. An illustration is given in Figure (16). Using the short notations

$$P_k = (B_k^-)^2 e^{2i\alpha_k^+}, \quad k = 1, ...N \tag{57}$$

$$Q_k = e^{-2\pi\nu_k+2i\sigma_k}, \quad k = 1, ...N \tag{58}$$

with

$$\sigma_k = \frac{1}{\hbar}\int_{\tilde{r}_{k-1}}^{\tilde{r}_k} p_{k-1}(r)dr, \quad k \geq 2 \tag{59}$$

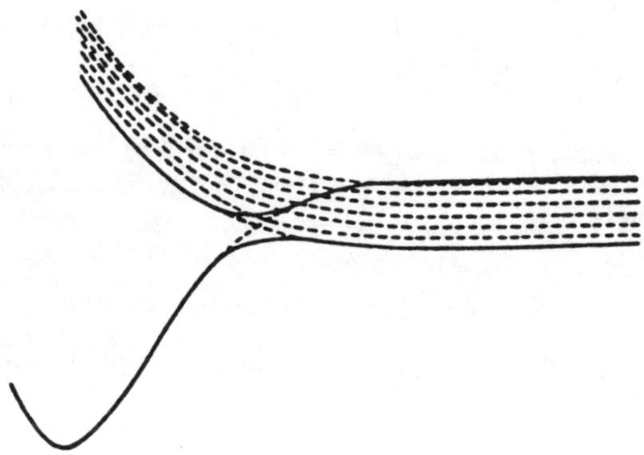

Figure 16.
N-state curve crossing.

and

$$\sigma_1 = \alpha_- \tag{60}$$

we obtain the final result for the quantum number function

$$N_{Nx}(E) = \frac{1}{\pi}\Big\{\beta^+ + \frac{1}{2i} ln\Big[P_N + Q_N[P_{N-1} + \ldots + Q_2[P_1 + Q_1]\ldots]\Big]. \tag{61}$$

The semiclassical complex energy resonances are again obtained from the familiar quantization condition

$$N_{Nx}(E_n) = n + 1/2. \tag{62}$$

7. Numerical Techniques And General Remarks

An application of the semiclassical methods requires only a few numerical techniques. An important requirement is, however, the analyticity of the potentials, and, moreover, in the curve-crossing case an adequate analytical fit, which takes the two-sheeted Riemann structure into account (for fitting techniques of this type see Reference 37). The basic numerical techniques are:

(i) Calculation of the complex turning points. Here the equation $E - V(r_t) = 0$ can be easily solved by complex Newton iteration.

(ii) Eventually one would like to check the Stokes line pattern. A method for following the Stokes lines emerging from a turning point has been developed [38].

(iii) The phase integrals can be easily computed by means of the Gauss-Mehler quadrature [39].

(iv) The solution of the semiclassical quantization condition $N(E_n) = n + 1/2$ can be solved for the E_n by complex Newton iteration. This method generally converges fast for a reasonable initial guess (it is also very fast if the resonances are traced as a function of a varying parameter). This method demands, however, an evaluation of the energy derivative of the quantum-number function N(E). This is straight forward (the energy derivation of the phase integrals can also be computed by Gauss-Mehler quadrature [39]), but admittedly the differentiation by hand is sometimes quite cumbersome. Here an efficient method for finding zeros of a complex function is highly desirable. The alternative methods we have tried so far have shown to be less efficient and reliable than the Newton iteration.

In closing we would like to make some general remarks: The semiclassical complex energy quantization developed recently can overcome most of the deficiencies of previous semiclassical treatments of resonances, like the breakdown of the approximation for shape resonances close to and above the potential barrier. The method can easily be implemented numerically and turned out to be in very good agreement with exact computations in all cases considered so far. (Note, that the methods described in this article are first order semiclassical approximations. Higher order methods are available [40–44], if a higher degree of accuracy is desired.) Some additional advantages of the semiclassical method should be underlined: First, the method allows an unambiguous assignment of a quantum number to a resonance state. This is not easily possible in exact quantum computations. Second, the semiclassical quantization formula of type

$$N(E_n) = n + 1/2 \tag{63}$$

is not restricted to integer values of n (the resonance states E_n, of course, appear <u>only</u> for integer n!), i.e. Equation (63) can be solved for arbitrary real values of n. This defines a (smooth) curve in the complex energy plane, which connects the resonance energies. This is sometimes desirable and not possible in the quantum case.

As a last point, we would like to note that the methods described above could be easily modified to give - instead of complex energy - a complex angular momentum quantization (keeping E fixed and solving $N(L_n) = n + 1/2$ for a complex angular momentum L_n), i.e. for Regge poles, or for a complex potential strength, i.e. a Weinberg resonance pole.

Acknowledgements

The author wishes to thank Dr. Ruth Möhlenkamp and Dr. Karl-Erik Thylwe for their important contributions to the work reviewed in the present article. This work was supported by the Deutsche Forschungsgemeinschaft (Sonderforschungsbereich 91).

References.

1. H. J. Korsch, H. Laurent and R. Möhlenkamp, Mol. Phys., **43**, 1441 (1981).
2. H. J. Korsch, H. Laurent and R. Möhlenkamp, J. Phys. B, **15**, 1 (1982).
3. H. J. Korsch, H. Laurent and R. Möhlenkamp, Phys. Rev., **26A**, 1802 (1982).
4. H. J. Korsch and R. Möhlenkamp, J. Phys. B, **15**, L559 (1982).
5. J. N. L. Connor and A. D. Smith, Chem. Phys. Lett., **88**, 559 (1982).
6. J. N. L. Connor and A. D. Smith, J. Chem. Phys., **78**, 6162 (1983).
7. H. J. Korsch and R. Möhlenkamp, Z. Phys. A, **314**, 267 (1983).
8. H. J. Korsch, *Resonances - Models and Phenomena*, S. Albererio et al. Eds., Lecture Notes in Phys., **211**, Springer-Verlag, Berlin (1984) p. 217-234.
9. R. Lefebvre, Can. J. Chem., **63**, 1635 (1985).
10. H. J. Korsch and K-E. Thylwe, J. Phys. B, **19**, 2139 (1986).
11. H. J. Korsch, R. Möhlenkamp and K-E. Thylwe, J. Phys. B, **19**, 2151 (1986).
12. R. Möhlenkamp and H. J. Korsch, Phys. Rev., **34A**, 4716 (1986).
13. A. F. Siegert, Phys. Rev., **56**, 750 (1939).
14. M. V. Berry and K. E. Mount, Rept. Progr. Phys., **35**, 315 (1972).
15. W. M. Miller, Adv. Chem. Phys., **25**, 65 (1974); Adv. Chem. Phys., **30**, 77 (1975).

16. M. S. Child, *Dynamics of Molecular Collisions*, Part B, W. M. Miller, Ed., Plenum, NY (1976) p. 171-216.

17. M. S. Child, J. Mol. Spectr., **53**, 280 (1974).

18. M. S. Child, *Semiclassical Methods in Molecular Scattering and Spectroscopy*, M. S. Child, Ed., Reidel, Dordrecht (1980) p. 127-154.

19. G. G. Stokes, Trans. Camb. Phil. Soc., **10**, 106 (1857).

20. J. Knoll and R. Schaeffer, Ann. Phys. NY, **97**, 307 (1976).

21. J. N. L. Connor, *Semiclassical Methods in Molecular Scattering and Spectroscopy*, M. S. Child, Ed., Reidel, Dordrecht (1980) p. 45-107.

22. J. N. L. Connor, Mol. Phys., **15**, 621 (1968); Mol. Phys., **23**, 717 (1973); Mol. Phys., **25**, 1469 (1973).

23. M. S. Child, J. Mol. Spect., **53**, 280 (1974).

24. D. S. F. Crothers, Adv. Phys., **20**, 405 (1971).

25. A. Bárány and D. S. F. Crothers, Phys. Scripta, **23**, 1096 (1981).

26. A. D. Bandrauk and M. S. Child, Mol. Phys., **19**, 95 (1970).

27. H. Nakamura, Phys. Rev., **26A**, 3125 (1982).

28. K. S. Lam and T. F. George, *Semiclassical Methods in Molecular Scattering and Spectroscopy*, M. S. Child, Ed. Reidel, Dordrecht (1980) p. 179-261.

29. O. Atabek, R. Lefebvre and M. Jacon, J. Phys. B, **15**, 2689 (1982).

30. E. Brändas, M. Rittby and N. Elander, These Proceedings, E. Brändas and N. Elander, Eds., Lecture Notes in Phys., Springer-Verlag, Berlin (1988).

31. M. Rittby, Phd Thesis, University of Stockholm, Stockholm (1985).

32. A. D. Bandrauk and M. S. Child, Mol. Phys., **19**, 85 (1970).

33. M. S. Child, *Molecular Spectroscopy*, Vol. 2, R. F. Barrow, D. A. Long and D. J. Millen, Eds., Chem. Soc. Spec. Period. Report, (1974).

34. M. S. Child, Can. J. Phys., **53**, 1838 (1975).

35. M. S. Child, Mol. Phys., **32**, 1495 (1976).

36. M. S. Child and R. Lefebvre, Mol. Phys., **34**, 979 (1977).

37. K-E. Thylwe and A. Bárány, J. Phys. B, **19**, L741 (1986).

38. H. J. Korsch, Unpublished.

39. F. J. Smith, Physica, **30**, 497 (1964).

40. N. Fröman, *Semiclassical Methods in Molecular Scattering and Spectroscopy*, M. S. Child, Ed., Reidel, Dordrecht (1980) p. 1-44.

41. P. Pajunen and J. Luppi, J. Chem. Phys., **76**, 4110 (1982).

42. J. Luppi and P. Pajunen, J. Chem. Phys., **76**, 4117 (1982); J. Chem. Phys., **77**, 1505 (1982); J. Chem. Phys., **78**, 4451 (1983), J. Chem. Phys., **81**, 1836 (1984).

43. J. Luppi, Phd Thesis, University of Helsinki (1987).

44. R. Paulsson, F. Karlsson and R. J. LeRoy, J. Chem. Phys., **79**, 4346 (1983).

REGGE POLES
AND
ATOM-MOLECULE DIFFRACTION

Karl-Erik Thylwe

Institute of Theoretical Physics
University of Uppsala
Thunbergsvägen 3
S-752 38 Uppsala
Sweden

Abstract

This paper partly reviews the basic development of non-relativistic Regge-pole theory in atom-molecule potential scattering. Formal aspects of the theory are emphasized, but computationally relevant semiclassical formulas are given. The application part of the paper deals with elastic and rotationally inelastic diffraction effects in sudden collisions of helium atoms by homonuclear nitrogen molecules. In particular, the sensitivity of the diffraction oscillations to variations of anisotropy parameters is investigated.

1. Introduction

In quantal collision theory one uses the notion of partial waves (i.e. waves with discrete angular momentum quantum numbers $\ell = 0, 1, 2, ...$) to compute and interpret the experimentally observed scattering cross sections. For light projectiles, such as electrons, only a small number of partial waves contribute to the scattering process at low energies and the notion is a useful one. It is, for example, used in the description of resonance phenomena observed in electron-atom and electron-molecule collisions. For heavy particles, such as atoms and molecules, a large number of partial waves contribute even at moderately low ($\sim$ eV) energies (often several thousand, seldom less than 50). This makes the partial wave approach a less suitable description of heavy particle collisions except at very low energies (thermal and below). At somewhat higher energies (keV), one can often treat the collisions classically, but the oscillations (rainbows, glories, diffractions) that are so prominent in the scattering differential cross sections at low energies are then lost.

Since the pioneering work of Ford and Wheeler [1] at the end of the fifties, the theoretical analysis of collision problems in the low energy regime has, to a large extent, been dominated by semiclassical considerations. In particular, the notion of interferences between semiclassical trajectories leading to oscillations in the differential cross section has proved extremely fruitful. The theory has been improved by the inclusion of complex trajectories [2,3,4] and extended to inelastic collisions [5,6,7].

Also at the end of the fifties, Regge [8] developed a theory representing the differential cross section in terms of the poles (Regge poles) of the scattering matrix in the complex angular momentum ℓ-plane. The key quantities that appear in the theory are the energy dependent Regge pole positions ℓ_m and their associated residues r_m, where m (= 0,1,2,...) distinguishes between different poles of the scattering matrix. The Regge pole (or complex angular momentum) representation consists of a (hopefully) small number of contributions from poles and a certain background integral that at first was neglected.

The main interest in the early development was in high-energy elementary particle physics [9–12]. During the seventies, this complex angular momentum theory began to find applications in nuclear physics [13–17] and in atomic and molecular physics.

In atomic and molecular scattering, Regge poles were first used as a tool to parametrize the differential cross section with a small number of complex pole parameters [18–22]. Later, potential scattering involving exact quantal [23–25] and semiclassical computations of Regge poles were also introduced [26–28].

At the end of the seventies, a process of incorporating the Regge pole theory into a systematic semiclassical complex angular momentum framework began. An important stride toward that goal was the identification of the background integral as the contribution from a classical trajectory reflected by the repulsive core of the potential [29]. Soon a new flexible CAM formalism was developed [30–33] in which the scattering amplitude is expressed in terms of a series of subamplitudes represented by complex contour integrals. Using the standard method of steepest descent, the subamplitudes reduce to well-known semiclassical trajectory amplitudes. If desired, an evaluation of some of the integrals by the residue theorem introduces a certain pole representation. It turns out that the original Regge pole description and a pure semiclassical trajectory description are only two (sometimes complementary) aspects of the more general CAM formulation.

The great asset of the CAM theory lies in its interpretational power. Differential cross sections can be analyzed in terms of a minimum of significant amplitudes where the series of partial wave amplitudes is slowly convergent or even divergent (see Coulomb scattering in [32]). Applications in different atomic and molecular systems have shown promising results, explaining some interesting features of differential cross sections in terms of interference between the 'background' amplitude and Regge pole (surface) waves propagating around the interaction region. Research has involved: the understanding of large-angle elastic scattering of chemically reactive systems (using parametrized S-matrix models [34–36] and complex optical potentials [33,37,38]), the theory of orbiting collisions [39–44], rainbow scattering [45], and the CAM theory of diffraction scattering in elastic [47]; and recently also rotationally inelastic collisions [48].

When it comes to inelastic processes, both complex angular momentum and semiclassical methods are not fully developed. There are formal generalizations of the CAM theory related to elementary particle physics, dealing with many channel two-body [8,9,67] and three-body collision[10], and scattering on non-local potentials[68]. However, lacking an appropriate analysis of the generalized background integral and the necessary computational tools, these results are not directly applicable to atom-molecule collisions. Here standard collision theories are presently in an expansive phase,and various quantum dynamical approximations has been introduced to simplify computations[69]. This being in part a result of the progress in experimental developments during the past two decades.

With the use of sudden approximations [49], a complex angular momentum theory of rotationally inelastic atom-molecule collisions has recently been developed [48]. Degeneracy averaged differential cross sections, which are in good agreement with coupled-channel calculations and recent measurements of He colliding with N_2, show diffraction oscillations in the elastic and inelastic channels [50]. These oscillations are interpreted as an interference of a (quantal) Regge pole amplitude and a (classical) repulsive trajectory amplitude. The theoretical approach in [48] provides a powerful tool to analyze diffraction oscillations. The present paper reports some new computational results which illustrate the influence of various anisotropy parameters in the $He-N_2$ potential on the oscillatory differential cross sections.

In Section 2 we review the basic relations in the complex angular momentum theory of elastic scattering. Section 3 discusses the diffraction mechanism in the Regge pole picture. We also introduce the so-called infinite-order-sudden approximation, and generalize the CAM formalism to describe rotationally inelastic collisions. Numerical results using an anisotropic potential modelled on the $He-N_2$ system are discussed. Conclusions are in Section 4.

2. Complex Angular Momentum Theory

The present section outlines the formal framework of elastic CAM theory. It should give the not initiated reader a useful theoretical background. Semiclassical formulas for accurate computation of key quantities of the theory are also given here.

2.1 Partial wave expansion

The CAM theory derives from the partial wave analysis by allowing the orbital angular momentum quantum number ℓ to take complex values. The key quantities of interest are the differential cross section $I(\theta)$ and the scattering amplitude $f(\theta)$. They are related through the equation

$$I(\theta) = |f(\theta)|^2 \tag{1}$$

where in the partial wave analysis $f(\theta)$ is given by the Legendre polynomial expansion

$$f(\theta) = ik^{-1} \sum_{\ell=0}^{\infty} (\ell + \frac{1}{2})(1 - S_\ell) P_\ell(\cos\theta). \tag{2}$$

In (2), k is the asymptotic wave number of relative motion and S_ℓ is the (elastic) scattering matrix (S-matrix) element. S_ℓ is determined from the (reduced) radial Schrödinger equation

$$d^2\Psi_\ell/dr^2 + (k^2 - U(r) - \ell(\ell+1)/r^2)\Psi_\ell = 0. \tag{3}$$

The regular solution of (3) (i.e. $\Psi_\ell = 0$, at $r = 0$) can be normalized to satisfy the asymptotic form

$$\Psi_\ell \sim e^{-ikr+i\ell\pi/2} - S_\ell e^{ikr-i\ell\pi/2}, \quad r \to \infty, \tag{4}$$

where the exponentials are the free radial waves of definite angular momentum. If the potential U(r) contains a Coulomb term, (2) and (4) must be modified appropriately [32]. Typically, U(r) is characterized by a dominant repulsion of short-range and a weak long-range (possibly van der Waals) attraction.

An accurate semiclassical approximation in the partial wave analysis employs the first-order WKB single turning point expression for the phase shift δ_ℓ, so that

$$S_\ell = e^{2i\delta_\ell^{WKB}}, \tag{5a}$$

$$\delta_\ell^{WKB} = \lim_{r\to\infty} \left[\int_{a_\ell}^{r} k_\ell(r')dr' - kr \right] + (\ell + \frac{1}{2})\pi/2, \tag{5b}$$

and

$$k_\ell(r) \equiv \left(k^2 - U(r) - (\ell + \frac{1}{2})^2/r^2\right)^{\frac{1}{2}}. \tag{6}$$

The classical turning point a_ℓ satisfies $k_\ell(a_\ell) = 0$ and is the only real and positive solution of this equation. The WKB phase shifts are computed by Gauss-Mehler quadrature [52,37].

2.2 S-matrix properties

A transformation to a complex angular momentum representation of the scattering amplitude $f(\theta)$ requires the analytic continuation of S_ℓ. It is standard to define S_ℓ by analytic continuation of the radial Schrödinger equation (3) and its regular solution satisfying (4), where now ℓ is complex valued.

It is well known that the S-matrix element for Lennard-Jones type potentials

$$U(r) = gr^{-n} - hr^{-p}. \tag{7}$$

with $n > p > 2$, is a meromorfic function of ℓ[11]. This class of potentials is sufficiently large for our purposes. Additional terms may be added in the expression (7) if desired [53].

The continuation of Legendre polynomials $P_\ell(\cos\theta)$ into the CAM plane is also standard [54,55]. The resulting Legendre function of the first kind of complex degree is entirely analytic in the half plane $\Re\ell \geq -\frac{1}{2}$ and the necessary relations for the treatment in Subsection 2.3 will be quoted there without derivation.

Essential for the transformation in subsection 2.3 is the establishment of bounds for the S-matrix in different regions of the complex ℓ plane, especially in the half plane $\Re\ell \geq -\frac{1}{2}$. It is straightforward to construct from the differential equation (3) and its complex conjugate the equality

$$\frac{d}{dr}\left(\Psi_\ell^* \frac{d}{dr}\Psi_\ell - \Psi_\ell \frac{d}{dr}\Psi_\ell^*\right) = 2i\Im\left(\frac{\ell(\ell+1)}{r^2}\right)|\Psi_\ell|^2. \tag{8}$$

With the aid of the regularity of the wave functions involved and their asymptotic forms given by (4) and its complex conjugate, (8) can be integrated. The result is a useful implicit expression for $|S_\ell|$:

$$\begin{aligned} |S_\ell|^2 = {} & \exp(-2\pi\Im\ell) + k^{-1}\exp(-\pi\Im\ell) \\ & \cdot 2(\Re\ell + \frac{1}{2})\Im\ell \int_0^\infty r^{-2}|\Psi_\ell|^2 dr. \end{aligned} \tag{9}$$

Since k is real and positive for scattering states, (9) implies the following behavior in the half plane $\Re\ell \geq -\frac{1}{2}$:

On the real axis

$$|S_\ell| = 1, \qquad \Im\ell = 0, \tag{10}$$

which reflects the unitarity of S_ℓ for real values of ℓ, provided the potential is real (if not, see [31]).

Below the real ℓ axis, $|S_\ell|^2$ is not larger than the first term on the right-hand side of (9), hence

$$|S_\ell| < \exp(-\pi\Im\ell), \quad \Im\ell < 0; \tag{11a}$$

$$|S_\ell| = \exp(-\pi\Im\ell), \quad \Re\ell = -\frac{1}{2}. \tag{11b}$$

Finally, above the real axis we obtain a lower bound for the scattering matrix, namely

$$|S_\ell| > \exp(-\pi\Im\ell), \quad \Im\ell > 0, \tag{12a}$$

$$|S_\ell| = \exp(-\pi\Im\ell), \quad \Re\ell = -\frac{1}{2}. \tag{12b}$$

It is evident from relations (10) - (12) that Regge poles (if any) and no zeros of S_ℓ must be located above the real ℓ axis in the half-plane $\Re\ell > -\frac{1}{2}$ and below the real ℓ axis in the half-plane $\Re\ell < -\frac{1}{2}$. We also see that the absolute value of S_ℓ on the line $\Re\ell = -\frac{1}{2}$ is a simple exponential function of $\Im\ell$, which tends to zero as $\Im\ell \to +\infty$ but diverges as $\Im\ell \to -\infty$. Our observations reflect the fact that S_ℓ satisfies certain symmetry relations and possesses an essential singularity at infinity. It is important, however, to know in more detail the distribution of poles and the asymptotic ($|\ell| \to \infty$) behavior of S_ℓ in regions where poles are not present. Before discussing how this can be done, it is useful to derive symmetry relations satisfied by the S-matrix.

We consider two symmetries of S_ℓ in the complex variable ℓ. The *unitarity symmetry* establishes a unique correspondence between the poles and zeros. It reads

$$S_\ell S^*_{\ell^*} = 1, \tag{13}$$

and derives from the fact that the regular wave functions $\Psi^*_{\ell^*}$ and Ψ_ℓ satisfy the same differential equation; hence they must be proportional to one another. The proportionality factor is determined by comparison of the asymptotic wave functions (see (4) which, by complex conjugation, also provides the expression for $\Psi^*_{\ell^*}$), and the relation (13) then follows. Each pole position ℓ_{pole} corresponds to a location $\ell_{zero} = \ell^*_{pole}$ of a zero.

Another symmetry of S_ℓ, the *reflection symmetry*, is valid only for strongly singular potentials of the type (7). For this class of potentials, Ψ_ℓ and $\Psi_{-\ell-1}$ are both regular solutions satisfying the same differential equation (3) and, therefore, must be proportional. Again, this implies their asymptotic forms and, consequently, S_ℓ and $S_{-\ell-1}$ must be proportional as well. One easily verifies

$$S_{-\ell-1} = \exp\left(-2\pi i(\ell + \frac{1}{2})\right) S_\ell, \tag{14}$$

which is the reflection symmetry. (14) shows that each pole position ℓ_{pole} corresponds to another pole position $\ell'_{pole} = -\ell_{pole} - 1$. Obviously, with the symmetries (13) and (14) available, it is sufficient to study S_ℓ in the first quadrant $\Re\ell \geq -\frac{1}{2}, \Im\ell \geq 0$. Finer details about the behavior of the S-matrix, especially as $|\ell| \to \infty$, can now be obained using reliable asymptotic (semiclassical) formulas [56–58]. Derivation of these formulas involves phase-integral type methods [14,56–59]. They are particularly powerful tools in many branches of elastic scattering, avoiding explicit use of wave functions.

There seem to be serious difficulties, however, in extending these methods to inelastic scattering where S_ℓ becomes a true matrix quantity. In the framework of the infinite-order-sudden approximation [49] such difficulties are effectively eliminated and results from the elastic theory are still meaningful (see Section 3).

A complete phase-integral analysis of S_ℓ pertaining to a potential of the Lennard-Jones type (7) requires a much more involved study of the complex turning points (as a function of ℓ) and of Stokes' and anti-Stokes lines [30,56]. This is beyond the scope of the present paper. Here we merely describe the main results.

It is well documented [53,57,58] that S_ℓ has an infinite number of poles $\ell_m, m = 0, 1, ..., \infty$ lying on a line (pole string) extending to infinity in the first quadrant. Both $\Re\ell_m$ and $\Im\ell_m$ increase without bound while the ratio $(\Im\ell_m/\Re\ell_m)$ tends to zero, as $m \to \infty$.

In the neighborhood of the pole string, two complex turning points influence the behavior of the S-matrix. One obtains the approximate phase-integral formula [56,59]

$$S_\ell = (1+\exp(-2i\gamma))^{-1} \, \exp(2i\bar{\delta}_\ell^{WKB}), \tag{15}$$

where γ, with $\Re\gamma > 0$, is the phase difference between the two complex turning points a_ℓ and e_ℓ (see references 33 and 56)

$$\gamma = \int_{a_\ell}^{e_\ell} k_\ell(r)dr, \quad |a_\ell| < |e_\ell|. \tag{16}$$

In (15) the asymptotic phase $\bar{\delta}_\ell^{WKB}$ is defined in the same way as (5b) except that the outer turning point e_ℓ is used in the phase integral.

The two turning points may be seen as forming a complex well between them in which the colliding particles are temporarily contained. Since the particles are material (probability) waves, represented by the wave function Ψ_ℓ of relative motion, one expects the build up of certain standing waves in the complex well. This is indeed the case. The phenomenon corresponds to the vanishing of the denominator in (15). Hence the scattering matrix has a pole (Regge pole) if the generalized Bohr-Sommerfeld quantization condition

$$\gamma = \left(m + \frac{1}{2}\right)\pi, \quad m = 0, 1, 2, ... \tag{17}$$

is satisfied [60]. The non-negative integer m obviously plays the role of a vibrational quantum number. Unlike particle states quantized in a real potential well, a Regge state with complex angular momentum satisfying (17) always has a purely outgoing radial wave component, which is leaking out from the complex well without tunneling through a barrier.

Numerical determination of Regge pole positions from the condition (17) is accomplished by complex Newton-Raphson iteration. The phase integral γ and its derivative with respect to ℓ are computed by complex Gauss-Mehler quadrature [37,52]. A reasonably accurate initial estimate of the leading Regge pole is required in order to keep track of the turning points a_ℓ and e_ℓ, moving about as functions of ℓ. Fortunately, for the class of potentials (7), an approximate closed-form expression for the leading pole positions has been derived:

$$\ell_m + \frac{1}{2} = \left(\frac{n}{n-2}\right)^{\frac{1}{2}} \left(\frac{n-2}{2}\right)^{\frac{1}{n}} k \left(\frac{g}{k^2}\right)^{\frac{1}{n}} e^{i\frac{\pi}{n}}$$
$$\left(1 + \frac{1}{n}\left(\frac{n-2}{2}\right)^{\frac{n-p}{n}} \frac{h}{g}\left(\frac{g}{k^2}\right)^{\frac{n-p}{n}} e^{-i\frac{p}{n}\pi}\right) + i(n-2)^{\frac{1}{2}}(m + \frac{1}{2}), \tag{18}$$

where k and g are large, h/g is much smaller than unity and m is not too large. The complex harmonic oscillator formula[61] generalized in (18), is retained in the limit $h \to 0$.

Knowing ℓ_m and the two turning point positions, it is then straight forward to calculate the residue r_m of the mth pole from the formula

$$r_m = [2id\gamma/d\ell]^{-1}_{\ell=\ell_m} \exp(2i\bar{\delta}^{WKB}_{\ell_m}), \tag{19}$$

which is obtained by expanding the denominator in (15).

Previous work [24,28,38] has shown that semiclassical calculations of ℓ_m and r_m compare well with accurate quantal results. Table 1 gives numerical values of ℓ_m and r_m for the isotropic $He - N_2$ potential model (see subsection 3.2), using (15) and (19).

Table I.

Regge pole positions and residues for the isotropic model of the $He - N_2$ interaction defined in Section 3.

m	$\Re(\ell_m)$	$\Im(\ell_m)$	$\Re(r_m)$	$\Im(r_m)$
0	26.45	6.37	1.40	2.81
1	26.41	9.42	2.44	-0.55
2	26.72	12.31	0.90	-0.97

Sufficiently far away from the pole string, the S-matrix is determined by the presence of a single turning point and is then given by the simple formula (5). The labeling of the single turning point is somewhat ambiguous. If we depart from the pole region (extending almost vertically upwards) by increasing $\Re\ell$, the term $\exp\{-2i\gamma\}$ in (15) rapidly becomes insignificant, and the dominating turning point should be labeled e_ℓ. On the other hand, by decreasing $\Re\ell$, the same term will soon dominate the denominator, and (15) reduces to (5) (since $\delta^{WKB}_\ell = \bar{\delta}^{WKB}_\ell + \gamma$) with the turning point properly labeled a_ℓ. This seems to contradict the fact that, in our cae, the single-turning-point region is simply connected (see Figure 1). However, turning points may interchange their positions. Thus, if we let ℓ vary arbitrarily <u>outside</u> the pole region we find that the dominating turning point indeed stays the same, indicating that a_ℓ and e_ℓ must be able to change place.

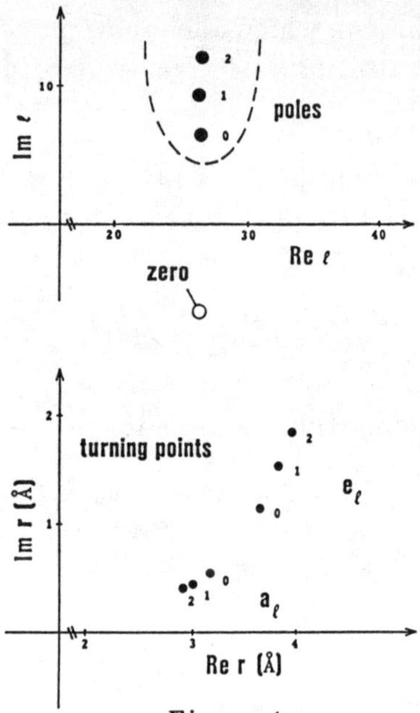

Figure 1.

Calculated positions of Regge poles and the corresponding complex turning points for the isotropic $He - N_2$ potential model in section 3. The dashed curve indicate the pole region where two complex turning points are important to explain the behaviuor of the S-matrix (see eq. 2.15) The numbers labelling the positions indicate the vibrational quantum numbers of the Regge state.

To the right of the pole string, one has found [57,62], using (5), that the S-matrix tends to unity, asymptotically. The phase shift is mainly influenced by the long-range part of the potential (7) and vanishes as

$$\delta_\ell \sim \frac{\sqrt{\pi}}{4}\frac{\Gamma(\frac{p}{2}-\frac{1}{2})}{\Gamma(\frac{p}{2})} h^p k^{p-2}\left(\ell+\frac{1}{2}\right)^{1-p}, \tag{20}$$

in the limit $|\ell| \to \infty$ in this region.

To the left of the pole string, it is sufficient for our purpose to recall (12b), which implies that the scattering matrix vanishes as $|\ell| \to \infty$ close to the line $\Re\ell = -\frac{1}{2}$ $(\Im\ell > 0)$. Here, the short-range part of the potential is important for the phase-shift and S_ℓ behavior.

2.3 Transformation of the scattering amplitude

In order to make use of the Regge poles in calculations and analyses of differen-

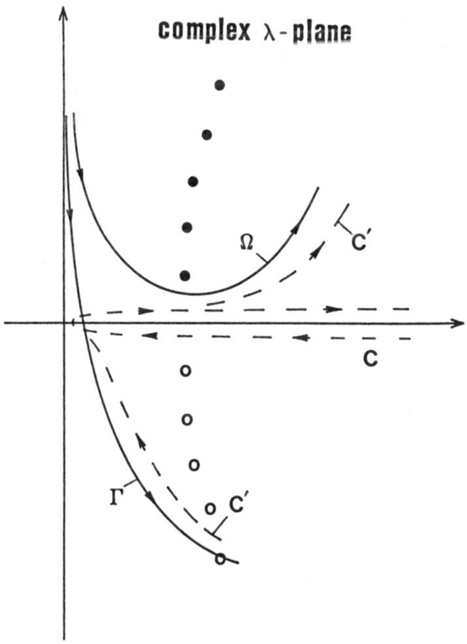

Figure 2

The contours C, C', Γ and Ω in the complex λ plane. Dots and circles indicate the positions of the poles and zeros of $S(\lambda)$, respectively.

tial cross sections, we transform the partial wave expansion (2) of the scattering amplitude. The starting point is the exact Sommerfeld-Watson transform [32,56] (see also an alternative approach in Reference 31)

$$f(\theta) = -\frac{1}{2k}\int_C (S(\lambda)-1)P_{\lambda-\frac{1}{2}}(-\cos\theta)\frac{\lambda d\lambda}{\cos(\pi\lambda)}, \tag{21}$$

using the alternative complex variable $\lambda \equiv \ell + \frac{1}{2}$ and the notation $S(\lambda) \equiv S_\ell$ for the scattering matrix. The contour C encloses the physical quantum numbers $\ell = 0, 1, \ldots$ in the clockwise sense (see Figure 2).

Next we deform the contour C to isolate the contribution from the Regge poles. First the asymptotic ends are moved away from the real λ axis. The lower end approaches infinity so that $\Im\lambda \to -\infty$ to the right of the zeros of $S(\lambda)$ in the fourth quadrant, where $S(\lambda) \sim 1$. A similar shift can be performed in the first quadrant if we ensure that all Regge poles stay to the left of the contour. The resulting contour C' is shown in Figure 2 The convergence of the integral in (21) is guaranteed by the following asymptotic behavior of the entire analytic Legendre function in the half-plane $\Re\lambda \geq 0$, i.e.

$$\begin{gathered} P_\ell(\cos\phi) \equiv P_{\lambda-\frac{1}{2}}(\cos\phi) \\ \sim (2\pi\lambda\sin\phi)^{-\frac{1}{2}}\Big(\exp[i(\lambda\phi-\pi/4)]+\exp[-i(\lambda\phi-\pi/4)]\Big), \\ |\lambda\sin\phi| \to \infty \quad . \end{gathered} \tag{22}$$

Second, we add and subtract a contribution along the positive imaginary axis. According to Figure 2, we then effectively split the integral into two parts; one part is defined on a non-closeable path Γ, and the other part is defined on a contour Ω which essentially surrounds the Regge poles in the first quadrant.

Third, having decomposed the scattering amplitude into two parts, the non-dynamical contribution (i.e. not including $S(\lambda)$) in each integrand can be eliminated. Along Ω, since the integrand has no poles enclosed by the contour, we have

$$\frac{1}{2k}\int_\Omega P_{\lambda-\frac{1}{2}}(-\cos\theta)\frac{\lambda d\lambda}{\cos(\pi\lambda)} \equiv 0.$$

Along Γ, since $P_{\lambda-\frac{1}{2}}(-\cos\theta) = P_{-\lambda-\frac{1}{2}}(-\cos\theta)$ [55], the odd symmetry of the integrand implies

$$-\frac{1}{2k}\int_\Gamma P_{\lambda-\frac{1}{2}}(-\cos\theta)\frac{\lambda d\lambda}{\cos(\pi\lambda)} = \frac{1}{2k}\int_{-i\infty}^{+i\infty} P_{\lambda-\frac{1}{2}}(-\cos\theta)\frac{\lambda d\lambda}{\cos(\pi\lambda)} \equiv 0.$$

Consequently, the scattering amplitude can be written as

$$\begin{aligned} f(\theta) = \frac{1}{2k}\int_\Gamma S(\lambda)P_{\lambda-\frac{1}{2}}(-\cos\theta)\frac{\lambda d\lambda}{\cos(\pi\lambda)} \\ -\frac{1}{2k}\int_\Omega S(\pi)P_{\lambda-\frac{1}{2}}(-\cos\theta)\frac{\lambda d\lambda}{\cos(\pi\lambda)}, \end{aligned} \tag{23}$$

where the first integral along Γ is the traditional background integral $f_B(\theta)$. The evaluation of the second integral in (23), using the residue theorem, gives us now the familiar Regge pole representation [10]. However, with semiclassical considerations in mind, it is preferable to slightly modify the background integral. We use the identity [31,55]

$$P_{\lambda-\frac{1}{2}}(-\cos\theta) = -2i\cos(\pi\lambda)Q^{(-)}_{\lambda-\frac{1}{2}}(\cos\theta) + i\exp(-i\pi\lambda)P_{\lambda-\frac{1}{2}}(\cos\theta), \qquad (24)$$

where the traveling wave (Jost-like) decomposition [31,63] of the Legendre functions

$$P_{\lambda-\frac{1}{2}}(\cos\theta) = Q^{(-)}_{\lambda-\frac{1}{2}}(\cos\theta) + Q^{(+)}_{\lambda-\frac{1}{2}}(\cos\theta),$$

with the asymptotic behaviors (large $|\lambda\sin\theta|$)

$$Q^{(\pm)}_{\lambda-\frac{1}{2}}(\cos\theta) \sim (2\pi\lambda\sin\theta)\exp\left(\pm i(\lambda\theta - \pi/4)\right) \qquad (25)$$

is introduced.

With the aid of (24) the background in (23) is split into two integrals; the second of which is convergent and unchanged also after shifting the path Γ to run along the imaginary axis. When using the fact that the Legendre function and the cosine are even functions of λ, the integral in question may be defined on the positive imaginary axis. We thus have for the background integral

$$f_B(\theta) = -\frac{i}{k}\int_\Gamma S(\lambda)Q^{(-)}_{\lambda-\frac{1}{2}}(\cos\theta)\lambda d\lambda$$
$$-\frac{i}{2k}\int_0^{+i\infty}\left(S(\lambda) - S(-\lambda)\exp(2i\pi\lambda)\right)P_{\lambda-\frac{1}{2}}(\cos\theta)\frac{\exp(i\pi\lambda)}{\cos(\pi\lambda)}\lambda d\lambda \ .$$

However, from the exact symmetry relation (14), satisfied by the S-matrix, we see that the last integral vanishes identically, i.e.

$$f_B(\theta) = -\frac{i}{k}\int_\Gamma S(\lambda)Q^{(-)}_{\lambda-\frac{1}{2}}(\cos\theta)\lambda d\lambda \ . \qquad (26)$$

The exact Regge pole representation of the full scattering amplitude $f(\theta)$ is now obtained by using the residue theorem for the second integral in (23)

$$f(\theta) = f_B(\theta) + f_P(\theta) \ , \qquad (27)$$

where $f_B(\theta)$ is given in (26), equivalent to the first integral in (23), and $f_P(\theta)$ is given by the residue series

$$f_P(\theta) = -\frac{i\pi}{k} \sum_{m=0}^{\infty} r_m \lambda_m \frac{P_{\lambda_m - \frac{1}{2}}(-\cos\theta)}{\cos(\pi\lambda_m)}. \tag{28}$$

In (28), $\lambda_m (= \ell_m + \frac{1}{2})$ is the position of the mth Regge pole and r_m is the corresponding residue.

The semiclassical approximation of $f_P(\theta)$ consists of formulas (17) and (19) of the preceding paragraph and the asymptotic expression (22) for the Legendre function. It is seen in (28) that each Regge pole contributes its characteristic angular distribution given by $|P_{\lambda_m - \frac{1}{2}}(-\cos\theta)|^2$. For a Regge pole situated near the real axis ($\Im\lambda_m << 1$), the angular distribution is oscillatory in the whole range of scattering angles, typical for a resonance. This is in contrast to diffraction scattering where $\Im\lambda_m$ is large (see Table 1) and where a typical Regge pole contribution is exponentially decaying with an increased scattering angle. More precisely, inserting $\phi = \pi - \theta$ into (22), we find

$$|P_{\lambda_m - \frac{1}{2}}(-\cos\theta)|^2 \sim |2\pi\lambda\sin\theta|^{-1} \exp(2(\pi - \theta)\Im\lambda_m), \quad \Im\lambda_m >> 1\,. \tag{29}$$

Such behavior reflects a situation where Regge states are short lived. A Regge state can be visualized as a rotating and vibrating quasi-molecule formed by the colliding constituents. During the time elapsed before decaying the quasi-molecule rotates slightly, with the effect that the escaping projectile has undergone a net negative deflection. One can define from (29) a characteristic angular life (or survival angle) $\Gamma_{\bar{\theta}_m}$ through the relation

$$\Gamma_{\bar{\theta}_m} = (2\Im\lambda_m)^{-1}\,. \tag{30}$$

The larger the imaginary part of the pole position the smaller the survival angle. The rotational period of the quasi-molecule is 2π on the same angular scale.

It should be emphasized that the quasi-molecules (Regge states) discussed here are formed without any support of a binding (attractive) force and they do not have a classical correspondence to a rotating pair of classical particles. Indeed, taking the limit $k \to \infty$, we find $\Gamma_{\bar{\theta}_m} \to 0$ (see (22)).

The background amplitude $f_B(\theta)$, however, has a classical correspondence. Using the asymptotic expressions (5) and (25) in the integrand of (26) and applying the standard method of steepest descent, we obtain

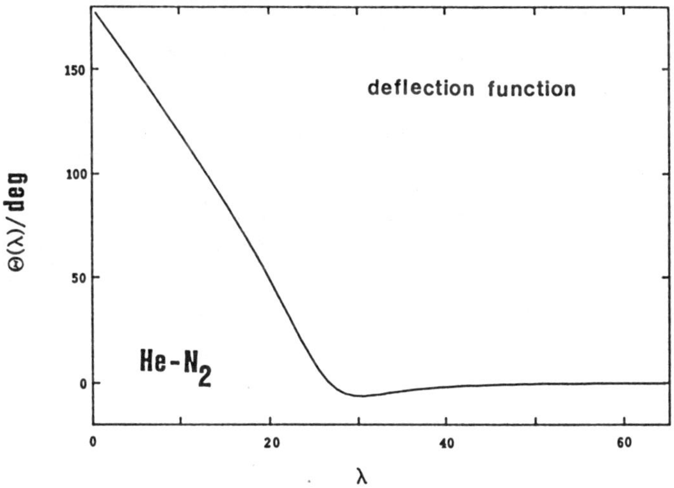

Figure 3

Calculated classical deflection function for the isotropic $He - N_2$ potential model is section 3.

$$f_B(\theta) = -\frac{i}{k}\left[\frac{\lambda_\theta}{\sin\theta|d\Theta(\lambda_\theta)/d\lambda|}\right]^{\frac{1}{2}} \exp\left\{i(2\delta^{WKB}(\lambda_\theta) - \lambda_\theta\theta)\right\} , \tag{31}$$

with the saddle point condition

$$\Theta(\lambda_\theta) = 2\frac{d\delta^{WKB}(\lambda)}{d\lambda}|_{\lambda=\lambda_\theta} = \theta , \tag{32}$$

determining the saddle point λ_θ. $\Theta(\lambda)$, in (31) and (32), is the classical deflection function. For real potentials, λ_θ is real (positive) and the corresponding phase shift is also real. Hence, the squared magnitude of $f_B(\theta)$ is identical to the classical differential cross section due to the classical trajectory scattered from the repulsive potential core. A complete classical analysis shows that this trajectory is the only one contributing to the cross section in the angular range considered.

The attractive force is simply too weak to result in a sufficiently large negative deflection of the projectile (see Figure 3.)

Numerically, the saddle point condition (32) is solved using Newton-Raphson iteration. The integrals $\delta^{WKB}(\lambda)(=\delta_\ell^{WKB})$, $\Theta(\lambda)$, and $d\Theta(\lambda)/d\lambda$ are computed by Gauss-Mehler quadrature [52].

3. Diffraction In $He - N_2$ Scattering

The present section utilizes the Regge pole theory in analyzing differential cross sections of a simple atom-molecule collision. In Subsection 3.1 we briefly discuss the diffraction mechanism in the Regge pole picture. Using the potential of [48], modeled on the $He - N_2$ system, we can then make the analysis more quantitative. The full aniosotropic potential has the Legendre expansion [48]:

$$V(r,\gamma) = V_o(r) + V_2(r)P_2(\cos\gamma) + V_4(r)P_4(\cos\gamma), \tag{33}$$

where

$$V_{0,2}(r) = \frac{\epsilon_i}{n-p}\left(p\left(\frac{r_{*i}}{r}\right)^n - n\left(\frac{r_{*i}}{r}\right)^p\right), \qquad n = 12, \quad p = 6, \quad \epsilon_0 = 1.65\ meV, \quad r_{*o} = 3.9\ \mathring{A}, \tag{34a}$$

$$\epsilon_2 = 0.087\ meV, \quad r_{*2} = 5.0\ \mathring{A} \tag{34b}$$

and

$$V_4(r) = \epsilon_4 e^{-\beta_4(r-r_{*4})}, \tag{35a}$$

$$\epsilon_4 = 63.37\ meV, \quad r_{*4} = 2.7\ \mathring{A}, \quad \beta_4 = 4.04\ \mathring{A}^{-1}. \tag{35b}$$

The angle γ in (33) specifies the orientation of the N_2 molecule relative to the He atom.

Retaining only the spherically symmetric term in the potential, we can directly apply the theory of Section 2. This is dealt with in Subsection 3.2 below. Finally, the full anisotropic potential is considered in the framework of the infinite-order-sudden approximation. The molecule is assumed initially in its rotational ground state.

3.1 Diffraction mechanism

The complex angular momentum theory of Section 2 has revealed that two competing mechanisms are present in the collision, represented by the two amplitudes $f_B(\theta)$ and $f_P(\theta)$. $f_B(\theta)$, the background integral, corresponds to the situation where the particles have undergone a classical collision. That is, the projectile has followed the one possible classical trajectory (with a net positive deflection) as illustrated in Figure 4. It is interesting to note that in Coulomb scattering this mechanism is the only one possible [32]. As a result, the classical and quantal differential cross sections are identical. However, in the present situation, the colliding particles may also form Regge states. Having escaped an intermediate collision complex, a Regge state, the projectile acquires a negative deflection as if a strong attractive force had been present. However, as we have pointed out in Section 2, Regge states may still exist for purely repulsive potentials.

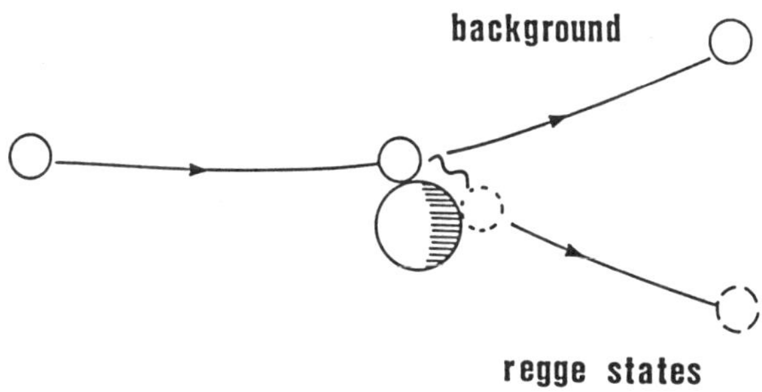

Figure 4

Schematic illustration of the two dynamical mechanisms giving raise to the diffraction interference.

The diffraction oscillations in the differential cross section are produced by quantum mechanical interference of the two amplitudes $f_B(\theta)$ and $f_P(\theta)$ (see Figure 4). Diffraction phenomena in material particle scattering are non-classical. They are directly associated with the probability wave nature of quantal particles.

3.2 Isotropic potential model

The model consists of the centrally symmetric part $V_o(r)$ of the potential given in (33). In Section 2, Figure 3 shows the classical deflection function $\Theta(\lambda)$ for that potential and Table 1 gives the positions and residues of the three leading Regge poles. Now we are interested in the elastic differential cross section, which for the isotropic potential model is analyzed in Figure 5.

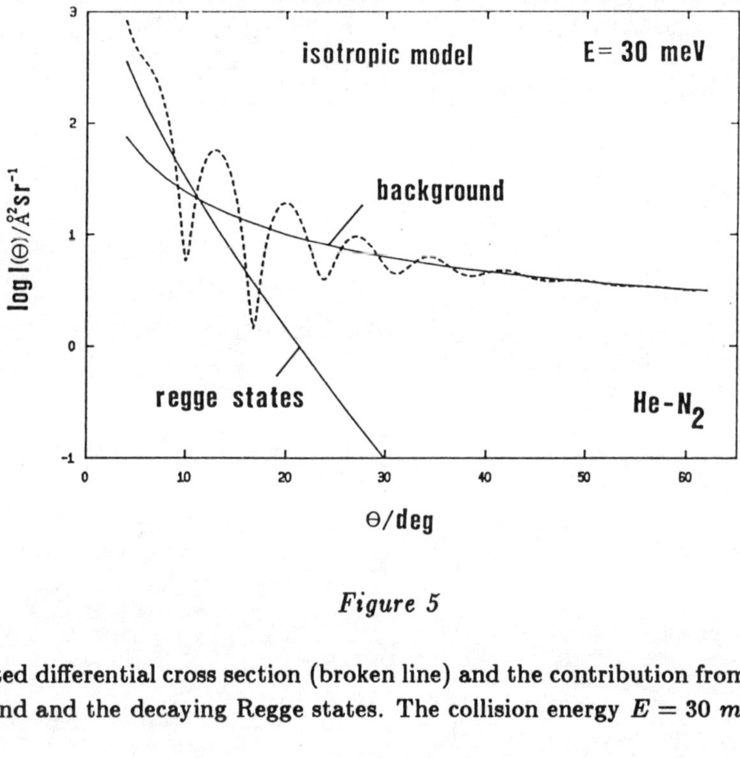

Figure 5

Calculated differential cross section (broken line) and the contribution from the classical background and the decaying Regge states. The collision energy $E = 30$ *meV*.

The broken curve in Figure 5 is the differential cross section, 30 meV center-of-mass energy, showing a Fraunhofer type diffraction pattern; a large peak in the forward direction and damped, regular oscillations with increasing scattering angle. Computations involved the standard partial wave series (2) and WKB phase shifts (5). The CAM result (not shown) is essentially equivalent. However, using just three Regge poles, the agreement becomes poor in the range of the forward diffraction peak. Consequently, we focus our attention on scattering angles beyond that range. With this restriction in the analysis even a single Regge pole approximation is quite accurate [48]. The CAM theory provides a decomposition of the differential cross section due to a (classical) background, a Regge states contribution and their interference. Shown in Figure 5 are the background and Regge states contributions. At scattering angles $\theta < 10$ degrees, most of the scattered intensity comes from decaying Regge states. As the scattering angle is increased beyond $\theta \approx 10$ degrees, it is seen that the quantum differential cross section more and more approaches the classical one. Although the contribution from the Regge states is 100 times smaller than the classical background at $\theta \approx 30$ degrees , it still produces significant effects in the cross section.

It is clear that the oscillatory pattern in the cross section in Figure 5 is due to an interference of the amplitudes $f_B(\theta)$ and $f_P(\theta)$. This result is not a prioritive. In resonance, and orbiting scattering, for example, the Regge states contribution alone shows prominent oscillations in a wide angular range [41,44]. Oscillations in the background (cusped rainbows) can also be found [65].

In the angular range where only the leading Regge pole (m=0) needs to be included in $f_P(\theta)$ (see (28)), one can, with two additional approximations, derive a simple expression for the period of the diffraction oscillations. First, the single pole amplitude simplifies for $\Im\lambda_o >> 1$:

$$f_P(\theta) \sim \frac{1}{k}\left(\frac{2\pi\lambda_o}{\sin\theta}\right)^{\frac{1}{2}} r_o \exp\left(i(\lambda_o\theta - \pi/4)\right) \tag{36}$$

Expression (36) is now similar in form to the semiclassical result (31) for the background amplitude $f_B(\theta)$. By neglecting the implicit θ-dependence of the exponential factor in (31), we finally find that the interference (and the relative phase) of the amplitudes $f_B(\theta)$ and $f_P(\theta)$ has the approximate angular period

$$\Delta\theta \approx \frac{2\pi}{\lambda_\theta + \Re\lambda_o} \tag{37}$$

Note that the imaginary part of the pole position does not influence the period. It mainly influences the rate of damping of the diffraction oscillations.

Figure 6 illustrates what happens at a significantly larger scattering energy (200 meV). In brief, Regge states decay much faster and the differential cross section has become more classical. Both amplitude and period of the oscillations have decreased. The position of the leading Regge pole is now $\lambda_o = 57.94 + i15.15$.

3.3 Anisotropic potential models

The use of an anisotropic potential has the consequence that orbital angular momentum is no longer conserved in the collision. Angular momentum can be transferred to the rotational motion of the molecule. Radial kinetic energy of the

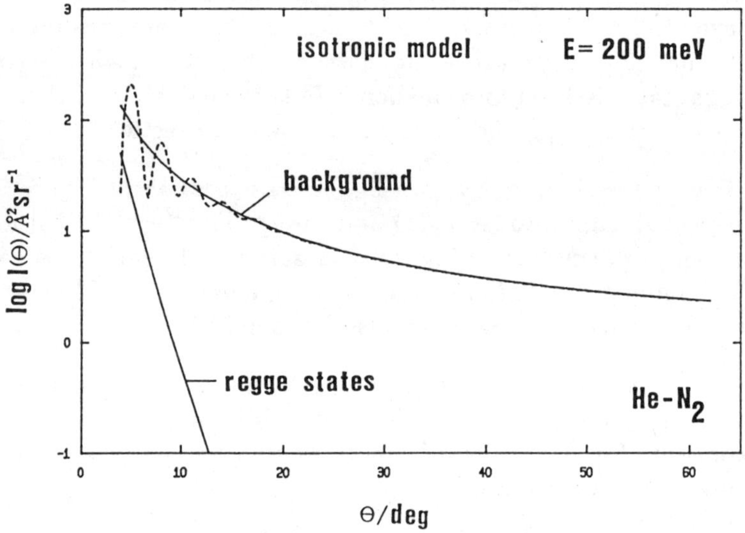

Figure 6

Same as in Figure 5, but with a larger collision energy $e = 200meV$

relative motion is similarly affected. However, the CAM theory has recently been generalized to this type of collision using the powerful infinite-order-sudden (IOS) approximation [48]. In this approximation, the rotational levels of the molecule are degenerate and the centrifugal potential is the same in all coupled channels. Hence, an effective asymptotic wave number k and an effective orbital quantum number ℓ can be specified. In what follows, k is identical to the initial channel wave number and ℓ refers to the so called post-collision identification [49].

With the molecule N_2 initially in its rotational ground state, the degeneracy averaged (over magnetic quantum numbers) differential cross sections are

$$I(0 \to j|\theta) = |f_B(0 \to j|\theta) + f_P(0 \to j|\theta)|^2 , \qquad (38)$$

where j(=0,2,...) is the final rotational quantum number, and where the generalized background and Regge state amplitudes are

$$f_{B,P}(0 \to j|\theta) = \frac{1}{2}(2j+1)^{\frac{1}{2}} \int_0^{\pi} f_{B,P}(\gamma|\theta) P_j(\cos\gamma) \sin\gamma d\gamma \; . \qquad (39)$$

γ is the orientation angle of the molecule appearing in the potential (33), and $f_{B,P}(\gamma|\theta)$ are the 'old' amplitudes of the single-channel theory, depending parametrically on γ through the potential. The integral in (39) projects a γ-dependent amplitude onto the relevant orientational distribution of the final rotational state of the molecule.

Below, we study differential cross sections for three anisotropic potential models (I-III) of the $He - N_2$ interaction. I is already defined in (33) to (35), while II-III are modifications of I. The purpose is to investigate the sensitivity of the diffraction to variations in the anisotropy. The modifications of I are specified as follows: in II the P_4 anisotropy is neglected, i.e. $V_4(r) \equiv 0$; and in III, $V_4(r) \equiv 0$ and the attractive part of $V_2(r)$ is neglected.

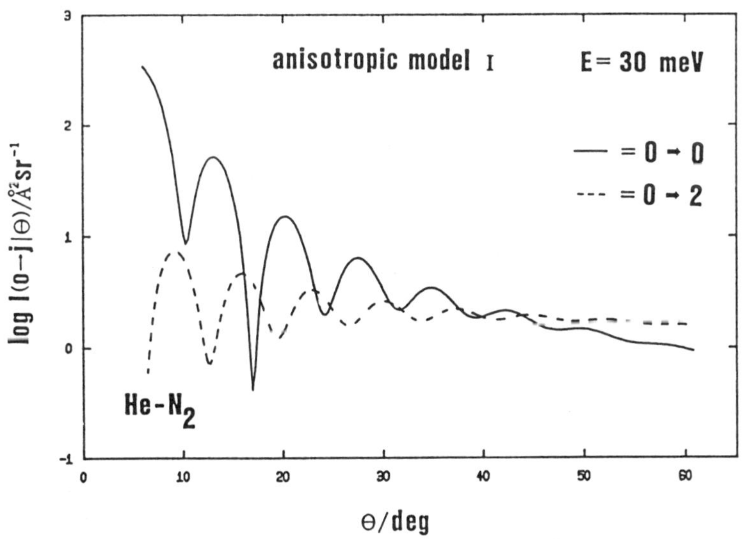

Figure 7

Calculated elastic and inelastic differential cross section for the anisotropic $He - N_2$ model I.

The two dominating differential cross sections, corresponding to the transitions $0 \rightarrow 0$ and $0 \rightarrow 2$, for the anisotropy model I, are shown in Figure 7. We see that diffraction oscillations of the same period appear in both cross sections, but the peaks are qualitatively out of phase. The elastic cross section is similar to that in the isotropic model (Figure 5), apart from the fact that its magnitude becomes increasingly lower at larger angles.

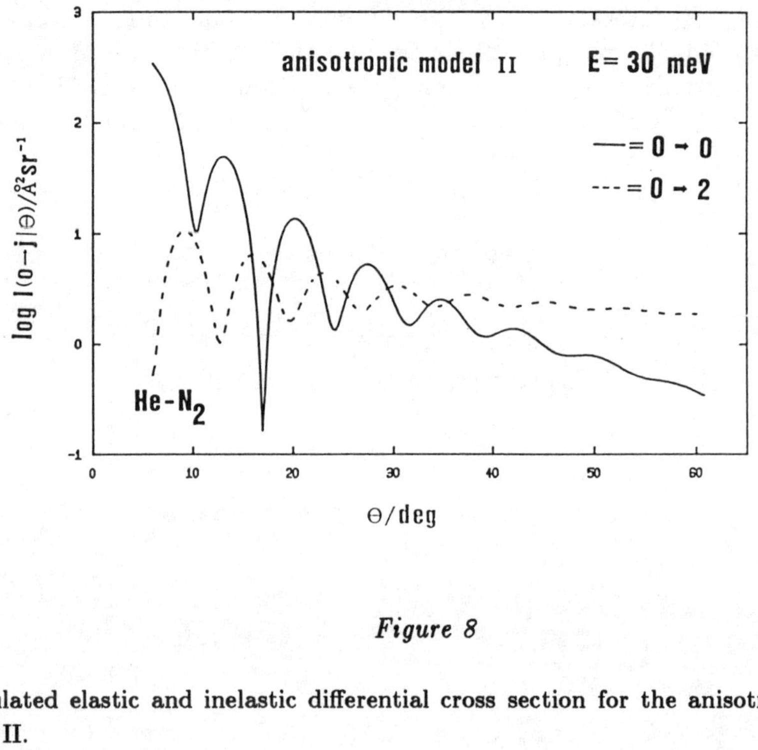

Figure 8

Calculated elastic and inelastic differential cross section for the anisotropic $He - N_2$ model II.

For the anisotropic model II in Figure 8, this effect is accentuated further. The elastic differential cross section at $\theta = 60$ degrees is now one order of magnitude smaller compared to the isotropic model. In II, the inelastic curve is slightly shifted upwards by a nearly angular independent amount (on the scale of the figures) compared to the anisotropic model I. The period and, damping of the diffraction oscillations are the same, or very nearly the same, in Figures 7 and 8. We may conclude that the P_4-anisotropy, present in the potential model I, does not influence the period, damping or relative peak shift (inelastic versus elastic) in

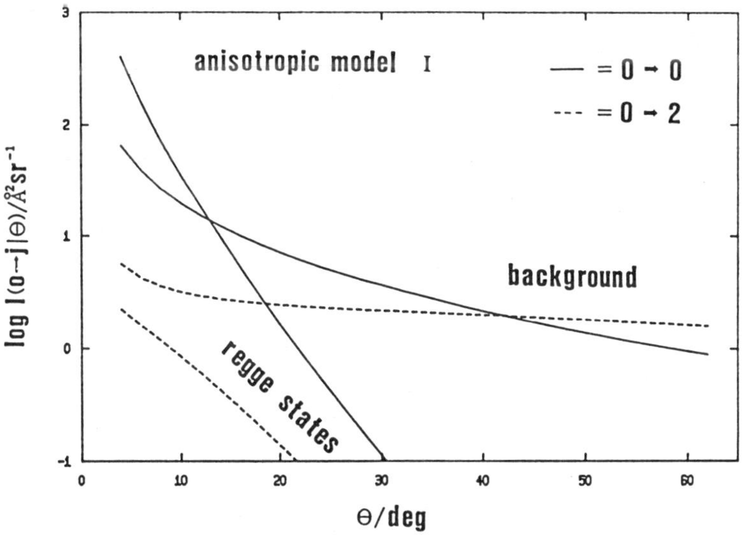

Figure 9

Calculated contributions form background and Regge states in the elastic and inelastic differential cross sections shown in figure 7.

the complete diffraction pattern. It merely modulates the magnitude of the cross sections at larger angles.

Figures 9 and 10 show the contributions from the classical background and Regge states mechanisms for the two potential models just discussed. Here we can see that the elastic Regge states contribution is not sensitive at all to the present P_4-anisotropy. The other components show minor changes; the most noticeable effect being the tendency of the elastic background to decrease at larger angles without the P_4-term. We also note a cross-over of the elastic background and Regge states contributions (see also Figure 5), which is responsible for a deep minimum in the differential cross section at a nearby scattering angle. This cross-over effect is not present in the inelastic channel.

Turning now to the anisotropic model III in Figure 11, we observe a continued lowering of the elastic differential cross section at, preferably, large angles, and a rise of the inelastic one toward lower angles. The elastic cross-over effect occurs

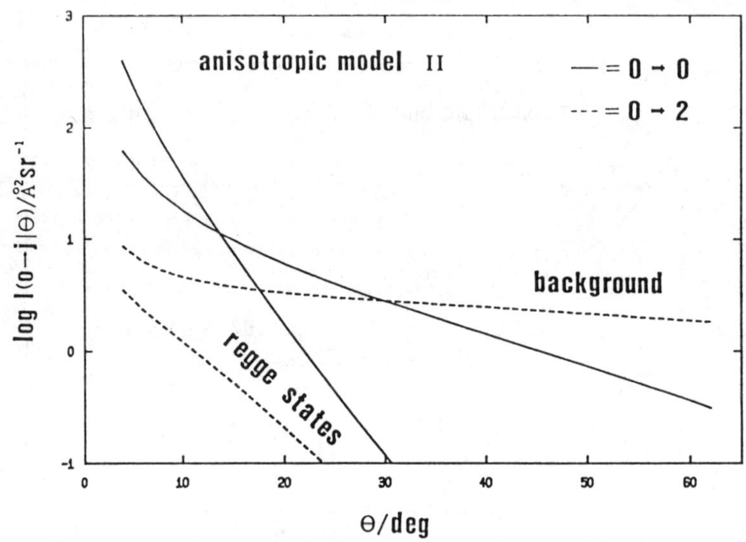

Figure 10

Calculated contributions form background and Regge states in the elastic and inelastic differential cross sections shown in figure 8.

at a somewhat larger scattering angle and the damping of the oscillations is much slower. By neglecting the attractive part of the P_2-anisotropy it is again the elastic background which has been affected the most; this is seen in Figure 12. The rapid intensity fall of the elastic background contribution results in a wider range of angles where it can interfere with the decaying Regge states. The shift of the cross-over effect is similarly explained.

A new effect can be seen in Figure 11. For the first time in our study we notice a small shift of the inelastic diffraction peaks toward smaller angles. This effect is interesting because other diffraction theories [50] of similar collisions predict inelastic peaks exactly out of phase with the elastic ones for our system $He - N_2$.

With the present theory, it is possible to analyze further the relative positions of elastic and inelastic diffraction peaks (see Reference 48). We define a local phase of the elastic/inelastic interference oscillations through the equation

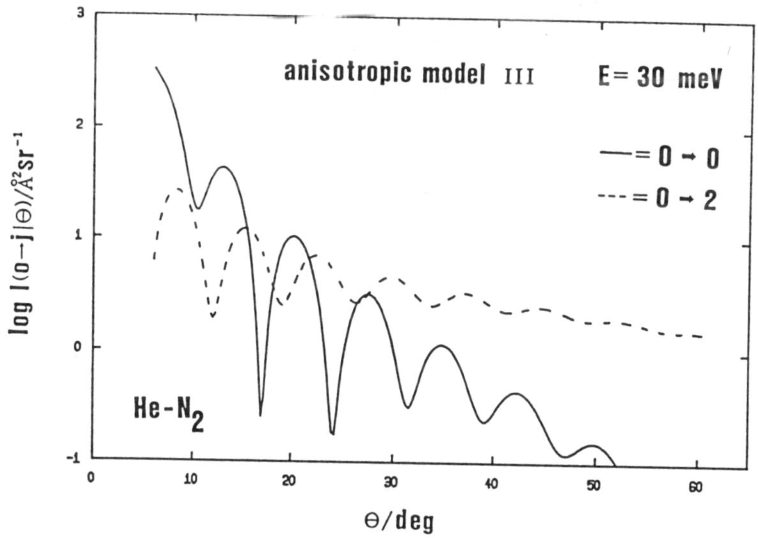

Figure 11

Calculated elastic and inelastic differential cross section for the anisotropic $He - N_2$ model III.

$$\varphi_j(\theta) = arg\{f_P(0 \to j|\theta)\} - arg\{f_B(0 \to j|\theta)\} \tag{40}$$

The quantity $\kappa_j(\theta)$, defined by

$$\kappa_j(\theta) \equiv \varphi_j(\theta) - \varphi_o(\theta) \quad (mod\ 2\pi), \tag{41}$$

then measures the shift in peak positions (normalized to a period of 2π). In Figure 13 we have plotted the reduced peak shift $\kappa_2(\theta)$ for all three anisotropic models (note that $\kappa_2(\theta)$ is defined here in the interval $(-2\pi, 0)$). The curves corresponding to II and III are well separated and fairly independent of the scattering angle. The peak shift produced by model I is similar to that of model II at small scattering angles. At larger angles, however, its value is closer to the model III result. For all three models, $\kappa_2(\theta)$ is significantly larger than (i.e. lies above) the out-of-phase result $-\pi$.

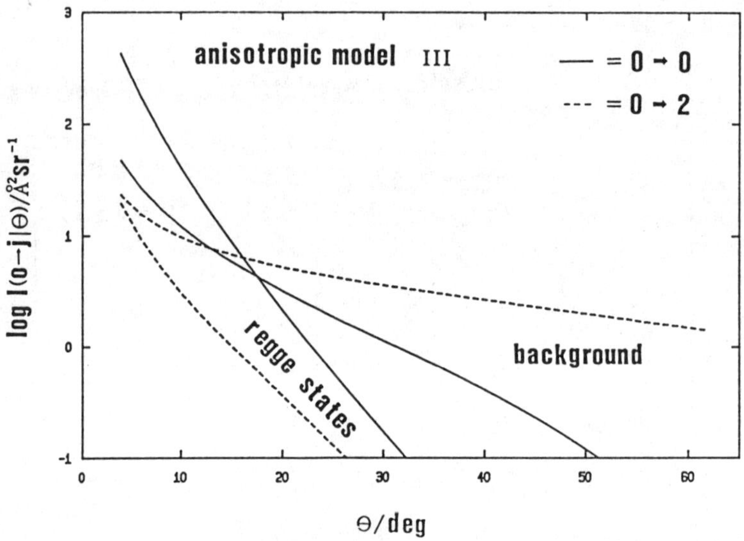

Figure 12

Calculated contributions form background and Regge states in the elastic and inelastic differential cross sections shown in figure 11.

4. Summary and Conclusions

In the present paper we have discussed a Regge pole picture of the diffraction mechanism in atom-molecule collisions. The Regge pole (or complex angular momentum) theory of elastic scattering was reviewed and diffraction oscillations are explained as an interference between a classically related background (or trajectory) amplitude and a non-classical amplitude due to decay of short-lived quasi-molecules (so called Regge states) formed in the collision.

Combined with the infinite-order-sudden approximation, the Regge pole theory was then applied in a study of rotationally inelastic diffraction scattering of He by homonuclear N_2 molecules at $E = 30\ meV$ center-of-mass energy. By neglecting various terms in the anisotropic part of the potential surface, we gained useful insights toward a general understanding of diffraction features in degeneracy averaged differential cross sections. For the $He - N_2$ system, the elastic differential cross section shows a diffraction pattern with two angular branches separated by a

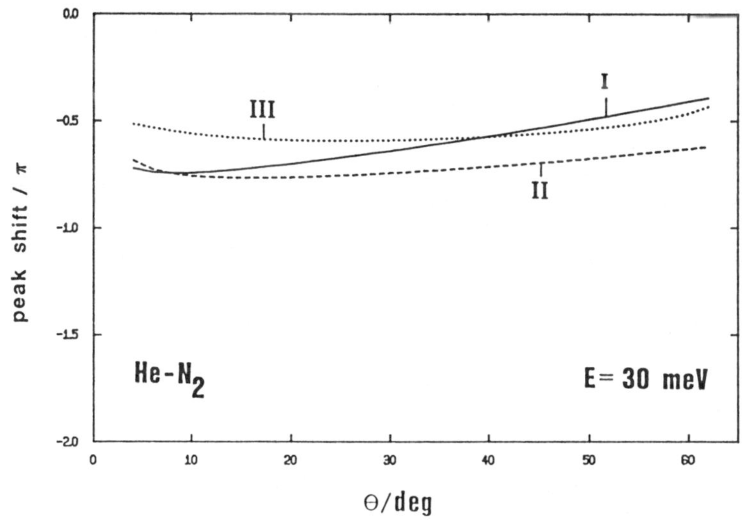

Figure 13

Calculated peak shifts for the anisotropic models I, II and III as a function of the scattering angle.

cross-over region. The small-angle branch, including the forward diffraction peak, is dominated by the Regge states contribution which is insensitive to our variations in the anisotropy. Indeed, in this angular region a centrally symmetric average potential (our isotropic model) yields accurate elastic cross sections. On the other hand, at larger angles where the classical mechanism dominates, the magnitude of the elastic cross section depends strongly on the variations in the anisotropy.

The rotationally inelastic $0 \rightarrow 2$ differential cross section shows no sign of any cross-over effect and is always dominated by the classical background. Here, in contrast to the elastic case, the sensitivity in its magnitude is found in smaller angles.

We can also conclude that the inelastic diffraction peaks are not exactly out-of-phase with the elastic ones. The peak shift is significantly influenced by details of the anisotropic part of the potential. It even seems to be (weakly) dependent on the scattering angle.

Acknowledgment

This work is supported by the Swedish Natural Science Research Council.

References

1. K. W. Ford and J. A. Wheeler, Ann Phys NY, **7**, 259, 287 (1959).

2. R. J. Gordon, J. Chem. Phys., **63**, 3109 (1975).

3. G. E. Zahr and W. H. Miller, Mol. Phys., **30**, 951 (1975).

4. H. J. Korsch and D. Leissing, J. Phys. B, **9**, 1857 (1976).

5. R. A. Broglia and A. A. Winther, Phys. Rep., **4C**, 153 (1972).

6. W. H. Miller, J. Chem. Phys., **53**, 3578 (1970).

7. J. Stine and R. A. Marcus, Chem. Phys. Lett., **15**, 536 (1972).

8. T. Regge, Nuovo Cim., **14**, 951 (1959).

9. E. J. Squires, *Complex angular momentum and particle physics*, Benjamin, New York 1963.

10. R. G. Newton, *The Complex j-plane*, Benjamin, New York 1964.

11. V. DeAlfaro and T. Regge, *Potential scattering*, North-Holland, Amsterdam 1965.

12. R. L. Omne's, Ann Rev Nucl Sci, **16**, 263 (1966).

13. K. W. McVoy, Phys. Rev., **3C**, 1104 (1971).

14. J. Knoll and R. Schaeffer, Ann. Phys. NY, **97**, 307 (1976).

15. N. Rowley and C. Marty, Nucl. Phys., **266A**, 494 (1976).

16. T. Tamura and H. H. Wolter, Phys. Rev., **6C**, 1976 (1972); T. Takemasa and T. Tamura, Phys. Rev., **18C**, 1282 (1978).

17. R. Anni, L. Renna and L. Taffara, Nuovo Cim., **45A**, 123 (1978); R. Anni and L. Renna, Nuovo Cim., **65A**, 331 (1981).

18. E. A. Remler, Phys. Rev., **3A**, 1949 (1971).

19. L. D. Doverspike, R. L. Champion, S. M. Bobbio and W. G. Rich, Phys. Rev. Lett., **25**, 909 (1970).

20. S. M. Bobbio, W. G. Rich, L. D. Doverspike and R. L. Champion, Phys. Rev., **4A**, 957 (1971).

21. W. G. Rich, S. M. Bobbio, R. L. Champion and L. D. Doverspike, Phys. Rev., **4A**, 2253 (1971).

22. J. N. L. Connor, Mol. Phys., **29**, 745 (1975).

23. C. V. Sukumar and J. N. Bardsley, J. Phys. B, **8**, 568 (1975).

24. C. V. Sukumar, S. L. Lin and J. N. Bardsley, J. Phys. B, **8**, 577 (1975).

25. S. Bosanac, Mol. Phys., **35**, 1057 (1978).

26. J. B. Delos and C. E. Carlson, Phys. Rev., **11A**, 210 (1975).

27. J. N. L. Connor, J. B. Delos and C. E. Carlson, Mol. Phys., **31**, 1181 (1976).

28. J. N. L. Connor, W. Jakubetz and C. V. Sukumar, J. Phys. B, **9**, 1783 (1976).

29. J. N. L. Connor and D. C. Mackay, Chem. Phys. Lett., **59**, 163 (1978).

30. K.-E. Thylwe, PhD Thesis, Acta Universitatis Upsaliensis, Uppsala 1981.

31. K.-E. Thylwe, J. Phys. A, **16**, 1141 (1983).

32. K.-E. Thylwe and J. N. L. Connor, J. Phys. A, **18**, 2957 (1985).

33. J. N. L. Connor and K.-E. Thylwe, J. Chem. Phys., **86**, 188 (1987).

34. J. N. L. Connor and W. Jakubetz, Chem. Phys. Lett., **36**, 29 (1975).

35. J. N. L. Connor and W. Jakubetz, Mol. Phys., **1619**, 33 (1977).

36. J. N. L. Connor and D. L. Mackay, Chem. Phys., **40**, 11 (1979).

37. J. N. L. Connor, D. C. Mackay and K.-E. Thylwe, J. Chem. Phys., **85**, 6368 (1986).

38. J. N. L. Connor, D. C. Mackay and C. V. Sukumar, J. Phys. B, **12**, L515 (1979).

39. S. Bosanac, Phys. Rev., **19A**, 125 (1979).

40. S. Bosanac, Phys. Rev., **24A**, 777 (1981).

41. S. Bosanac, Phys. Rev., **28A**, 1344 (1983).

42. S. Bosanac, Phys. Rev., **30A**, 153 (1984).

43. S. Bosanac and K. Knesaurek, Phys. Rev., **28A**, 2173 (1983).

44. K.-E. Thylwe, J. Phys. B, **16**, 1915 (1983).

45. J. N. L. Connor and W. Jakubetz, Mol. Phys., **35**, 949 (1978).

46. See discussion in Report Number 34, Reference 30.

47. J. N. L. Connor, D. Farrelly and D. C. Mackay, J. Chem. Phys., **74**, 3278 (1981).

48. K.-E. Thylwe and J. N. L. Connor, J. Chem. Phys., to be published (1988).

49. G. A. Parker and R. T. Pack, J. Chem. Phys., **68**, 1585 (1978).

50. M. Faubel, Adv. At. Mol. Phys., **19**, 345 (1983).

51. A. Messiah, *Quantum Mechanics*, North-Holland, Amsterdam 1970.

52. R. T. Pack, J. Chem. Phys., **60**, 633 (1974).

53. W. M. Frank, D. J. Land and R. M. Spector, Rev. Mod. Phys., **43**, 36 (1971).

54. A. Erdélyi, (Ed.) *Higher Transcendental Functions Vol I*, McGraw-Hill, New York 1953.

55. L. Robin, *Fonctions Sphérigues de Legendre et Fonctions Sphéroïdales Vol II*, Gauthier-Villars, Paris 1958.

56. J. N. L. Connor, *Semiclassical Methods in molecular scattering and spectroscopy*, Ed., M. S. Child, Reidel, Dortrecht 1980, p. 45.

57. O. Brander, Ark. Fys., **32**, 131 (1966).

58. N. Dombey and R. H. Jones, J. Math. Phys., **9**, 986 (1968).

59. K.-E. Thylwe, J. Phys. A, **16**, 3325 (1983).

60. K.-E. Thylwe, J. Phys. A, **18**, 3445 (1985).

61. J. N. L. Connor, W. Jakubetz, D. C. Mackay and C. V. Sukumar, J. Phys. B, **13**, 1823 (1980).

62. A. Paliov and S. Rosendorff, J. Math. Phys., **8**, 1829 (1975).

63. R. C. Fuller, Phys. Rev., **12C**, 1561 (1975).

64. J. N. L. Connor, J. Phys. B, **15**, 1683 (1982).

65. J. N. L. Connor and D. Farrelly, J. Chem. Phys., **75**, 2831 (1981).

66. D. Brink, *Semiclassical methods in nucleus-nucleus scattering*, Cambridge University Press, Cambridge 1985.

67. J.M. Charap and J.E. Squires, Ann. Phys., **25**, 143 (1963).

68. Dai Yaun-ben, Scienta Sinica, **13**, 1496(1963); Dai Yaun-ben, Scienta Sinica, **14**, 1496(1965); Chou Lon-Shiang and Dai Yaun-ben, Acta Physica Sinica, **21**, 1552 (1965).

69. R. B. Bernstein, (Ed.) " *Atom-Molecule Collision Theory*", Plenum Publ. Co. New York (1979).

ENERGY SHIFTS AND WIDTHS IN ATOMIC AND MOLECULAR PHYSICS: MULTICHANNEL APPROACH

Roland Lefebvre

Laboratoire de Photophysique Moléculaire
du CNRS,
Campus d'Orsay,
91405 Orsay,
FRANCE

Abstract

The description of dynamical processes in atomic and molecular systems often leads to a set of coupled differential equations. Resonances within such species can be studied from a step by step propagation of solutions of these equations satisfying appropriate boundary conditions at both ends of the integration range. This is combined with a matching relation which plays the role of a quantization condition yielding a complex energy. Many variants are possible: the integration path can be real or complex; the addition of an optical potential is analyzed in some detail. This perturbation may, under some conditions, be completely equivalent to the use of a complex integration path. When these conditions are not fulfilled, the energy suffers a complex shift which may be eliminated through analytic continuation.

1. Introduction

Energy quantization in wave mechanics is a consequence of imposing some conditions on the wave function for two different ranges of the coordinate r for relative motion of two particles. A familiar example is the bound states of the hydrogen atom: the indicial equation for $r \simeq 0$ selects the behavior $f \simeq r^{\ell+1}$ (the regular function). Far from the origin the growing component in f can disappear only for energies $E_n = -1/2n^2$ (in a.u., n integer). The complex energies which are associated with atomic and molecular resonances, that is to states with a decaying character, are similar in this respect since conditions have to be fulfilled both for small or large distances between two entities. We note that the methodology has to take into account a number of special features:

a) The two entities (atoms, molecular fragments or molecules) have internal degrees of freedom. There are as many ways of describing the relative motion of these entities as there are relevant internal states at the selected energy. Each of these ways results in a channel. The dependence of the interentity potential energy on the internal variables produces a coupling between the channels. Instead of the single radial second order differential equation for the hydrogen atom, there is generally a set of coupled second order differential equations (coupled channel description). The wave function is replaced by a vector wave function (with one component per channel) and matrices made of these vector wave functions have to be built to take into account the fact that a number of independent solutions may fulfill given boundary conditions.

b) When two molecular species approach each other, very strong repulsive forces develop so that the range $r \simeq 0$ is classically forbidden. The boundary condition as one reaches small values of r is that the components of the wave function should be exponentially vanishing. This is, as we shall see below, a very favorable circumstance for the numerical integration of the coupled channel equations.

c) For large values of r, as in all fields of physics, the Siegert criterion applies for a resonance [1], which is also equivalent to stating that we are on a pole of the scattering matrix: the wave function should be of outgoing character only in all open channels; that is, in all channels allowing for free relative motion of the two entities.

d) The argument usually given to illustrate the effect of the operation called complex rotation on the resonance wave functions of a single channel [2] is easily extended to the multichannel case. The wave number in the nth channel is defined (in a.u.) by $k_n^2 = 2(E - \epsilon_n)$ where ϵ_n is the contribution of the internal states to the total energy. Open channels are such that $E - \epsilon_n > 0$. The Siegert criterion turns E into the complex quantity $E_r = \bar{E} - i\Gamma/2$ and k into $k_{n0} - ik_{n1} = K_n \exp[-i\beta_n]$, with k_{n0}, k_{n1}, K_n and β_n all positive. We assume that for all open channels the condition $Re(E_r) - \epsilon_n > 0$ is satisfied. Let θ_c be the largest of all β_n's. If we substitute $\rho \exp(i\theta)$ (with ρ real) instead of r in the coupled system of equations, with $\theta > \theta_c$ we ensure vanishing of the components of the wave function as $\rho \to \infty$, since the outgoing asymptotic wave $\exp[ik_n r]$ is turned into $\exp[iK_n\rho\, cos(\theta - \beta_n)] \exp[-K_n\rho\, sin(\theta - \beta_n)]$. We have now asymptotic vanishing of the wave functions in all channels, both open and closed. The situation is exactly that prevailing in the bound region of the energy spectrum.

We describe below a methodology for resonance state determination based on a numerical integration of the coupled channel equations supplemented by a matching condition to ensure continuity of the wave function and its derivative over the entire integration range [3]. Emphasis will be put on the analogy between one channel and multichannel situations. We will also point out that there is considerable flexibility in choosing an integration path which may even be a curve in the complex plane. This flexibility will be used to study in some detail the relation beween complex rotation and the addition of an optical potential which acts as a sink in the asymptotic region. For certain characteristics of this potential it is possible to show that there is a complete equivalence if the path is properly chosen. When the optical potential produces a breakdown of this equivalence, the associated complex eigenenergies are no longer the resonance energies but are still significant. An analytic continuation based on Padé approximants can be performed to 'deperturb' these energies and to recover unambiguously the resonance energies.

2. Multichannel Complex Energy Quantization

The complete wave function of the system is written:

$$\Psi(q,r) = \sum_n \varphi_n(q,r)U_n(r) \tag{1}$$

where q denotes collectively all coordinates except r which measures the inter-fragment distance. The φ_n's collect the information on the internal states (for example vibration and rotation of a diatomic molecule) and also on the overall rotation of the system (as in the steps leading to the radial equation for the hydrogen atom). These functions could depend explicitly on r (adiabaticity). It is simpler for numerical purposes to choose them as independent from r (crude adiabaticity), or possibly different for different ranges in r (adiabaticity by sectors [4]). Introduction of (1) into the wave equation yields, for the U_n 's , a system of coupled differential equations:

$$\sum_n \langle\varphi_m|H|\varphi_n\rangle\ U_n(r) = EU_n(r) \tag{2}$$

The complete Hamiltonian H contains the relative radial kinetic energy operator which we now write $-d^2/dr^2$. (2) can be reduced to the standard form

$$[+d^2/dr^2\mathbf{1} + E\mathbf{1} - \mathbf{V}(\mathbf{r})]\mathbf{U}(\mathbf{r}) = 0 \tag{3}$$

where $\mathbf{U}(\mathbf{r})$ is the vector of N components $U_n(r)$ and $\mathbf{V}(\mathbf{r})$ the potential matrix of elements $< \varphi_m|H + d^2/dr^2|\varphi_n >_q$. In an N-channel situation it is possible to build N independent vectors satisfying some prescribed boundary condition in a given range for r. It is therefore convenient to group these N vectors into a matrix of solutions $\mathbf{U}(\mathbf{r})$[5,6] and to perform a step by step propagation of such matrices or of quantities derived from them [3,6]. One such quantity is the Fox-Goodwin 'ratio' **P**(r) defined [7] for two adjacent points of the integration grid r-h and r as:

$$\mathbf{P}(r) = \mathbf{U}(r-h)\mathbf{U}^{-1}(r) \tag{4}$$

This quantity is stable even in classically forbidden regions where the growing of closed channel components makes it difficult to propagate **U**(r) itself. We now prepare the ground for a discussion of integration grids which may be along any path in the complex plane, so as to allow for an extension to the complex rotation method and some variants of it. Consider three adjacent points along a path, of coordinates $r-h_1$, r and $r+h_2$, with r, h_1 and h_2 real or complex. We define the generalized Numerov matrices [8]:

$$\boldsymbol{\alpha}(r-h_1) = h_2\{1 + 1/12(h_1^2 + h_1h_2 - h_2^2)(E\mathbf{1} - \mathbf{V}(r-h_1))\}$$

$$\boldsymbol{\beta}(r) = (h_1+h_2)\{1 - 1/12(h_1^2 + 3h_1h_2 + h_2^2)(E\mathbf{1} - \mathbf{V}(r))\} \tag{5}$$

$$\boldsymbol{\gamma}(r+h_2) = h_1\{1 + 1/12(-h_1^2 + h_1h_2 + h_2^2)E\mathbf{1} - \mathbf{V}(r+h_2))\}$$

P(r) can be propagated with the help of:

$$\mathbf{P}(r+h_2) = \mathbf{U}(r)\mathbf{U}^{-1}(r+h_2) = [\boldsymbol{\beta}(r) - \boldsymbol{\alpha}(r-h_1)\mathbf{P}(r)]^{-1}\boldsymbol{\gamma}(r+h_2) \tag{6}$$

with $\mathbf{P}(r) = \mathbf{U}(r-h_1)\mathbf{U}^{-1}(r)$. With proper choices of the $h_i's$ this can be done either <u>outward</u> from the internal region or <u>inward</u> from the external region. In both cases one has to consider choosing the initial values of $\mathbf{P}(r)$ and matching the results of both integrations at some intermediate point r_m. There are several possibilities:

a) For bound states <u>and</u> resonant states treated within the complex rotation method and for which the range $r \simeq 0$ is classically forbidden, the choice can be made the same at both ends of the integration range $r_1 - r_2$:

$$\mathbf{P}(r_1) = \mathbf{P}(r_2) = 0 \tag{7}$$

This is not because the desired **P**(r) is really zero in forbidden regions (for example for a single channel with real r and $U(r) \sim \exp(-Kr)$, P(r) assumes the constant value $\exp[-Kh]$), but simply because the correct matrices are automatically produced by the propagation from a forbidden region toward an allowed one [6]: the channel components, which decay as one penetrates into a forbidden region, become the dominant contribution as one leaves such a region. Numerical experiments show that the best choice for the matching point r_m is to take it, whenever this is possible, in a region which is classically allowed for all channels. The matching condition expresses the fact that one may combine linearly the independent solutions present in both $\mathbf{P}^o(r)$ (outward integration) and $\mathbf{P}^i(r)$ (inward integration) so as to produce the same vector solution $\mathbf{U}(r)$. It reads [3,9]:

$$det|\mathbf{P}^i(r_m) - \mathbf{P}^o(r_m+1)^{-1}| = 0 \tag{8}$$

b) When the range $r \simeq 0$ is accessible, outward integration can be started from $\mathbf{P}^o(h) = \mathbf{U}(0)\ U^{-1}(h) = 0$, except when the potential is singular at $r = 0$ (for instance for the atomic static Stark effect [10]) since (6) would require the evaluation of $\mathbf{V}(0)$. Assuming a constant step h in this neighborhood, it is possible to start from $\mathbf{P}(2h)$, the correct wave function close to the origin being obtained from series expansions [6]. The matching condition can again be given the form (8). As a variant one may perform only the inward integration starting from the asymptotic region. For a potential with no singularity at the origin, the matching (or quantization) condition is simply:

$$p^i(0)^{-1} = 0 \tag{9}$$

For a singular potential $P^i(r)$ can be matched to the matrix $\mathbf{P}(2h)$ already alluded to.

c) The treatment of resonant states with integration along the real axis requires a change in the initial inward ratio $p^i(r)$. If the channels become asymptotically decoupled, the initial form given to $\mathbf{P}^i(r)$ at $r = r_2$ can be derived from the expected Siegert behavior of the U_n's, that is $\exp[ik_n r]$.

Whatever particular variant we choose within the present scheme, the method is of the 'shooting' type: a trial energy has to be chosen and the quantization conditions (8) or (9) are fulfilled iteratively, using, for instance, the complex Newton-Raphson procedure. As an example of the efficiency of the method, we quote a study [11] of the bound states and of the first five resonances of a potential studied by Ginocchio [12]. This is a one-channel case for which bound and resonance energies admit exact analytic expressions. The potential V(r) is expressed with an auxiliary variable y:

$$V(r) = \lambda^2\ \nu(\nu+1)(1-y^2) + (1-y^2)(1-\lambda^2)[5(1-\lambda^2)y^2 - (7-\lambda^2)y^2 + 2]/4 \tag{10}$$

with, between r and y the relation:

$$r = \frac{1}{2\lambda^2}\left\{\log\left(\frac{1+y}{1-y}\right) + iC\log\left(\frac{i+Cy}{i-Cy}\right)\right\} \tag{11}$$

and $C = (\lambda^2 - 1)^{1/2}$. The case examined here corresponds to $\lambda = 26$ and $\nu = 5.5$. The peculiarity of this potential (see Reference 13 for a discussion of this point) is that a singularity in the complex plane prevents the use of ordinary complex rotation (that is integration along a rotated axis making an angle with the real axis sufficient to ensure localization of the resonance wave-function). In Reference 11 we have used a path suggested by Simon [14] which consists of following the real axis up to some point of abcissa r_o, after which r is changed into

$$r_o + (\rho - r_o)\ \exp[i\theta] \quad (\rho\ real). \tag{12}$$

This 'exterior scaling' method is obviously within reach of the present procedure. In an inward propagation, the step just before r_o is complex (i.e. $h\ \exp[i\theta]$), while it is real (i.e. h) just after r_o. The quantities defined in (5), which reduce to scalars for a one channel problem, are precisely built to deal with such a situation. Table 1 compares analytic and numerical results. The agreement is excellent since it goes from 12 figures for the lowest bound state to 7 or 8 figures for the resonances.

Before ending this discussion, it should be noted that the three-point Numerov technique is just one among a large family of numerical integration procedures (see Reference 15 for an interesting approach to a generation of the members of this family). Other algorithms are possible to reduce the error made at each propagation step. We would rather comment on a procedure which is simpler than the Numerov one. For a one-channel case we rewrite (6) with $h_1 = h_2 = h$ as:

$$P(r+h) = \left\{2\ \frac{1 - \frac{5h^2}{12}(E - V(r))}{1 + \frac{h^2}{12}(E - V(r+h))} - \frac{1 - \frac{h^2}{12}(E - V(r-h))}{1 + \frac{h^2}{12}(E - V(r+h))}P(r)\right\}^{-1} \tag{13}$$

Taylor expanding $V(r \pm h)$ around r and keeping only terms in h^2 at most gives

$$P(r+h) = \{2[1 - h^2/2(E - V(r))] - P(r)\}^{-1} \tag{14}$$

This is nothing but the Schrödinger equation with second derivative of the wave function replaced by its finite difference approximation:

$$d^2U/dr^2 \simeq h^{-2}[U(r-h) - 2U(r) + U(r+h)] \tag{15}$$

Table I.

Bound state energies (B_i) and resonance energies (R_i) calculated analytically (left column) or numerically (right column) with the exterior scaling procedure for the Ginocchio potential [12]. The imaginary contributions to the numerical bound state energies are due to the use of a program manipulating complex quantities to produce numbers which should in principle be real.

Analytical	Numerical
B_1 -24012.0418063	-24012.0418061 + i 0.22(-6)
B_2 -22366.3386049	-22366.3386045 + i 0.11(-5)
B_3 -19414.3235410	-19414.3235405 + i 0.33(-5)
B_4 -15192.2034036	-15192.2034043 + i 0.95(-5)
B_5 -9757.95858080	-9757.95859139 + i 0.22(-4)
B_6 -3259.03268051	-3259.03281672 + i 0.90(-4)
R_1 4140.5 - i 840.7644	4140.4993 - i 840.7631
R_2 13576.5 - i 1751.3755	13576.499 - i 1751.3900
R_3 24360.5 - i 2657.2664	24360.500 - i 2657.2670
R_4 36492.5 - i 3634.0574	36492.500 - i 3634.0577
R_5 49972.5 - i 4699.6269	49972.500 - i 4699.6269

The fact that E is an eigenvalue of a tridiagonal matrix is expressed in (14), with the eigenvectors being the values of U(r) at the points of the integration grid. If integration goes from r_1 to r_2, $U(r_1 - h)$ and $U(r_2 + h)$ are missing and therefore implicitly assumed to be zero. This is equivalent to imposing the boundary conditions for a bound state. This procedure has been called by Truhlar [16] the finite boundary value method and shown by him to be efficient to find all at once the eigenvalues and eigenvectors of the bound states of the one-dimensional Schrödinger equation. It can be extended [17] to the search for resonant states if the integration is performed along a rotated axis (complex rotation) or a mixed path (exterior scaling). No trial energies are needed. However the outcome consists of both the resonant states and spurious states with a node at both ends of the integration range. Examination of stability versus rotation angle and integration range is necessary to distinguish the significant eigenenergies. As an example of the efficiency of this procedure, Table 2 (taken from Reference 18) gives the resonant energies associated with the potential $A\ \exp(-\alpha r)$ for various values of the parameter $\Lambda = 2\sqrt{A}/\alpha$ with the tridiagonal method, the matching method and a direct search from the analytic scattering amplitude [19].

The multichannel complex energy quantization scheme described here has been applied to various atomic and molecular problems. The static Stark effect in the hydrogen atom [10] offers an interesting situation: in a coupled channel formulation the interchannel coupling increases indefinitely with r. Some rearrangement of the channels would be necessary to apply explicitly the Siegert outgoing-wave only condition. Complex rotation leads to considerable simplification since bound state boundary conditions are applicable. Van der Waals complexes have also been studied in this context [20]. Rotational predissociation in the $Ar - H_2$ system and interference effects in the photofragmentation of $Ar - HD$, with respectively 2 and 3 channels, have been examined with a potential which is defined in a piece-wise manner [21]. Such potentials cannot be treated with ordinary complex rotation because two potentials which match at some value of r, say r_o, no longer match when r is made complex. Exterior scaling [14] is the natural answer to this difficulty since complex rotation and analytic continuation of the potential can be performed only after r_o.

3. Complex Potentials and Complex Rotation

Complex (or optical) potentials are found in various contexts. They may be the result of partitioning out the continua in a situation involving both closed and open channels (Feshbach projection operator method [22]). They may also be simply added to a physical potential to act as a sink in the asymptotic region to avoid reflection of the waves when the system is enclosed in a box. There is, as we shall now show, an interesting connection between this function of the optical potential and the method of complex rotation. This relation was first stated by Jolicard and Austin [23] for wave functions developed in bases of square integrable functions and can be shown to hold in the context of propagation and matching procedures for both one and multichannel situations [24].

We first examine how a complex potential may affect a resonance wave function in a single channel case if it is applied progressively in the asymptotic region. We assume this potential goes from zero at $r = r_o$ to some constant imaginary value of the form $-iV_o$ when r reaches r_2, the end of the integration range. A potential with these properties is:

$$V_{opt}(r) = -i(V_o/2)[tanh\ \alpha(r - r_c) \ - \ tanh\ \alpha(r_o - r_c)], r_0 \leq r \leq r_2 \qquad (16)$$

with $\alpha(r_c - r_0)$ and $\alpha(r_2 - r_c)$ both much larger than unity. It is well known that a smooth potential guarantees the validity of the JWKB approximation [25]. The

Table II.

The complex energies obtained for various values of $\Lambda(\alpha = 2)$ with: (a) the finite-difference boundary value method; (b) a matching procedure based on the propagation of the ratio of amplitudes of the wavefunction at two adjacent points of the integration grid; and (c) the search for the poles of the analytic scattering amplitude. Upper value, real part; lower value, imaginary part. The integration conditions (methods (a) and (b)) were: step size h = 0.01 on an interval 0-40 and a rotation angle given in radians in the column θ.θ_c the critical rotation angle given by $\theta_c = 1/2\ Tan^{-1}(Im(E)/Re(E))$.

Λ	$\theta_c(rad)$	$\theta(rad)$	(a)	(b)	(c)
2.4	1.08	1.3	-3.064234 -4.515600	-3.064397 -4.615494	-3.064398 -4.615494
2.6	1.03	1.3	-2.917356 -5.466206	-2.917563 -5.466100	-2.917562 -5.466100
	1.55	1.6	-13.749040 -0.523997	-13.752445 -0.518461	-13.729326 -0.524841
2.8	0.99	1.3	-2.703302 -6.355462	-2.703558 -6.355359	-2.703557 -6.355358
	1.48	1.6	-14.158597 -2.563463	-14.159992 -2.562189	-14.159991 -2.562186
3.0	0.95	1.3	-2.421519 -7.280778	-2.421830 -7.280682	-2.421833 -7.280682
	1.44	1.6	-14.625898 -3.916485	-14.627353 -3.915471	-14.627351 -3.915473

Siegert outgoing wave $\psi_s(r) = \exp[ikr]$ corresponding to a resonant state of energy E_r remains of outgoing character (the JWKB method is equivalent to assuming that the potential produces no reflection of the wave [26]) but is changed into:

$$\psi_s(r) \propto (k_{opt.}(r))^{-1/2} \exp\left[i \int_{r_o}^{r} k_{opt.}(r')dr'\right] \tag{17}$$

with $k_{opt.}(r) \propto [E_r - V_{opt.}(r)]^{1/2}$. For r approaching r_2 such that

$$V_{opt.}(r) \to -iV_o,$$

the function assumes the form

$$\psi_s(r) \to k^{-1/2} \exp(ikr) \tag{18}$$

with $k \propto [E_r + iV_o]^{1/2}$. The resonance energy E_r has a negative imaginary part $\Gamma_r/2$. If the optical potential satisfies $V_o > \Gamma_r/2$, we have $\Im(k) > 0$: this is sufficient to produce attenuation of the wave function. The purpose of the optical potential is to change the sign of the imaginary part of the wave number. This is also the purpose of complex rotation, since the Siegert wave becomes $\exp|[k \exp i(\theta - \beta)]\rho$, with $\Im[K \exp i(\theta - \beta)] > 0$ if $\theta > \beta$. The two methods are producing localization of an otherwise divergent function.

It is possible to go one step further by showing that there can be complete equivalence [24] between the two procedures, provided we take advantage of the possibility offered by the formulation embodied in (5) and (6) of performing the integration along any path in the complex plane. We first formulate this equivalence and then illustrate it for a well known model potential. The validity of the JWKB approximation is assumed in both cases (optical potential or complex path).

3.1. Optical potential

The discussion is no longer limited to a range where the physical potential $V_{ph.}(r)$ has vanished and the optical potential $V_{opt.}(r)$ is switched on. The phase integral of the JWKB function can be approximated as

$$\int^{r} k(r)dr \simeq \sum_{p} h_p[E_r - V_{ph.}(r_p) - V_{opt.}(r_p)]^{1/2} \tag{19}$$

The h_p's are steps along the real axis and $r_p = \sum^{p} h_i$.

3.2. Complex Path (or generalized complex rotation)

The phase integral is calculated along a complex path approximated by a series of complex steps h_p^c:

$$\int^{r_p^c} k(r^c)dr^c \simeq \sum_p h_p^c[E_r - V_{ph.}(r_p^c)]^{1/2} \tag{20}$$

with $r_p^c = \sum^p h_i^c$.

The lower limits of integration or summation in all these expressions depend on the choice of a reference point. If the optical potential is such as to lead to the resonance energy, the complex path built from the h_p^c's, obtained by equating term by term the contributions to the two summations in (19) or (20), should also produce this energy.

We study this equivalence for the potential $V(r) = 15r^2 \exp(-r)$ which has a resonance (with the kinetic energy operator written as $-d^2/dr^2$) with the energy [2,27]

$$E_r = 6.85278062 - i0.02554976 \tag{21}$$

The optical potential is chosen to be:

$$V_{opt.}(r) = -i\ 10^{-n} r^n \tag{22}$$

It has been shown in previous work [23,24] that with n sufficiently large this potential can produce the same effects as those described above. Table 3 gives the eigenenergies obtained with either the addition of the optical potential with various values of n or those corresponding to quantization along an 'equivalent' complex path. We can observe that for n=8 or 7 there is very close agreement between the two procedures. As the value of n is lowered the optical potential departs from the properties which are required to correctly produce the resonance wave function and energy: there is increasing overlap of the ranges where the two potentials (optical and physical) are effective. The complex eigenenergy is shifted progressively until for n=1 its imaginary part ($\simeq 0.124$) is no longer reminiscent of the correct value ($\simeq 0.026$). On the other hand, complex equivalent paths always produce the expected resonance energy for n as low as 3. This can be understood with Figure (1) which gives these paths: they are all of the 'exterior scaling' type, being first very close to the real axis and then departing from it with an angle sufficient to produce localization of the resonance wave function. In this series of calculations we have, therefore, the two situations of optical potentials which produce effects equivalent to those of complex rotation (or rather of 'complexization,' since the path is not characterized by a single angle) and of optical potentials which significantly perturb the eigenenergy.

Table III.

Column (a): eigenenergies obtained for the potential $15r^2 \exp(-r)$ with the addition of an optical potential of the form $-i\ 10^{-n} r^n$ for various values of n. The real steps along the real axis are h = 0.01 over the range 0-20. Column (b): eigenenergies along 'equivalent' complex paths. The procedure fails to converge for n = 1,2. In both cases the quantization is based on inward propagation and (6).

n	(a)	(b)
8	6.8527808 - i 0.0255487	6.8527806 - i 0.0255490
7	6.8527829 - i 0.0255491	6.8527806 - i 0.0255490
6	6.8527906 - i 0.0255533	6.8527806 - i 0.0255490
5	6.8528212 - i 0.0255855	6.8527806 - i 0.0255490
4	6.8529385 - i 0.0258003	6.8527806 - i 0.0255490
3	6.8533743 - i 0.0272311	6.8527806 - i 0.0255490
2	6.8549144 - i 0.0375450	6.8527806 - i 0.0255490
1	6.8595946 - i 0.1237970	- -

We are therefore led to a further question: is there any information left in the energies when the optical potential is perturbing them? This problem has been examined elsewhere [27]. One may, for instance, repeat a series of calculations with an optical potential in the form $-i\ A_p r$ with A_p going from 0.1 (the energy is that reported in Table 3 in row n=1) to 1.0 by steps of 0.1. These ten energies $E(A_p)$ are used to build a point-Padé representation [28] of the analytic function E(A) in the form of a ratio of two polynomials with their coefficients determined from the fits

$$E(A_p) = \frac{P(A_p)}{Q(A_p)} \qquad p = 1, 2, ...10. \tag{23}$$

With P being a polynomial of order 4 and Q a polynomial of order 5 there is

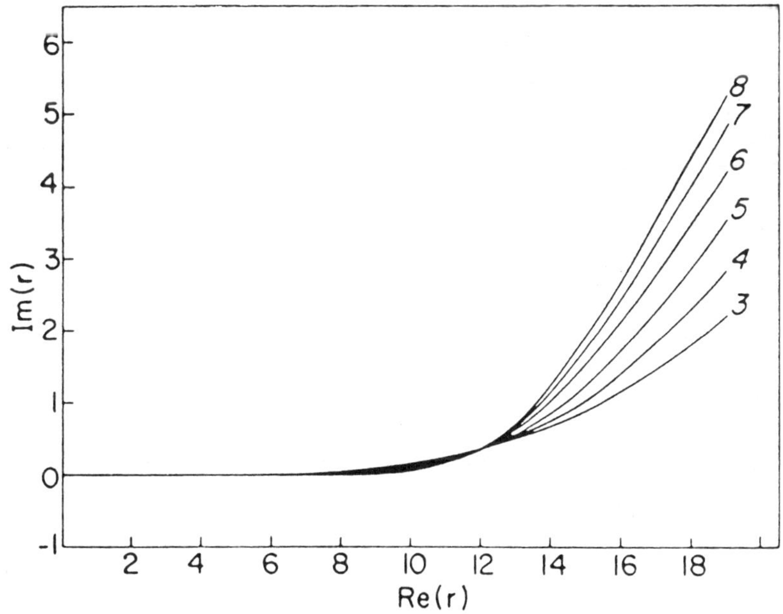

Figure 1.

The complex paths equivalent to the use of an optical potential of the form $-i\ 10^{-n} r^n (n = 3 - 8)$ added to the potential $15r^2 \exp(-r)$.

complete determination of the unknown coefficients. Elimination of the optical potential amounts to evaluation of $E(0)$ which is found to be $6.85277 - i0.025521$, in fairly reasonable agreement with the correct value.

Such an analytic continuation is reminiscent of the procedures used in the study of the resonance energies associated with the atomic Stark effect [29]: the perturbation series giving the energy as a function of powers of the electric field intensity is divergent. Summation, however, can be performed for imaginary values of the field intensity (this step yields real energies). A point-Padé representation is used to go back to real values of the field intensity and to produce the correct complex resonance energies. The analogy between this situation and that described in the present survey can be strengthened by the use of a multichannel quantization procedure to calculate the eigenvalues for an imaginary or complex field intensity[30]

References

1. A. F. J. Siegert, Phys. Rev., **56**, 750 (1939).
2. R. A. Bain, J. N. Bardsley, J. N. Junker and C. V. Sukumar, J. Phys. B, **7**, 2189 (1974).
3. O. Atabek and R. Lefebvre, Phys. Rev., **22A**, 1817 (1980).
4. C. L. Shoemaker, N. Abusalbi and D. J. Kouri, J. Phys. Chem., **87**, 5389 (1983).
5. R. G. Gordon, J. Chem. Phys., **51**, 14 (1969).
6. B. R. Johnson, J. Chem. Phys., **69**, 4678 (1978).
7. D. W. Norcross and M. J. Seaton , J. Phys. B, **6**, 614 (1973).
8. G. Bergeron, X. Chapuisat and J. M. Launay, Chem. Phys. Lett., **38**, 349 (1976).
9. R. Lefebvre *Intramolecular Dynamics*, J. Jortner and B. Pullman ,Eds., Reidel (1982), p. 55.
10. O. Atabek and R. Lefebvre, Int. J. Quant. Chem., **19**, 901 (1981).
11. R. Lefebvre and M. Garcia-Sucre, Ann. Fond. L. de Broglie, **12**, 195 (1987).
12. J. N. Ginocchio, Ann. Phys., **152**, 203 (1984).
13. D. T. Colbert, R. Mayrhofer and P. R. Certain, Phys. Rev., **33A**, 3560 (1986).
14. B. Simon, Phys. Lett., **71A**, 211 (1979).
15. R. Guardiola and J. Ros, J. Comp. Phys., **45**, 374 (1982).
16. D. G. Truhlar, J. Comp. Phys., **10**, 123 (1972).
17. O. Atabek and R. Lefebvre, Chem. Phys. Lett., **84**, 233 (1981).
18. O. Atabek, R. Lefebvre and M. Jacon, J. Phys. B, **15B**, 2689 (1982).
19. R. Jost, Helv. Phys. Acta **20**, 256 (1947).
20. R. Lefebvre, J. Phys. Chem. **88**, 4839 (1984).

21. R. J. Le Roy and J. S. Carley, Adv. Chem. Phys. **42**, 353 (1980).

22. H. Feshbach, Ann. Phys. **19**, 287 (1962).

23. G. Jolicard and E. J. Austin, Chem. Phys. Lett. **121**, 106 (1985).

24. M. Garcia-Sucre and R. Lefebvre, J. Chem. Phys. **85**, 4753 (1986); Int. J. of Quant. Chem. **20**, 541 (1986).

25. A. Messiah, *Mécanique Quantique* (Dunod, Paris, 1969), p. 197.

26. H. Bremmer, Physica **15**, 593 (1949).

27. R. Lefebvre and M. Garcia-Sucre, in *Stochasticity and Intermolecular Redistribution of Energy*,(Eds. R. Lefebvre and S. Mukamel Reidel, 1987).

28. L. Schlessinger, Phys. Rev. **167**, 1411 (1968).

29. W. P. Reinhardt, Int. J. of Quant. Chem. **21**, 133 (1982).

30. M. Garcia-Sucre and R. Lefebvre, Mol. Phys. **60**, 941 (1987).

TWO STUDY CASES IN THE CALCULATION OF RESONANCES USING THE MULTICHANNEL SCHRÖDINGER EQUATION

Mario A. Natiello and Alejandro R. Engelmann

Department of Quantum Chemistry
Uppsala University
Box 518
S-751 20 Uppsala
Sweden

Abstract

The numerical solution of the Multichannel Schrödinger Equation is discussed as a way to introduce dynamical effects beyond the Born-Oppenheimer approximation. The algorithmic procedure is described and two model examples are displayed where resonances are present. Resonances are calculated using different scaling procedures.

Introduction

The Born-Oppenheimer approximation (BO)[1] has been the usual starting point for molecular and atomic calculations since the early days of Quantum Mechanics. The total wave function for a system of nuclei and electrons is assumed to be a sum of products of nuclear wavefunctions times electronic ones. The electronic functions are calculated as eigenfuctions of the electronic hamiltonian, namely the one obtained from the total hamiltonian neglecting the nuclear kinetic energy terms. The nuclear coordinates are considered to be parameters. The approximation consists further in taking these electronic wavefunctions (for a certain choice of nuclear coordinates, depending on the particular problem) as the quantum state of the system in study. Even in systems where the partition nuclei/electrons may not be evident, there are ways of regarding some coordinates of the system as relatively slow or fast. A partition analog to BO can be considered in these cases[2].

This approximation and the model hence obtained have explained successfully a large number of atomic and molecular properties. The computation of electronic structures in this context has a history of more than 20 years, and it is still evolving. However, there exists a number of problems where the approximation breaks down. This is the case when for example a single electronic wavefunction for a single choice of nuclear coordinates, no longer can describe the system in study within the required accuracy.

The natural step forward within the context of non-relativistic quantum mechanics is to solve the Multichannel Schrödinger Equation (MSE), namely to keep to the partition of the wave function into a sum of products, but taking now into account the nuclear kinetic energy to a higher extent. The high quality of the solutions for the electronic problem reinforces the choice, since the electronic functions needed to go beyond BO can now be calculated with a high level of precision.

Of the many dynamical problems where the MSE is relevant, only the calculation of bound state- and resonance energies will be addressed. Resonances are used to describe peaks in the scattering cross section and also poles of the analytic continuation of the resolvent beyond the first Riemann sheet[3]. Although these concepts are not completely equivalent, we will adopt for the purposes of the paper the latter definition. Resonance states correspond to long-lived "quasi-stable" states, which can be identified with confined classical orbits in configuration space in a potential that allows escape to infinity (the movement "inside" a potential barrier with an energy larger than $V(\infty)$). In the MSE context, resonances can appear as the effect of a coupling between the bound states of a channel with the continuum of another channel[4].

In this paper we describe the algorithms used to solve the MSE, we discuss the use of Complex Scaling[5] within the MSE to calculate resonances and we show results for two model problems: Two Morse potentials coupled with a gaussian function and the Helium atom. The former can be regarded as a reasonable model for intermolecular interactions. A study of the variation of the states with the strength of the coupling is performed.

Other works relating coupled differential equations and Complex Scaling can be found in this volume. The reader is especially encouraged to consult the contributions of Lefèbvre and Elander et al., among others.

The Multichannel Schrödinger Equation

We start with the Schrödinger equation:

$$H\Psi = E\Psi, \tag{1}$$

where

$$H = T_N + T_e + V_{NN} + V_{Ne} + V_{ee} = T_N + H_e \tag{2}$$

and the wave function is assumed to be

$$\Psi(R,r) = \sum_i \Omega_i(R)\phi_i(R,r). \tag{3}$$

The $\phi_i(r,R)$ are the electronic wave functions and they are obtained as the solutions of:

$$H_e(R)\phi_i = E_i(R)\phi_i. \tag{4}$$

The solution of this equation has been a central aim of computational quantum chemistry for many decades. There exist nowadays a number of methods to calculate accurately ϕ_i and E, most of them using the Ritz variational principle. Without loss of generality, we will assume for convenience, that the ϕ_i's are orthogonal under integration of the r-variables, what we indicate as: $<\phi_i|\phi_j>_r = \delta_{ij}$, for each R.

The MSE is the resulting equation for the $\chi_i(R)$ after insertion of expansion (3) into equation (1). Under the assumption that there exists only one relevant nuclear (or slow) coordinate, and assuming angular separability (namely that the coupling parameters depend only on $|R|$), we obtain, for R-spherically symmetric systems (atomic units are assumed):

$$\frac{-1}{2M}\frac{d^2}{dR^2}\left(\sum_i \chi_i\phi_i\right) + \sum_i \chi_i\phi_i E_i(R) = E\sum_i \chi_i\phi_i, \tag{5}$$

where $\chi_i = R\Omega_i$.

After expansion of the product derivative, eq. (5) is left-multiplied by ϕ_j and integrated over the r variables. The MSE is thus obtained:

$$\begin{aligned} &\frac{-1}{2M}\chi_j'' + (E_j(R) - E)\chi_j - \frac{1}{M}\sum_i \chi_i' < \phi_j|\frac{d\phi_i}{dR} >_r - \\ &- \frac{1}{2M}\sum_i \chi_i < \phi_j|\frac{d^2\phi_i}{dR^2} >_r = 0, \quad j = 1, \ldots \end{aligned} \tag{6}$$

where the prime denote derivation with respect to R.

This is a set of coupled second order differential equations. The assumption of spherically symmetric (*"S"*) solutions gives the boundary condition $\chi_k(0) = 0$, as in the analogous one-electron case.

The information of the "electronic" (or fast) coordinates is carried by three sets of functions of R derived from the quantities obtained in the solution of eq. (4), namely the potentials $E_j(R)$ and the two arrays of derivatives $<\phi_j|\frac{d\phi_i}{dR}>_r$ and $<\phi_j|\frac{d^2\phi_i}{dR^2}>_r$. Each E_j represents a "channel". Due to the r-orthogonality, the first derivative functions can be regarded as the entries of an antisymmetric matrix. The analogous second derivative matrix can be written as a sum of a symmetric matrix and an antisymmetric one. The entries of the latter are the derivatives of the entries in the first derivatives matrix[6].

The Algorithm

The procedures used in this article are based on the method of Gordon[7]. The numerical algorithm to solve the MSE is chosen under the assumption that the degree of coupling between the different channels decreases with the energy separation between them, and therefore the set of equations can be truncated to include only N close-lying channels. The choice of N depends on the nature of the problem.

The potentials and derivatives cannot, in general, be obtained in closed analytic form. Rather, a table of values can be computed by solving eq. (4) for different electronic configurations. The numerical integration of the MSE will require these functions to be evaluated at arbitrary values of R, and therefore some interpolation procedure has to be applied to the table of values, to obtain an estimate of the functions for the desired R.

The technique proposed here is to use splines[8], adequating the powers of R used in the interpolation to the behaviour of the function to be approximated: If the function has a R^{-n} behaviour, splines can be performed to $R^n f(R)$, which will now have a polynomic behaviour, and scale it back with R^{-n} when it is to be used.

To obtain the energy eigenvalue and eigenfunction of eq.(6) the actual procedure is as follows:

- Turn the truncated set of second order differential equations into a double-sized set of first order differential equations.

- Decide an integration interval $[x_0, x_1]$.

At this point, the aim is to obtain a value of E and a wave function that is continuous, with continuous derivative and is zero at the boundaries of the integration interval. This wave function can be expressed as a column matrix with the χ_i as entries.

For a given value of E, N linearly independent integrations can be produced that fulfill the boundary condition at x_0 (left integrations). Also, N linearly independent "reverse" integrations (right integrations) can be produced integrating from x_1 (towards x_0) and satisfying the boundary condition at x_1. Only for the value of E corresponding to an eigenstate of the hamiltonian, there will be a linear combination of the left solutions that will be equal to a linear combination of the right solutions at all points in the integration interval. In fact, it is only necessary to assure that at one point inside the interval (the matching point) a linear combination of the left integrations is equal to a combination of the right integrations together with their derivatives. The next steps are, hence:

- Choose a trial value for E.

- Choose a point (x_m) somewhere inside the interval (matching point).
- Integrate in $[x_0, x_m]$, with the initial condition $\varphi^{l\prime}_{ij}(x_0) = \epsilon\delta_{ij}$ (for the i-th integration only one of the components –also the i-th– has non-zero initial derivative).
- Integrate backwards in $[x_m, x_1]$ with similar initial condition in x_1 (also N linearly independent integrations, giving ψ^r_{ij}).

The index i labels the entries in the column vector solution and the index j labels the different linearly independent solutions.

- Build the following matrix with the values of the linearly independent integrations at the matching point x_m (l and r stand for left and right).

$$\mathbf{M}(E) = \begin{pmatrix} \varphi^l_{11}(x_m) & \dots & \varphi^l_{1N}(x_m) & \varphi^r_{11}(x_m) & \dots & \varphi^r_{1N}(x_m) \\ \vdots & \ddots & \vdots & \vdots & \ddots & \vdots \\ \varphi^l_{N1}(x_m) & \dots & \varphi^l_{NN}(x_m) & \varphi^r_{N1}(x_m) & \dots & \varphi^r_{NN}(x_m) \\ \varphi^{l\prime}_{11}(x_m) & \dots & \varphi^{l\prime}_{1N}(x_m) & \varphi^{r\prime}_{11}(x_m) & \dots & \varphi^{r\prime}_{1N}(x_m) \\ \vdots & \ddots & \vdots & \vdots & \ddots & \vdots \\ \varphi^{l\prime}_{N1}(x_m) & \dots & \varphi^{l\prime}_{NN}(x_m) & \varphi^{r\prime}_{N1}(x_m) & \dots & \varphi^{r\prime}_{NN}(x_m) \end{pmatrix} \tag{7}$$

If the value of E used is such that an eigenfunction exists, there must exist a linear combination of the left columns which is equal to another combination of the right columns. In other words, a certain election of the derivatives at the boundaries yields a solution which is continuous and with continuous derivative at x_m, and hence everywhere. This condition is equivalent to requesting the determinant of $\mathbf{M}(E)$, or one of the eigenvalues, to be zero. Hence the procedure is finished as follows:

- Find the zero eigenvalues (or determinant) of $\mathbf{M}$ as a function of E.

$$\mathbf{M}(E)\,\mathbf{c} = 0, \qquad \mathbf{c} = \begin{pmatrix} c^l_1 \\ \vdots \\ c^l_N \\ c^r_1 \\ \vdots \\ c^r_N \end{pmatrix} \tag{8}$$

The corresponding eigenvector will yield the relative weights of the different components in the solution:

$$\psi = \begin{pmatrix} \chi_1 \\ \vdots \\ \chi_N \end{pmatrix} \tag{9}$$

and

$$\chi_i(R) = \begin{cases} \sum_j \varphi^l_{ij}(R)c^l_j, & x_0 \leq R \leq x_m \\ \sum_j \varphi^r_{ij}(R)c^r_j, & x_m \leq R \leq x_1 \end{cases} \tag{10}$$

The integration and the search for zero eigenvalues can be performed with standard algorithmic packages. The applications discussed below were carried out using the *NAg*[8] program library.

The approximations involved can be summarised as:

1. The number of equations is truncated to a finite value N.
2. The electronic potentials are not exact, but just interpolated ones.
3. The integration limits are not the exact ones (usually $[0, \infty)$).
4. The integration procedure is approximated. This is probably the less critic problem, since the procedure can be checked against standard model equations with solutions known in closed-form.

The method looses accuracy when E is close to a threshold. In this case, the wavefunction is significantly different from zero for large values of R and it is therefore necessary to use large integration domains. This in turn might lead to numerical overflows.

Case 1: A molecular model

The Morse potential has been widely used to describe intermolecular interactions. It seems thus relevant the study of two similar Morse potentials, separated by a gap in energy, and coupled by a gaussian function describing the off-diagonal part of the second derivative matrix. This can be thought of as representing the interaction of two systems with two electronic states ($N = 2$). The antisymmetric first derivative matrix is zero and thus the second derivative matrix becomes symmetric. This is a necessary condition for the hermiticity of the model hamiltonian. Hence, we have:

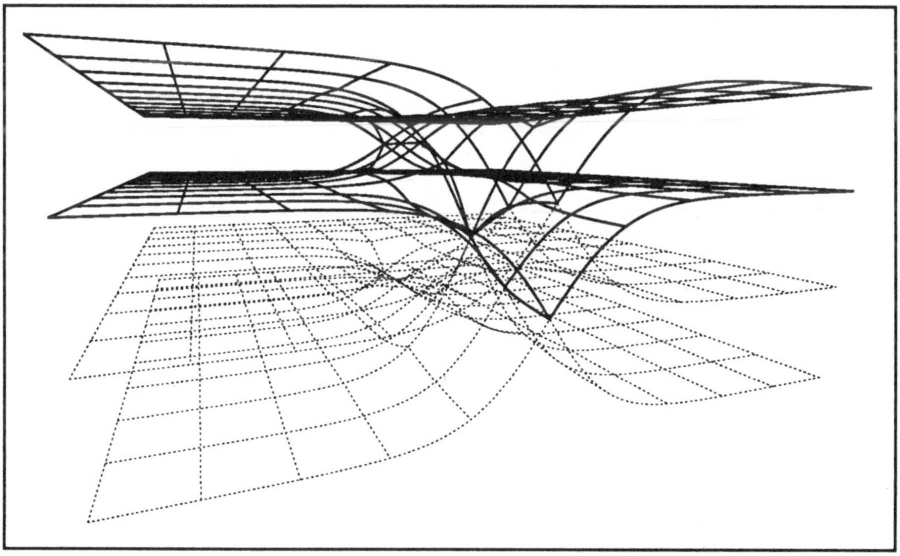

Figure 1

A schematic view of a two-threshold Riemann surface. Connections between different sheets corresponding to the first (lower) threshold are shown in full line.

$$E_1(R) = D[\left(1 - e^{-\beta(R-R_0)}\right)^2 - 1]$$

$$E_2(R) = C + E_1(R)$$

$$< \phi_1 | \frac{d\phi_2}{dR} >= 0, \quad \forall i, j$$

$$< \phi_1 | \frac{d^2\phi_2}{dR^2} >= Ae^{-\left(\frac{(R-r_1)}{\sigma}\right)^2}$$

where the constants used are $C = 4$, $D = 10$, $R_0 = 1$, $\beta = 1.5$, $\sigma = 4$, $r_1 = 2.25$. The mass is chosen to be 1.

The dependence of the eigenvalues of this model problem with the amplitude A of the coupling was studied. For $A = 0$, the system has 6 bound states, three of them corresponding to the lower Morse channel, and the other three shifted correspondingly in C. The Riemann surface of this model is more complicated than that of a single particle problem (which can be imagined as the Riemann surface of the square root in the field of complex numbers). It consists of the

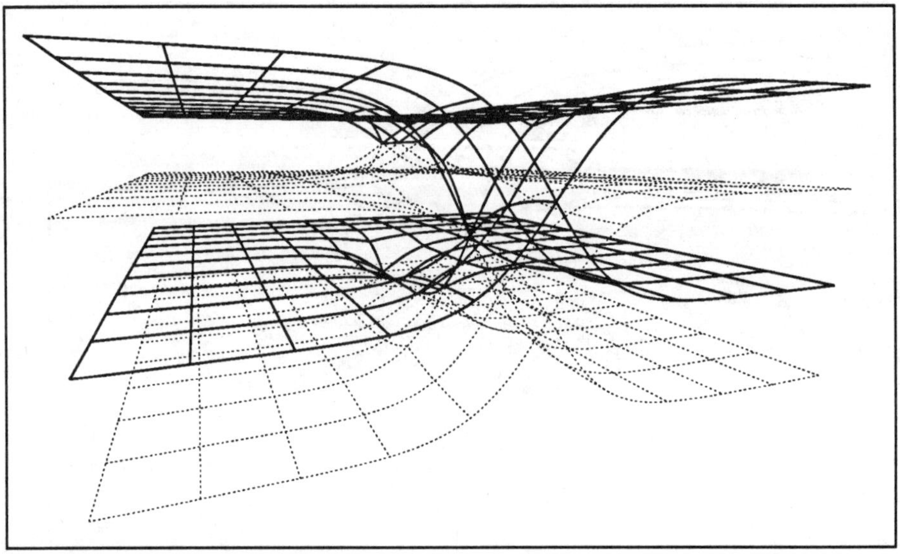

Figure 2

A schematic view of a two-threshold Riemann surface. Connections between different sheets corresponding to the second (higher) threshold are shown in full line.

superposition of two continua, starting at the thresholds $E = 0$ and $E = C$, and four different sheets can be distinguished. A pictorial description of this surface can be seen in figs. 1 and 2. The "vertical" axis in those figures is chosen so that a better appreciation of the connections between surfaces at both continua can be obtained. Different Riemann sheets are displayed in full lines in each figure to better appreciate the crossings. Fig. 1 stresses the structure of the first (lower) threshold, and fig. 2 concentrates in the second threshold.

Two of the eigenvalues of the higher channel lie in the continuum of the lower one, thus becoming resonances when the coupling is switched on ($A \neq 0$) [4].

Using the method of Complex Scaling[5], resonances can be regarded as certain "bound" states of a modified hamiltonian problem. Complex Scaling has been discussed in other contributions to this Conference. See particularly the contribution of Balslev, Elander and Brändas.

From the algorithmic point of view, this technique requires to perform the following transformation in the integration coordinate R:

$$R \to e^{i\theta} R, \qquad (CS)$$

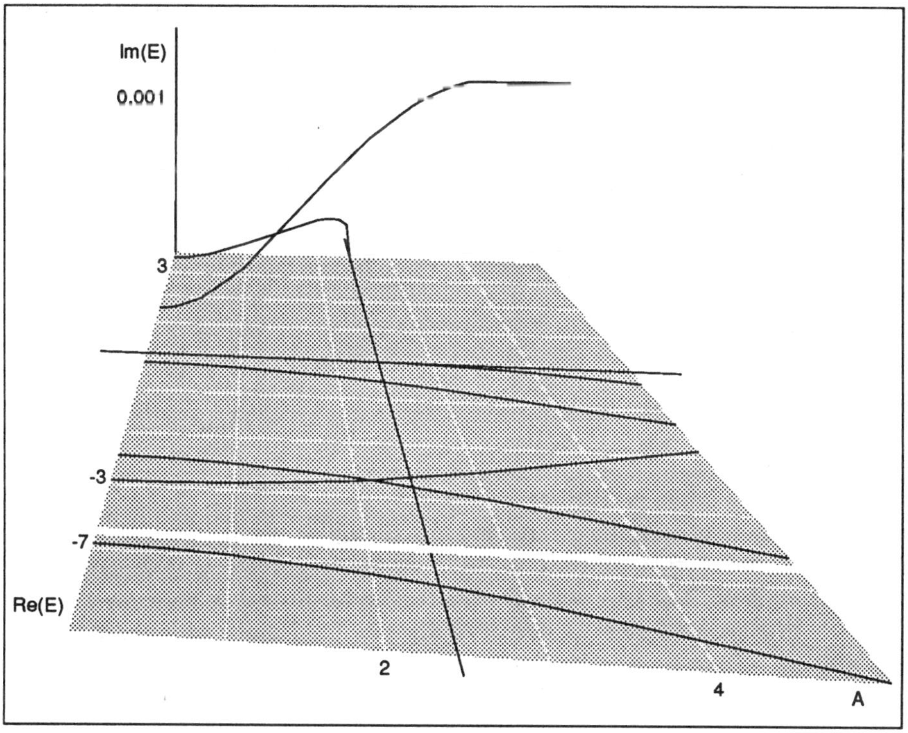

Figure 3

Variation of the energy eigenvalue of the discrete states of Case 1 with the amplitude A.

where $\theta \in [0, 2\pi]$.

Eq. (6) becomes then:

$$\frac{-1}{2M}\chi_j'' + e^{2i\theta}(E_j(e^{i\theta}R) - E)\chi_j - \frac{e^{i\theta}}{M}\sum_i \chi_i' < \phi_j | \frac{d\phi_i}{d(e^{i\theta}R)} >_r$$
$$- \frac{e^{2i\theta}}{2M}\sum_i \chi_i < \phi_j | \frac{d^2\phi_i}{d(e^{i\theta}R)^2} >_r = 0, \quad j = 1, \ldots N \qquad (6.CS)$$

The quantities in eq. $(6.CS)$ are now complex. By separating into real and imaginary parts, this equation can be transformed into a double-sized set of real equations, which can be solved in the same way as the real (non-scaled) case.

The variation of the six eigenvalues with the amplitude A is displayed in fig. 3. Some remarks are relevant. First, it is to be noticed that the eigenvalues corresponding to the lower channel in the uncoupled case decrease their energy with increasing A, while the remaining real eigenvalue increases, as well as the real part of the two resonance eigenvalues. This can be interpreted saying that the action of the coupling is somewhat opposite in each channel. The inspection of the relative signs of each component of the three lowest states at a given value of A supports this hypothesis: The states 0 and 2, (coming from the lower channel) have both components with roughly the same relative sign, while state 1 does not (see fig. 5).

When the real part of the energy of the resonance eigenvalues approaches the threshold value $E = C$, the imaginary part diminishes, so that the eigenvalue is "absorbed" by the threshold. At the same value of A (roughly 2.5) a "new" bound state arises. The connection between these two phenomena cannot be *proved* by this numerical calculation. At most, it can be said that the calculation suggests that a resonance absorbed by the threshold at $E = C$ apparently coincides with the appearance of a bound state, in such a way that the number of discrete states seems to be constant throughout the variation of A.

Finally, it has to be noted that two of the eigenvalues cross. In fig. 4 and 5 the wave functions before and after the crossing are plotted. It is apparent that each state somehow keeps its identity throughout the crossing, with no apparent transfer of information from one state to the other. This is quite different to the situation where a so-called "avoided crossing" is present[9].

Case 2: The Helium atom

The calculation of the states of Helium has been a subject of great interest since the early days of Quantum Mechanics. The purpose of this study is to use this problem as an example of the possibilities of the joint use of the MSE, hyperspherical coordinates and Exterior Scaling, rather than attempting to improve the very accurate calculations known since 25 years ago.

This example contains a different partition of the Schrödinger equation, namely that of the hyperspherical coordinates[10]. This consists in the election of a coordinate system to describe the problem having one radial variable and $3n-1$ "angular" ones. In the case of the Helium atom ($n = 2$ after separation of the global translational motion), the radial variable is taken to be $R = \left(r_1^2 + r_2^2\right)^{1/2}$, where the r_i are the nucleus-electron distances. R is the "slow" variable.

Following a procedure described by Klar & Klar [11], using a partition similar to that of eq. (3), the equation corresponding to eq. (4) becomes:

$$\left(\frac{\Lambda^2 + 15/4}{2R^2} + \frac{V}{R}\right) \phi_n(R,\omega) = E_n(R)\phi_n(R,\omega), \tag{11}$$

where Λ is the hyperangular momentum operator and V collects the hyperangular part of the three coulombic interactions present.

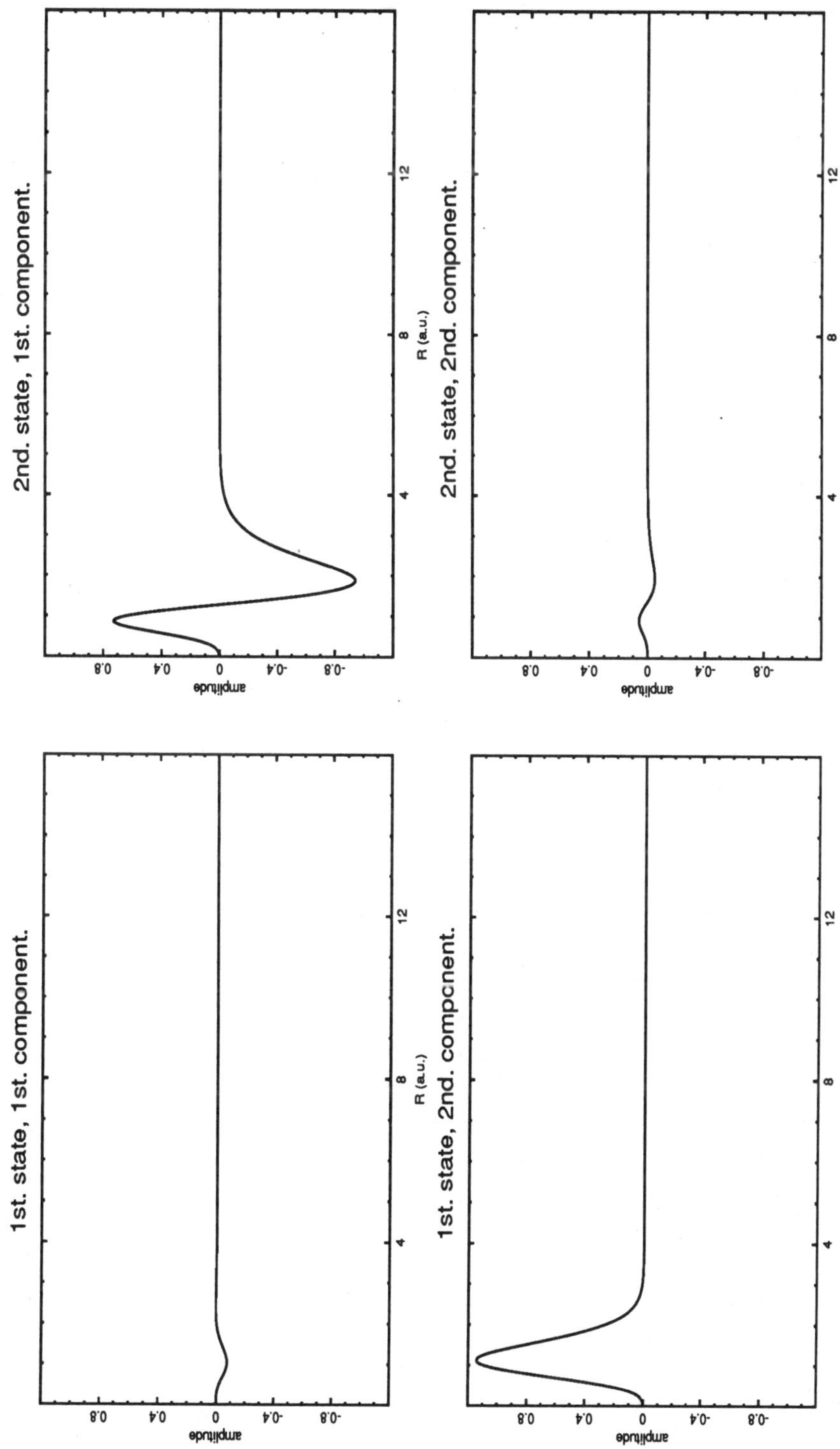

Figure 4: Wavefunctions of states 1 & 2 of Case 1 before the crossing. A=0.5

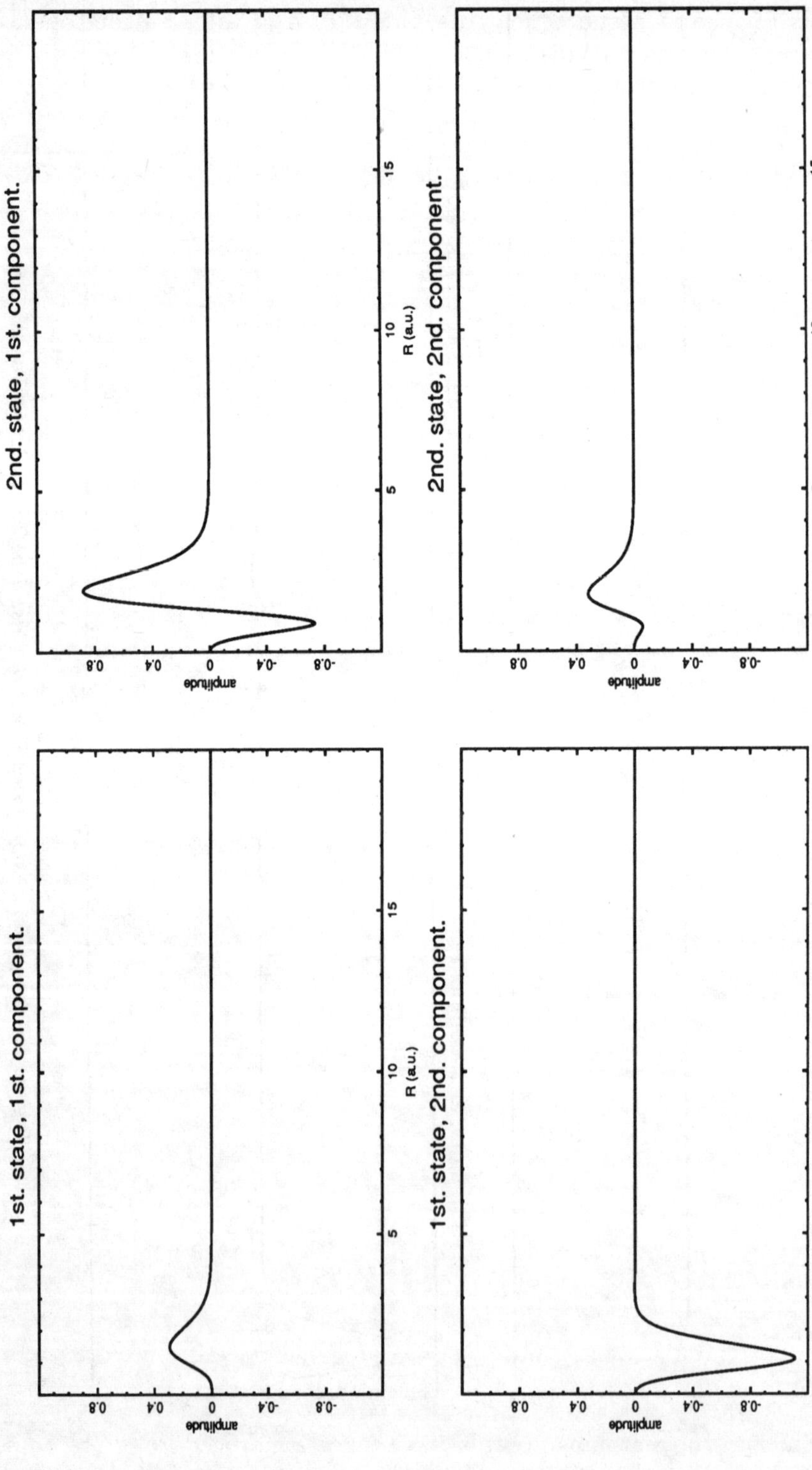

Figure 5: Wavefunctions of states 1 & 2 of Case 1 after the crossing. A=2.5

This equation can be solved by truncation and expansion on the basis of eigenfunctions of Λ, and a subsequent diagonalization of the resulting operator matrix. The remaining hyperradial equation is formally similar to Eq.(6), where $M \simeq 1 a.u.$.

One of the difficulties with this approach is that the energies arising from the angular equation, eq. (11), have wrong asymptotic behaviour, because of the truncation to a finite matrix problem. This energies should fulfill:

$$E_k(R) \quad \stackrel{R\to\infty}{\longrightarrow} \quad -\frac{2}{k^2},$$

since one of the possible realizations of the limit $R \to \infty$ is the configuration with one electron free and the other attached to the nucleus; i.e. a free electron plus a hydrogenic atom with charge $Z = 2$. However, the limit values for the E_k's obtained from eq. (11), are equal to 0.

Thus, the integration limits x_0, x_1 are forced to be very inaccurate, since x_1 must be small enough to still have a reliable approximation to the E_k's for all R in the integration interval, and x_0 cannot be set to zero because the energies diverge and numeric problems appear. Moreover, the lack of an analytic expression for the angular functions prevents the use of the standard framework of Complex Scaling to calculate the possible resonances.

The procedure proposed in this work is to use an estimated analytic tail for the large R behaviour of the angular functions, $(E_k(R) \sim a/R + b/R^2 - 2/k^2)$[11] setting a and b so that the tail has continuous matching with the tabulated function, with continuous derivative. The derivative matrices are matched as well to decaying exponentials $(f_i(R) \sim (a + br)e^{-c(R-R_0)}$, where a and b are determined as above, and the decaying rate c is still a free parameter. At this stage, the possible resonances can be calculated using Exterior Scaling[12]. The description of this procedure can also be found in other contributions to this volume. We briefly summarise its algorithmic consequences:

The transformation in R is now:

$$\begin{cases} R \to R, & R \leq R_0; \\ R \to R_0 + e^{i\theta}(R - R_0), & R > R_0. \end{cases} \qquad (ES)$$

The numeric procedure consists in performing a left integration following eq.(6) and a right integration according to eq.(6.CS). Again, the solutions will be complex-valued. The matching conditions in $\mathbf{M}(E)$ have to be changed, due to the discontinuity of the scaling transformation (ES): The solution ψ has now a jump in $e^{-i\theta}$ in the derivative at the matching point $x_m = R_0$.

For convenience, the matching point of the integration procedure can be chosen as the scaling starting point R_0. Since the analytic tails do not fit exactly the true curves, the calculation might now be spuriuosly dependent on the matching point.

The energy of the ground state of Helium was calculated by Pekeris[13] and more recently by Morgan[14] both using a variational approach. The first few significant figures for this value are $E = -2.90372 a.u.$ (calculated with 440 variational parameters).

The present method is, on the contrary, non variational. It contains nevertheless parameters (x_0, x_1, a, b, etc.) which are not determined variationally, but rather by the numeric constraints that the procedure has. It is worth noting, however, that in this particular example the solution of eq. (11), was chosen to be variational. This is not actually essential to the integration of the MSE, this being just one of the possible methods to produce the R functions required to integrate the hyperradial equation.

Table 1

Some calculations of the ground state of He.[1−3]

Λ_{max}	x_m	Energy ($a.u.$)
48	4.50	-2.90233
48	4.75	-2.90234
48	5.00	-2.90234
58	4.50	-2.90272
58	4.75	-2.90271
58	5.00	-2.90272

Notes:

1 Only the three lowest $E_k(R)$ curves were included in the calculation.
2 c was set equal to unity.
3 The accepted value[13,14] is $E = -2.90372 a.u.$

Some calculations of the ground state energy of Helium are displayed in Table I. The integration interval is $I_1 = [0.01, 40]$ when using hyperangular functions up to $\Lambda_{max} = 48$ and $I_2 = [0.005, 40]$ when Λ max $= 58$. In both cases only the 3 lowest E_k curves are coupled ($N = 3$). The agreement with the accepted value is better than 0.05%, having invested only a moderate amount of computational resources.

The $(2s^2)$ resonance of Helium has also been an interesting topic for many years. It can be interpreted as the coupling (through the MSE) of the "eigenstate" that can be associated to the second potential curve alone, with the continuum of the first (lowest) potential. In table II some results concerning this resonance are displayed. The coincidence with the accepted value [15] is quite satisfactory. The large value of x_m required to obtain a good agreement indicates, however, that an accurate representation of the $E_k(R)$ and the derivative coupling functions is crucial for the description of this resonance. This representation is better accomplished with eq.(11) than with the analytic tails, enven for moderately large values of R (ca. $10 a.u.$).

Table 2

Some calculations of the $(2s^2)$ resonance in He.[1,2]

Λ_{max}	c	x_m	Energy eigenvalue $(a.u.)$
48	0.5	4.5	$-0.7922 \pm i1.64 \times 10^{-3}$
48	1.0	4.5	$-0.7925 \pm i1.39 \times 10^{-3}$
48	2.0	4.5	$-0.7929 \pm i0.85 \times 10^{-3}$
58	1.0	4.5	$-0.7932 \pm i1.40 \times 10^{-3}$
58	1.0	4.7	$-0.7883 \pm i1.54 \times 10^{-3}$
58	1.0	8.0	$-0.7738 \pm i2.27 \times 10^{-3}$
58	1.0	9.0	$-0.7739 \pm i2.25 \times 10^{-3}$
58	1.0	10.0	$-0.7739 \pm i2.28 \times 10^{-3}$

Notes:

1 An accurate calculated value is $E = -0.7784 \pm i\, 2.26 \times 10^{-3} a.u.$ It was obtained in Ref. 15 with a basis set method. Other calculations are cited there as well.

2 Only the three lowest $E_k(R)$ curves were included in the calculation.

Concluding remarks

The purpose of this work was to contribute to this Conference describing the actual procedure to calculate resonances in a context somewhat less elementary than the study of one-electron model potentials. As such, the effort was concentrated in the methodological approach, using some simple examples to better illustrate what can be calculated.

Other contributions to this volume describe a different approach to the study of resonant states, namely to integrate the time-dependent Schrödinger equation (see the paper of Yngve Öhrn). The computational effort invested in this method can be adapted with little extra work to implement this approach on a discretised R, t mesh. This would require to assume a shape for the wave function at $t = 0$ (as a discretised table of values in R) and to integrate the first-order time-dependent Schrödinger equation, studying the resulting wavefunction for different values of t. This amounts roughly to perform similar tasks to those already achieved in this work, namely to integrate a first order differential equation, to operate with derivation and multiplication operators and to evaluate inner products and integrals. Work on this line is in progress.

Acknowledgments

One of us (M.A.N.) acknowledges C.O.N.I.C.E.T. (Consejo Nacional de Investigaciones Científicas y Técnicas de la República Argentina) for partial financiation of this work through a scholarship. He also thanks the organisers of the Conference at Lertorpet (Värmland, Sweden) for the invitation to participate. The authors acknowledge also the valuable comments of Prof. Jan Linderberg.

References

[1] M. Born and R. Oppenheimer, Ann. d. Phys. 84, 457 (1927).

[2] J. Linderberg Int. J. Quantum Chem. Q.C.S. 15, 559 (1981).

[3] See a discussion in Ref. 4, chapter 12-6.

[4] M.Read and B.Simon Methods of Modern Mathematical Physics, vol.4, Academic Press, N.York, 1979.

[5] S. Aguilar and J. Combes, Commun. Math. Phys. 22, 265 (1971) and E. Balslev and J. Combes, Commun. Math. Phys. 22, 280 (1971).

[6] H. Ågren, A. Flores-Riveros, H. J. Aa. Jensen, Phys. Rev. A 34, 4606 (1986).

[7] R. G. Gordon, J. Chem. Phys. 51, 14 (1969), and A. M. Dunker and R. G. Gordon, J. Chem. Phys. 64, 4984 (1976).

[8] Fortran Library Manual Mark 9, Numerical Algorithms Group, Oxford, U.K. (1982), and references therein.

[9] M.A. Natiello, E.J. Brändas and A.R. Engelmann, Int. J. Quantum Chem. Q.C.S. 21, 555 (1987).

[10] J. Linderberg in Intermolecular Dynamics, pp. 235, D. Reidel, 1982. See also his contribution in this volume.

[11] H. Klar and M. Klar, J. Phys. B 13, 1057 (1980) and Phys. Rev. A 17,1007 (1978).

[12] B. Simon, Phys. Lett. 71A, 211 (1979).

[13] C. L. Pekeris, Phys. Rev. 126, 1470 (1962).

[14] J. Baker, D. E. Freund, R. Nyden Hill and J. D. Morgan III, Phys. Rev. A, in press.

[15] A. Macías and A. Riera, Phys. Lett. A. 119, 28 (1986).

MULTICHANNEL COMPLEX SCALED TITCHMARSH WEYL THEORY A MODEL FOR DIATOMIC FRAGMENTATION

Erkki Brändas

Department. of Quantum Chemistry
University of Uppsala
Box 518
S-751 20 Uppsala
SWEDEN

Magnus Rittby[†] and Nils Elander

Manne Siegbahn Institute of Physics
Frescativägen 24
S-014 05 Stockholm
SWEDEN

Abstract

A single channel Titchmarsh-Weyl resonance scattering theory is extended to a coupled equations formulation. It is demonstrated that the intrinsic simplicity of the theory persists also in the multichannel case. Explicit theorems in connection with the multichannel case are proved, and simple Wronskian expressions for resonant scattering wave functions and related physical quantities are derived. The computational features are illustrated by an application to a non-trivial curve crossing problem. The numerical results display the detailed behaviour of resonance poles in the complex plane. The relevant diabatic and/or adiabatic limits as well as the effect of the strength of the interaction on the resonance structures are studied. Applications to the modeling of laser photofragment spectroscopy data are finally mentioned

1. Introduction

Several physical processes can be described in terms of a set of coupled differential equations. The purists formulation of the Hartree-Fock equations and the diabatic close coupling formulations for the vibrational motions in a diatomic molecule are examples from atomic and molecular physics. The methods for finding bound eigenvalues to these problems are standard in a theorists toolbox. When it comes to dynamical problems involving, in particular resonance phenomena, each theorist has his/her own tricks which all have the common feature of trying to avoid, one way or the other, the terrible divergency associated with the wellknown Gamow waves. The formally correct treatment of these problems originated in the classical works of Weyl[1] and Titchmarsh[2]. The development of the complex rotation theory due to Aguilar, Balslev and Combes [3–6] led to new insights regarding resonance formation. Besides its mathematical rigour this theory is also of considerable practical value in connections with computations in scattering theory.

In the treatment of atomic collisions one is often confronted with the problem of solving a set of coupled radial equations. These equations result after an integration over electronic coordinates and an appropriate separation of the angular coordinates of the system, see e.g. refs. 7-9. The numerical integration of these coupled equations provides us with information concerning cross sections, possible bound states, etc.

An important phenomenon in atomic collisions is the occurence of resonances, i.e. quasi-bound states of the diatomic system. These resonances can also be observed in molecular spectroscopy where the phenomenon is usually referred to as predissociation. From the spectroscopist's point of view, the resonance states are bound states that are somehow perturbed by a continuum state. These perturbations are represented as off-diagonal matrix elements of the Hamiltonian, see e.g. ref. 10.

In a general theoretical description of the dynamical processes in atomic and molecular physics we need all possible interactions between bound and continuum states to be taken into account. Depending on the representation used to describe the electronic many-body problem, different potential energy curves and couplings are obtained. In this way the adiabatic and diabatic pictures can be defined. Ignoring the couplings and chosing the adiabatic states as the zeroth order problem we obtain the Born-Oppenheimer approximation, wellknown to molecular physisists. For diatomic molecules these curves obey the, so called, non-crossing rule. The separation of electronic and nuclear motion leading to the Born-Oppenheimer approximation gives a highly simplified but in many cases relevant picture of the actual situation.

Phenomena such as predissociations or charge transfer processes in atomic collisions, however, can not be formulated in a strict Born-Oppenheimer picture. For instance, an *avoided crossing*, is a typical feature associated with such a breakdown. To correct the deficiency of this approximation an analysis incorporating the relevant interaction elements is necessary. Furthermore, in addition to a detailed study of the actual potential energy curves in the non-crossing region (as functions of *real* internuclear distances), one may try to locate the corresponding crossing point in the complex plane (for a *complex* internuclear distance!). This crossing point is of fundamental importance in a semi-classical formulation of the problem [11,12].

Alternatively, one may try to extract the dynamical properties of the system without a detailed knowledge of the interaction elements. The transition properties of the system are in this picture dependent on the narrow non-crossing region on the real axis. The situation is formally related to the analysis of avoided crossings in the study of stabilization graphs. Studies along these lines have been presented by Simons,[13] see also supporting analysis by Löwdin augmented to ref. 13.

In the theory of resonances, stabilizations graphs occur as an indication of some form of *spectral concentration.* We have previously[14–16] discussed the connection between Weyl's theory for singular second order differential equations and the continuous spectrum with regards to the occurence of predissociating levels in diatomics. We have also presented attempts to formulate a multichannel generalization of the Titchmarsh-Weyl theory [17,18].

Recently the case of coupled equations were considered in connection with charge transfer processes in atomic collisions [19]. In the following section, we present a more detailed description of the Titchmarsh-Weyl's theory for coupled equations. For clarity we have divided it into subsections. Section *2.1* contains the theoretical generalization of the single channel theory the corresponding multichannel version, section *2.2* relates this developments to conventional multichannel scattering theory, and section *2.3* treats the complex scaled analogue of the theory.

The utility of the present development is illustrated in section *3* by an application to a curve crossing model problem describing the lowest electronic states in the diatomic radical *MgH.* For completeness, formal relations between adiabatic and diabatic potential energy curves are given in an appendix.

In section *4* we emphasize some of the advantages offered by the present theory used as a tool to analyze photofragment spectroscopy data. A summary with conclusions are given in section *5.*

2. Theory

2.1 Titchmarsh-Weyl's Theory for a Set of Singular Second Order Differential Equations - A Way of Analyzing the Coupled Equations for a Continuum Problem

In the single channel case, the Titchmarsh-Weyl theory assumes a very simple and condensed form, see e.g. refs. 20 and 21. In particular, key quantities as spectral densities, Green's functions, Jost functions and scattering matrices are cornerstones in the formulation. As we have stressed earlier these fundamental building blocks can be analytically continued into higher order Riemann sheets and, most importantly, computed numerically there. In this section we will show that essentially the same features, as in the single equation case, appears also for a set of coupled second order differential equations.

Consider a physical problem including coupling between different zeroth order states $|\Psi_i\rangle$ of a one dimensional, usually radial, variable r, which in the non-relativistic framework can be described by the Schrödinger equation:

$$\Big(-\frac{d^2}{dr^2}\mathbf{1} + \mathbf{V}(r) - E\mathbf{1}\Big)\mathbf{\Psi} = \mathbf{0}$$

Here the diagonal potentials, $V_{ii}(r)$, of the potential matrix $\mathbf{V}(r)$ represents the zeroth order, uncoupled, description whereas the off-diagonal potentials, $V_{ij}(r)$, $i \neq j$, in a perturbation sense, give the interaction between the zeroth order states to all orders.

We rewrite this set of coupled equations as

$$\mathbf{Y}'' = \{\mathbf{V} - \lambda\mathbf{1}\}\mathbf{Y} \tag{1a}$$

where

$$\lambda = E + \imath\varepsilon \tag{2}$$

defines a complex energy in which the positive quantity ε will be taken to the limit $\varepsilon \to 0$ before analytically continuing λ to $E - \imath\Gamma/2$ thereby fullfilling resonance boundary conditions to be defined later. Here the real part, E, is the location of the resonance on the real energy axis and Γ is the width of the resonance. The potential matrix $\mathbf{V} \equiv \mathbf{V}(\mathbf{r})$ is a (usually) symmetric interaction matrix (satisfying standard boundary conditions for scattering theory and /or Weyl's limit point[2])

$$\mathbf{V} = \tilde{\mathbf{V}} \tag{3}$$

supposed to be diagonal as $r \to 0$ and as $r \to \infty$. In what follows all selected values of the coordinate belong to the positive half axis R^+. We further assume that the potential is less singular than r^{-2} as $r \to 0$ and goes to zero faster than r^{-2} at infinity. With minor modifications, extensions to more general situations are possible within Weyl's limit-point classification[2].

To the set of N coupled equations there exist $2N$ linearly independent solutions. Below we will make use of the solution matrix $\mathbf{Y}$,

$$\mathbf{Y} = (\mathbf{y}^{(1)}\mathbf{y}^{(2)}\ldots\mathbf{y}^{(N)}) \tag{4}$$

defined by :

$$\mathbf{y}^{(i)} = \begin{pmatrix} y_1^{(i)} \\ y_2^{(i)} \\ \vdots \\ y_N^{(i)} \end{pmatrix} \tag{5}$$

Because of eq.(3) we also have

$$\tilde{\mathbf{Y}}'' = \tilde{\mathbf{Y}}(\mathbf{V} - \mathbf{1}\lambda) \tag{1b}$$

The individual solution vectors satisfy

$$(\mathbf{y}^{(\mathrm{i})})'' = (\mathbf{V} - \mathbf{1}\lambda)\mathbf{y}^{(\mathrm{i})} \tag{6a}$$

$$(\tilde{\mathbf{y}}^{(\mathrm{i})})'' = \tilde{\mathbf{y}}^{(\mathrm{i})}(\mathbf{V} - \mathbf{1}\lambda) \tag{6b}$$

We further define the logarithmic derivative solution matrix

$$\mathbf{R} = \mathbf{Y}'\mathbf{Y}^{-1} \tag{7a}$$

$$\tilde{\mathbf{R}} = \tilde{\mathbf{Y}}^{-1}\tilde{\mathbf{Y}}' \tag{7b}$$

The logarithmic derivative in eq.(7) is a solution of the matrix Riccati equation

$$\mathbf{R}' = (\mathbf{V} - \mathbf{1}\lambda) + \mathbf{R}^2 \tag{8a}$$

or since V is symmetric

$$\tilde{\mathbf{R}}' = (\mathbf{V} - \mathbf{1}\lambda) + \tilde{\mathbf{R}}^2 \tag{8b}$$

For the development below we will need the following theorems.

Theorem 1: If $\boldsymbol{\chi}$ and $\tilde{\boldsymbol{\chi}}$ are solutions to eq.(1a) and eq.(1b) respectively, then the matrix

$$\mathbf{T} = \tilde{\boldsymbol{\chi}}\boldsymbol{\chi}'$$

satisfies $\mathbf{T} - \tilde{\mathbf{T}} =$ constant matrix, independent of the coordinate r

Proof: Define

$$\boldsymbol{\chi}^{(i)} = \begin{pmatrix} \chi_1^{(i)} \\ \chi_2^{(i)} \\ \cdots \\ \chi_n^{(i)} \end{pmatrix}$$

such that :

$$\mathbf{T}_{\mathrm{ij}} = \{\tilde{\chi}\chi'\}_{\mathrm{ij}} = \tilde{\chi}^{(\mathrm{i})}(\chi^{(\mathrm{j})})'$$

$$\tilde{\mathbf{T}}_{ij} = \{\tilde{\chi}'\chi\}_{ij} = (\tilde{\chi}^{(i)})'\chi^{(j)}$$

Since $\boldsymbol{\chi}^i$ and $\tilde{\boldsymbol{\chi}}^i$ satisfy eq.(6a) and eq.(6b) respectively it follows that

$$(\mathbf{T}_{ij})' = (\tilde{\chi}^{(i)})'(\chi^{(j)})' + \tilde{\chi}^{(i)}(\mathbf{V} - \mathbf{1}\lambda)\chi^{(j)} = (\tilde{\mathbf{T}}_{ij})'$$

or $\mathbf{T}' = \tilde{\mathbf{T}}'$ $\forall$ r. Hence $\mathbf{T} - \tilde{\mathbf{T}}$ is a constant matrix independent of r. **Q.E.D.**

Corollary 1.1: If $\mathbf{T} = \tilde{\mathbf{T}}$ for some $r = a$ then $\mathbf{T} = \tilde{\mathbf{T}}$ $\forall$ r.

Corollary 1.2: If $\boldsymbol{\chi}$ and $\boldsymbol{\phi}$ are solutions to eq(1) then:

$$\mathbf{W}\{\boldsymbol{\chi}, \boldsymbol{\phi}\} = (\tilde{\chi}\phi' - \tilde{\chi}'\phi) = \text{ constant matrix independent of } r$$

Theorem 2: The logarithmic derivative matrices

$$\mathbf{R} = \mathbf{Y}'\mathbf{Y}^{-1} \ ; \ \tilde{\mathbf{R}} = (\tilde{\mathbf{Y}})^{-1}\tilde{\mathbf{Y}}'$$

satisfying the Riccati equations eq.(8) are

a) invariant with respect to any arbitrary linear (non-singular) coordinate independent transformation.

b) they fulfil $\mathbf{R} = \tilde{\mathbf{R}}$ $\forall$ r if $\mathbf{R}(a) = \tilde{\mathbf{R}}(a)$ for some $r = a$.

Proof:

a) Let $\chi \equiv \mathbf{Y}\mathbf{d}$ with the determinant $det(\mathbf{d}) \neq 0$.

Hence $\mathbf{d}^{-1}$ exists, and we obtain:

$$\chi'\chi^{-1} = \mathbf{Y}'\mathbf{d}(\mathbf{Y}\mathbf{d})^{-1} = \mathbf{Y}'\mathbf{d}\mathbf{d}^{-1}\mathbf{Y}^{-1} = \mathbf{R}$$

b) From $\mathbf{R}(a) = \tilde{\mathbf{R}}(a)$ follows

$$\mathbf{T}(\mathrm{a}) = \tilde{\chi}(\mathrm{a})\mathbf{R}(\mathrm{a})\chi(\mathrm{a}) = \tilde{\chi}(\mathrm{a})\tilde{\mathbf{R}}(\mathrm{a})\chi(\mathrm{a}) = \tilde{\mathbf{T}}(\mathrm{a})$$

Hence $\mathbf{T} = \tilde{\mathbf{T}}$ $\forall$ r.

From this follows that $\forall$ r:

$$\mathbf{R} = (\tilde{\chi})^{-1}\mathbf{T}\chi^{-1} = (\tilde{\chi})^{-1}\tilde{\mathbf{T}}\chi^{-1} = \tilde{\mathbf{R}}$$

We will now proceed to define the necessary solution matrices to eq.(1a) as defined by their boundary conditions.

The solution matrix regular at the origin is given by :

$$\boldsymbol{\Psi} = (\, \boldsymbol{\psi}^{(1)} \quad \boldsymbol{\psi}^{(2)} \quad \ldots \quad \boldsymbol{\psi}^{(\mathrm{N})} \,) \tag{9a}$$

where:

$$\boldsymbol{\psi}^{(\mathrm{i})} = \begin{pmatrix} \psi_1^{(i)} \\ \psi_2^{(i)} \\ \vdots \\ \psi_N^{(i)} \end{pmatrix} \tag{9b}$$

and the components are defined by:

$$\psi_k^{(i)} = \lim_{r \to 0} \delta_{ki} c_\ell r^{\ell+1} \tag{10}$$

or by the boundary conditions (useful e.g. when $\ell = 0$):

$$\begin{cases} \psi_k^{(i)} = \delta_{ki} \cos \alpha_i \\ (\psi_k^{(i)})' = \delta_{ki} \sin \alpha_i \end{cases} \quad ; \text{ at } \quad r = a > 0 \quad \text{and} \quad 0 \le \alpha_i < \tfrac{\pi}{2} \tag{11a}$$

Here the diagonal matrix $\boldsymbol{\alpha}$ with $(\boldsymbol{\alpha})_{ij} = \alpha_i \delta_{ij}$ is chosen so that $\boldsymbol{\psi}$ is regular at the origin.

A third choice also is possible:

$$\begin{cases} \psi_k^{(i)} = \delta_{ki} \frac{a^{\ell+1}}{(2\ell+1)!!} \\ (\psi_k^{(i)})' = \delta_{ki} \frac{(\ell+1)a^\ell}{(2\ell+1)!!} \end{cases} \quad ; \text{ at } \quad r = a > 0 \tag{11b}$$

In general it could happen that different channels exhibit different quantum numbers ℓ. This would lead to the replacement of ℓ by ℓ_i in eqs.(10) and (11b). See also eq.(16) below. The Jost solution matrices:

$$\mathbf{F}^{\pm} = (\, \mathbf{f}^{(1)\pm} \quad \mathbf{f}^{(2)\pm} \quad \ldots \quad \mathbf{f}^{(\mathrm{N})\pm} \,) \tag{12}$$

where :

$$\mathbf{f}^{(\mathrm{i})\pm} = \begin{pmatrix} f_1^{(i)\pm} \\ f_2^{(i)\pm} \\ \vdots \\ f_N^{(i)\pm} \end{pmatrix} \tag{13a}$$

and the components are defined by the conditions:

$$f_k^{(i)\pm} = \lim_{r\to\infty} \delta_{ki} \imath^{\ell} \exp(\pm \imath k_i r) \tag{13b}$$

$$k_i = \lim_{r\to\infty} \sqrt{\lambda - V_{ii}(r)}; \tag{13c}$$

$$(\mathbf{k})_{ij} = k_i \delta_{ij}$$

To establish a connection between $\mathbf{\Psi}$ and $\mathbf{F}$ we define the irregular solution matrix:

$$\mathbf{\Phi} = (\boldsymbol{\phi}^{(1)} \quad \boldsymbol{\phi}^{(2)} \quad \ldots \quad \boldsymbol{\phi}^{(N)} \ ) \tag{14a}$$

with

$$\boldsymbol{\phi}^{(i)} = \begin{pmatrix} \phi_1^{(i)} \\ \phi_2^{(i)} \\ \vdots \\ \phi_N^{(i)} \end{pmatrix} \tag{14b}$$

by the boundary conditions:

$$\begin{cases} \phi_k^{(i)} = \delta_{ki} \sin \alpha_i \\ \\ (\phi_k^{(i)})' = -\delta_{ki} \cos \alpha_i \end{cases} ; \text{ at } \quad r = a > 0 \text{ and } \quad 0 \le \alpha_i < \tfrac{\pi}{2} \tag{15}$$

and the alternative definition:

$$\begin{cases} \phi_k^{(i)} = \delta_{ki} \frac{(2\ell-1)!!}{a^{\ell}} \\ \\ (\phi_k^{(i)})' = \delta_{ki} \frac{-\ell(2\ell-1)!!}{a^{\ell+1}} \end{cases} ; \text{ at } \quad r = a > 0 \tag{16}$$

Note that the definitions (11a) and (15) as well as (11b) and (16) fulfil

$$\mathbf{W}\{\mathbf{\Phi}, \mathbf{\Psi}\} = \mathbf{1} \tag{17}$$

By combining $\mathbf{\Phi}$ and $\mathbf{\Psi}$ via the matrix $\mathbf{M}$ we obtain for the asymptotic solution

$$\boldsymbol{\chi} = \mathbf{\Phi} + \mathbf{\Psi}\mathbf{M} = \mathbf{F}\mathbf{d} \tag{18a}$$

Since $\mathbf{\Phi}$, $\mathbf{\Psi}$ and $\mathbf{M}$ are given by eqs.(11), (15) and (16), $\mathbf{M}$ and $\mathbf{d}$ can be found uniquely.

From Theorem 2 we now obtain the following:

Corollary 2.1: Solutions to the Riccati equation (8), fulfilling the boundary conditions (11a), (11b), (13b), (15) and (16) have the property that

$$\mathbf{R} = \tilde{\mathbf{R}} \qquad \forall\ r.$$

From eq.(7) also follows that

$$\mathbf{R}' = \tilde{\mathbf{R}}' \qquad \forall\ r.$$

Theorem 3:

a)

$$\mathbf{M} = -[\mathbf{W}\{\mathbf{F},\boldsymbol{\Psi}\}]^{-1}\mathbf{W}\{\mathbf{F},\boldsymbol{\Phi}\} = (\mathbf{R}_J\boldsymbol{\Psi} - \boldsymbol{\Psi}')^{-1}(\boldsymbol{\Phi}' - \mathbf{R}_J\boldsymbol{\Phi}) = \tilde{\mathbf{M}}$$

b)

$$\mathbf{d} = -(\mathbf{W}\{\boldsymbol{\Psi},\mathbf{F}\})^{-1}$$

Proof:

a) From (18*a*) and (18*b*) given by

$$\boldsymbol{\chi}' = \boldsymbol{\Phi}' + \boldsymbol{\Psi}'\mathbf{M} = \mathbf{F}'\mathbf{d} \tag{18b}$$

one obtains:

$$\mathbf{W}\{\mathbf{F},\mathbf{F}\}\mathbf{d} = \mathbf{W}\{\mathbf{F},\boldsymbol{\Phi}\} + \mathbf{W}\{\mathbf{F},\boldsymbol{\Psi}\}\mathbf{M}$$

From Corollary 1.1 and 1.2 $\mathbf{W}\{\mathbf{F},\mathbf{F}\} = \mathbf{0}$. Hence

$$\begin{aligned}\mathbf{M} &= -[\mathbf{W}\{\mathbf{F},\boldsymbol{\Psi}\}]^{-1}\mathbf{W}\{\mathbf{F},\boldsymbol{\Phi}\}\\ &= -(\tilde{\mathbf{F}}\boldsymbol{\Psi}' - \tilde{\mathbf{F}}'\boldsymbol{\Psi})^{-1}(\tilde{\mathbf{F}}\boldsymbol{\Phi}' - \tilde{\mathbf{F}}'\boldsymbol{\Phi})\\ &= (\tilde{\mathbf{R}}_J\boldsymbol{\Psi} - \boldsymbol{\Psi}')^{-1}(\boldsymbol{\Phi}' - \tilde{\mathbf{R}}_J\boldsymbol{\Phi})\end{aligned}$$

From Corollary 2.1 we have $\mathbf{R}_J = \tilde{\mathbf{R}}_J$. It follows trivially that M is symmetric for all initial boundary conditions where $\boldsymbol{\Psi}^{(1)}$ and $\boldsymbol{\Phi}^{(1)}$ are proportional to the unit matrix. In particular this is fulfilled by the boundary conditions (11b) and (16) and by (11a) and (15) provided $\alpha_i = \alpha$, for all i.

b) Similarly

$$\mathbf{W}\{\boldsymbol{\Psi},\boldsymbol{\Phi}\} + \mathbf{W}\{\boldsymbol{\Psi},\boldsymbol{\Psi}\} = \mathbf{W}\{\boldsymbol{\Psi},\mathbf{F}\}\mathbf{d}$$

From Corollary 1.1 and 1.2, $\mathbf{W}\{\boldsymbol{\Psi},\boldsymbol{\Psi}\} = \mathbf{0}$ and from (11), (13) and (15)

$$\mathbf{W}\{\boldsymbol{\Psi},\boldsymbol{\Phi}\} = -\mathbf{1}$$

Thus

$$\mathbf{d}^{-1} = -(\mathbf{W}\{\mathbf{\Psi},\mathbf{F}\})$$

Q.E.D.

Now define the differential operator:

$$\mathbf{L}(\mathrm{r}) = \{-\mathbf{1}\frac{d^2}{dr^2}+\mathbf{V}(\mathrm{r})\} \tag{19}$$

In eq.(19) we allow $\mathbf{V}(r)$ to be completely general in the sense that it may include both the case $\ell = 0$ as well as the case of different ℓ–values for different channels. The special case of a unit matrix boundary condition (11b), $\ell > 0$ will be treated briefly at the end of section 2.2. Then the matrix Green's function corresponding to $\mathbf{L}$ is given by:

$$\mathbf{G}(\mathrm{r},\mathrm{r}') = \begin{cases} \mathbf{\Psi}(\mathbf{k},r)(\tilde{\mathbf{W}}\{\mathbf{\Psi},\mathbf{F}\})^{-1}\tilde{\mathbf{F}}(\mathbf{k},r') & r < r' \\ \mathbf{F}(\mathbf{k},r)(\tilde{\mathbf{W}}\{\mathbf{\Psi},\mathbf{F}\})^{-1}\tilde{\mathbf{\Psi}}(\mathbf{k},r') & r > r' \end{cases} \tag{20}$$

where $\mathbf{k^2} = \lambda\mathbf{1} - \mathbf{V}(\infty)$, see e.g. eq.(13c).

Theorem. 4:

$$(\lambda - \mathbf{L}(r))\mathbf{G}(r,r') = \delta(r-r')\mathbf{1}$$

Proof: Consider an arbitrary matrix function $\mathbf{h}(\mathrm{r})$ which is *not* a solution of eq. (1), i.e.

$$\mathbf{H}(r) = \int_0^\infty \mathbf{G}(r,r')\mathbf{h}(r')dr'$$

exists. Taking the first derivative of $\mathbf{H}$ one obtains

$$\mathbf{H}'(r) = \mathbf{F}'(\mathbf{k},r)\int_0^r \mathbf{W}\{\mathbf{\Psi},\mathbf{F}\}\tilde{\mathbf{\Psi}}(\mathbf{k},r')\mathbf{h}(r')dr' + \mathbf{\Psi}'(\mathbf{k},r)\int_r^\infty \tilde{\mathbf{W}}\{\mathbf{\Psi},\mathbf{F}\}\tilde{\mathbf{F}}(\mathbf{k},r')\mathbf{h}(r')dr'$$

Note that from Corollary 2.1

$$\begin{aligned}&\mathbf{F}(\mathbf{k},r)\mathbf{W}\{\mathbf{\Psi},\mathbf{F}\}\tilde{\mathbf{\Psi}}(\mathbf{k},r) - \mathbf{\Psi}(\mathbf{k},r)\mathbf{W}\{\mathbf{\Psi},\mathbf{F}\}\tilde{\mathbf{F}}(\mathbf{k},r)\\ &= [(\mathbf{R}_J - \tilde{\mathbf{R}}_R)^{-1} - (\tilde{\mathbf{R}}_J - \mathbf{R}_R)^{-1}] = 0\end{aligned} \tag{20b}$$

To prevent $(\mathbf{R}_J - \tilde{\mathbf{R}}_R)^{-1}$ from becoming singular, one usually keeps $\boldsymbol{\lambda} = \mathbf{k}^2 + \mathbf{V}(\infty) + \imath\varepsilon\mathbf{1}; \varepsilon \neq 0$ and let ε approach zero provided there is no bound state at

$\lambda = \mathbf{k}^2 + \mathbf{V}(\infty)$. We will return to this limit later on. Hence if $\varepsilon \neq 0$ the above term vanishes.

Proceeding to the next derivative

$$\begin{aligned}\mathbf{H}''(r) =& \mathbf{F}''(\mathbf{k},r)\int_0^r \mathbf{W}\{\mathbf{\Psi},\mathbf{F}\}\tilde{\mathbf{\Psi}}(\mathbf{k},r')\mathbf{h}(r')dr' \\ &+ \mathbf{\Psi}''(\mathbf{k},r)\int_r^\infty \tilde{\mathbf{W}}\{\mathbf{\Psi},\mathbf{F}\}\tilde{\mathbf{F}}(\mathbf{k},r')\mathbf{h}(r')dr' \\ &+ \{\mathbf{F}'(\mathbf{k},r)\mathbf{W}\{\mathbf{\Psi},\mathbf{F}\}\tilde{\mathbf{\Psi}}(\mathbf{k},r) - \mathbf{\Psi}'(\mathbf{k},r)\tilde{\mathbf{W}}\{\mathbf{\Psi},\mathbf{F}\}\mathbf{F}(\mathbf{k},r)\}\mathbf{h}(r)\end{aligned}$$

or from eq.(1a)

$$\mathbf{H}''(r) = (\mathbf{V} - \lambda)\mathbf{H}(r) + \mathbf{h}(r)$$

Thus

$$(\lambda - \mathbf{L}(r))\mathbf{H}(r) = \mathbf{h}(r)$$

or since $\mathbf{h}(r)$ is arbitrary in the domain of $\mathbf{G}$

$$(\lambda - \mathbf{L}(r))\mathbf{G}(r,r) = \mathbf{1}\delta(r - r')$$

Using the notation $\langle r|\mathbf{L}\rangle = \mathbf{L}(r)$ and $\langle r|\mathbf{G}|r'\rangle = \mathbf{G}(r,r')$, we can also write **Theorem. 4** as:

$$\mathbf{G} = (\lambda - \mathbf{L})^{-1}$$

$$\mathbf{H} = \mathbf{G}\mathbf{h}$$

Again we consider

$$\langle r|\mathbf{H}\rangle = \langle r|(\mathbf{k} - \mathbf{L})^{-1}|\mathbf{h}\rangle = \int_0^\infty \langle r|\mathbf{G}|r'\rangle\mathbf{h}(r')dr'$$

and we also use (18) and **Theorem 3b** to write :

$$\mathbf{G}(r,r') = \begin{cases} -\mathbf{\Psi}(\mathbf{k},r)\tilde{\boldsymbol{\chi}}(\mathbf{k},r') & r < r' \\ -\boldsymbol{\chi}(\mathbf{k},r)\tilde{\mathbf{\Psi}}(\mathbf{k},r') & r > r' \end{cases} \tag{21}$$

since $\mathbf{W}\{\mathbf{\Psi},\chi\} = -\mathbf{1}$.

For $\varepsilon \to 0^+$, $\mathbf{k}^2 = \lambda - \mathbf{V}(\infty)$, not in the bound part of the spectrum of $\mathbf{L}(r)$ one gets via (18) and **Theorem 3a**

$$\Im(\mathbf{L} - \lambda)^{-1} = \mathbf{\Psi}\mathbf{M}_\imath\tilde{\mathbf{\Psi}} \quad ; \quad \mathbf{M} = \mathbf{M}_r + \imath\mathbf{M}_\imath \tag{22}$$

Theorem 5: The matrix spectral density

$$\pi\left(\frac{d\boldsymbol{\rho}}{d\omega}\right)_{\omega=E} = \mathbf{M}_\imath$$

satisfies the relation

$$\mathbf{M}_\imath = \tilde{\mathbf{d}}^+\mathbf{k}_r\mathbf{d}^-$$

where

$$\mathbf{k}_r = \frac{1}{2}(\mathbf{k}+\mathbf{k}^*)$$

and

$$\mathbf{d}^\pm = -[\mathbf{W}\{\boldsymbol{\Psi},\mathbf{F}^\pm\}]^{-1}$$

and where $\mathbf{F}$ has two branches according to eq.(12).

Proof: According to (18) we obtain

$$\begin{aligned} -2\imath\mathbf{M}_\imath &= \mathbf{M}^- - \mathbf{M}^+ = \mathbf{W}\{\boldsymbol{\chi}^+,\boldsymbol{\chi}^-\} \\ &= \mathbf{W}\{(\mathbf{F}^+\mathbf{d}^+),(\mathbf{F}^-\mathbf{d}^-)\} \\ &= \tilde{\mathbf{d}}^+(-2\imath\mathbf{k}_r)\mathbf{d}^- \end{aligned}$$

Here

$$\begin{aligned} \mathbf{M}^+ &= -[\mathbf{W}\{\mathbf{F}^+,\boldsymbol{\Psi}\}]^{-1}\mathbf{W}\{\mathbf{F}^+,\boldsymbol{\Phi}\} \\ &= [\mathbf{R}_J^+\boldsymbol{\Psi} - \boldsymbol{\Psi}']^{-1}[\boldsymbol{\Phi}' - \mathbf{R}_J^+\boldsymbol{\Phi}] \\ &= \mathbf{M}_r + \imath\mathbf{M}_\imath \end{aligned}$$

The spectral distribution matrix $\boldsymbol{\rho}$ thus takes the form

$$\boldsymbol{\rho}(E) = \lim_{\delta\to 0}\ \lim_{\varepsilon\to 0}\frac{1}{\pi}\int_{\delta}^{E+\delta}\mathbf{M}_\imath(E+\imath\varepsilon)dE \tag{23a}$$

eq.(23a) can also be written

$$\left(\frac{d\boldsymbol{\rho}}{d\omega}\right)_{\omega=E} = \frac{1}{\pi}\left(\tilde{\mathbf{W}}\{\boldsymbol{\Psi},\mathbf{F}^+\}\right)^{-1}\mathbf{k}_r\left(\mathbf{W}\{\boldsymbol{\Psi},\mathbf{F}^-\right)^{-1}\} \qquad E \geq E_{thd} \tag{23b}$$

and

$$\boldsymbol{\rho}(E) = \sum_{i=1}^{N}\delta(E-E_i)(-\mathrm{Res}(\mathbf{M}^+)) \qquad E < E_{thd} \tag{23c}$$

where

$$E_{thd} = k^2 + \min\{V_{ii}(\infty)\}$$

From (23a - c) the matrix identity $\mathbf{1}$ can be represented as

$$\mathbf{1} = \int_{-\infty}^{+\infty}|\boldsymbol{\Psi}(\mathbf{k})\rangle d\boldsymbol{\rho}(\omega)\langle\boldsymbol{\Psi}(\mathbf{k})| = \int_{-\infty}^{+\infty}\boldsymbol{\Psi}(\mathbf{k})d\boldsymbol{\rho}(\omega)\tilde{\boldsymbol{\Psi}}(\mathbf{k}) \tag{24}$$

In principle one could of course have bound states above the threshold energy, E_{thd}, provided there is no coupling between the closed channels and any of the other open ones. We will not consider this problem further here.

2.2 Connections with Scattering formulations

We will now show that the frequently appearing Wronskian (for $\ell = 0$)

$$\mathbf{J}^+ \equiv -\mathbf{W}\{\mathbf{\Psi}, \mathbf{F}^+\} = (\mathbf{d}^+)^{-1}$$

is nothing but the wellknown Jost function matrix appearing in standard scattering theory.

Theorem 6: The partial wave $\mathbf{S}$ -matrix is given by

$$\mathbf{S} = [\mathbf{W}\{\mathbf{\Psi}, \mathbf{F}^+\}]^{-1}\mathbf{W}\{\mathbf{\Psi}, \mathbf{F}^-\} = (\mathbf{J}^+)^{-1}\mathbf{J}^-$$

Proof:

$$\boldsymbol{\chi}^- - \boldsymbol{\chi}^+ = \mathbf{\Psi}(\mathbf{M}^- - \mathbf{M}^+)$$

or

$$\frac{\imath}{2}(\boldsymbol{\chi}^- - \boldsymbol{\chi}^+)\mathbf{M}_\imath^{-1} = \mathbf{\Psi} = \frac{\imath}{2}(\mathbf{F}^- - \mathbf{F}^+\mathbf{d}^+(\mathbf{d}^-)^{-1})\mathbf{k}_r^{-1}(\tilde{\mathbf{d}}^+)^{-1}$$

Hence

$$\mathbf{\Psi} = \frac{\imath}{2}(\mathbf{F}^- - \mathbf{F}^+\mathbf{S})\mathbf{k}_r^{-1}(\tilde{\mathbf{d}}^+)^{-1}$$

with

$$\mathbf{S} = \mathbf{d}^+(\mathbf{d}^-)^{-1} = (\mathbf{J}^+)^{-1}\mathbf{J}^- \qquad \text{(independent of } r\text{)}$$

Corollary 6.1: Since $\mathbf{S} = (\mathbf{F}^+)^{-1}(\mathbf{R}_J^+ - \mathbf{R}_R)^{-1}(\mathbf{R}_J^- - \mathbf{R}_R)\mathbf{F}^-$ one obtains at $r = b$, where $\mathbf{F}^+$ has taken its asymptotic form (13),

$$\mathbf{S} = -[(\mathbf{k}^{1/2}\exp(\imath b\mathbf{k})]^{-1}(\imath\mathbf{1} - \mathbf{D})^{-1}(\imath\mathbf{1} + \mathbf{D})\mathbf{k}^{1/2}\exp(-\imath b)$$

$$\mathbf{D} = \mathbf{k}^{-1/2}\mathbf{R}_R\mathbf{k}^{-1/2}$$

Hence:

$$\mathbf{S}\mathbf{S}^\dagger = \mathbf{S}^\dagger\mathbf{S} = \mathbf{1}$$

and

$$\mathbf{S} = \exp(2\imath\boldsymbol{\delta})$$

where $\boldsymbol{\delta}$ is a Hermitian matrix according to Stone's theorem

Note that formula (23b) in general , i.e. $\ell > 0$, can be taken in the limit $a \to 0$. The boundary condition for a multichannel problem $\boldsymbol{\Psi}$ then reads assuming unit matrix boundary conditions (see eq.(11b))

$$\lim_{r\to 0} \psi_k^{(i)}(2\ell+1)!!r^{-(\ell+1)} = \delta_{ki}\mathbf{1} \tag{25}$$

Redefining the Jost function as

$$\mathbf{J}_\ell^{\pm} = (-\mathbf{k})^\ell \mathbf{W}\{\boldsymbol{\Psi}_\ell, \mathbf{F}_\ell^{\pm}\} \tag{26}$$

with

$$f_k^{(i)\pm} = \lim_{r\to\infty} (\imath)^\ell \exp(\pm\imath k_i r)\delta_{ik} \tag{27}$$

one obtains

$$\pi\left(\frac{d\boldsymbol{\rho}}{d\omega}\right)_{\omega=E} = \left(\mathbf{W}\{\boldsymbol{\Psi}_\ell, \mathbf{F}_\ell^{+}\}\right)^{-1} \mathbf{k}_{\mathrm{r}}^{2\ell+1} \left(\mathbf{W}\{\boldsymbol{\Psi}_\ell, \mathbf{F}_\ell^{-}\}\right)^{-1} \tag{28}$$

For the sake of completeness we also give the equations for solutions $\boldsymbol{\Psi}$ and $\mathbf{F}$ at a resonance eigenvalue $\varepsilon = \varepsilon_0$. Note that the solution $\boldsymbol{\Psi}$ can always be integrated numerically for the (complex) resonance energy, $\varepsilon = \varepsilon_0$. Since we already have determined $\boldsymbol{\Psi}$, $\mathbf{F}$, $\mathbf{d}$ and $\mathbf{M}$ we can also obtain a direct formula for the appropriate outgoing solution (Gamow wave). Knowing that $\mathbf{M} = \tilde{\mathbf{M}}$, for unit matrix boundary conditions as well as the connection between $\mathbf{d}$ and the Jost function $\mathbf{J}(\varepsilon)$ (see above) we obtain straight forwardly the following relations for $\det\{\mathbf{J}'(\varepsilon_0)\} \neq 0$

$$\begin{aligned}\mathbf{d}^{-1} = \mathbf{J}(\varepsilon) &= (\varepsilon - \varepsilon_0)\mathbf{J}'(\varepsilon_0) \\ &\quad + \text{higher Taylor expansion terms in } (\varepsilon - \varepsilon_0)\end{aligned}$$

$$\mathbf{M} = \tilde{\mathbf{M}} = \mathbf{W}\{\boldsymbol{\Phi}, \mathbf{F}\}\mathbf{J}^{-1}$$

and

$$\begin{aligned}\lim_{\varepsilon\to\varepsilon_0} (\varepsilon - \varepsilon_0)\boldsymbol{\chi}(\varepsilon) &= \mathbf{F}[\mathbf{J}'(\varepsilon)]^{-1}|_{\varepsilon=\varepsilon_0} \\ &= \boldsymbol{\Psi}\left[\mathbf{W}\{\boldsymbol{\Phi}, \mathbf{F}\}(\mathbf{J}'(\varepsilon))^{-1}\right]|_{\varepsilon=\varepsilon_0} \\ &= \boldsymbol{\Psi}\mathrm{Res}(\mathbf{M}(\varepsilon_0))\end{aligned}$$

From the above equations partial widths and other physical parameters can be easily deduced. Note the central role of the derivative of the Jost function matrix in these equations. The same quantity appears in partial wave S-matrix expansion in the paper by Rittby et al. in this book. We will now proceed to derive the complex scaled analogue of the present theory.

2.3 The Complex Dilated Analogue of the Titchmarsh-Weyl's Theory for a Set of Coupled Singular Second Order Differential Equations

In analogy with previous studies of the one dimensional case, we now study the effects of complex dilation of eq.(1) and the concomitant development. For recent reviews of the field of complex scaling see refs. 22-24.

Consider the case of a uniform complex scaling of the coordinates

$$r \to \eta r; \qquad \eta = \exp(\imath\theta); \qquad \theta \text{ real} \tag{29}$$

For the definition of complex exterior scaling see eq.(34) below. Two versions of complex scaling transformations are illustrated in figure 1.

The uniform complex scaling presupposes that we can analytically continue the full potential into the complex plane. Treating η as a complex parameter, with r real, we rewrite the set of coupled differential equations as

$$\mathbf{Y}_\eta'' = \eta^2(\mathbf{V}_\eta - \lambda\mathbf{1})\mathbf{Y}_\eta \tag{30}$$

The solution matrices are correspondingly replaced by their rotated analogues, and specifically we obtain the logarithmic derivative

$$\mathbf{R} \to \mathbf{R}_\eta = \mathbf{Y}_\eta'\mathbf{Y}_\eta^{-1} \tag{31}$$

fulfilling the dilated Riccati equation

$$\mathbf{R}_\eta' = \eta^2(\mathbf{V}_\eta - \lambda\mathbf{1}) - \mathbf{R}_\eta^2 \tag{32}$$

From eq.(20a) and (20b) we find that the diagonal element ($r = r'$) of the dilated Green's matrix can be expressed as

$$\mathbf{G}_\eta(k, r, r) = (\mathbf{R}_{R,\eta} - \mathbf{R}_{J,\eta})^{-1} \tag{33}$$

Note that the symmetries of appropriate matrix quantities defined in the previous sections prevails. A symmetric matrix stays symmetric (although of course complex symmetric) and unitary ones become star-unitary under the complex scaling transformations. Apart from having singularities for bound state energies, the analytic continuation of the Green's matrix has simple or higher order poles at resonant energies in the lower half of the energy plane. The complex scaling method provides in effect an analytic continuation of the Green's function to the second Riemann sheet. According to the formal theory of complex dilations, the result of the scaling eq.(29) amounts to a rotation of the possible continua of the system around their respective thresholds, exposing resonances in appropriate sectors in the complex energy plane. In a multichannel system we further have to consider the possibility of resonances becoming absorbed by the rotated continuum of another channel. This is illustrated in fig. 2.

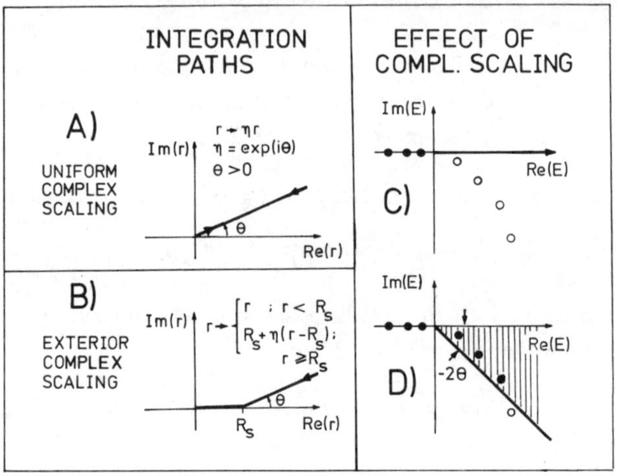

Figure 1

In the theory of complex scaling the original integration path on the real axis is distorted out into the complex plane. **A)** For *uniform* complex scaling the path follows a ray through the origin at an angle θ relative to the real axis. **B)** For *exterior* complex scaling the path first follows the real axis out to some exterior scaling point R_s, and then goes out in the first quadrant on a ray at an angle θ relative to the real axis. **C)** In the unscaled formulation of the problem we are only able to compute the bound levels of the problem (•) whereas the resonances (∘) remain out of reach. **D)** Both of the transformations above will uncover the same sector, defined by the angle -2θ (dashed in the figure), of the complex energy plane. The bound eigen energies of the unscaled Hamiltonian H_1 associated with the potential are unaffected by the scaling but a new set of eigenenergies show up in the uncovered sector. Within this formulation we are thus able to compute the appearing resonances. Resonance which are not uncovered by the transformation (∘) do not belong to the discrete spectrum of H_η.

The expression (33) may thus have singularities for complex energies provided the rotation angle is in the appropriate interval for the resonance to be exposed. We have seen that the **M**−function is intimately connected with the Green's function, and from the unique definition of **M** in Theorem 3a, we find that **M** has the same poles as **G**. From the definition of the Jost function, it also follows that the zeroes of $\mathbf{J}^+$ correspond to the same resonance energies. Since a pole in **G**, given by eq.(33), provides the necessary and sufficient condition for the excistence of resonance poles as well as for bound state poles, we may e.q. use the logarithmic derivative matrices obtained from a numerical integration of the Riccati equation

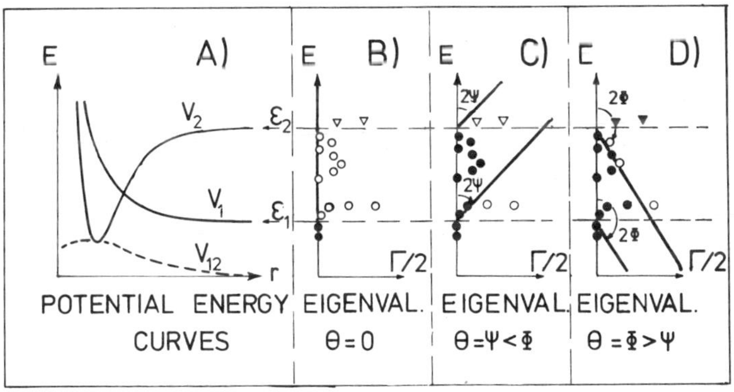

Figure 2

The effects of complex scaling of a decaying two-body system is illustrated here by the nuclear motion of a diatomic molecule in a two-channel curve crossing problem. For more than one channel, ν, we will have different thresholds, ε_ν as indicated in the figure. Complex thresholds can also appear. The filled symbols represent levels which can be obtained computationally in the different scaling modes. **A)** shows the potentials for the unscaled problem. **B)** For the unscaled ($\theta = 0$) formulation of the problem we are able to calculate two bound levels (filled symbols) but none of the resonances (indicated with open symbols). **C)** If the problem is scaled with an angle $\theta = \Psi$ we will be able to compute some resonances in addition to the bound levels. **D)** If the scaling angle is increased to $\theta = \Phi$ we will be able to derive even more resonance positions but some of the previously uncovered ones are now absorbed by the rotated continuum.

(8) to obtain bound and resonance eigenvalues.

Accurate interaction potentials for short radial distances are usually only defined for a discrete set of points. The long-range part of the potential can be described by some analytic expression obtained from considerations of the approximate long-range behaviour of the interactions. In these cases, we find that the above described complex scaling procedure is not defined since the potential is not given in a form which permits an analytic continuation for all radial distances. We are thus led to more general coordinate transformations.

The exterior complex scaling method [25–27] defined by

$$r \rightarrow \begin{cases} R_s + \eta(r - R_s) & ; r \geq R_s \\ r & ; r < R_s \end{cases} \tag{34}$$

can conveniently be used to treat problems with partly numeric potentials. This transformation is illustrated in figure 2 C). Exterior complex scaling enables one to treat almost any type of problem since the asymptotic form of a potential or an interaction can be derived, or if necessary, approximated analytically. We also have the experience that, by using the exterior scaling transformation, we can reach better numerical stability than with the uniform scaling transformation. A many channel problem is more complicated. Figure 2 illustrates some of the important features. For every channel, ν, we will have a threshold, ε_ν. Additional features do however appear. Complex thresholds, which corresponds to resonances of subsystems [27,28] can also be present.

3. Application to a model of a Diatomic Curve Crossing of the $X^2\Sigma^+ - B'\ ^2\Sigma^+$ States in the MgH Radical

As an illustration of the present formulation of the partial wave scattering problem we have chosen a problem which relates both to atomic scattering and diatomic molecular spectroscopy. We have investigated the resonance structure associated with a curve crossing of two electronic potential energy curves of the same symmetry. In the adiabatic representation one obtainings an avoided crossing, since potential energy curves of the same symmetry are not allowed to cross. (The formal connections between the adiabatic and the diabatic frameworks are given as an appendix). In the diabatic representation of the problem, the curves do cross but are coupled by an off-diagonal, r-dependent, matrix element of the Hamiltonian. This matrix element describes the mixing of the diagonal, "pure states". We have previously [19] used this theory to study a model for the charge transfer reaction:

$$N^{3+}(1s^2 2s^2)^1S + H(1s)^2S \rightarrow N^{2+*}(1s^2 2s^2 3s)^2S + H^+ \tag{35}$$

In this investigation we were able to relate the resonant peaks in the total low energy reactive cross section with partial wave S-matrix poles as discussed above. In the present application we will focus on the spectroscopic aspects of the problem, i.e. molecular predissociation. This application is justified by the development of new experimental techniques in the study of diatomic predissociation phenomena such as the HFD method [30] and especially photofragment spectroscopy [31]. Theoretical studies of this kind have been treated by several authors. The methods can be divided into semiclassical ones and quantum mechanical ones. Childs[32] reviewed the status of the field up to 1973 discussing the essential structure of the level positions as well as of their widths. In certain cases these semiclassical methods seem to work well, but in general different approximations have to be made for Mulliken's different predissociation cases. In other cases one has to be careful when judging the results. To our knowledge semiclassical methods have not yet been applied to more than two coupled channels whereas quantum me-

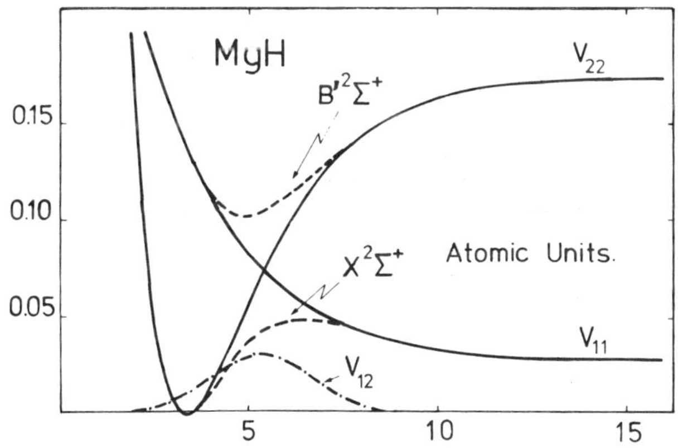

Figure 3

The diabatic and adiabatic potential energy curves in our model of the $X^2\Sigma^+$ and $B'\ ^2\Sigma^+$ states in MgH. The curves are generated with eqs.(37 - 39) and correspond to $J = 0$. The numeric values are given in table I. We will refer to this as MgH (table I).

chanical methods have proven quite successful in general, e.g. the complicated predissociation phenomena in the BeH/BeD radical[33].

Quantum mechanical calculations have usually been focused on determining the phase shift, see e.g. the classical work by Levine and Bernstein[34]. Conventional phase shift methods do however lack the property of describing complicated resonance structures originating in several close lying quasibound levels. The existence of such structures have been clearly demonstrated previously[19,35] Furthermore, Atabek and Lefebvre [36–38] have recently undertaken a series of detailed studies of coupled channel problems in small molecules. In the first paper in this series [36] the Goodwin-Fox method for a numerically stiff problem was analysed. The Siegert boundary condition was applied in the second paper in the series[37] which also contains a discussion of the partial width concept. Multichannel quantization performed with the complex rotated Goodwin-Fox propagator was described in the third paper[38] together with a study complex rotations of a Hamiltonian depending on more then one coordinate. Various schemes were proposed, which for the model problem treated, gave a final result independent of the schemes applied.

As a comparison, see e.g. the previous section, we found that singularities of

Table 1

Potential parameters for the curve crossing application as defined in eqs. (37 - 39) The reduced mass used was $\mu = 1763.0698$ (a.u.)

Parameter	Spectroscopic units	Atomic units
A	95817.22 cm^{-1}	0.4365754 Hartree
b	0.7872477 $Å^{-1}$	0.4165934 $Bohr^{-1}$
D_e^{11}	5859.824 cm^{-1}	0.02669932 Hartree
D_e^{22}	38339.19 cm^{-1}	0.1746862 Hartree
T_e^{22}	-100.2809 cm^{-1}	-4.569134×10^{-4} Hartree
β	0.9766027 $Å^{-1}$	0.5167957 $Bohr^{-1}$
r_e	1.736154 Å	3.280857 Bohr
A_{int}	6822.694 cm^{-1}	0.03108649 Hartree
c	2.824000 Å	5.336588 Bohr
d	1.0 Å	0.2880284 Bohr

the Green's matrix could be calculated by a matching procedure of logarithmic derivative matrices. The numerical integration of the associated coupled set of Riccati-equations eq. (8) is a very convenient way to obtain these solution matrices particularly utilizing the efficient technique due to Johnson [39]. This alternative is certainly a competitive numerical technique both in scope and accuracy. The conclusion from these recent investigations as well as our own previous studies, is clearly that quantum mechanical treatments, including the use of appropriate analytic continuation procedures are necessary to resolve complicated spectral resonance structures.

The optical rovibronic spectra of alkaline earth radicals have recently been studied and it was found that several curve crossings are present in the potential energy curve description of these molecules[40]. We have here chosen to study a simple model potential based on experimental data optimized so as to describe the curve crossing between the X and the $B'\ ^2\Sigma^+$ states of MgH. The curve crossing which occurs between these states is due to the mixing of the asymptotically ionic

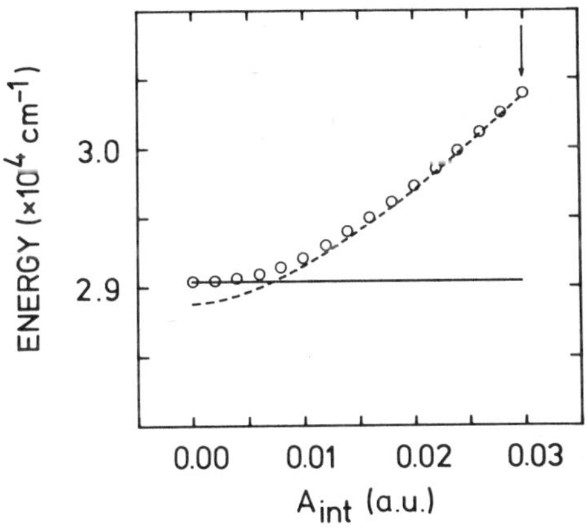

Figure 4

Variation of the energy for a resonance pole lying well above the curve crossing as a function of the interaction strength. Each ring represents an increase of the interaction strength (A_{int}) by 0.002 a.u. The solid line shows the diabatic limit for the level. The dotted line shows the variation of the energy for the same level in the adiabatic approximation. The arrow indicates the ring corresponding to the MgH (table 1) case.

excited molecular eigenstates and the valence type molecular ground state

$$Mg(1s^2 2s^2 2p^6 3s^2)\ ^1S + H(1s)\ ^2S \longleftrightarrow Mg^+(1s^2 2s^2 2p^6 3s)\ ^2S + H^-(1s^2)\ ^1S \quad (36)$$

Note that this is a rather simplified model and in order to obtain a good description of experimental data one would obviously need to include electronic states of other symmetries than these two $^2\Sigma^+$ states.

The diabatic potential energy curves are constructed such that the repulsive potential has a pure exponential behaviour:

$$V_{11}(r) = A\exp(-br) + D_e^{11} \quad (37)$$

The attractive potential (V_{22}) was assumed to be of Morse type:

$$V_{22}(r) = D_e^{22}(1 - \exp\{\beta(r - r_e)\})^2 + T_e^{22} \quad (38)$$

whereas the interaction potential was represented by a gaussian type function:

$$V_{12}(r) = A_{int}\exp\{(r - c)d\}^2 \quad (39)$$

The diabatic potential energy are based on the parameters given in table 1, and they are displayed in figure 3. In the present investigation however, we have allowed

Table 2

Vibrational energies and non-radiative lifetimes for the $X^2\Sigma^+$ of MgH as obtained with the model potentials presented by eqs.(37 - 39) with the parameters in table I. The energies are referred to the v = 0 level of the X state. The difference between the values obtained from the solution of the full coupled problem and the values in the table is 513.5 cm^{-1}.

v	E(v,0) (cm^{-1})	τ (nsec)
0	0.0	-
1	1481.2	-
2	2903.0	-
3	4257.8	-
4	5534.4	-
5	6717.5	$< \infty$
6	7785.8	$< 10^{+11}$
7	8713.4	$4.0\ 10^{+4}$
8	9478.7	0.8
9	10066.0	$5.1\ 10^{-4}$

the interaction strengths, A_{int}, to vary, in order to be able to study the behaviour for different interaction strength. In all cases below, the remaining parameters of table 1 were kept fixed.

Following of Levine and Bernstein[34] we first investigate the quality of the diabatic and adiabatic limits as the interaction strength varies with $J = 0$. Figure 4 displays the variation of the energy as a function of the interaction strength for a level situated well above the crossing point. We find that the level exhibits a relatively quick departure from the diabatic domain followed by a slow approach to the adiabatic region. Note the appearance of a transition zone, where neither the adiabatic nor the diabatic formulation is satisfactory. However, MgH (table 1) (see arrow), is well in the adiabatic limit.

The simultaneous variation of the width and energy as functions of (A_{int}) is shown in figure 5. In parallel with figure 4 we find that there exists an intermediate region where the width becomes very large, while for MgH (table 1) the level will acquire only a very small width. Our present technique now allows us to study the intermediate region in more detail.

Figure 6 shows the distribution of energies and widths for $A_{int} = 0.015 a.u.$

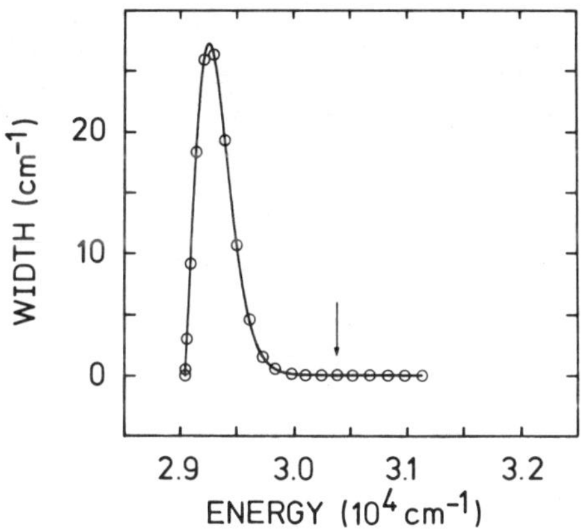

Figure 5

Variation of energy and width for the same level as in figure 4 as a function of the strength of the interaction at the crossing point. Each ring represents an increase of the interaction strength (A_{int}) by 0.002 a.u. The arrow indicates the ring corresponding to the MgH case.

In this case, the lower adiabatic potential will show a substantial barrier, and we find vibrational levels with an increasing width as the barrier maximum is approached. Above the crossing point an oscillatory pattern emerges. These effects have previously been discussed by Childs [32] and Robbe[41] as well as by Childs and Lefebvre [42] in connection with a semiclassical treatment of the IBr molecule[43].

In figure 7 we study the behaviour of the energy and width as a function of J for a specific level in the upper adiabatic potential. Each ring here represents an increase of five units of J. Again the width oscillates with a very large amplitude. The full variation of the width for the energy levels is thus to be interpreted as an interference between the continuum and the bound states. The rotational and vibrational effects are not additive and hence we have to solve the total matrix problem eqs.(30,32) for each given energy level $E(v, J)$, to obtain reliable results.

Figures 6 and 7 suggest that we may to find regular variations of line intensities, and thus lifetimes of the transitions originating from vibrational levels in the upper adiabatic potential energy curve. For an interaction strenth such as the one above, these variations will be extremely dramatic with only a few visible transitions. Such a band would be very hard to analyse. For MgH(table 1), the variations predicted by this simple model will of course be much less.

In experimental investigations[44–46] transitions have been observed from $v' =$

Table 3

Vibrational energies and non-radiative lifetimes for the $B'\ ^2\Sigma^+$ of MgH as obtained with the model potentials defined in eq.(37-39) with the parameters given in table I. The energies are referred to the $v = 0$ level of the X state.

v	E(v,0) (cm^{-1})	τ (nsec)		v	E(v,0) (cm^{-1})	τ (nsec)
0	22326.5	14747		16	32705.6	6.6
1	23170.8	6320		17	33197.1	3.3
2	23962.9	10835		18	33667.2	2.1
3	24712.9	40839		19	34114.9	1.7
4	25430.7	856		20	34539.5	1.6
5	26124.8	161		21	34940.4	1.7
6	26800.3	54		22	35317.0	2.0
7	27460.2	24		23	35668.6	2.9
8	28105.7	14		24	36994.7	4.8
9	28736.8	10		25	36294.7	11
10	29353.9	9.3		26	36568.0	38
11	29955.7	11		27	36814.0	2794
12	30541.4	18		28	37032.1	177
13	31110.2	72		29	37221.7	38
14	31661.2	804		30	37512.9	16
15	32193.3	22				

$0-6$ to $v'' = 0-13$ in emission. We would like to stress that we do not claim that our set of model potentials should be able to represent the experimental data to any high accuracy, but rather that our results suggest possible effects that might be detectable in systems similar to MgH.

Tables II and III give the energies and widths for the X and $B'\ ^2\Sigma^+$ states of MgH as obtained from our calculation. Note that the data predicts that, for $J = 0$, only a few vibrational levels are stable enough to be observed in emission. Apart from the more or less bound states for lower v, there exists occasional tightly

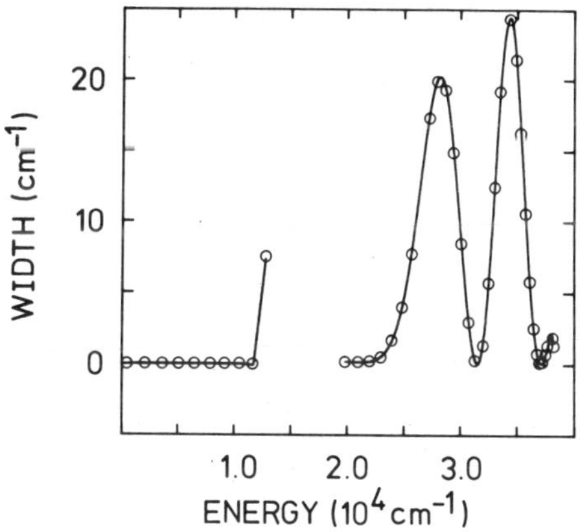

Figure 6

Distribution of resonances and bound state poles for a value of the interaction strength of 0.015 a.u.

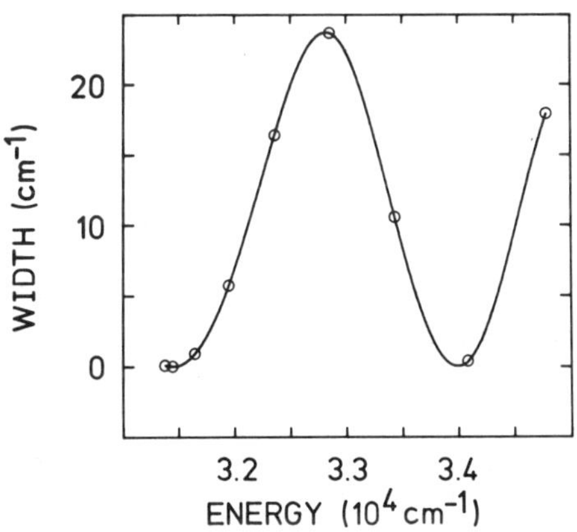

Figure 7

Variation of energy and width, for the same level as in figures 3 and 4, as a function of the rotational quantum number with the same interaction strength as in figure 6. Each ring represents an increase of J by five units.

bound states at higher v. If Franck-Condon factors are favourable one should be able to observe transitions from these states. We must, however, take the result

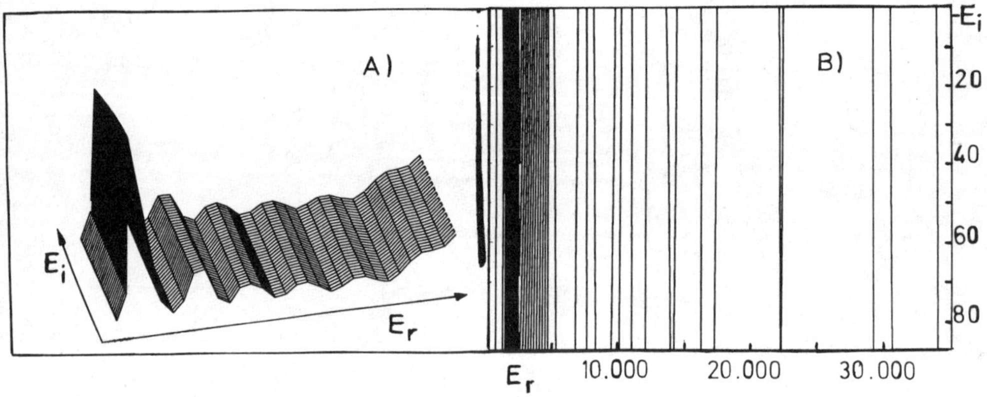

Figure 8

The diagonal part of the complex rotated Green's function in the complex energy region $\Re(E) = 0$ to 35 000 cm^{-1} $\Im(E) =$ - 85 to 0 cm^{-1}. A) 3-D hidden line plot of the region. The peaks correspond to the poles of the $J = 0$ partial wave S-matrix or Green's function. B) Contour plot of the same energy region.

of figure 5 into account. Thus, it is possible that very few lines can be observed in the band, consequently leading to problems in detecting such a system.

After the appearence of the present data[47] Korsch and coworkers[48,49] have used the present model as well as the previously mentioned charge transfer study[19] as a test case for their semiclassical method. The general agreement between the two entirely different approaches is quite good but some comments are warranted. As can be seen from their figures 3 and 6, in ref. 49, (see also discussion in Korsch's contribution in this book) the differences between the quantal data of this work and the semiclassical results of Korsch and Möhlenkamp might at first sight seem rather big. One should note, however, that we are here presenting a full treatment of coupled (ordinary) differential equations dealing with two *coupled quanteties*, i.e. a real and an imaginary part of a complex energy. Here the real part is of the order of 10^5 whereas the imaginary part is of the order of 10. From the difference between the two methods only it is not possible to say whether we have obtained all the complex poles of this problem.

It is furthermore interesting to note a couple of additional surprises concerning the variation of the width of the adiabatic $v = 18$ level with the interaction strength, A_{int}, as found by Korsch and Möhlenkamp [49]. To spread some light on this two problems we have used our new CRAY version of the program to scan the diagonal part of the Green's function in eq.(33). The result is given in figure 8. A number of peaks, corresponding to Smatrix poles, are found in the upper left half of the figure. When comparing with the results of Korsch and Mölenkamp[49] one would expect to find poles also in the remaining part of the figure. The scan of the dilated Green's function, $\mathrm{G}_\eta(k,r,r)$,(see eq. (33)), however, does not seem to show our usual pole structure. Instead we see a number of parallell ridges where $|\mathrm{G}_\eta(k,r,r)|$ is large, typically 10^7 to 10^{10}. We do not at the present time fully understand these features and further studies of these effects are presently under way.

4. Possible Applications to Photofragment Spectroscopy

The applicability of the present method to the analysis of photofragment data[31] is in principle obvious although some practical problems need to be investigated. Freed and cooworkers[51–56] have recently presented a series of careful studies describing methods for analyzing photofragment spectra as well as applications. In the analysis of a typical photofragment situation, it is likely that one might need to include around ten channels of which several have the same or almost the same dissociation limit. This means that the boundary conditions given by e.g. eq. (13) have to be modified to include possible asymptotic couplings between degenerate or almost degenerate molecular eigenstates. In a recent study of the states leading to the photodissociation of the CH^+ ion[51,55]

$$CH^+ + h\nu \rightarrow C^+(^2P_{\frac{1}{2},\frac{3}{2}}) + H(^2S)$$

the coriolis and spin-orbit coupling were found responsible for the asymptotic coupling. The potential matrix is then *not* diagonal in its asymptotic limit.

$$\mathbf{V}(r) = \begin{pmatrix} V_{11}(r) & V_{12}(r) & \dots & V_{1n}(r) \\ V_{21}(r) & V_{22}(r) & \dots & V_{2n}(r) \\ \vdots & \vdots & \ddots & \vdots \\ V_{n1}(r) & V_{n2}(r) & \dots & V_{nn}(r) \end{pmatrix} \tag{40}$$

The diagonal elements describe the asymptotic dissociation limits of the fragments. Some of the off-diagonal interaction potential elements $V_{ij}(r)$ represent for example the spin-orbit splitting between the $C^+\ ^2P_{3/2}$ and the $C^+\ ^2P_{1/2}$ in the CH^+ dissociation. Freed and coworkers were able to find methods to set up the asymptotic potential and the corresponding asymptotic wave function[51]. As can be seen from eqs. (12), (13) and (27) the outgoing Jost solution matrix can no longer be considered separable since off-diagonal coupling elements are present in the potential matrix (eq.(40)). Rather the asymptotic Jost solution should be written as:

$$\mathbf{F}^{+}(k,r) = \begin{pmatrix} f_{11}^{+}(k,r) & f_{12}^{+}(k,r) & \ldots & f_{1n}^{+}(k,r) \\ f_{21}^{+}(k,r) & f_{22}^{+}(k,r) & \ldots & f_{2n}^{+}(k,r) \\ \vdots & \vdots & \ddots & \vdots \\ f_{n1}^{+}(k,r) & f_{n2}^{+}(k,r) & \ldots & f_{nn}^{+}(k,r) \end{pmatrix} \tag{41}$$

where the diagonal elements no longer will have the simple form of eq.(13b). Even if the diagonal and off diagonal elements are not as simple as in the standard case (since they represent the interference between the diabatic states) they still exhibit exponential (in- and outgoing) character.

The complex scaling transformation of eq. (29) or eq. (34) does however in practice save us here. The effect of these transformations is to bring the growing amplitude of the outgoing Gamow wave to become exponentially decreasing thereby forcing the total wave function matrix to become square integrable. If we want to avoid explicitly calculating the non-diagonal Jost solution matrix (eq. (41)) we may simply resort to the complex scaling technique of Engdahl et al.[21] appropriately extended to the multichannel case.

Freed et al. computed experimental observables like total or differental cross sections for the photofragmentation more or less directly. The underlying structure of the partial wave S-matrix or the corresponding Green's function is thus not available in their papers. With the present method we have adopted a more rigorous attitude emphasizing the importance of fundamental quantities like the partial wave S-matrix, the Green's function and the Titchmarsh-Weyl **M**-matrix. Based on this we can for instance get a deeper and simplifying insight of the behavior of cross section in terms of underlying molecular electronic states and their couplings.

We can in principle go one step further and treat the whole photofragment process selfconsistently by using the dressed molecule formalism [57] in which the laser excitation is treated simultaneously with the the dissociation process. The dressed molecule framework is illustrated in figure 9. The idea here is to consider the excitation as a formation of a collision complex between the initial state and the incoming photon. The potential energy curve of the dressed state is given by the sum of the original potential energy curve and the energy of the laser photon :

$$V_{X-dressed}(r) = V_X(r) + h\nu.$$

This state interacts with the excited states through the usual transition moments between the original initial state and the excited states. The latter states in turn interact with each other through their original coupling matrix elements.

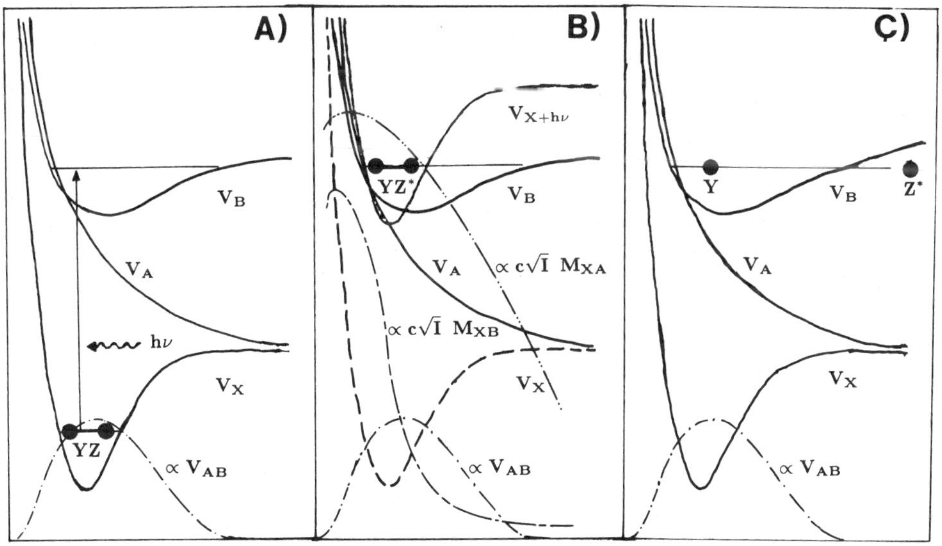

Figure 9

Schematics of the dressed molecule picture of a photodissociation involving two excited states. **A)** The inital state (in this case ground state X) of a diatomic molecule absorbs a resonant photon, $h\nu$. **B)** It is thereby raised by the energy of the photon to the dressed initial state (full line) "$X + h\nu$". The dressed state will interact with the excited states A and B **C)** Due to the coupling of the dressed state and the repulsive A state our molecule will then dissociate.

Within this framework we are thus back to the same type of coupled differential equation (eq. (1)). The size of the problem has just increased from $N \times N$ to $(N+1) \times (N+1)$. By the synthesis of the dressed molecule picture and the present method we have in principle achieved a simplified formulation which will embody the radiative and non-radiative channels on essentially the same footing. The method is in this sense selfconsistent, since it allows for appropriate interferences between these two kinds of processes[57]. The price we have paid is not too high. We have just added one more channel to the problem.

Although the present theory has so far not been tested to a photofragment spectroscopy application, our experience with complex rotated coupled channel theory tells us that including the dressed molecule picture with the possibility to have a selfconsistent model for photofragmentation of diatomic molecules is indeed an encouraging avenue.

The present MgH model potential calculation may in some sense be viewed as simulating a photofragmentation process. The bound fully diabatic curve V_{11} in figure 3 does in this framework represent the *dressed ground state* $X + h\nu$, which interacts with a fully repulsive state represented by the potential energy curve V_{22}. The interaction potential $V_{12}(r)$ in this context is the transition moment multiplied by the the electric field strength:

$$V_{12}(r) = const\sqrt{(I)}M(r) \tag{43}$$

where I is the laser intensity in atomic units and $M(r)$ can be taken as the transition moment between the non-dressed initial state and the final state. The effect of the variation of level position and widths studied above may here be interpreted as a model of possible variations of widths and positions with the laser intensity. One must, however, observe that *only those levels which correspond to direct vertical excitations of the initial state are possible to observe.* We believe that this effect should be investigated in more detail.

5. Summary and Conclusions

In this paper we have presented a simple generalization of Titchmarsh-Weyl's theory to a set of coupled differential equations as well as its numerical realization. By the use of the complex coordinate technique, the formulation can be analytically continued to the complex plane alloowing the explicit calculation of key quantities like the Titchmarsh-Weyl **M**-matrix (eqs. (18) (22) and (23)) or the S-matrix in the complex plane. The coupled analogue displays the same intrinsic simplicity as the one-channel formulation with respect to the derivation of relevant quantities, their analytic continuation and their actual numerical computation.

Before proceeding with the conclusions, we note that that our development has centered around Titchmarsh-Weyl's theory, whereas conventional scattering theory develops essentially from the theory of integral equations in terms of Jost functions and S-matrices. Even if both approaches should finally give the same answer, we believe that the present formulation may prove more transparent in connection with analytic continuation and rigourous classification of allowable potentials. This viewpoint is formed as a result of several investigations.

The classical versions of Titchmarsh-Weyl's theory can be found in in chapter nine of the book by Coddington and Levinson[2]. The treatment of the Stark effect by Hehenberger et al.[58] shows how this theory is applicable to a physical problem by the computation and analysis of the (one channel) $m-$function and its poles in the complex energy plane. Complementary features of the $m-$function were given by Brändas et al.[59] and by Hehenberger et al [60,61]. The early work was based on numerically computing the complex valued $m-$function on the real axis. The position of the complex poles of the $m-$function were then computed by analytic continuation of a Padé approximant representing the $m-$function on the second Riemann sheet. In later development[15,29] we concentrated on determining the poles of the $m-$function directly on the second Riemann sheet by using a synthesis of Titchmarsh-Weyl's theory and the complex scaling method. Recently the spectral theorem has been analyzed via the analytically continued $m-$function[20] and its implication for the expansion of the associated Green's functions in the complex momentum or energy planes[62].

In the present contribution we have generalized the Titchmarsh-Weyl's theory, see references 1,2,58-60, to the multichannel case. The main assumption has been that the interaction of a system can be described with an effective *symmetric* interaction matrix. Although any matrix can be transformed to (complex) symmetric form, the non-symmetric case can be treated analogously with some modification. Hence we have shown the following:

1. The one dimensional $m-$function can be generalized to an $\mathbf{M}$-matrix defined as the energy dependent mixing matrix which allows us to have a square integrable solution matrix $\boldsymbol{\chi}$ in terms of the regular and irregular initial value solution matrices $\boldsymbol{\Phi}$ and $\boldsymbol{\Psi}$. For unit matrix boundary conditions $\mathbf{M}$ is symmetric.

2. The square integrable solution matrix $\boldsymbol{\chi}$ can be expressed as the Jost solution matrix $\mathbf{F}$ times inverse of the Wronskian matrix of $\boldsymbol{\Psi}$ and $\mathbf{F}$.

3. The spectral density and spectral distribution matrices can be defined as a direct generalization of the one channel formulation.

4. The explicit connections between the multichannel Titchmarsh-Weyl's theory and the multichannel S-matrix formulation were explored and the single channel relations can essentially be straight forwardly extended to a matrix formulation in which $\mathbf{S}$ is unitary.

5. The matrix theory just presented allows for a direct analytic continuation onto the second Riemann sheet via the complex dilation method without resorting to any kind of analytic fitting procedures. Apart form yielding direct convenient computation formulae for fundamental scattering quantities the theory also exhibits, in a simple and direct way, properties such as complex symmetry of the spectral density matrix $\mathbf{M}$, star unitarity of $\mathbf{S}$ etc.

As an illustration to our theoretical formulation we presented and discussed some curve crossing situations in a diatomic molecule. In particular we focused on the calculation of resonance poles and associated structures, and found that the knowledge of the behaviour of resonance spectra with interaction strength and J-values may help to resolve complicated band patterns. Although our $MgH-$model represents a rather clean adiabatic case, we found that the calculated energies and lifetimes may change drastically under parameter variation. The occurrence of interferences between waves in different channels may lead to long or short lived resonances, which may or may not be expected in a more simplified analysis. In general our discussion is not limited only to two channel problems where semiclassical methods may work sufficiently well. The general agreement with the work of Korsch et al.[48,49] gives additional support to our results.

Although we have here considered a potential scattering problem we are confident that energy dependent potentials should be no problem as our studies for the one channel case indicates[66,67]. One can furthermore generalize our Green's function and S-matrix expansions like the Mittag-Leffler expansion and related methods[66,67] also to the multichannel energy dependent case.

In our numerical work we have made use of the log-derivative integration technique[39] as well as the renormalized Numerov method[39]. In related studies the Goodwin-Fox method has been advocated[68–70].

Finally, we stress that the present technique of describing molecular predissociation processes and atom-ion collisions, within a non-relativistic formulation, in terms of a coupled set of ordinary singular differential equations can in principle be employed numerically to any desired accuracy.

Acknowledgements

We are grateful to Drs. Korsch, Möhlenkamp and Thylwe for supplying us with their results prior to publications as well as for discussions and to Drs. Lefebvre, Atabek and Krylstedt for enlightning discussions. M.S. Carlsund has given constructive criticism concerning the numerical procedures. One of us (E.B.) thanks the University of Florida and in particular the Office of Academic Affairs for the warm hospitality that was shown to him during the time he was an Uppsala Visiting Professor at the Quantum Theory Project at the University of Florida.

Appendix: Adiabatic and Diabatic States of a Diatomic Molecule

As an introduction to the theoretical development we outline the derivation of the coupled equations for the radial motion of a diatomic molecule. For a more detailed treatment we refer the reader to Pack and Hirschfelder[7,8] and a recent review by Macias and Riera [9]. Note that the Hamiltonian obtained below is expressed in terms of so-called separated atom coordinates[7,8], while the coordinates in ref. 9 refer to center of mass of the nuclei. In the latter case the various contributions to H_{tot} will be radically different[71].

Consider a diatomic molecule with nuclei A and B. The respective nuclear charges are denoted Z_A and Z_B respectively. The distance between the two nuclei is R. An electron i is positioned at a distance r_{iA} relative to the nuclei A. and the two electrons i and j are located at a distance r_{ij} relative to each other. The nonrelativistic Hamiltonian then reads:

$$H_{tot} = T + V =$$

$$-\frac{1}{2M_A}\nabla_A^2 - \frac{1}{2M_B}\nabla_B^2 - \frac{1}{2}\sum_{i=1}^{N}\nabla_i^2 - \sum_{i=1}^{N}\{\frac{Z_A}{r_{iA}} + \frac{Z_B}{r_{iB}}\} + \sum_{i<j}^{N}\frac{1}{r_{ij}} + \frac{Z_A Z_B}{R} \quad (A.1)$$

which after separation of the total center of mass motion, and reference to separated atom system relative coordinates, dropping the cross term between electron and nuclear momenta as well as mass polarization reads

$$H = -\frac{1}{2\mu}\nabla_R^2 + H_{el}(\vec{r}, R) \quad (A.2a)$$

$$H_{el} = -\sum_i \frac{1}{2}\nabla_i^2 + V \quad (A.2b)$$

where μ now is the reduced mass of the nuclei $\mu = \frac{M_A M_B}{M_A + M_B}$. It is convenient to make a transformation to the molecular fixed frame and to express the total wave function as a product of the radial part of the nuclear wave function $F_{nJ\Lambda}(R)$, the electronic part $\Phi_{n\Lambda}(\vec{r}, R)$ and the angular part of the nuclear wave function $Y_{J\Lambda}(\theta, \phi)$ as (we here chose $\phi = 0$) :

$$\Psi(\vec{r}, \vec{R}) = \frac{1}{R}\sum_{nJ\Lambda} F_{nJ\Lambda}(R)\Phi_{n\Lambda}(\vec{r}, R)Y_{J\Lambda}(\theta, 0) \quad (A.3)$$

This construction leads to a separation of the wave function into an electronic and a radial nuclear part. We are thus able to describe the nuclear motion in terms of a set of coupled differential equations written as:

$$\left[\frac{d^2}{dR^2} + 2\mu\{E - \frac{J(J+1) - \Lambda^2}{R^2}\}\right]F_{nJ\Lambda} =$$

$$= \sum_{n'\Lambda'}\Big[2\mu F_{n'J\Lambda'}\langle\Phi_{n\Lambda}|H_{el}|\Phi_{n'\Lambda'}\rangle\delta_{\Lambda\Lambda'} - F_{n'J\Lambda'}\langle\Phi_{n\Lambda}|\frac{\partial^2}{\partial R^2}|\Phi_{n'\Lambda'}\rangle\delta_{\Lambda\Lambda'}$$

$$- 2\frac{dF_{n'J\Lambda'}}{dR}\langle\Phi_{n\Lambda}|\frac{\partial}{\partial R}|\Phi_{n'\Lambda'}\rangle\delta_{\Lambda\Lambda'} \quad (A.4)$$

$$- \frac{F_{n'J\Lambda}}{R^2}\{\langle\Phi_{n\Lambda}|L_x^2 + L_y^2|\Phi_{n'\Lambda'}\rangle\delta_{\Lambda\Lambda'}$$

$$- \sqrt{J(J+1) - \Lambda(\Lambda+1)}\langle\Phi_{n\Lambda}|L_+|\Phi_{n'\Lambda'}\rangle\delta_{\Lambda\Lambda'+1}$$

$$- \sqrt{J(J+1) - \Lambda(\Lambda-1)}\langle\Phi_{n\Lambda}|L_-|\Phi_{n'\Lambda'}\rangle\delta_{\Lambda\Lambda'-1}\}\Big]$$

We are still free to choose the complete set of electronic wave functions $\{\Phi\}$ and to provide such a set is one of the main objectives of quantum chemistry. These functions are usually the ones diagonalizing the electronic Hamiltonian defining the so called adiabatic potential energy curves.

It is sometimes convenient to work with a different representation than the one obtained from the diagonalization of H_{el}. This forms the set of diabatic representations. Such a choice would be electronic wave functions for which the $\partial/\partial R$ coupling vanishes for all internuclear distances [72]. This definition has since then been criticized as unphysical[73] especially for low energy close coupling scattering formulations. Delos has [74] recently presented a more appropriate formulation. For our present purposes Smith's definition is sufficient. The transformation to such a diabatic form can be made rather easily for a two-channel problem. We will ignore some of the couplings above and restrict our discussion to Σ−states (i.e. $\Lambda = 0$). We now rewrite the coupled equations as

$$\frac{d^2F_{nJ}}{dR^2}+2\sum_m P_{nm}\frac{dF_{mJ}}{dR}+\sum_m Q_{nm}F_{mJ}+2\mu\{E-U_n(R)-\frac{J(J+1)}{R^2}\}F_{nJ}=0 \quad (A.5)$$

with

$$P(R)_{nm} = \langle\phi_n|\frac{\partial}{\partial R}|\phi_m\rangle \quad (A.6a)$$

$$Q(R)_{nm} = \langle\phi_n|\frac{\partial^2}{\partial R^2}|\phi_m\rangle \quad (A.6b)$$

and

$$U_n(R) = \langle\Phi_n|H_{el}|\Phi_n\rangle \quad (A.6c)$$

where ϕ_k are the eigenfunctions of H_{el}. Following Garqaud et al [75] we write

$$F_{nJ} = \sum_m D_{nm}G_{mJ} \quad (A.7)$$

where $\mathbf{D}$ satisfies :

$$\frac{d\mathbf{D}}{dR} + \mathbf{PD} = \mathbf{0} \; ; \quad \lim_{R\to\infty} \mathbf{D} = \mathbf{1} \quad (A.8)$$

With this substitution we find :

$$\frac{d^2G_{nJ}}{dR^2} - 2\mu\sum_{m=1} V_{nm}G_{mJ} + \{2\mu E - \frac{J(J+1)}{R^2}\}G_{nJ} = 0 \quad (A.9)$$

$\mathbf{V}$ is the diabatic potential defined by :

$$\mathbf{V} = \mathbf{D}^{-1}\mathbf{UD} \quad (A.10)$$

For a two channel problem the matrix $\mathbf{D}$ can be defined through :

$$\mathbf{D} = \begin{pmatrix} \cos\omega(R) & \sin\omega(R) \\ -\sin\omega(R) & \cos\omega(R) \end{pmatrix}$$

with

$$\omega(R) = \int_{R}^{\infty} P_{12}(R')dR' \tag{A.11}$$

For more then two channels we do not have such simple expressions. In practice the limitation to two channels is not such a large restriction since one rarely finds curve crossing problems where more than two curves cross at the same internuclear distance.

† Permanent adress :Quantum Theory Project
Williamson Hall
University of Florida
Gainesville, Florida 32611
U.S.A.

References

1. H. Weyl, Ann. Math. **68**, 220 (1910).

2. E. A. Coddington and N. Levinsson, *Theory of Ordinary Differential Equations,* McGraw-Hill New York, 1955.

3. J. Aquilar and J. M. Combes, Commun. Math. Phys. **22**, 269 (1971).

4. E. Balslev and J. M. Combes, Commun. Math. Phys. **22**, 280 (1971).

5. B. Simon, Ann. Math. **97**, 247 (1973).

6. C. van Winter, J. Math. Anal. Appl. **47**, 633 (1974).

- C. van Winter, J. Math. Anal. Appl. **47**, 368 (1974).

7. R. T. Pack and J. O. Hirschfelder, J. Chem. Phys. **49**, 4009 (1970).

8. R. T. Pack and J. O. Hirschfelder, J. Chem. Phys. **52**, 521 (1970).

9. A. Macias and A. Riera, Phys. Rep. **90**, 299 (1982).

10. H. Lefebvre-Brion, in *Atoms, Molecules and Lasers,* International Atomic Energy, Vienna (1974), p.411.

11. K. Morokuma and T. F. George, J. Chem. Phys. **59**, 1959 (1973).

12. P. Pechukas, T. F. George, K. Morokuma, F. J. McLafferty and J. R. Laing, J. Chem. Phys. **64**, 1099 (1976).

13. J. Simons, J. Chem. Phys. **75**, 2465 (1981) ; P. O. Löwdin, Int. J. Quantum Chem. **27**, 494 (1985).

14. M. Rittby, N. Elander and E. Brändas, Phys. Rev. **A24**, 1636 (1981).

15. M. Rittby, N. Elander and E. Brändas, Mol. Phys. **45**, 553 (1982).

16. M. Rittby, N. Elander and E. Brändas, Chem. Phys. **87**, 55 (1984).

17. M. Hehenberger, E. Brändas and N. Elander, Int. J. Quantum Chem. Symp. **12**, 67 (1978).

18. N. Elander, M. Rittby and E. Brändas Int. J. Quantum Chem. **22**, 445 (1982).

19. M. Rittby, N. Elander, E. Brändas and A. Barany, J. Phys. **B17**, L677 (1984).

20. E. Brändas, M. Rittby and N. Elander, J. Math. Phys. **26**, 2648 (1985).

21. E. Engdahl, E. Brändas, M. Rittby and N. Elander, Phys. Rev A. **37**, 377 (1988).

22. B. R. Junker, Adv. Atom. and Molec. Phys. **18**, 207 (1982).

23. W. P. Reinhardt, Ann. Rev. Phys. Chem. **33**, 223 (1982).

24. Y. K. Ho, Phys. Rep. **1**, 99 (1983).

25. B. Simon, Phys. Lett. A **71**, 211 (1979).

26. J. D. Morgan and B. Simon, J. Phys. **B14**, L167 (1981).

27 E. Balslev, in *Resonances - Models and Phenomena* Lectures Notes in Physics **211**, (Eds. S. Albeverio, L. S. Ferreira and L. Streit, Springer Verlag , Heidelberg 1984) page 27.

28 M. Reed and B. Simon, *Methods of Modern Mathematical Physics IV, Analysis of Operators* (Academic Press, New York, 1978) page 183 -191.

29. M. Rittby, N. Elander and E. Brändas, Int. J. Quant. Chem. **23**, 865 (1983).

30. P. Erman in *Specialist Periodic Reports* Vol **6** (Chem. Soc. ,London, 1979) page 501.

31. C. P. Edwards, C. S. MacLean and P. J. Sarre, Mol. Phys. **52**, 1453 (1984).

32. M. S. Childs in *Molecular Spectroscopy*,Vol. **2** (Eds. R. F. Barrow, D. A. Long and D. J. Millen Chem. Soc. Specialist Periodical Report London 1974) page 466.

33. H. Lefebvre-Brion and R. Colin J. Mol. Spectrosc. **65**, 33 (1977).

34. R. D. Levine, B. R. Johnson and R. B. Bernstein, J. Chem. Phys. **50**, 1694 (1969).

35. G. Sillén, M. Rittby, N. Elander and E. Brändas, J. Chem. Phys. submittted (1988).

36. O. Atabek and R. Lefebvre, Chem. Phys **52**, 199 (1980).

37. O. Atabek and R. Lefebvre, Chem. Phys **55**, 395 (1981).

38. O. Atabek and R. Lefebvre, Chem. Phys **56**, 195 (1981).

39. B. R. Johnson, J. Chem. Phys. **69**, 4678 (1978).

40. L. Klynning and H. Martin , Physica Scripta **20**, 594 (1979).

41. J. M. Robbe, *Thesis*, L'Universite des Sciences et Techniques de Lille (1978).

42. M. S. Childs and R. Lefebvre, Chem. Phys. Lett. **55**, 213 (1978).

43. L. E. Selin Ark. Fys. **21**, 529 (1962).

44. W. J. Balfour and H. M. Cartwright, Can. J. Phys. **54**, 1898 (1976).

45. W. J. Balfour and H. M. Cartwright, Astron. Astrophys. Suppl. **26** 389 (1976).

46. W. J. Balfour and B. Lindgren, Can. J. Phys. **56**, 767 (1978).

47. M. Rittby, *Thesis*, University of Stockholm (May 1985).

48. H. J. Korsch, R. Möhlenkamp and K. E. Thylwe, J. Phys. **B19**, 2151 (1986).

49. H. J. Korsch and R. Möhlenkamp, Phys. Rev. **A34**, 4716 (1986).

50. O. Atabek, R. Lefebvre and A. Jacon, J. Phys. **Bxx**, yy (198ZZ).

51. C. J. Williams, K. F. Freed, S. J. Singer and Y. B. Band, Faraday Discuss. Chem. Soc. **82**, *paper 6* (1986).

52. S. J. Singer K. F. Freed and Y. B. Band, J. Chem. Phys. **79**, 6060 (1983).

53. S. J. Singer, K. F. Freed and Y. B. Band, J. Chem. Phys. **81**, 3064 (1984).

54. S. J. Singer, K. F. Freed and Y. B. Band, J. Chem. Phys. **81**, 3091 (1984).

55. C. J. Williams and K. F. Freed, J. Chem. Phys. **85**, 2699 (1986).

56. S. Lee, C. J. Williams and K. F. Freed, Chem. Phys. Lett. **130**, 271 (1986).

57. X. He, O. Atabec and A. Suzor Submitted to Phys. Rev. A (1988).

58. M. Hehenberger, H. V. McIntosh and E. Brändas, Phys. Rev. **A10**, 1494 (1974).

59. E. Brändas, M. Hehenberger and H. V. McIntosh, Int. J. Quantum Chem. **9**, 103 (1975).

60. M. Hehenberger, B. Laskowski and E. Brändas, J. Chem. Phys. **65**, 4559 (1976).

61. M. Hehenberger, P. Froelich and E. Brändas, J. Chem. Phys. **65**, 4571 (1976).

62. E. Engdahl, E. Brändas, M. Rittby and N. Elander, J. Math. Phys. **27**, 2629 (1986).

63. P. Krylstedt, N. Elander and E. Brändas, Int. J. Quantum Chem Symp. **20**, 253 (1986).

64. P. Krylstedt, M. Rittby, N. Elander and E. Brändas, J. Phys. **B** in press 1987.

65. P. Krylstedt, N. Elander and E. Brändas, submitted to J. Phys. **B** 1987.

66. E. Brändas, E. Engdahl, M. Rittby and N. Elander, Int. J. Quantum Chem. **S19**, 457 (1986).

67. M. Rittby, N. Elander and E. Brändas , this book page 133.

68. D. W. Norcross and M. J. Seaton, J. Phys. **B6** 614,1973.

69. R. Lefebvre in *Intramolecular Dynamics* (Eds. J. Jortner and B. Pullman, D. Reidel Publ. Comp., Paris, 1982).

70. R. Lefebvre, in ACS SYMPOSIUM SERIES **263***Resonances in Electron-Molecule Scattering, van der Waals Complexes and Reactive Chemical Dymamics* (Ed. D. Truhlar, American Chem. Soc. Washington D.C. 1984).

71. A. Fröman, J. Chem. Phys. **36**, 1490 (1962).

72. F. T. Smith Phys. Rev. **118**, 349 (1960).

73. B. Andresen and S. E. Nielsen, Mol. Phys., **21**, 523 (1971).

74. J. B. Delos Rev. Mod. Phys. **53**, 287 (1981).

75. J. Garquad, Phys. **B14**, 2259 (1981).

ON THE WAY TO A MULTICONFIGURATION TREATMENT OF RESONANCE PHENOMENA IN ATOMS

Nils Elander, Christina Carlsund, Peter Krylstedt

Manne Siegbahn Institute of Physics
Frescativägen 24
S - 104 05 Stockholm
SWEDEN
and

Peter Winkler

Department of Physics
University of Nevada at Reno
Reno, NV 89557
U.S.A.

Abstract

Some features in current experimental and theoretical electron - atom (ion) collision physics are pointed out. The problems are given a formal analysis and some previous methods are discussed. The close coupling formalism of Burke and Seaton[1] is reviewed in some detail. A synthesis of this method and the theory of exterior complex dilation is suggested. The formal properties of this novel approach are discussed. Two preparatory steps for the full treatment, are discussed. The first one concerns the reduction of the problem to a single channel equation - the distorted wave approach or the continuum local density approach introduced by Hara[2]. Several improvements of the Hara theory are reported. The second model problem has the full close coupling formalism, but uses expansions in a small set of hydrogenic orbitals to approximate the full machinery. Calculation of cross sections and other experimentally measurable collision data are discussed and results from single channel partial-wave S-matrix pole expansion calculations are reported. The possible use of many channel S-matrix pole expansions is discussed.

1. Introduction

1.1. Experimental Motivation

Collisions between electrons and neutral atoms or atomic ions lead to the simplest kind of atomic particle exchange processes one can think of. Early experimental studies revealed few details of the electron-atom scattering cross section. It was only in the beginning of the 1960th that resonances in electron-atom collisions were found. Through the introduction of new experimental methods such as ultra high vacuum technology, particle counting, low-current measurements and signal averaging the experimental study of electron-atom or electron-ion collisions reached an explosive, blooming period with pioneering work of Schulz[3] during the sixties. Schulz's use of the word ***resonances*** as equivalent to ***compound states*** or ***temporary negative ions*** emphasis the particular interest at the time. The negative ions of hydrogen and some other light atoms were particularly theoretically well studied. Experimental results were obtained for several other atomic as well as diatomic species[4].

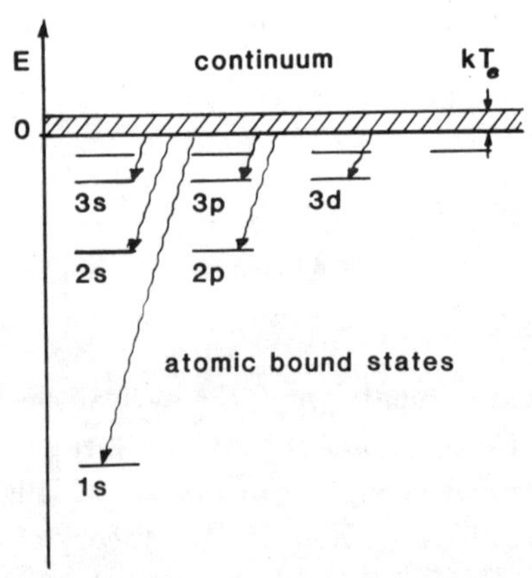

Figure 1

Direct Radiative capture of an electron to an atomic ion

Schulz defined *resonances* in more restrictive way than we intend in this paper: " *For example, doubly excited states of atoms are nonstationary states which decay by electron emission, yielding positive ions. Nevertheless, such states are excited*

states of the neutral atom, and therefore do not meet our definition, namely that the states be temporary negative ions".

While this early work mostly relates to rather accurate measurements of negative ions of fundamental atoms and diatomic molecules another problem of atomic collision physics began to receive some attention. Studies of stellar spectra of atomic ions revealed the importance of recombination processes. There are two basic reactions for free electron-ion recombination in dilute astrophysical or magnetically confined laboratory plasmas.

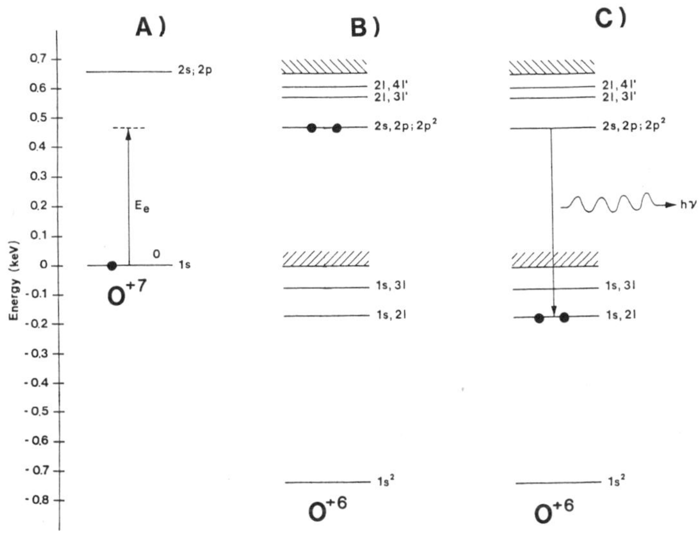

Figure 2.

Schematic of the dielectronic recombination in O^{7+} .

In direct radiative recombination the continuum electron is captured by the ion into a state of the new ion. The energy and momentum are conserved by the simultaneous emission of a photon.

This process is drastically enhanced in the presence of resonances in the electron-ion collision complex. If the continuum electron has an appropriate kinetic energy it may orbit around the ion for some short time instead of just pass by. The chance of getting captured is thus considerably increased. Direct radiative recombination processes may be found in nude ions as well as in ions that still have a number of electrons left. We note that even though this process is expected to be enhanced by shape resonances, it is *not an energy resonant process* as such.

This is quite different in the *dielectronic recombination* process. Here the energy balance is handeled by a simultaneous excitation of an electron in the target ion. This means that DR occur only be found in non-nude ions. DR is the invers of excitation autoionization.

$$O^{7+}(1s) + e^- \rightarrow O^{6+}(2s2p) \rightarrow \begin{cases} O^{7+} + e^- & \text{(a)} \\ O^{6+}(1s2s)^{3,1}S + h\nu & \text{(b)} \end{cases} \quad (1.1.1)$$

In order for the recombination to occcur, the resonant capture of the electron into a doubly excited state of the ion has to be followed by deexcitation to a lower electronic state which does not autoionize. Even though the process had been proposed as early as 1939, it was not until 1964 that the importance of the DR process was fully appreciated. At this time Burgess[5] showed that the DR rate in the solar corona (for $e^- + He^+$) can, in fact, dominate the direct radiative rate by orders of magnitude.

Until the present date (1987) electron-ion collision experiments do not yet have enough energy resolution. It is obvious from recent model calculations that a rich structure, sensitive to various external fields, will appear once the experimentalists will be able to bring the energy resolution to an accuracy enough to be able to measure the signatures of different atomic terms.

Such accurate experiments will hopefully be possible in the new generation of heavy ion storage rings[6], which are under construction. By cooling the circulating ion beam in a bath of cold electrons[7], it is possible to reduce the energy spread of the ions to a about $\delta E/E \sim 10^{-4}$ or at best 0.1 eV. It has even been suggested that it is possible to go even further. By using cooled electrons, emitted by photoemission from a cold cathode[8], one might be able to cool the ions by another factor of ten to $\delta E/E \sim 10^{-5}$ or at best around 0.01 eV. Schuch[9] has very recently suggested to use capture into autoionizing states with negligible radiative decay rates as a way to cool an ion beam. The advantage of this method is that the beam loss, via direct radiative capture, which is large at low relative kinetic energies, is avoided. If these proposals turn out to hold in reality one will be able to do experiments with ion beams with an energy definition better that 10^{-5}.

Experimental studies of single and, particularly, double ionization of free atoms by synchrotron radiation constitutes rather a challenge to atomic theory. Photoionization experiments aiming at just the opposite of the above capture experiments, have recently been proposed [10]. Here a low charged ion is hit by a synchrotron photon which makes it ionize.

$$A^{q+} + h\nu \rightarrow A^{(q+1)+} + e^- \quad (1.1.2)$$

Still large discrepancies between theory and experiment exists. This is especially true close to thresholds where large resonances occur.

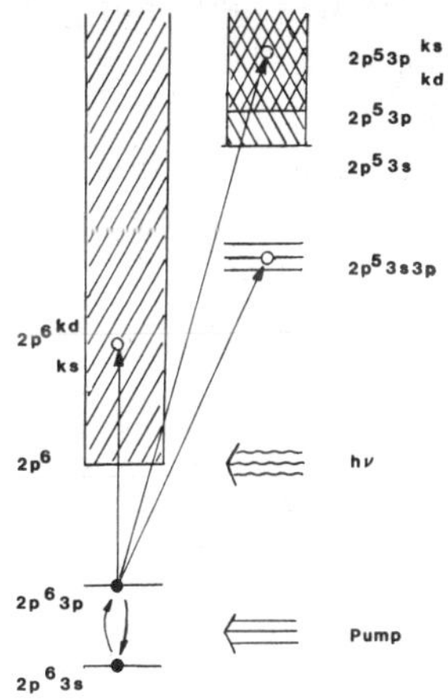

Figure 3.

Example of photoionization by a UV photon. Neutral sodium is laser excited to $2p^6 3p$. From this excited state the electron is either put into a continuum ks or kd orbital, or for example the autoionizing $2p^5 3s3p$ state.

1.2. Theoretical approaches to electron-atom scattering phenomena

The experimental progress reviewed by Schulz[3,4] encourages theoretical developments. The status of theoretical approaches reviewed by Nicolaides[11] who distinguishes four categories:

(a) Theories using the scattering viewpoint.

(b) Theories treating resonances as *decaying states.*

(c) Theories treating resonances as *special types of bound states using projection operators.*

(d) Theories treating resonances as *special types of bound states* without *using projection operators.*

In his own work, Nicolaides[11] uses a description with the wave function of the system taken at $t = 0$. He can thereby obtain a formalism where the Gamow states in some sense are square integrable. A recent monograph on autoionization edited by Temkin[12] supports such a division into categories.

Present experiments on dielectronic recombination have so far been done with rather moderate energy resolution, thus favouring not termspecific theoretical methods. In particular the average configuration methods of Griffin et al.[13], as well as the studies of Hahn et al.[14,15], have proven to compare well with experimental results. In the work of Griffin et al. energies and wave functions for a small number of configurations of the target ion A^{q+}, and a larger numbers of doubly excited configurations of the capture ion $(A^{(q-1)+})^{**}$ and singly excited configurations $(A^{(q-1)+})^{*}$ are calculated with a modified Hartree-Fock code[16]. Hydrogenic Rydberg configurations are added and the DR cross section is calculated according to the original formalism of Bates and Massey[17]. The cross section for DR, which is derived from detailed balance considerations is given as:

$$\sigma_j^{DR}(E_r) = \frac{\pi^2 \hbar^3}{m_e E_r} \sum_i \frac{g_j}{2g_i} A_a(i \to j) \frac{\sum_f A_r(j \to f)}{\sum_{i'} A_a(j \to i') + \sum_f A_r(j \to f)} L_j(E_r), \tag{1.2.1}$$

where g_i and g_j are the statistical weights for the inital, (i), N-electron system and the resonance level (j), of the *(N+1)*- electron system, respectively. E_r is the asymptotic relative energy of the projectile electron and the target ion, $A_a(i \to j)$ is the autoionization rate for capture, $A_r(j \to f)$ is the radiative stabilisation rate and $L_j(E_r)$ is a Lorentzian line profile of the DR transition.

The basic idea behind eq(1.2.1) is that A_a and A_r are computed independently of each other. A rigorous analysis, which accounts for overlapping resonances, interference effects etc. has been given by Bell and Seaton[18]. This leads to a unified treatment of electron-ion scattering and DR. Within the exact Bell and Seaton formulation of the theory of dielectronic recombination, accurate configuration interaction close coupling calculations may in principle be able to estimate observable quantities within 10%. The unified treatment of DR, autoionization resonances and direct radiative recombination is also represented in recent work by Winkler[18a] who emphasizes a propagator approach with the construction of an optical potential to serve all competing processes.

Recent experimental work made it necessary to incorporate the effects of external fields, that are present in the reaction region, in the theoretical calculations to obtain good agreement between the experimental and theoretical results [19,20,21]. An exact theory of DR needs to consider the eigenstates and dynamics of the electron and ion systems and thus derive the individual autoionization and radiative transition probabilities in the presence of an electric field. Harmin[22] and Sakimoto[23] have recently tried to formulate theories which includes the Stark field in the DR framework of Bell and Seaton[18]. In contrast to earlier work these theories are able to describe a smooth transition between high and low field regimes.

A complete description should also take radiation damping into account. In the electron-ion continuum, DR will in principle compete with radiative recombination and cause interference. Continuum contributions to DR, reflect the strength of the radiative coupling between the electron-ion continuum and the final bound state. It is of the order $\alpha^3 Z^2$ and could therefore be quite significant for heavy, highly charged ions.

Still another set of channels need to be taken into account to describe DR properly, namely, the fine structure of an excited state of the A^{q+} ion in the process. Transitions between different fine structure levels with the same term has been found to enhance the DR rate. This has for instance been reported for the Ba^+ ion[24]. Theoretical models furthermore need to employ intermediate coupling. This is true also for the more highly charged ions, that dominate in the thin stellar or laboratory fusion plasmas. Badnell[25] has during the last few years presented a series of papers where DR cross section has been calculated for individual DR lines. The calculations were made in conventional LS coupling as well as intermediate coupling. With his code *AUTOSTRUCTURE*, which is based on the Cambridge group code *SUPERSTRUCTURE*, Badnell has been able to calculate DR cross sections and rates for low as well as highly charged ions. The code uses multiconfiguration wave functions and different ionization limits for different term series are taken into account. Recent results show how states, which are forbidden to stabilize via DR in pure LS-coupling, in the *intermediate* description are participating in the DR process.

While early workers like Schulz[3,4] pay most of their attention to negative ions created by low energy electron impact. Experiments have now proceeded to study indirect processes in electron impact ionization[26] of atoms and atomic ions. Following a recent comment by Pindzola et al.[27], the major indirect contribution is the excitation-autoionization

$$\begin{array}{l} e^- + A^{q+} \rightarrow [A^{q+}]^* + e^- \\ \qquad\qquad\qquad\qquad \searrow \\ \qquad\qquad\qquad\qquad\quad A^{(q+1)+} + e^- \end{array} \tag{1.2.2}$$

where q is the charge of the ion. Pindzola et al.[27] compare the agreement between experiment and the distorted wave and the close coupling approaches. Their conclusion is that, for comparison with present low energy resolution crossed beam experiments, both methods are capable of giving a fair understanding of the process. The close coupling is certainly more accurate but also more complicated and time consuming in actual computations.

The situation in the photoionization business, discussed in the previous section, is in some sense quite similar to the indirect electron impact autoionization. In some cases the appearence of the large resonant contributions in the photoionization cross section can be understood as follows: The originally bound inner electron is first excited by a synchrotron photon to an excited state. Due to correlations effects etc. this state may be quasibound. In a simple distorted wave formalism the potential of the excited configuration can host quasibound eigenstates. These so-called shape resonances are localized eigenstates and are like bound states described by orbitals with large amplitudes close to the atomic core. They do however also have a non-negligible ionization rate. The fact that quasibound orbitals have a large overlap with some bound states, makes the transition probability for the absorption of a photon, larger for these states as compared with normal non-localized continuum states. The net result is that the quasibound states appear as large resonances in photoionization cross sections and for that matter also in electron impact ionization cross sections. This kind of resonances have recently been discussed by Younger in connection with electron-impact ionization of Cs^+ [28]. Large "unidentified resonances in Ca^{++} and Ba^{++}" [29] are most likely of this nature. Our group is currently engaged in the calculations of such resonances[30].

Comparison of experimental results with theoretical data from different methods has, as usual, helped the progress of both fields. From this point of view the net result of Schulz review[3] is that: (a) Accurate comparisons can only be made from studies of hydrogic systems. (b) The configuration interaction close coupling (CCC) method seems to give the best agreement with experiments, especially when one is interested in both the position and the width of the resonance. (c) Computationally the CCC method is the most time consuming of all four listed above. Even if several year have past since these statements were made, these three observations still seem to be true.

2. Exterior Complex Dilation

The basic idea of the present contribution is to *localize* the resonant, close coupling, many channel, Gamow type of wave function by exterior complex rotation techniques. The theory of this and other dilatation or regularization methods is the topic of several papers in this book[31]. Here we accordingly limits ourselves to a quite brief discussion. The specific transformation used here is the exterior complex rotation affecting only the radial coordinates of the spherically symmetric system.

$$r \to \xi(r) = \begin{cases} r & ; \quad 0 \leq r < R_s \\ R_s + \exp(i\theta)(r - R_s) & ; \quad r \geq R_s \text{ and } 0 < \theta < \theta_{crit} < \pi \end{cases} \tag{2.1}$$

Here θ_{crit} is an angle that is dependent on the asymptotic analytic properties of the potential[32].

The close coupling approach is formally an effective many channel framework. The exterior complex rotation on a many channel problem is discussed by Brändas et al. in this book[33]. The effect of the transformation is, in some sense, to *gently suppress* the amplitude of the system outside the exterior rotation radius R_s (see figure 4). This means, vaguely speaking, that the probability of finding the system outside this radius is suppressed. We are deliberately localizing the system inside this sphere and studying it there. The amplitude of the exterior rotated Gamow wave is decreasing exponentially with the radius outside R_s. It will thus have as many nodes and antinodes as the unrotated Gamow wave in contrary to the box quantization wave function which has a finite number of nodes depending on the

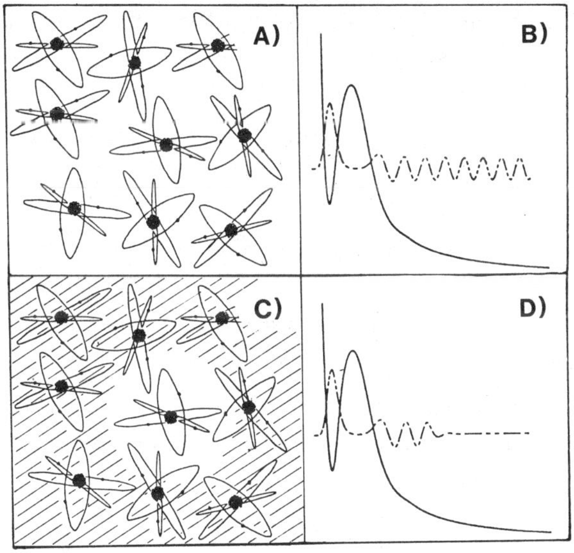

Figure 4.

The effect of exterior complex dilation. **(a)** In the non-dilated space we need in principle to consider a whole ensemble of atoms, in particular when we are interested in environmental effects. **(b)**A resonant quasibound orbital in one atom is here represented by a purely outgoing wave (here we display the real part of ψ_{out}). **(c)** By introducing the exterior complex dilation transformation with an exterior rotation radius R_s around the center of our atom we focus on the internal structure of our quasibound system. **(d)** The orbital wave function for the quasibound resonant level has now turned into a square integrable one for which the modified bound orbital methods that are discussed in this contribution can be used.

box wall distance. The complex rotated system has thus never lost its contact with its environment, it is an open localized system. The ordinary Hilbert space scalar product is here generalized to to the one discused by Elander and Brändas[34] (see also the paper by Brändas and Dreismann[34a] in this book), where the bra and ket spaces are chosen to be the pure incoming and the pure outgoing states respectively. Both spaces are properly complex rotated in order to obtain a well defined square integrable integrand.

$$\langle f|g\rangle_{\xi} = \int_{0}^{\xi(b)} \mathcal{GD}(f,g)d\xi(r) = \int_{0}^{\xi(b)} f^{*}(\xi^{*}(r))g(\xi(r))d\xi(r), \tag{2.2}$$

where $\mathcal{GD}(f,g)$ is a generalizable density derived out of the generalized inner product[34]. The integration path is in the present exterior dilated framework defined by eq(2.1). The *norm* of the complex rotated wave function is then naturally associated with our generalized scalar[34]

$$N_f = \langle f|f\rangle_\xi = \int_0^{\xi(b)} \mathcal{GD}(f,f)d\xi(r) = \int_0^{\xi(b)} f^*(\xi^*(r))f(\xi(r))d\xi(r). \tag{2.3}$$

A normalized resonance eigenfunction

$$\psi_N(\xi(r)) = \frac{1}{\sqrt{N_\psi}}\psi(\xi(r)) \tag{2.4}$$

can thus be defined for the time independent formulation. In the full time dependent formulation it can be shown to be a function of time, reflecting that we are dealing with decaying levels

$$N_f(t) = \langle f|f\rangle_\xi \exp(-\lambda t). \tag{2.5}$$

Here λ can be related to the complex energy of the resonance eigenfunction.

The complex dilation transformation has thus transformed our open system to a time dependent open localized one.

3. Selfconsistent Complex Dilated Approach - Previous Work

Complex dilated selfconsistent field approximations appeared early in the literature after the physicists and chemists became aware of the rigour of the method. Balslev and Combes and others[35,36,37] addressed the atomic Schrödinger problem in their early contributions. The manybody Hamiltonian, too, is feasable to dilate basis set methods. We shall however remember that the Slater type orbitals[38] were introduced long after Hartree and Fock had invented their method(s)[39]. Indeed, the STO and Gaussian basis functions were invented to mimic the numerical orbital wave functions for *bound* orbitals. Their simple analytical form and properties shortened the computational labor considerably. One bottleneck in basis set applications is the inflexibility of the basis set used. Even for bound states, several basis functions are needed to accurately represent a true analytic or a given numerical function. Too large a basis set can easily, on the other hand, introduce severe linear dependences among the basis functions. The literature is full of tricks and methods to avoid such problems. The fact that we are dealing with resonant orbitals as well as bound ones, does not make the situation simpler. Formal as well as numerical integration methods, show that the Gamow functions, albeit tamed to be square integrable by the complex scaling transformation, has infinitely many nodes. The boundary condition as well as the part close to the outer boundary has to be represented correctly by the basis set expansion. It is thus easy to realize why STO or Gaussian basis sets grow considerably when leaving their secure bound orbital home. Several problems beyond HF and few electron problems, were in the early complex dilation days in practice not possible to treat.

The use of the complex variational principle[40] has turned out to be an important step. Junker[41] and McCurdy[42] have advocated a version in which a real valued (unscaled) Hamiltonian and a complex-valued wave function are used to determine the stationary complex resonance energies E,

$$E[\Psi] = \frac{\int \Psi H \Psi d\tau}{\int \Psi \Psi d\tau}. \tag{3.1}$$

One disadgantage with the complex variational method of Junker and and McCurdy is, as far as we can understand, that they loose contact with the boundary conditions. The dilated Gamow waves can not be roatated back[42a]

The introduction of the exterior complex scaling gave in some sense support for this complex wave function variational calculations. Basically they deal with formalisms that can have complex eigenvalues, but in which the stability of the roots with respect to the complex dilation parameters (for example the complex scaling parameter $\eta = \exp(\imath\theta)$), is not fulfilled by the rigourous mathematical theory. Still the method has been successfully applied to several problems in the literature[43]. The reason for this success is that one mainly has studied narrow resonances which, accordingly, are close to the real axis. Following Brändas et al.[44] we see that the roots are not delta functions in the complex plane. Rather the peak of the Green's function around a pole has a width. If the true resonance modulus Green's function, $|G_\ell(r,r,\lambda)|$, peak is close to the real axis the width will give a peak on the real energy axis cut through the edges of the modulus Green's function structure.

The idea behind the complex basis method is still a good one. One want to be able to make use of the exterior scaling idea[45,46]. Loosely speaking, the core part of the system is not transformed, only the outer parts. This is however not an argument for not transforming the core orbitals at all and only transforming the valence or Rydberg ones. The correct form of the orbitals at large distances from the center of the system is especially important for low energy processes. Another approach to tame the stationarity problem of complex roots in an expansion method by tyhe use of sum-rules has been given by Winkler[46a].

Froelich et al.[47] has presented a rigorous method that models the exterior scaling idea in a basis set method. By partitioning the total dilated Hamiltonian matrix according to a division of the basis set in "bound" and "scattering" subspaces, which interact via off-diagonal blocks, the authors were able to mimic the exterior scaling ideas. Only the scattering part of the basis need to be complex dilated. It was shown that this improved the stability of the complex variation procedure considerably.

Although some small incompleteness exists, the basic ideas behind it are good. Its success and basic stability is proven by its ability to be the foundation for a complex multiconfiguration SCF theory[42]. With this work McCurdy and coworkers have gone further, than for instance Chung and Davis[48] or Nicolaides et al.[49]. The former, in their rigorously dilated saddle point method, do a two step optimization - one for the bound orbitals and one for the continuum ones. Good results for various systems have been reported. Nicolaides et al.[50] have, by using Fano theory[51], proven the validity of this approach for narrow resonances.

Froelich et al.[52] have recently presented another approach in which matrix elements are first calculated with a real scaling factor applied to the problem. Each matrix element is then fitted to a polynomial in the real scale factor after which each fit is analytically continued. The resonance eigenvalues can now be obtained by diagonalising the Hamiltonian matrix containing the now by fitting continued matrix elements. This method avoids the complications of complex SCF and CI as such during the iterations, while they give the analytic continuation from the fitted polynomials. The method works excellently for narrow resonances since the fit is valid close to the real axis.

While complex dilated selfconsistent field theories for basis functions are reported in the literature, little progress on numerical integration methods are found. To our knowledge only one successful calculation has been reported. Amstrong and coworkers[53] have, using the Froese-Fischer numerical Hartree-Fock codes, developed a complex dilated version. They are currently trying to generalize the Froese-Fischer real valued MCHF-codes to complex dilated ones. Several problems, arising from the fact that the original codes are constructed for real eigenvalues have shown up.

We belive that the close coupling equations of Burke and Seaton[1] are the most natural starting point for obtaining a complex dilated continuum multiconfiguration Hartree-Fock method.

4. Formulation of the Theory

As such, the present theory is not limited to resonant atomic processes, rather it can be applied to any problem in many-body theory where effective central symmetric potentials can be used. In cases where we need to go beyond spherical symmetry in the inner reaction region, but still can define some spherical radius around the "reaction region", as is done in exterior complex dilation, it will also be possible to treat systems without manifest spherical symmetry.

4.1. Formulation of the Equations

Following Burke and Seaton[1] we assume that the wave function for the entire system can be written as:

$$\Psi = \sum_{\Gamma} C_{\Gamma} \Theta_{\Gamma} \tag{4.1.1}$$

where

$$\Theta_\Gamma = \frac{1}{\sqrt{N_{total}!}} \mathcal{A} \psi_\Gamma(p)$$

is the antisymmetrized product of possible different target and projectile states:

$$\psi_\Gamma(p) = \sum_{M_{\ell_T}, m_\ell, M_{S_T}, m_s} (L_T M_{L_T} \ell m_\ell | L M_L)(S_T M_{S_T} s m_s | S M_S)$$
$$\times \phi_\gamma(S_T L_T M_{S_T} M_{L_T} | \vec{x}_p) \chi_{m_s} Y_{\ell m_\ell}(\hat{r}_p) \frac{1}{r_p} F_{k\ell}(r_p)$$

Here $\vec{x}_p$ designates the collection of the position and spin variables of all electrons *except the p-th.* ϕ_γ is, in the general case, a normalized and antisymmetric target state coupled to $(L_T M_{L_T} S_T M_{S_T})$ whereas the overal product function is coupled to $(L M_L S M_S)$.

4.2. A Simple Example : Coupling of Two Configurations - Hydrogenic Target

If channel I and channel II are represented by the following sets of quantum numbers: $(a,k)^S L_{M_L M_S}$ and $(a',k')^S L_{M_L M_S}$ we have :

$$\psi_I(1,2) = \sum_{m_{\ell_a} m_{\ell_k} m_{s_a} m_{s_k}} (\ell_a m_{\ell_a} \ell_k m_{\ell_k} | L M_L)(s_a m_{s_a} s_k m_{s_k} | S M_S) \times \phi_a(1)\phi_k(2) \tag{4.2.1}$$

and

$$\psi_I(1,2) = \sum_{m_{\ell_a} m_{s_a}} (\ell_a m_{\ell_a} \ell_k (M_L - m_{\ell_a}) | L M_L)(\frac{1}{2} m_{s_a} \frac{1}{2} (M_S - m_{s_a}) | S M_S) \times \phi_a(1)\phi_k(2) \tag{4.2.2}$$

which gives the antisymmetrized function :

$$\Theta_I(1,2) = \frac{1}{\sqrt{2}} \Big\{ \sum_{m_{\ell_a} m_{s_a}} (\ell_a m_{\ell_a} \ell_k (M_L - m_{\ell_a}) | L M_L)(\frac{1}{2} m_{s_a} \frac{1}{2} (M_S - m_{s_a}) | S M_S)$$
$$\times [\phi_{n_a \ell_a m_{\ell_a} m_{s_a}}(1) \phi_{k \ell_k (M_L - m_{\ell_a})(M_s - m_{s_a})}(2)$$
$$- (-1)^{\ell_a + \ell_k - L} (-1)^{1-S} \phi_{k \ell_k (M_L - m_{\ell_a})(M_s - m_{s_a})}(1) \phi_{n_a \ell_a m_{\ell_a} m_{s_a}}(2)] \Big\} \tag{4.2.3}$$

$\Theta_{II}(1,2)$ will be the same except that a,k will be replaced with a',k' This gives the collision function for the two configuration case:

$$\Psi = C_I \Theta_I + C_{II} \Theta_{II} \tag{4.2.4}$$

The key equation to derive the close-coupling equations from is:

$$\langle\Psi|\hat{H}|\Psi\rangle = E\langle\Psi|\Psi\rangle \tag{4.2.5}$$

where the Hamiltonian for the total scattering system is :

$$\hat{H} = \sum_j \hat{h}_j + \sum_{i>j} v(i,j) \tag{4.2.6}$$

Assume for simplicity that we only consider two configurations: Θ_I and Θ_{II}, each of which has only one term different from zero in the $m_\ell - m_s$ sums. Such an example is the coupling of two s–functions to $L=0, M_L=0, S=1, M_S=1$ or an s–function with a p–function to $L=1, M_L=1, S=1, M_S=1$ etc.

$$\langle\Psi|\hat{H}|\Psi\rangle = E\langle\Psi|\Psi\rangle \tag{4.2.7}$$

will then read :

$$\begin{aligned} &\langle C_I\Theta_I + C_{II}\Theta_{II}|\hat{H}|C_I\Theta_I + C_{II}\Theta_{II}\rangle \\ &= E\langle C_I\Theta_I + C_{II}\Theta_{II}|C_I\Theta_I + C_{II}\Theta_{II}\rangle \end{aligned} \tag{4.2.8}$$

If $H_{I,J} = \langle\Theta_I|\hat{H}|\Theta_J\rangle$ and $\langle\Theta_I|\Theta_J\rangle$ is written as $\langle I|J\rangle$ this equation can be expanded into :

$$\begin{aligned} &C_I^*C_IH_{I,I} + C_{II}^*C_{II}H_{II,II} + C_I^*C_{II}H_{I,II} + C_{II}^*C_IH_{II,I} = \\ &E\{C_I^*C_I\langle I|I\rangle + C_{II}^*C_{II}\langle II|II\rangle + C_I^*C_{II}\langle I|II\rangle + C_{II}^*C_I\langle II|I\rangle\} \end{aligned} \tag{4.2.9}$$

Resolved in orbitals, not assuming orthonormality we have :

$$\begin{aligned} \hat{H}_{I,I} =K^2\Big\{&h_{aa}(1)s_{kk}(2) + h_{kk}(2)s_{aa}(1) + v_{ak,ak} \\ &- (-1)^{\ell_a+\ell_k-L+1-S}[h_{ak}(1)s_{ka}(2) + h_{ka}(2)s_{ak}(1) + v_{ak,ka}]\Big\} \end{aligned} \tag{4.2.10}$$

with the one electron operator matrix element:

$$h_{ij}(1) = \int \phi_i^*(1)\hat{h}(1)\phi_j(1)d\tau_1 \tag{4.2.11}$$

and the overlap intergral:

$$s_{ij}(1) = \int \phi_i^*(1)\phi_j(1)d\tau_1 \tag{4.2.12}$$

and finally

$$K = (\ell_a m_{\ell_a} \ell_k (M_L - m_{\ell_a})|LM_L)(\frac{1}{2}m_{s_a}\frac{1}{2}(M_s - m_{s_a})|SM_S) \tag{4.2.13}$$

The relation of the channel indices to the orbital indices is like above $I \to (a,k)$ and $II \to (a',k')$

In the same way we obtain :

$$H_{II,II} = K'^2 \Big\{ h_{a'a'}(1)s_{k'k'}(2) + s_{a'a'}(1)h_{k'k'}(2) + v_{a'k',a'k'} - (-1)^{\ell_{a'}+\ell_{k'}-L+1-S}[h_{a'k'}(1)s_{k'a'}(2) + s_{a'k'}(1)h_{k'a'}(2) + v_{a'k',k'a'}] \Big\} \quad (4.2.14)$$

with

$$K' = (\ell_{a'} m_{\ell_{a'}} l_{k'} (M_L - m_{\ell'_a}) | L M_L)(\frac{1}{2} m_{s_{a'}} \frac{1}{2} m_{s_{k'}} | S m_S) \quad (4.2.15)$$

analogously :

$$H_{I,II} = KK' \Big\{ h_{aa'}(1)s_{kk'}(2) + s_{aa'}(1)h_{kk'}(2) + v_{ak,a'k'} - (-1)^{\ell_a+\ell_k-L+1-S}[h_{ak'}(1)s_{ka'}(2) + s_{ak'}(1)h_{ka'}(2) + v_{ak,k'a'}] \Big\} \quad (4.2.16)$$

and

$$H_{II,I} = KK' \Big\{ h_{a'a}(1)s_{k'k}(2) + s_{a'a}(1)h_{k'k}(2) + v_{a'k',ak} - (-1)^{\ell_a+\ell_k-L+1-S}[h_{a'k}(1)s_{k'a}(2) + s_{a'k}(1)h_{k'a}(2) + v_{a'k',ka}] \Big\} \quad (4.2.17)$$

The overlap integral of the configuration functions are expressed as :

$$\langle I|I\rangle = K^2 \Big\{ s_{aa}(1)s_{kk}(2) - (-1)^{\ell_a+\ell_k-L+1-S} s_{ak}(1)s_{ka}(2) \Big\} \quad (4.2.18)$$

$$\langle II|II\rangle = K'^2 \Big\{ s_{a'a'}(1)s_{k'k'}(2) - (-1)^{\ell_{a'}+\ell_{k'}-L+1-S} s_{a'k'}(1)s_{k'a'}(2) \Big\} \quad (4.2.19)$$

$$\langle I|II\rangle = KK' \Big\{ s_{aa'}(1)s_{kk'}(2) - (-1)^{\ell_a+\ell_k-L+1-S} s_{ak'}(1)s_{ka'}(2) \Big\} \quad (4.2.20)$$

$$\langle II|I\rangle = KK' \Big\{ s_{a'a}(1)s_{k'k}(2) - (-1)^{\ell_a+\ell_k-L+1-S} s_{a'k}(1)s_{k'a}(2) \Big\} \quad (4.2.21)$$

Note that we have not assumed orthonormality between the orbitals. Many terms will vanish when some or all orbitals are orthonormal since, in general, resonant orbitals are not orthonormal upon each other or on bound states.

4.3. Transformation to a Set of Coupled Differential Equations - One Step Generalized Closed-Coupling Equations

The equations above represent a set of multigrated second order differential equations. We want to transform them back to the not multigrated set. This can be done if we follow the strategy below:

(a) Collect all terms with $\int \phi_k(2) \ldots d\tau_2$ and undo the multigration and equate to zero.

(b) Do the same with respect to the multigration $\int \phi_{k'}(2) \ldots d\tau_2$.

At this point each term has been collected either into (a) or (b). We have in this fashion obtained two coupled integro-differential equations yielding the functions ϕ_k and $\phi_{k'}$. These equations contain terms which depend on the choice of the remaining functions ϕ_a and $\phi_{a'}$. In this way we thus only improve the orbitals with labels k and k'. The first step is thus restricted in the sense of Nicolaides[50].

4.4. *Iterative Generalized Closed-Coupling Equations*

In order to obtain two coupled integro-differential equations for ϕ_a and $\phi_{a'}$ we regroup the terms. This time we

(c) collect all terms with $\int \phi_a(1) \ldots d\tau_1$ and

(d) collect all terms with $\int \phi_{a'}(1) \ldots d\tau_1$ and
again undo the corresponding multigrations.

We now have four integro-differential equations to solve. Consider for example :

$$\begin{aligned}
&C_I^* C_I K^2 [h_{aa} + s_{aa}(\hat{h} - E) + \hat{V}_{aa}]\phi_k + C_I^* C_{II} K K' \hat{V}_{aa'} \phi_{k'} \\
&C_I^* C_{II} K K' [h_{aa'} + s_{aa'}(\hat{h} - E)]\phi_{k'} \\
&(-)^{\Sigma} C_I^* C_{II} K K' [h_{ak'} + s_{ak'}(\hat{h} - E)]\phi_{a'} \\
&(-)^{\Sigma} C_I^* C_I K^2 [h_{ak} + s_{ak}(\hat{h} - E)]\phi_a = 0
\end{aligned} \tag{4.4.1}$$

with K and K' defined by equations (4.2.13) and (4.2.15) and

$$\Sigma = \ell_a + \ell_k - L - S$$

and

$$\hat{V}_{ij}\phi(1) = \int \phi_i^*(2) \frac{1}{r_{12}} \phi_j(2) d\tau_2 \phi(1) + (-)^{\Sigma} \int \phi_i^*(2) \frac{1}{r_{12}} \phi(2) d\tau_2 \phi_j(1) \tag{4.4.2}$$

If it is understood that only terms with $K, K' \neq 0$ are included, we can divide by $C_I^* C_I K^2$ and use the definition

$$\mu = \frac{C_{II}}{C_I} \frac{K'}{K} \tag{4.4.3}$$

to rewrite (4.4.1) as

$$\begin{aligned}
&[h_{aa} + s_{aa}(\hat{h} - E) + \hat{V}_{aa}]\phi_k + \mu \hat{V}_{aa'} \phi_{k'} \\
&\mu [h_{aa'} + s_{aa'}(\hat{h} - E)]\phi_{k'} \\
&(-1)^{\Sigma} \mu [h_{ak'} + s_{ak'}(\hat{h} - E)]\phi_{a'} \\
&(-1)^{\Sigma} [h_{ak} + s_{ak}(\hat{h} - E)]\phi_a = 0
\end{aligned} \tag{4.4.4}$$

Using analogous definitions for the primed quantities:

$$\Sigma' = \ell_{a'} + \ell_{k'} - L - S \tag{4.4.5}$$

$$\mu' = \frac{C_I}{C_{II}} \frac{K}{K'} (-\mu^{-1}) \tag{4.4.6}$$

We obtain the other three of the four coupled integro-differential equations:

$$\begin{aligned}
&[h_{a'a'} + s_{a'a'}(\hat{h} - E) + \hat{V}_{a'a'}]\phi_{k'} + \frac{1}{\mu}\hat{V}_{a'a}\phi_k \\
&\frac{1}{\mu}[h_{a'a} + s_{a'a}(\hat{h} - E)]\phi_k \\
&(-1)^{\Sigma'}\frac{1}{\mu}[h_{a'k} + s_{a'k}(\hat{h} - E)]\phi_a \\
&(-1)^{\Sigma'}[h_{a'k'} + s_{a'k'}(\hat{h} - E)]\phi_{a'} = 0
\end{aligned} \tag{4.4.7}$$

$$\begin{aligned}
&[h_{kk} + s_{kk}(\hat{h} - E) + \hat{V}_{kk}]\phi_a + \mu\hat{V}_{kk'}\phi_{a'} \\
&\mu[h_{kk'} + s_{kk'}(\hat{h} - E)]\phi_{a'} \\
&(-1)^{\Sigma}\mu[h_{ka'} + s_{ka'}(\hat{h} - E)]\phi_{k'} \\
&(-1)^{\Sigma}[h_{ka} + s_{ka}(\hat{h} - E)]\phi_k = 0
\end{aligned} \tag{4.4.8}$$

$$\begin{aligned}
&[h_{k'k'} + s_{k'k'}(\hat{h} - E) + \hat{V}_{k'k'}]\phi_{a'} + \frac{1}{\mu}\hat{V}_{k'k}\phi_a \\
&\frac{1}{\mu}[h_{k'k} + s_{k'k}(\hat{h} - E)]\phi_a \\
&(-1)^{\Sigma'}\frac{1}{\mu}[h_{k'a} + s_{k'a}(\hat{h} - E)]\phi_k \\
&(-1)^{\Sigma'}[h_{k'a'} + s_{k'a'}(\hat{h} - E)]\phi_{k'} = 0
\end{aligned} \tag{4.4.9}$$

In these equations the relevant quantities are defined below. The orbitals:

$$\phi(\mathbf{x}) = \chi_{m_s} Y_{\ell m_\ell}(\hat{\mathbf{r}}) \frac{\phi_{n\ell}(r)}{r} \tag{4.4.10}$$

where

(1) χ_{m_s} is the spin function.

(2) $Y_{\ell m_\ell}$ is the spherical harmonic function.

(3) $\phi_{n\ell}(r)$ is the radial wave function. In the continuum case n is replaced by k.

The one electron operator is composed of the kinetic energy term, the nuclear attraction coulomb term and the centrifugal term:

$$\hat{h} = -\frac{1}{2}\frac{d^2}{dr^2} - \frac{Z}{r} + \frac{\ell(\ell+1)}{2r^2} \tag{4.4.11}$$

The one electron matrix elements :

$$h_{ij} = \int \phi_i^*(r)\hat{h}\phi_j(r)dr \tag{4.4.12}$$

The overlap integrals :

$$s_{ij} = \int \phi_i^*(r)\phi_j(r)dr \tag{4.4.13}$$

The coupling potential terms : $\hat{V}_{ij}\phi_k(r)$ will be discussed below.

The angular integration was already carried out in the key equations to give:

$$\mu = \frac{C_{II}}{C_I}\frac{(\ell_{a'}m'_{\ell_a}l_{k'}(M_L - m_{\ell_a})|LM_L)(\frac{1}{2}m_{s_{a'}}\frac{1}{2}m_{s_{k'}}|Sm_S)}{(\ell_a m_{\ell_a}\ell_k(M_L - m_{\ell_a})|LM_L)(\frac{1}{2}m_{s_a}\frac{1}{2}(M_s - m_{s_a})|SM_S)}$$

It is understood that those terms for which the Clebsch-Gordan coefficients vanishes are not included. C_I and C_{II} are included for completeness. In the conventional close coupling scheme[1] (Burke and Seaton) they are set equal to unity.

The Two-Electron Matrix Elements in a representation coupled to LM_LSM_S are evaluated as follows:

$$|\gamma\ell_a\ell_k LM_L\rangle = \sum_{m_{\ell_a}m_{\ell_k}} |\gamma\ell_a m_{\ell_a}\ell_k m_{\ell_k}\rangle(\ell_a m_a\ell_k m_k|LM_L) \tag{4.4.14}$$

the matrix element is thus :

$$\begin{aligned}
&\langle\ell_a\ell_k LM|v(1,2)|\ell_{a'}\ell_{k'}L'M'\rangle = \\
&\sum_{\kappa=0}^{\kappa_{max}} F^{(\kappa)}(n_a\ell_a n_{a'}\ell_{a'}n_k\ell_k n_{k'}\ell_{k'})(-1)^{\ell_k+\ell_{k'}-L}\delta_{LL'}\delta_{MM'} \\
&\sqrt{(2\ell_a+1)(2\ell_{a'}+1)(2\ell_k+1)(2\ell_{k'}+1)}\begin{pmatrix}\ell_a\kappa\ell_{a'}\\ 0\,0\,0\end{pmatrix}\begin{pmatrix}\ell_k\kappa\ell_{k'}\\ 0\,0\,0\end{pmatrix}\begin{Bmatrix}\ell_a\ell_k L\\ \ell_{a'}\ell_{k'}\kappa\end{Bmatrix}
\end{aligned} \tag{4.4.15}$$

where $\kappa_{max} = \min[(\ell_a + \ell_{a'}), (\ell_k + \ell_{k'})]$

Here the interation is a coulomb interaction $v(1,2) = \frac{1}{r_{12}}$, and thus:

$$F^{(\kappa)} = \int_0^\infty dr_1 r_1^2 \int_0^\infty dr_2 r_2^2 R_{n_a \ell_a}(r_1) R_{n_{a'} \ell_{a'}}(r_1) R_{n_k \ell_k}(r_2) R_{n_{k'} \ell_{k'}}(r_1) \frac{r_<^\kappa}{r_>^{\kappa+1}}$$

$$= \int_0^\infty dr_1 R_{n_a \ell_a}(r_1) R_{n_{a'} \ell_{a'}}(r_1) r_1^{1-\kappa} \int_0^{r_1} dr_2 R_{n_k \ell_k}(r_2) R_{n_{k'} \ell_{k'}}(r_1) r^{2+\kappa}$$

$$+ \int_0^\infty dr_2 R_{n_k \ell_k}(r_2) R_{n_{k'} \ell_{k'}}(r_2) r_1^{1-\kappa} \int_0^{r_2} dr_1 R_{n_a \ell_a}(r_1) R_{n_{a'} \ell_{a'}}(r_1) r^{2+\kappa} \tag{4.4.16}$$

where

$$R_{n\ell}(r) = \frac{1}{r}\phi_{n\ell}(r).$$

In terms of the $\phi_{n\ell}$ we write :

$$F^{(\kappa)} = \int_0^\infty dr \phi_a(r) \phi_{a'}(r) \frac{1}{r^{\kappa+1}} \int_0^r dr' \phi_k(r') \phi_{k'}(r') r'^\kappa$$
$$+ \int_0^\infty dr \phi_a(r) \phi_{a'}(r) r^\kappa \int_r^\infty dr' \phi_k(r') \phi_{k'}(r') \frac{1}{r^{\kappa+1}} r'^\kappa \tag{4.4.17}$$

With this formula we define the antisymmetrized interaction potential $\hat{V}_{kk'}$ with the components :

$$V_{kk'}^{(\kappa)}(r)\phi_{a'}(r) = \int_0^r dx \phi_k(x) \phi_{k'}(x) x^\kappa \frac{\phi_{a'}(r)}{r^{\kappa+1}} + \int_r^\infty dx \phi_k(x) \phi_{k'}(x) \frac{1}{x^{\kappa+1}} r^\kappa \phi_{a'}(r)$$
$$+ (-1)^{\ell_{a'}+\ell_{k'}-L-S} \left\{ \int_0^r dx \phi_k(x) \phi_a(x) \frac{x^\kappa \phi_k(r)}{r^{\kappa+1}} + \int_r^\infty dx \phi_k(x) \phi_{a'}(x) \frac{r^\kappa}{x^{\kappa+1}} \phi_{k'}(r) \right\} \tag{4.4.18}$$

The full interaction potential is given by :

$$V_{kk'}(r)\phi_{a'}(r) = (-1)^{\ell_k+\ell_{k'}-L} \sqrt{(2\ell_a+1)(2\ell_{a'}+1)(2\ell_k+1)(2\ell_{k'}+1)}$$
$$\times \sum_{\kappa=0}^{\kappa_{max}} \begin{pmatrix} \ell_a \kappa \ell_{a'} \\ 0\,0\,0 \end{pmatrix} \begin{pmatrix} \ell_k \kappa \ell_{k'} \\ 0\,0\,0 \end{pmatrix} \begin{Bmatrix} \ell_a \ell_k L \\ \ell_{a'} \ell_{k'} \kappa \end{Bmatrix} V_{kk'}^{(\kappa)}(r)\phi_{a'}(r)\} \tag{4.4.19}$$

where κ_{max} is defined as above at equation (4.4.15). Another way of calculating the antisymmetrized interacting of eq. (4.4.2) is given in the following subsection.

4.5. A Way of Solving the Non-local Problem

The non-local exchange potential is a consequence of the Pauli principle and its calculation is the main obstacle in atomic and molecular theory. The textbooks usually speak of the coulomb operator as a density operator constructed from all the fellow electrons. An electron a naturally feels the charge of all its peers. In real valued, bound state, conventional descriptions this electron density can be found by solving Poisson's equation for each participating electron:

$$\frac{1}{r}\frac{d^2}{dr^2}\bar{V}^b_{coul}(r) = -\mathcal{G}\mathcal{D}(\phi_b, \phi_b)/r^2 \tag{4.5.1}$$

In the case of exterior rotation, which is used here, the scalar product is defined by eq(2.2) above. Note that the functions ϕ_b must be normalised for eq(4.5.1) to be valid. Then the total direct coulomb potential, experienced by electron a, is given by:

$$V^a_{coul}(r) = \sum_{b\neq a}^{N} f(\gamma)V^b_{coul}(r) \tag{4.5.2}$$

where γ describes the angular coefficients that are found above.

In the Hartree-Fock theory for a closed shell system the energy is given as:

$$E = 2\sum_{a}^{N}\varepsilon_a - \sum_{ab}(2J(a,b) - K(a,b)) \tag{4.5.3}$$

where ϵ_a is defined below. The summations extend over all spin and space quantum numbers and:

$$J(a,b) = \int\int \phi^*_a(1)\phi^*_b(2)(1/r_{12})\phi_a(1)\phi_b(2)dr_1dr_2 \tag{4.5.4}$$

$$K(a,b) = \int\int \phi^*_a(1)\phi^*_b(2)(1/r_{12})\phi_b(1)\phi_a(2)dr_1dr_2 \tag{4.5.5}$$

The Hartree-Fock equations for a closed shell system are:

$$F(1)\phi_a(1) = \varepsilon_a\phi_a(1) \tag{4.5.6}$$

The Fock operator is given by:

$$F(1) = H_0 + \sum_{a}^{N}(2J_a(1) - K_a(1)) \tag{4.5.7}$$

$$H_0 = -\frac{1}{2}\frac{d^2}{dr^2} - \sum_{p}\frac{Z_p}{r_p} \tag{4.5.8}$$

$$J_a(1)\phi_b(1) = V^a_{coul}(r_1)\phi_b(1) = \left\{\int \phi^*_a(2)(1/r_{12})\phi_a(2)dr_2\right\})\phi_b(1) \tag{4.5.9}$$

$$K_a(1)\phi_b(1) = V^{b,a}_{exch}(r_1)\phi_a(1) = \left\{\int \phi^*_a(2)(1/r_{12})\phi_b(2)dr_2\right\}\phi_a(1) \qquad (4.5.10)$$

While $V^a_{coul}(r)$ was defined as a solution of the Poisson equation above in eq(4.5.1), the exchange potential usually has been derived in a less straight forward manner. Slater[54] and later, independently, Laksonen et al.[55] noticed that since

$$\nabla^2(1/r_{12}) = -4\pi\delta(r_{12}) \qquad (4.5.11)$$

From (5.5.5) and (4.5.10) follows:

$$K(a,b) = \int dr_2\phi^*_b(2)\phi_a(2)V^{a,b}_{exch}(2) \qquad (4.5.12)$$

Now operating on $V^{a,b}_{exch}(2)$ with ∇^2_2 gives :

$$\nabla^2_2 V^{a,b}_{exch}(2) = \int dr_1\phi_a(1)^*\phi_b(1)\nabla^2_2(1/r_{12}) = -4\pi\phi^*_a(2)\phi_b(2) \qquad (4.5.13)$$

which is a Poisson like equation. For the dilated problem we need to use the natural integrand of the generalized scalar product[34] (see above) instead of the usual scalar product analogue, when forming the source of the Poisson like equation:

$$\nabla^2_2 V^{a,b}_{exch} = -4\pi\,\mathcal{GD}(\phi_a(\xi),\phi_b(\xi))/r^2 \qquad (4.5.14)$$

We are thus in the position to find the full exchange potential for a given electron a expressed as a sum of the exchange terms with all its fellow electrons:

$$V^a_{exch}(r) = \sum_{b\neq a}^{N} g_{a,b}(\gamma)V^{a,b}_{exch}(r) \qquad (4.5.15)$$

where $g_{a,b}(\gamma)$ are appropriate angular coefficients discussed below.

5. On the Exterior Complex Dilation of the Close Coupling Equations

We are now in the position to combine the close coupling equations and the exterior complex dilation theory. The coulomb and exchange potentials are calculated as solutions of eq. (4.5.1) and (4.5.14) above.

The potentials felt by an electron in the set of coupled equations, are a sums of potentials in the sense of Balslev[57]. The nuclear attractive coulomb containes i h. The coulomb and exchange potentials $V^a_{coul}(r)$ and $V^{a,b}_{exch}(r)$ are asymptotically of exponential since they reflect electron-electron distances wihtin the atom. The full weighted sums, where angular momentum coupling coefficients are the weight-factors, are thus sums of exponential functions. Following Balslev, we can thus state that this formulation is complex dilation analytic, and we can in particular apply the exterior complex rotation transformation.

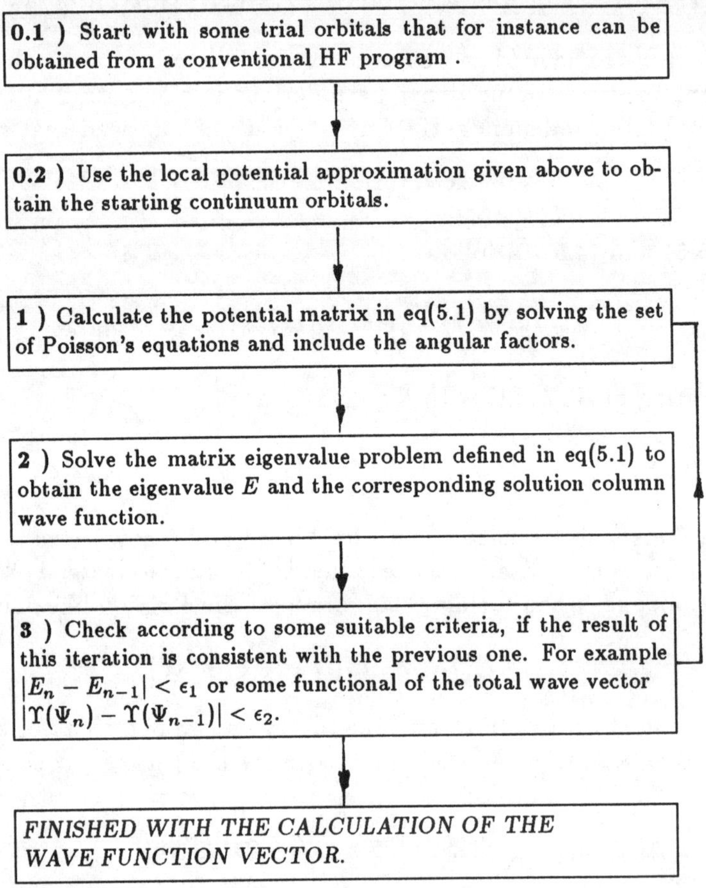

Figure 5.

Tentative iterative scheme for the *ATOMIC RESONANCE MCSCF*.

Now all orbitals, originally bound as well as those of Gamow type, have exponentially decaying modulus. Hence there is no need to distinguish between the two kinds.

Since we are only interested in finding resonances, each channel has its appropriate boundary condition defined by the exterior transformation of the usual many channel boundary conditions for resonance states, i.e. purely outgoing waves.

Our set of coupled equations can in be written as a matrix equation:

$$\mathbf{Y}'' = \{\mathbf{V} - \mathbf{\Lambda}\}\mathbf{Y} \tag{5.1}$$

where Λ is a diagonal matrix with eigenvalues $\lambda^i = E^i - \frac{i\Gamma^i}{2}$ We have thus obtained a set of N coupled equations to which there are $2N$ linearly independent solutions. Below we will make use of the solution matrix $\mathbf{Y}$,

$$\mathbf{Y} = (\mathbf{y}^{(1)}\mathbf{y}^{(2)}\ldots\mathbf{y}^{(N)}) \tag{5.2}$$

defined by

$$\mathbf{y}^{(i)} = \begin{pmatrix} y_1^{(i)} \\ y_2^{(i)} \\ \vdots \\ y_N^{(i)} \end{pmatrix} \tag{5.3}$$

With the two boundary conditions at the origin as well as at infinity taken into account the resonance wave function, is obtained as a column vector i.e. is a linear combination of all $\mathbf{y}^{(i)}$. The tentative, iterative, selfconsistent scheme we will employ is displayed in figure 5.

5.1. A Local Energy Dependent Study

The aim of our present work is to accomplish a reliable computational method based on numerical integration. It is thus of interest to gain experience in numerical determination of resonances for electron-atom(ion) scattering processes. Since we have accumulated considerable knowledge on single channel as well as multichannel analysis of potential scattering problems for molecules, it seems wise to us to see how we can adapt these algorithms to the present problems. The most straight forward extension is the local density approximation to the non-local exchange potential. The work cited below has been done by Krylstedt and co-workers[59–63]. These ideas can also be viewed in another, formal, way. Assuming that the coupling between the different channels in the close coupling formalism are so small that we can neglect them, we then arrive at what is known as the distorted wave approximation[27]:

$$\left\{-\frac{d^2}{dr^2} + \frac{\ell(\ell+1)}{r^2} + 2\mu V_{DW}(k^2, r)\right\}\phi_\ell(k,r) = k^2\phi_\ell(k,r), \tag{5.1.1}$$

where μ is the reduced mass of the collisional complex, and $k^2 = 2\mu E$. The orbital wave function is for a given energy, E, calculated as a single channel problem.

$$V_{tot}^{DW}(k^2, r) = V_{stat}(r) + V_{cex}(r) + V_{pol}(r) + V_{cen}(r) \tag{5.1.2}$$

Since the non-local exchange potential in principle has to be derived from knowledge of the continuum wave function itself we need to use approximate methods for this interaction. Hara[2] is the father of a local static exchange approximation for the continuum electron exchange. Using the free electron gas approximation (see e.g. Slater[58]), we obtain

$$V_{EX}^{0}(r) = -\frac{2}{\pi}k_F(r)f(\gamma_0(r)), \tag{5.1.3}$$

with the distribution function

$$f(\gamma_0(r)) = \frac{1}{2} + \frac{(1-\gamma_0^2)}{4\gamma_0}\ln\left|\frac{\gamma_0+1}{\gamma_0-1}\right|, \tag{5.1.4}$$

where

$$\gamma_0(r) = k_0(r)/k_F(r), \tag{5.1.5}$$

and the Fermi momentum in the free electron gas

$$k_F(r) = (3\pi^2\rho(r))^{1/3} \tag{5.1.6}$$

Hara defined the momentum of the continuum electron as

$$k_{0,H}^2(r) = k_F^2(r) + 2|Emax| + k_\infty^2, \tag{5.1.7}$$

where $|Emax|$ is the difference between the Fermi level of the bound electrons and the zero point level, while $k_\infty^2/2$ is the kinetic energy of the continuum electron at infinite radial distance. The exterior complex dilated potential has the form:

$$V_{CEX,\eta}(\xi(r)) = -\frac{2}{\pi}k_F(\xi(r))$$
$$\left\{\frac{1}{2} + k_F(\xi(r))\left[1-\left[\frac{k_{0,H}(\xi(r))}{k_F(\xi(r))}\right]^2\right]\left[4k_{0,H}(\xi(r))\right]^{-1}\ln\frac{k_{0,H}(\xi(r))+k_F(\xi(r))}{k_{0,H}(\xi(r))-k_F(\xi(r))}\right\}, \tag{5.1.8}$$

where $k_F(\xi(r))$ and $k_{0,H}(\xi(r))$ are given by the dilated analogues to equations (5.1.6) and (5.1.7) respectively.

In the original synthesis[59] (CEXI) of the Hara formulation and the exterior complex dilation theory we took the ordinary standpoint and assumed that the projectile electron only interacts with the outermost, valence, electrons

$$\rho(r) = w_N|P_N(r)|^2 \Rightarrow V_{CEXI}(r). \tag{5.1.9}$$

Here w_N is the occupation number for the outermost orbital N. This choice is unnatural and was later improved in the CEXII paper[59] to contain all electrons

$$\rho(r) = \sum_{i=1} w_i|P_i(r)|^2 \Rightarrow V_{CEXII}(r). \tag{5.1.10}$$

It is well known that the Hara local continuum electron momentum has incorrect asymptotic behaviour. This deficiency was overcome by introducing a tempered $|Emax|$:

$$|Emax(r)| = \begin{cases} |Emax| & ; \quad r < r_0 \\ |Emax| e^{-\lambda(r-r_0)} & ; \quad r \geq r_0, \end{cases} \tag{5.1.11}$$

The tempering parameter λ is the same as the fitting parameter for the static part of the potential at large radial distances

$$V_{stat}(R) = A \exp(-\lambda r) \tag{5.1.12}$$

which is used to derive the exterior scaled form of the asymptotic analogue of the static atomic potential $V^{\eta}_{stat}(\xi(r))$

The off diagonal elements of the close coupling potential matrix represent correlation. In the DW representation this is represented by the polarization term:

$$V_{pol}(r) = -\frac{\alpha_0}{2r^4}\left[1 - \exp\left(-\left(\frac{r}{r_0}\right)^6\right)\right], \tag{5.1.13}$$

where α_0 is the static polarizability of the target atom. This expression also contains a real and arbitrary cut-off parameter r_0. By fixing the calculated resonance peak maximum to an observed one, it is possible to determine r_0 for all resonances of the system, irrespectively of the different angular momenta ℓ. The physical meaning of this cutoff parameter is to switch from an outer region, where electron-electron exchange is neglected and where the interaction can be represented by virtual excitations i.e. polarization, to an inner region where the continuum electron exchange potential is important. Since different atoms, belonging to the same group in the periodic table have similar electron configurations in the outermost valence shells we assume that the cut-off parameters for different atoms of the same group are related according to their average outer orbital radius (Krylstedt[59,61,62]).

$$\frac{r_0^a}{r_0^b} = \frac{\langle P_N^a | r | P_N^a \rangle}{\langle P_N^b | r | P_N^b \rangle} = \frac{\int_0^\infty r |P_N^a(r)|^2 dr}{\int_0^\infty r |P_N^b(r)|^2 dr}. \tag{5.1.14}$$

Thus, once r_0 has been determined for one resonance in one atom, it is approximately determined for all related resonances in the other atoms within the same group in the periodic table.

The four parts building the potential in equation (5.1.2) were analysed and found to be rigourously exterior dilation analytic[62].

The strength of these local density models was demonstrated through numerical studies of shape resonances in the negative ions of the three first elements in the

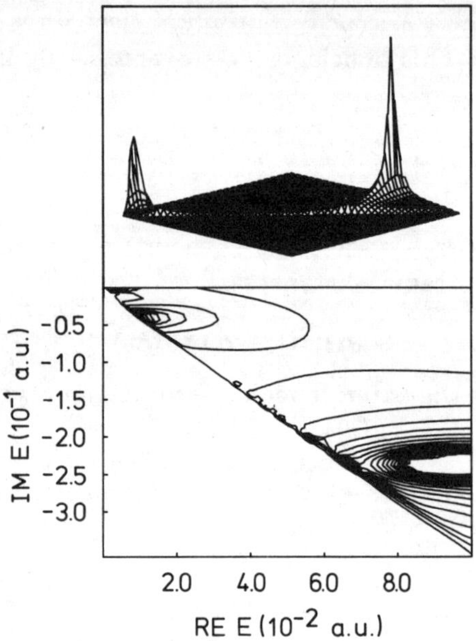

Figure 6.

Hidden and contour plots of the diagonal element of the e^- – Be partial p-wave Green's function, $|G_\ell(a, a, \lambda)|$ and $\ell = 1$, for the CEXII model (of continuum electron exchange). The value of $r_0 = 3.45$ Bohr is obtained from our method to fix the cut-off parameter for various symmetry related scattering systems. Shown are the first and second resonance poles.

alkaline earth group. The $(1s^2 2s^2)2p\ ^2P$ resonance in Be^- and the corresponding $(1s^2 2s^2 2p^6 3s^2)3p\ ^2P$ and the(...)$3d\ ^2D$ resonances in Mg^-,were studied in the CEXI and CEXII papers.

The simultaneous poles of the partial-wave S-matrix and the corresponding Green's function were found by conventional Newton-Raphson search procedures which we have used in most of our numerical work[65,33]. It is illustrated by a plot of the absolute value of the diagonal part of the Green's function in figure 6, which shows the signatures of the two poles for the Be^- collision complex: (...)$2p\ ^2P$ and (...)$3p\ ^2P$.

The difference of the two CEX-approaches is clearly seen, in figure 7, from the two exchange potentials as well as the total continuum electron potentials. The success of Hara formulation, used in CEXI, can in some sense be understood by considering figure 7a. It is not likely that a classical low energy projectile electron will penetrate deepely into the target atom. Neglecting the $1s$ orbital compared to the $2s$ is, for a low energy electron feeling the electron repulsion, not as bad as

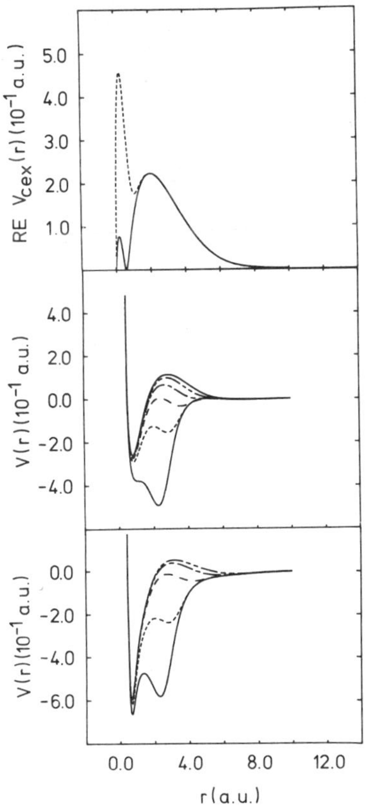

Figure 7.

(a): Real part of the continuum electron exchange potential for the $e^- -$Be scattering system. The solid curve represents the CEXI model and the broken curve the CEXII case. Since the potential is energy dependent we have chosen to show the case corresponding to the first partial p-wave S-matrix resonance pole. (b): The real part of the total CEXII potential felt by the continuum electron in the e^- – Be (ground state) scattering for various values of the cut-off parameter r_0. The parameters selected are : $r_0 = 2.5 - 5.0$ in steps of 0.5 Bohr. Notice the building up of a barrier as r_0 is increased. (c): The same as in (b) but for CEXI ($r_0 = 2.5,\ 3.0 - 6.0$ in steps of 1.0 Bohr). Note the obvious difference between the CEXI and CEXII potentials. The figure is taken from[61].

one first would think. That does not mean that the CEXI form is justified. From a formal point of view the CEXI $\Rightarrow$ CEXII development is certainly a natural one. Fig 7b and 7c shows the difference between the CEXI and the CEXII forms of the total continuum electron potential. The difference between the two potentials is not negligible.

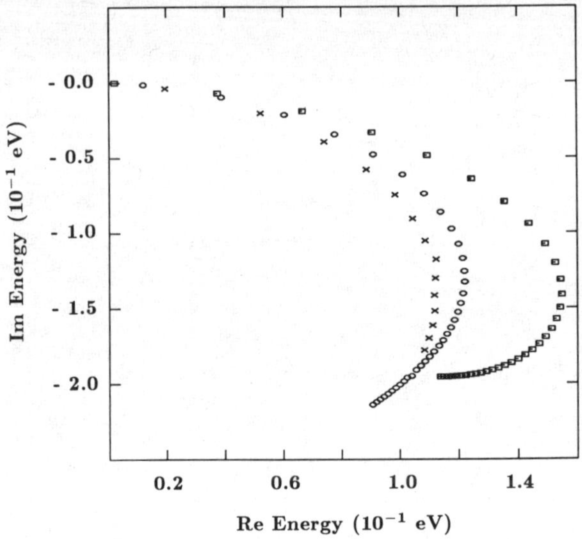

Figure 8.

First partial p-wave S-matrix complex resonance pole (as a function of the cut-off parameter) assigned to the ^{2}P shape resonance occurring in the e^- – Be scattering system. (X) CEXI, (○) CEXII and (□) AC model of continuum exchange. The figure is taken from [62] .

The position of the different poles as a function of the cut-off parameter was thoroughly studied. The results are displayed in figures 8 and 9. The resulting partial-wave S-matrix poles are highly dependent on the cut-off parameter, r_0. The variations are smooth, some even without changing the sign of the derivative. The CEXI and CEXII first Be^- pole trajectories are closer to each other than to the AC poles reflecting the importance of the tempered Fermi level ionization potential, $|Emax|$. For the second Be^- pole we find the opposite. The CEXII and AC pole trajectories have similar forms and move close to each other. The CEXI second pole trajectory, on the other hand, is close to the real axis and may accordingly have stronger influence on real axis properties, in particular the elastic cross section. The CEXI $\Rightarrow$ CEXII improvement is thus essential in the understanding of the shape of the $(1s^2 2s^2)2p\ ^2P\ Be^-$ elastic cross section resonance peak. The CEXI description indicates a more noticable dependence on the second p-wave S-matrix pole whereas the CEXII and AC descriptions seems to give a more single particle description of this resonance peak. Considering the corresponding model dependence of the cross section, we observe in figure 10 that the peak gets sharper when going from CEXI over CEXII to the AC formulation. The $(1s^2 2s^2 2p^6 3s^2)np\ ^2P\ Mg^-$ resonances, studied in the same papers support the above Be^- observations.

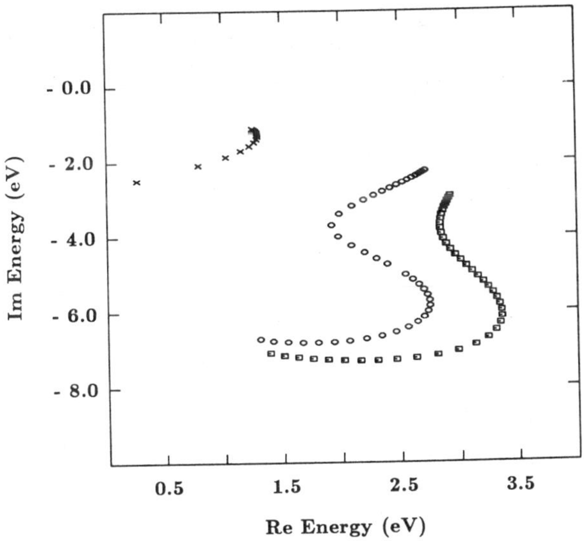

Figure 9.

Second partial p-wave S-matrix complex resonance pole (as a function of the cut-off parameter) assigned to the ^{2}P shape resonance occurring in the e^- – Be scattering system. (X) CEXI, (○) CEXII and (□) AC model of continuum exchange.The figure is taken from [62] .

Besides a ^2P resonance in the Ca^- ion we found a bound state, thereby verifying some very recent experimental work of Pegg et al.[66] The positions of the ^2P and the two ^2D resonance poles as a function of the cut-off parameter r_0 for Ca^- is illustrated in figure 11. This resonance peak is different from the others studied. It is not unlikely that we need more than one pole to build this spectrum. The influence of the position of the poles can, although not rigorously, be associated with the form of the corresponding elastic scattering cross sections.

The pole trajectories in figure 12 supports this ansats. The first and the second poles are different from the rest of the pole pairs in this study, in the sense that they are closer to each other than all the rest.

Krylstedt et al.[30] studied so called "giant resonances " which experimentally are found in electron impact ionization or indirect photoionization cross sections. Highly excited quasibound shape resonances, usually with high ℓ-quantum number are here acting as intermediate states of the process. The single particle nature of these states justifies the use of the DW approximation. HF orbitals were used to construct the bound ion orbitals.

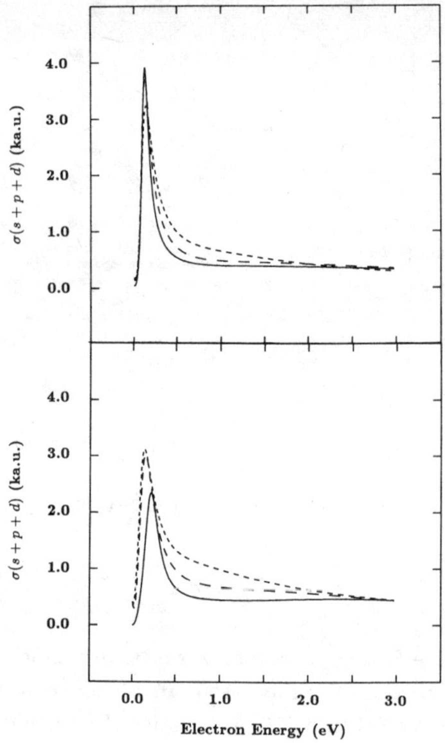

Figure 10.

(a) Total (s + p + d) cross section for elastic $e^- - Be$ scattering as calculated by the previous CEXI (dotted curve; $r_0 = 5.25$) and CEXII (dashed curve; $r_0 = 3.45$) models, and the present AC (solid curve; $r_0 = 3.45$) model of continuum exchange. The various values of the cut-off parameter are obtained from the equation (2.4.2) using Mg^- as reference. (b) The same as in (a), but for the $e^- - Mg$ collisional system. The figure is taken from [62] .

$$V_{HF}(r) = -\frac{Z}{r} + V_{EE}(r), \qquad (5.1.15)$$

Since we are interested in relatively heavy elements Cs^+, Ba^{++} and Ca^{++} a mass-velocity correction to the potential was introduced.

$$V_{MV}(k_\infty^2, r) = -\frac{(E - V_{HF}(r))^2}{2c^2}. \qquad (5.1.16)$$

The total DW potential was constructed using the AC formalism of Krylstedt[62] (see above)

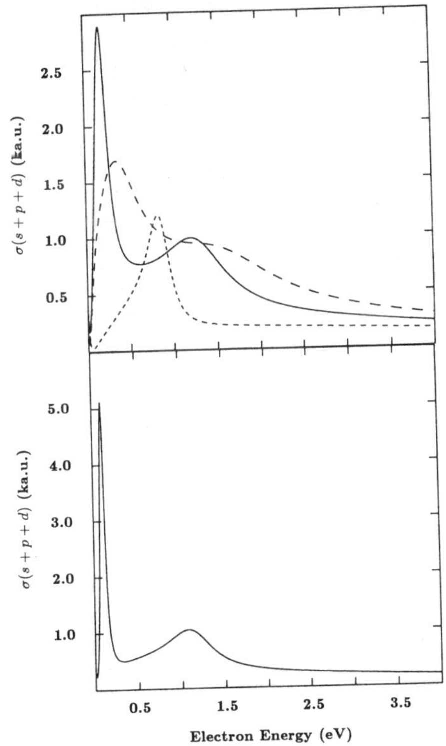

Figure 11.

(a) Total (s + p + d) cross section for elastic electron-Ca scattering as calculated by the present AC model of continuum exchange. The r_0 values chosen are 4.00 (dotted curve), 4.25 (solid curve) and 4.50 (dashed curve) Bohr. (b) The same as in (a), but here $r_0 = 4.15$ Bohr. The figure is taken from[62] .

$$V_{DW}(k_0^2, r) = V_{HF}(r) + V_{MV}(k_\infty^2, r) + V_{CEX}(k_0^2, r). \tag{5.1.17}$$

In order to compare the AC continuum exchange with the non-local one, we are currently studying the Poisson-like equation (3.5.14) derived above. Solving this equation with the local density eigenfunction of equation (5.1.1) as an input, we will be able to compare the two, and also judge the quality of $V_{AC}(r)$. Some previous DW calculations have utilized the Riley Truhlar[67] semiclassical exchange potential (RTSCE).

$$V_{SCE}(E, r) = \frac{E - V_{stat}(r)}{2} - \frac{\sqrt{(E - V_{stat}(r))^2 + 4\pi\rho(r)}}{2}, \tag{5.1.18}$$

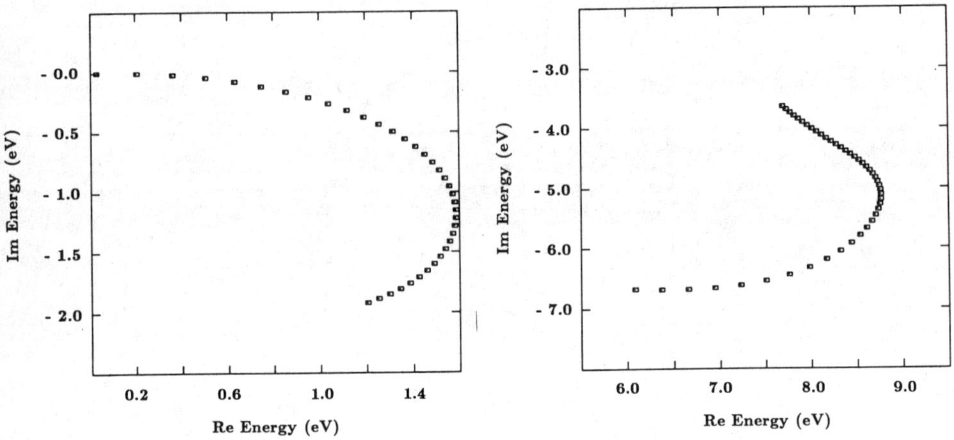

Figure 12.

First and second partial d-wave S-matrix complex resonance pole (as a function of r_0) assigned to the ^{2}D shape resonance occurring in the elastic $e^- - Ca$ scattering problem. (a) For the first pole the r_0 values ranges from 3.70 to 5.50 in steps of 0.05 Bohr while (b) for the second pole the cut-off parameter ranges from 3.30 to 5.50 in steps of 0.05 Bohr.

which is valid for targets with double occupied spatial orbitals. In this contribution we made a comparison of the results obtained from the two exchange schemes as well as investigated the effects of relativistic corrections to the models.

This series of papers [59,61,62,30] taught us that the DW approximation is a reliable and stable one. In the more general exterior complex rotated schemes, it will be used as a first step to obtain the zero order continuum electron wave function.

5.2. A Model Problem Study using Hydrogenic Basis

In order to check our set-up of generalized coupled equations we resorted to a basis set study employing hydrogenic orbitals which have the correct number of nodes. Since we further want to describe doubly excited states it is likely that the screening, that is important for low lying bound states, will be of less importance here.

Our model problem was further chosen to be the most simple non-trivial problem : the *He* like atom. First a series of bound states were calculated in order to check the angular and formal part of the formalism. The results for a number of the bound state - here restricted to real energies only - calculations and the lowest doubly excited state are given in table 1.

Table 1.

Energies from Exterior Rotated Close Coupling Hydrogenic Basis Model Calculation for the Helium atom. *N* is the number of basis functions used.

Configuration	N = 2	N = 4	N = 6	Exp.
$(1s^2)\ ^1S$	-2.8236	-2.8336	-	-2.9036
$(1s2s)\ ^3S$	-	-2.1250	-	-2.175
$(1s2s)\ ^1S$	-2.012	-2.1471	-	-2.146
$(1s2p)\ ^3P$	-2.1241	-	-	-2.133
$(1s2p)\ ^1P$	-2.1233	-2.1245	-	-2.124
$(2s^2)\ ^1S$	-0.69	-0.7145	-0.7119	-0.777

The absolute difference between the calculated energies and the experimental data[68] is shown in table 2.

The accuracy of the present model calculation is surprisingly good. This is especially true for those that are likely to be good independent particle terms i.e. the $(1s2p)\ ^{3,1}P$ terms. We accordingly find the largest difference where correlation effects are typically large i.e. the $(1s^2)\ ^1S$ and the $(2s^2)\ ^1S$ terms. The present results displayed in this subsection are thus encouraging for the development of the current method. With very small inflexible basis sets and with a form that has only radial correlation, we have thus obtained already acceptable results. What will we then be able to obtain with fully numerical orbitals and a larger configuration space?

6. Partial-Wave S-matrix Poles and Reality - A Mittag-Leffler Expansion

The fundamental notions in resonant quantum scattering theory: *the phase shift*, *time delay* or *mean lifetime* of a quasibound state are based on the Breit

Table 2.

| Experimental- Calculated | Energies from Exterior Rotated Close Coupling Hydrogenic Basis Model for the Helium atom. N is the number of basis functions used.

Configuration	N = 2	N = 4	N = 6
$(1s^2)\ ^1S$	0.08	0.07	-
$(1s2s)\ ^3S$	-	0.05	-
$(1s2s)\ ^1S$	0.134	-0.0011	-
$(1s2p)\ ^3P$	0.0089	-	-
$(1s2p)\ ^1P$	0.0007	-0.0005	-
$(2s^2)\ ^1S$	0.087	0.0625	0.0651

Wigner isolated resonance formula:

$$S_\ell(E) = \left(1 + \frac{i\Gamma}{(E_r - E) - i\Gamma/2}\right) \exp(i2\gamma) \tag{6.1}$$

$\mathcal{E} = E_r - i\Gamma/2$; γ is the background phase shift

This is why the complex resonance energy concept, which is computationally strongly supported by the complex dilation ideas, is attractive. Still, the complex eigenvalues are as such, only meaningful when they are interpreted in some observable quantity. Accordingly, complex eigenvalue methods are only meaningful when they can be used to describe observable phenomena

-i conceptionally simpler than competing methods or

-ii give more insight than competing methods or

-iii with less computational efforts compared to competing methods.

The elastic cross section discussed above is such an observable. It is shown elsewhere in this book[69,70,71] how one can utilise certain analytical properties of the partial-wave S-matrix $S_\ell(E)$ or the corresponding Green's function $G_\ell(r,r,E)$ to derive expansions from which these functions can be calculated parametrically. Rittby et al.[71] have in particular shown how one can:

-i through a generalization of Green's formula to a complex dilated integration path show that a wronskian also here is independent of is evaluation point

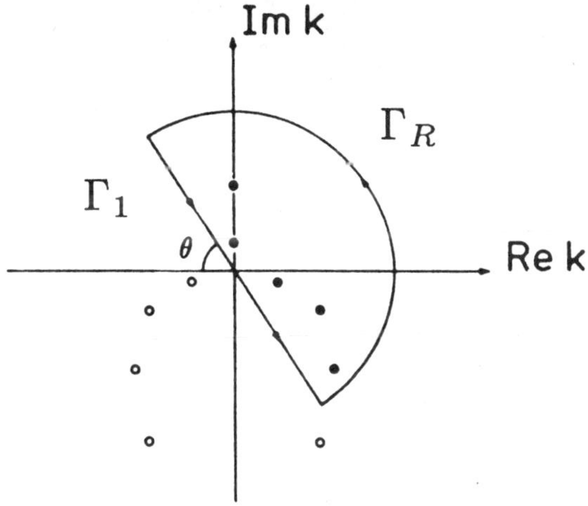

Figure 13.

The partial-wave S-matrix integral contribution term is evaluated on the contour Γ by individual complex dilations of the Jost functions, that defines $S_\ell(k)$

-ii then use this dilated wronskian to analytically continue the partial-wave S-matrix away from the real energy or momentum axis and

-iii finally Mittag-Leffler expand this S-matrix in its residues, poles and a background - cut integral. The resulting expression for the formal expansion is:

$$S_\ell(k) = S_\ell(a) + \sum_{j=1}^{N} Res(S_\ell(k_j)) \left[\frac{1}{k-k_j} + \frac{1}{k_j - a} \right] + \frac{k-a}{2\pi i} \oint_\Gamma \frac{S_\ell(k')}{(k'-k)(k'-a)} dk', \tag{6.2}$$

where Γ is some closed contour enclosing a set of $\{k_j\}_{j=1}^N$ poles, with corresponding residues $Res(S_\ell(k_j))$ (see figure 13). a is an "arbitrary" expansion point, which should be chosen conveniently far away from any resonance pole. The k-residues are formally given as:

$$Res(S_\ell(k_j)) = \lim_{k \to k_j} (k - k_j) S_\ell(k) = \left. \frac{F_\ell^-(k)}{\left(\frac{dF_\ell^+(k)}{dk} \right)} \right|_{k=k_j}. \tag{6.3}$$

Krylstedt et al.[63] have very recently shown how the uniform dilation approach of Rittby et al. can be extended to an exterior dilated analogue, thereby making the basic idea applicable on realistic problems, which are only asymptotically analytic continuable. The original method of Rittby et al.[71] contains an evaluation of the contour integral part in equation(6.2). This means in principle that one would have to evaluate the S-matrix on the rotated cut instead of on the real axis i.e. making a mountain out of a mole hill.

In practice Krylstedt et al.[63] have chosen to first calculate the cross section at a small number of points, between 5 to 10. The difference between the conventional, for instance Riccati-Bessel integration method, and the present pure pole expansion expression, i.e. the first two terms in eq (5.2), is then assumed to be a slowly varying function which is fitted by a third order spline. This fit ,$\mathcal{G}_\ell^{RD}(k)$, is then added to the Mittag-Leffler pole expansion with the integral term omitted to compute the full partial-wave S-matrix.

$$\begin{aligned} S_\ell^{RD}(k) &= S_\ell^R(k) + \mathcal{G}_\ell^{RD}(k) \\ &= S_\ell(0^+) + \sum_{j=1}^{N} Res(S_\ell(k_j))\left[\frac{1}{k-k_j} + \frac{1}{k_j - 0^+}\right] + \mathcal{G}_\ell^{RD}(k) \end{aligned} \tag{6.4}$$

A unitary deficiency Λ is defined for the purpose of checking the quality of the approximate reduced partial-wave S-matrix. It is defined through:

$$S_\ell(k) = S_\ell^R(k) + \mathcal{G}_\ell(k) \Longrightarrow S_\ell^R(k) = S_\ell(k) - \mathcal{G}_\ell(k)$$

$$\begin{aligned} S^R(S^R)^\dagger &= (S - \mathcal{G})(S - \mathcal{G})^\dagger = SS^\dagger - S\mathcal{G}^\dagger - \mathcal{G}S^\dagger + \mathcal{G}\mathcal{G}^\dagger = \\ &= \mathbf{1} - (S\mathcal{G}^\dagger + \mathcal{G}S^\dagger - \mathcal{G}\mathcal{G}^\dagger) = \mathbf{1} - \Lambda, \end{aligned} \tag{6.5}$$

Krylstedt et al[62] chose the poles derived in the previous AC study on the negative alkaline earth ions Be^-, Mg^- and Ca^- to illustrate the formal theory. Here we pick the Ca^- ion as an example.

The p-wave S-matrix decomposition around the ^2P resonance in Ca^-, displayed in figure 14a,b, shows close agreement between the full and the approximate single pole approximation calculation. The corresponding background function, $\mathcal{G}_\ell^{RD}$ in figure 14c,d, shows some variation around the real position of the resonance. The quality of the one pole approximation is further illustrated by the unitary deficiency defined in eq(6.5) above. The quality of the reduced partial-wave S-matrix formulation is clearly demonstrated in figure 15 for the ^2P resonance peak in the elastic $e^- + Ca$ scattering cross section. No deviation between the "exact" and the reduced single pole approximation is noticed in the figure. The ^2D resonance peak, illustrated in figure 16, requires a two pole reduced partial-wave S-matrix formulation to give a description with the same quality. The deviation between the reduced form and the "exact" cross section is possible to observe. The difference is however very small. When comparing to experiments, the statistical relative error

$$\frac{\Delta\sigma}{\sigma} \propto \frac{\sqrt{n}}{n} = \frac{1}{\sqrt{n}} \tag{6.6}$$

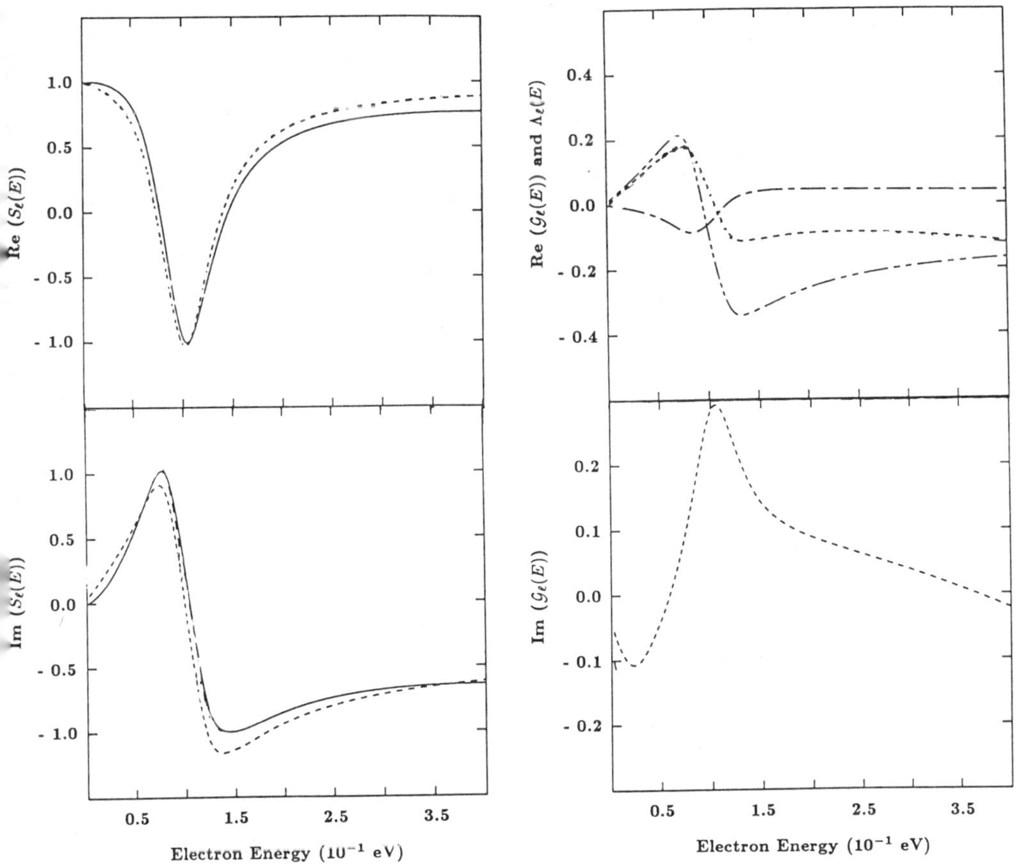

Figure 14.

(a) The real part of the partial p-wave S-matrix for $e^- - Ca$ scattering as a function of the kinetic energy. The solid curve represents the correct result of the present model. The dotted curve shows the results using a one pole approximation without the background function. (b) The same as in (a), but for the imaginary part of the partial p-wave S-matrix. (c) The real part of the partial p-wave background function $\mathcal{G}_\ell(E)$ for $e^- - Ca$ scattering as a function of the kinetic energy. The dotted curve shows the results with one pole approximation. The long-short curve gives the unitary deficiency defined in equation (6.5) (d) Here the dotted and the dashed curve shows the corresponding imaginary part of the background function $\mathcal{G}_\ell(E)$.

we find that we need at least to the order of $n = 10.000$ counts to observe this difference.

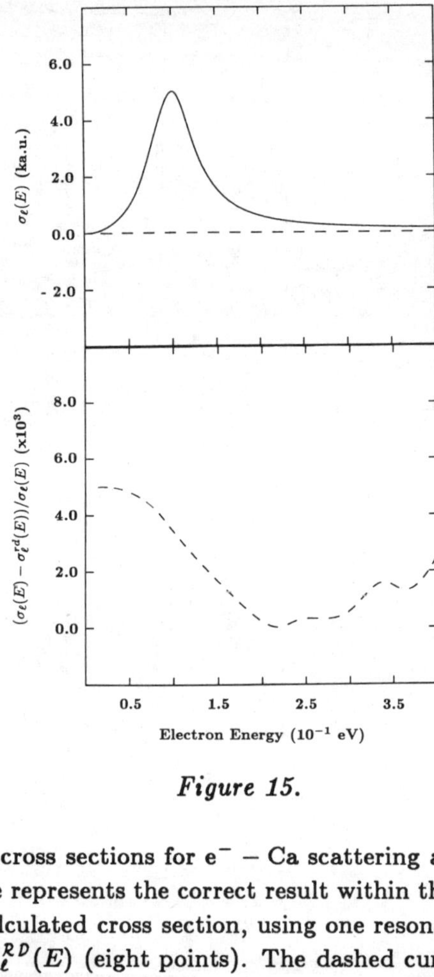

Figure 15.

(a) Partial p-wave cross sections for $e^- - Ca$ scattering as a function of the kinetic energy. The solid curve represents the correct result within the present model. The dotted curve shows the calculated cross section, using one resonance pole plus the reduced background function $\mathcal{G}_\ell^{RD}(E)$ (eight points). The dashed curve gives the difference between the two cases. (b) Here we show the relative deviation from the correct result of the reduced approximation.

Now consider :

$$\begin{aligned} \Lambda^{RD} &= \mathbf{1} - S_\ell^{RD} S_\ell^{RD\dagger} = \\ &= \mathbf{1} - \left(S^R + \mathcal{G}_\ell^{RD}\right)\left(S^R + \mathcal{G}_\ell^{RD}\right)^\dagger = \\ &= \mathbf{1} - S^R S^{R\dagger} - S^R \mathcal{G}_\ell^{RD\dagger} - \mathcal{G}_\ell^{RD} S^{R\dagger} - \mathcal{G}_\ell^{RD} \mathcal{G}_\ell^{RD\dagger}. \end{aligned} \tag{6.7}$$

Unitarity is a sensitive property of the partial-wave S-matrix. It can be used to analyze the quality of the reduced partial-wave S-matrix of eq(6.4). All terms

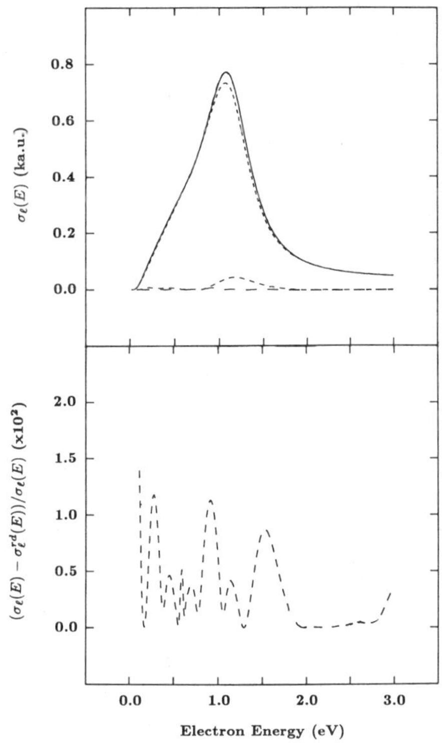

Figure 16.

(a) Partial d-wave cross sections for $e^- -$ Ca scattering as a function of the kinetic energy. The solid curve represents the correct result of the present model. The dotted and dashed curves show the calculated cross sections as well as the differences, using one and two resonance poles plus the reduced background function $\mathcal{G}_\ell^{RD}(E)$ (eight points) respectively. Note that the dashed curve is right on top of the exact results. (b) Here we display the relative deviation from the correct result of the reduced two pole approximation.

in eq(6.7) are part of the reduced partial-wave S-matrix since they are inexpensive to compute. Thus eq(6.7) lends itself to be a good test of the reduced partial-wave S-matrix. Naturally this test is only meaningfull outside the small set of calibration points that is used to define the background function.

The advantage of these Mittag-Leffler expansions and, in particular, the reduced form introduced by Krylstedt et al.[63] , is that

1) it introduced exact quantization into scattering theory by allowing an exact separation of the partial-wave S-matrix into resonant states plus a smooth background.

2) The reduced form is computationally about 50 times more powerful than a direct quantum mechanical calculation.

3) The unitary deficiency allows for a computationally inexpensive quality control of the reduced form.

7. Future Perspectives

The Swiss philosopher Pestalozzi had as his motto the sentence

" *to start with the simple and move towards the complicated*"

In our work cited above as well as in other contributions in this book we have demonstrated a number of results.

-i Within the single channel formalism it is possible to use the exterior complex dilation method to calculate stable poles of the partial-wave S-matrix or the corresponding Green's function.

-ii The results for a single channel can be extended to a many channel, close coupling problem in potential scattering. This particular result is for example applicable to atom-atom collisions or their inverse process of diatomic fragmentation reactions.

-iii The single channel methods do also apply to an energy dependent potential.

-iv The local density, distorted wave approximation is an example of such an energy dependent potential, which is exterior dilation analytic.

-v The exterior complex dilation method can also be used to analytically continue the partial-wave Jost functions such that they are simultaneously defined in a sector of the complex energy or momentum planes.

-vi The partial-wave S-matrix can accordingly be defined in the same sector and there being a meromorphic function.

-vii The partial-wave S-matrix can be expanded in terms of its poles and corresponding residues within the uncovered sector.

-viii The resonant eigenvalues and eigenfunctions be given a quantized meaning on the same footing as bound states and their eigenfunctions.

-ix Observable quantities like cross sections, mean lifetimes etc., which can be described in terms of the partial-wave S-matrix, can be interpreted and analyzed in terms of contributions from, poles respective residues and background.

-x Through the simplicity of the partial-wave S-matrix expansion formula (eq (6.4) reduced) it can be used to interpret experiments even when theoretical data is not available. This can be done by assuming the validity of eq(6.4), assigning at least one pole and residue to each peak plus a smoothly varying background and then by a least square fitting, obtain a consistent physical description of a set of relevant different, partial-wave, S-matrices, say $\ell = 1, 2, 3...$ The experimentalist and the theoretician can thus meet and compare a discrete relatively small set of numbers which describes the scattering process of their common interest.

So far we have only demonstrated these latter expansion properties for single channel problems, albeit some of the examples we have used are single channel approximations, to true many channel forms. Our impressions from the review of other methods above, makes it plausible that the full close coupling method gives the most detailed results of all non-dilated ones. We view the complex dilation theory not as an alternative to other regularization theories but rather a complement. These two statements are thus the arguments for this progress report.

The question to be discussed below is now :

"What can we expect if we succeed with this synthesis ?"

7.1. On the Possibility of Pole Expansions

The most obvious step in the development is to look for possible partial-wave S-matrix of Green's functions expansions for the many channel problem. Is the many channel S-matrix analytically continuable in the same manner as the single channel has shown to be? The form of the exterior complex dilated close coupling methods is such that the effective potential matrices $\mathbf{V}(r)$ with matrix elements $V_{i,j}(r)$, which are weighted sums of individual "primitive" coulomb $V_{n,l}^{coul}(r)$ and exchange potentials $V_{n\ell,i}^{exch}(r)$. These individual potential matrix elements are sums of potentials which as such are dilation analytic. They are thus, according to Balslev[57,72] dilation analytic. The next question concerns the integration contour. The different thresholds which exists in many channel scattering is likely to introduce problems and accordingly make the integration contour complicated. Figure 17 gives a taste of what kind of contours one might have to deal with.

The degree of complication is still increased when noticing that so called complex thresholds[73] also will generate complex cuts. An illustration of complex dilation for a many channel problem is given by Brändas et al.[33] in this book.

It we assume that this can be done successfully we can speculate further. Since we then are able to obtain partial-wave S-matrix poles and residues for a many channel problem, we assume that this is also possible for a problem where the potentials are effective, selfconsistent field ones. If this is true we will at the end

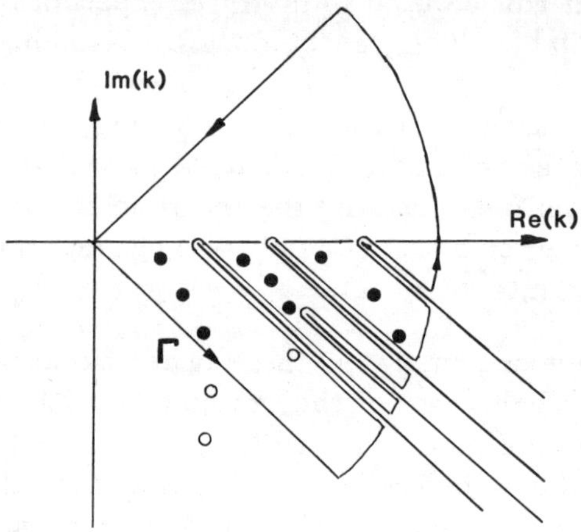

Figure 17.

Example of contour for evaluation of the integral part of a many channel partial-wave S-matrix in the k-plane. Only the poles which are uncovered by the dilation are summed over in the discrete many channel form of eq(6.2). The rest of the S-matrix is absorbed in the contour along the cuts.

be able to use the framework of exterior complex dilated close coupling to obtain poles and residues which in turn will give S-matrix expansions. These will then, as above, supply us with sets of numerical parameters through which we can describe observable quantities which can be expressed in terms of the S-matrix.

7.2. On the Applications to Electron Atom Collisions

Assuming we are able to make the above complex dilated close coupling method computationally functional, we now turn to some possible applications by returning to the subject of the first section - the experimental motivation.

Starting from the end of the first section, it is obvious that we can use the single channel DW formulation of Krylstedt[62,30] to study some of the intermediate states in excitation-autoinization. A rigorous formalism for radiative decay from a quasibound state is here of interest, since the radiative and autoionization transition probabilities are sometimes of the same order of magnitude. Such a formal study is in progress. Considering the present exterior dilated DW approximation we would thus be able to describe the excitation-autoionization process as a bound-bound state transition followed by a non-radiative decay of the excited state. The photoionization spectrum could then be described in terms of the "responsible" partial-wave S-matrix poles.

The same can be said about the dielectronic recombination. DW theory is presently successful in modeling DR and it is accordingly expected that this, in some sense deeper formulation may improve the understanding of the DR process.

Direct radiative recombination is another process, we may be able to give some different aspects on, even with the simple DW theory. We may for instance be capable of better modelling the interaction of the atom with external fields.

The kind of experiments reviewed by Schulz[3,4] have already been treated sucessfully with the complex dilation methods. The work of Krylstedt[59–64] discussed above is an example of what has already been done. To us it is also a direction post which shows us where to go.

The theoreticians usually keep their atoms, ions or molecules far away from anything else. Preferably they are also kept at zero temperature. They are easy to treat then. Experimentalists have a hard time to make experiments with this kind of species. Even if the atom or ion is essentially free, they most of the time feel some kind of environment, for example an electromagnetic field.

$\langle {}^1S \vert H \vert {}^1S\rangle$	$\langle {}^1S \vert D \vert {}^1P\rangle$		
$\langle {}^1P \vert D \vert {}^1S\rangle$	$\langle {}^1P \vert H \vert {}^1P\rangle$	$\langle {}^1P \vert D \vert {}^1D\rangle$	
	$\langle {}^1D \vert D \vert {}^1P\rangle$	$\langle {}^1D \vert H \vert {}^1D\rangle$	$\langle {}^1D \vert D \vert {}^1F\rangle$
		$\langle {}^1F \vert D \vert {}^1D\rangle$	$\langle {}^1F \vert H \vert {}^1F\rangle$

Figure 18.

Equation (7.2.3) has a tridiagonal block structure in which the originally *LS*-coupled two-electron blocks of the field free Hamiltonian with ${}^1S, {}^1P, {}^1D$ etc. structure are coupled together by the usual dipole operator

The static electric field is the simplest, and an example from Wendoloski and Reinhardt on the ${}^1P^o$ resonance in H^- will[75] serve as an example. The static electric field will introduce a term

$$\sum_{i=1}^{N} \mathcal{E} z_i = \sum_{i=1}^{N} \mathcal{E} r_i \cos \alpha_i, \tag{7.2.1}$$

into the field free hamiltonian. $\mathcal{E}$ represents the electric field while α_i is the angle between the field direction and the position vector $\mathbf{r}_i$ of the electron i. This form is at first sight not dilation analytic. Averon et al.[76] has recently shown that the complex rotation methods provide a correct analytic continuation of the resolvent associated with this formulation. The total two-electron dilated Hamiltonian is given as

$$H_\eta^{Stark} = H_\eta^{atomic}(\mathbf{r}_1, \mathbf{r}_2) + \sum_{i=1}^{2} \mathcal{E} r_i \cos \alpha_i \eta \;\; ; \;\; \eta = \exp(i\theta) \tag{7.2.2}$$

This equation can formally be written as :

$$H^{Stark}(\theta) = \exp(-2i\theta) T^{atomic} + \exp(-i\theta) V^{atomic} + \exp(i\theta) \mathcal{E} V^{field} \tag{7.2.3}$$

where T^{atomic} and V^{atomic} are the field free kinetic and potential matrices respectively. Equation (7.2.3) has a tridiagonal block structure in which the originally *LS*-coupled two-electron blocks of the field free Hamiltonian with $^1S, ^1P, ^1D$ etc. structure are coupled together by the usual dipole operator. The eigenvalue problem can for relatively week fields, where we can truncate to a finite number of blocks, be solved as a conventional configuration interaction problem.

The static magnetic field could in principle be treated in the same way. However, we have not been able to find that in the literature the varying electromagnetc field is well treated. In a recent review by Chu[77] the non-hermitian Floquet is discussed as a tool to include bound-free transitions such as multiphoton excitation ionization processes. The result of generalizing the standard bound-bound theory to one that also include part of the continuum is that one is able to describe the intensity-dependent width (the imaginary part of the eigen energy) in addition to the usual field induced shift.

References.

1. P. G. Burke and M.J. Seaton, Meth. Comp. Phys., **10**, 1-80 (1971).

2. S. Hara, J. Phys. Soc. Jpn., **22**, 710 (1967).

3. G. J. Schulz, Rev. Mod. Phys., **45**, 378 (1973).

4. G. J. Schulz, Rev. Mod. Phys., **45**, 423 (1973).

5. A. Burgess, Astrophys. J., **139**, 776 (1964).

   - ibid, **141**, 1589 (1965).

6. C. J. Herrlander, A. Bárány, K. G. Rensfeldt and J. Starker , Physica Scripta, **T22**, In press. (1988).

7. G. I. Budker and A. N. Skrinsky, Sov. Phys. Uspeki, **21**, 277 (1987).

8. D. Habs, Private communication.

9. R. Schuch, In the ABISKO workshop on,"*Ion Cooler Ring Experiments*", May 1988.

10. E. Källne, Private communication.

11. C. A. Nicolaides, Phys. Rev., **A6**, 2078 (1972).

    - **ibid.** **141**, 1588 (1965).

12. A. Temkin, Editor of *Autoionization, Recent Development and Applications*, (Plenum Press, New York 1985).

13. D. C. Griffin, M. S. Pindzola and C. Bottcher, Phys. Rev., **A31**, 568 (1985).

14. Y. Hahn, Adv. At. Mol. Phys., **21**, 1 (1985).

15. Y. Hahn, Comments in At. Mol. Phys., **19**, 99 (1987).

16. R. D. Cowan, *The Theory of Atomic Structure and Spectra*, (University of California, Berkeley 1981).

17. D. R. Bates and H. S. W. Massey, Phil. Trans. Roy. Soc. A, **239**, 269 (1943).

18. R. H. Bell and M. J. Seaton, J .Phys. B, **18**, 1588 (1985).

18a P. Winkler, Int. J. Quantum Chem. Symp **19**,201 (1986).

19. V. L. Jacobs, J. Davis and P. C. Kepple, Phys. Rev. Lett., **37**,1390 (1976).

20. K. LaGattuta, and Y. Hahn, Phys. Rev. Lett., **51**, 558 (1983).

21. C. Bottcher, D. C. Griffin and M. S. Pindzola, Phys. Rev., **A43**, 860 (1986).

22. D. Harmin, Phys. Rev. Lett., **57**, 1570 (1986).

23. K. Sakimoto, J. Phys. B, **19**, 3011 (1986).

    - **ibid.**, **20**, 807 (1987).

24. R. Kachru, N. H. Tran, H. B. van Linden, vand der Heuvel and Gallageher, Phys. Rev. A, **30**, 667 (1984).

25. N. Badnell, *The influence of core fine-structure interactions on Dielectronic re-*

combination at low temperatures: B-like C, N, O recombined ions. Submitted to J. Phys. B, (1987).

- J. Phys. B, **20**, 2081 (1987).

- J. Phys. B, **19**, 3827 (1986).

- J. Phys. B, **19**, 1541 (1986).

26. R. A. Phaneuf, In*Physics of Electron-Ion and Ion-Ion Collisions, NATO ASI,* Ed. F. Brouillard (Plenum, New York 1987) p. 117.

27. M. S. Pindzola, D. C. Griffin and C. Bottcher, Comments At. Mol. Phys., **20**, 337 (1987).

28. S. Younger, Phys. Rev. Lett., **56**, 2618 (1986).

29. J. B. West, In *New Results in Atomic Photoionization at the Daresbury Synchrotron Radiation Source*, Daresbury Preprint Jan 1987 to appear in Z. Phys. D: At. Mol. Clusters.

30. P. Krylstedt, N. Elander and M.S. Pindzola S, Manuscript to be submitted to J. Phys. B. (April 1987).

31. See for example the contributions in this book:

    T. Berggren: *Resonance State Expansions in Nuclear Physics,*

    B. Gyarmati: *Calculation of Resonant Wave Functions in Nuclear Physics,*

    B. Helffer: *Comparison between Different Notions of Resonances*

32. M. Rittby, N. Elander and E. Brändas, Phys. Rev. A, **26**, 1804 (1982).

33. E. Brändas, M. Rittby and N. Elander, *Multichannel Complex Scaled Titchmarsh-Weyl's Theory, a Model for Diatomic Fragmentation*, In this book.

34. N. Elander and E. Brändas, In *The Lertorpet Conversion for The Scalar Product of Resonant States*, In this book.

35. J. Aguilar and J. M. Combes, Commun. Math. Phys., **22**, 269 (1971).

    - E. Balslev and J. M. Combes, Commun. Math. Phys., **22**, 280 (1971).

36. B. Simon, Ann. Math., **97**, 247, (1973).

37. C. van Winter, J. Math. Anal. Appl., **47**, 633, (1974).

- C. van Winter, J. Math. Anal. Appl., **47**, 368, (1974).

38. J. C. Slater, *Quantum Theory of Atomic Structure*, (McGraw-Hill, New York 1960) and references therein.

39. D. R. Hartree, *The Calculation of Atomic Structure* (Wiley, New York 1957) and references therein.

40. N. Moiseyev and C. T. Corcoran, Phys. Rev., **A20**, 814 (1979).

41. B. R. Junker, Adv. At. Mol. Phys., **18**, 207 (1982).

42. S. Yabushita and C. W. McCurdy, J. Chem. Phys., **83**, 3547 (1985).

42a. E. Balslev, this book.

43. C. W. McCurdy, In *Autoionization*, Ed. A. Temkin (Plenum Press, New York 1985) p. 135.

44. E. Brändas, P. Froelich, C. Obcemea, N. Elander and M. Rittby, Phys. Rev., **A26**, 3656 (1982).

45. B. Simon , Phys. Lett. A, **71**, 211 (1979).

46. J. D. Morgan and B. Simon, J. Phys. B, **14**, L167 (1981).

46a. P. Winkler, Int. J. Quantum Chem. Symp. **20**, 147 (1986).

47. P. Froelich, O. Goscinski and N. Moiseyev, J. Chem. Phys., **84**, 3931 (1986).

48. K. T. Chung and B. F. Davis, Phys. Rev., **A26**, 3278 (1982).

- B. F. Davis and K. T. Chung, Phys. Rev., **A31**, 3017 (1985).

49. Th. Mercouris and C. A. Nicolaides, J. Phys. B, **17**, 4127 (1984).

50. C. A. Nicolaides, Y. Komninos and Th. Mercouris, Int. J. Quant. Chem. Symp, **15**, 355 (1981).

51. U. Fano, it Phys. Rev., **124**, 1866 (1961).

52. P. Froelich, E. Sangfelt and B. Weiner, Int. J. Quant. Chem., **31**, 823 (1987).

53. D. Frye and L. Armstrong, Phys. Rev., **34**, 1682 (1986).

54. Slater J C, *Quantum Theory of Atomic Structure*, Vol. 1, (MaGraw-Hill, New York, 1960), p. 486.

55. L. Laaksonen, P. Pyykkö and D. Sundholm, Chem. Phys. Lett., **96**, 1 (1983).

- L. Laaksonen, *PhD Thesis*, Åbo Akademi, Finland (1983).

56. M. A. Crees, M. J. Seaton and P. M. H. Wilson, Comput. Phys. Commun., **15**, 23 (1978).

57. E. Balslev, J. Math. Ana. Appl., **123**, 339 (1987).

58. J. C. Slater, Phys. Rev., **81**, 385 (1951).

59. P. Krylstedt, M. Rittby, N. Elander and E. Brändas, J. Phys. B, **20**, 1295 (1987).

60. P. Krylstedt, N. Elander and E. Brändas, Int. J. Quant. Chem., **31**, 755 (1987).

61. P. Krylstedt, N. Elander and E. Brändas, *"A Complex Rotated Approach to Electron scattering on Atoms in a Static Exchange Plus Polarisation Approximation, Part II. Full Continuum Electron Exchange." J. Phys. B*, Submitted (1988).

62. P. Krylstedt, N. Elander and E. Brändas, "A Complex Rotated Approach to Electron scattering on Atoms in a Static Exchange Plus Polarisation Approximation, Part III. Asymptotically Completed Form of Continuum Electron Exchange." J. Phys. B, Submitted (1988).

63. P. Krylstedt, C. Carlsund and N. Elander, *"On the Calculation of Electron-Atom Collision Properties Using Exterior Complex Dilatated S-matrix Expansions."* Submitted to J. Phys. B.

64. P. Krylstedt, C. Carlsund and N. Elander, *"On the Validity of Local Density Based Continuum Exchange for the Description of Electron-Atom Scattering Processes and its Connection With the Inverse Scattering Problem Studied by Means of Complex Dilatation Techniques."*, Manuscript to be submitted to J. Phys. B.

65. M. Rittby, N. Elander and E. Brändas, Molec. Phys., **45**, 553 (1982).

66. D. J. Pegg, J. S. Thompson, R. N. Compton and G. D. Alton, Phys. Rev. Lett., **59**, 2267 (1987).

67. M. E. Riley and D. G. Truhlar, J. Chem. Phys., **63** , 2182 (1975).

68. P. Winkler, Priv. com.

69. T. Berggren : *Resonance State Expansions in Nuclear Physics*, In this book and references therein.

70. J. Bang:*Mittag-Leffler Expansions in Nuclear Physics*, In this book and references therin.

71. M. Rittby, N. Elander and E. Brändas :*On the Complex Dilation of the Partial Wave S-matrix*, In this book.

72. E. Balslev and E. Skibsted, Matematisk Institut, Aarhus Universitet, Preprint Series, **7**, 1-35 (1987).

73. E. Balslev, In *Resonances Models and Phenomena*, Lecture Notes in Physics**211**, (Springer-Verlag, Heidelberg 1984), p. 27.

74. R. Lefebvre, In *Selected Topics in Physics, Astrophysics and Biophysics*, Proceedings of the XIVth Latin American School of Physics, Caracas 1972, *"Radiative and Non-radiative Processes in Molecules"* (D. Reidel Publishing Comp, Dordrect-Holland 1973) p. 87.

75. J. J. Wendoloski and W. P. Reinhardt, Phys. Rev., **A17**, 195 (1978).

76. Priv. Commmun. in ref. 75.

77. S. I. Chu, Adv. At. Mol. Phys., **21**, 197 (1985).

RESONANT STATES IN THE MICROSCOPIC CLUSTER MODEL

András T. Kruppa

Institute of Nuclear Research of the
Hungarian Academy of Sciences,
Debrecen, P.O. Box 51,
H-4001, Hungary

Abstract

Two equivalent formulations of the microscopic cluster model, the resonating-group model and the generator-coordinate method, are reviewed. Within these frameworks, the position and widths of resonances are usually determined by calculating the phase shift function. The purpose of this lecture is to show some alternative methods for calculation of resonance parameters. Two approaches are based on imposing explicitly the Gamow asymptotics for the wave function. In a third method the resonance solution of the resonating-group equation is searched for by applying the complex scaling technique. Numerical results concerning low lying resonances in ^{8}Be are given.

1. Introduction

In nuclear physics, resonance structures frequently appear in cross sections. Experimentally the parameters, the position E_r and width Γ, of a resonance can be defined through the pronounced peaks in the cross section: E_r is the energy at which the peak has the maximum and Γ is the width of the peak at its half height. The bumps in the cross section are related to steep rises of the phase shift function $\delta(E)$, which gives another way of defining E_r and Γ. The position and width of a resonance may be defined as the inflexion point of $\delta(E)$ and $\Gamma = 2/(d\delta/dE)_{E=E_r}$, respectively. In keeping with this, the theoretical resonance parameters are to be determined by solving the Schrödinger equation with the scattering boundary condition, which allows both incoming and outgoing waves. However, there is a direct way of obtaining E_r and Γ. Mathematical consideration shows that steep rises of π in $\delta(E)$ may be associated with poles of the S-matrix at complex energies $E_{res} \approx E_r - i\Gamma/2$. At these energy values the solution Ψ, of the Schrödinger equation $H\Psi = E_{res}\Psi$, satisfies the boundary condition prescribing a purely outgoing wave. These eigensolutions are called Gamow or Siegert wave functions, and the method for describing resonances in terms of them are called direct methods.

The microscopic cluster model has been extensively used in the description of the collision of nuclei since the sixties. A number of methods have been developed for the calculation of scattering wave functions. Much less attention has been paid to the direct description of resonance states. The main purpose of the present talk is to survey the methods used in these few works.

The two alternative realizations of the microscopic cluster model, the resonating-group model (RGM) and the generator-coordinate method (GCM), have been reviewed several times [1]. Our sketch of the RGM and GCM in Section 2 follows the line of [2]. In 1976, Fiebig and Weiguny introduced the Gamow-state into a GCM calculation of the alpha decay width of ^8Be and ^{20}Ne [3]. In 1984, Okhrimenko investigated the resonances of ^8Be using the algebraic version of the RGM [4]. These two methods are reviewed in Section 3. We recently applied the complex scaling technique, developed in atomic and molecular physics, to the cluster model [5]; Section 4 is devoted to our procedure. Some numerical results concerning ^8Be and the summary are contained in Section 5.

2. The Microscopic Cluster Model

2.1 The Resonating-group Model

The picturesque concept behind the microscopic cluster model is very simple. We imagine that two or more nuclei are able to form one nucleus in a way that the constituent nuclei more or less preserve their identities. The participant nuclei are called clusters. The RGM and GCM are two alternative quantum mechanical formulations of this idea. Both descriptions assume that an A-nucleon system can be described by the Hamiltonian

$$H = -\frac{\hbar^2}{2m}\sum_{i=1}^{A} \Delta_{\mathbf{r}_i} + \sum_{i<j}^{A}(V_{ij}^N + V_{ij}^C) - T_{cm}, \tag{1}$$

where T_{cm} is the kinetic energy of the center-of-mass motion, V_{ij}^N and V_{ij}^C are the nuclear and Coulomb interaction, respectively. In the following we discuss the simplest case, in which the A nucleons are grouped into two spin-zero clusters A_1 and A_2.

The RGM wave function for this system looks like

$$\Psi_{RGM} = A_{12}(\Phi_1^{int}\Phi_2^{int}g(\mathbf{r}_{12})), \tag{2}$$

where A_{12} is the inter-cluster antisymmetrization operator, Φ_i^{int} are wave functions in the cluster center-of-mass frame, and $\mathbf{r}_{12}$ is the relative coordinate between the centers of mass of A_1 and A_2. The function Φ_i^{int} is usually taken to be the (normalized and antisymmetrized) translation-invariant harmonic-oscillator shell-model ground state of the cluster A_i. The unknown relative motion function $g(\mathbf{r}_{12})$ can be determined through the variational principle for the energy,

$$\delta(\langle\Psi_t|H|\Psi_t\rangle/\langle\Psi_t|\Psi_t\rangle) = 0. \tag{3}$$

With a trial function of the form of (2), Equation (3) leads to the RGM equation of motion

$$\int d\mathbf{r}'\,(H(\mathbf{r},\mathbf{r}') - EN(\mathbf{r},\mathbf{r}'))\,g(\mathbf{r}') = 0, \tag{4}$$

where the kernel functions are

$$H(\mathbf{r},\mathbf{r}') = \left\langle \Phi_1^{int}\Phi_2^{int}\delta(\mathbf{r}_{12}-\mathbf{r})|H|A_{12}\left(\Phi_1^{int}\Phi_2^{int}\delta(\mathbf{r}_{12}-\mathbf{r}')\right)\right\rangle, \tag{5}$$

$$N(\mathbf{r},\mathbf{r}') = \left\langle \Phi_1^{int}\Phi_2^{int}\delta(\mathbf{r}_{12}-\mathbf{r})|A_{12}\left(\Phi_1^{int}\Phi_2^{int}\delta(\mathbf{r}_{12}-\mathbf{r}')\right)\right\rangle. \tag{6}$$

The kernels may be made more explicit. The antisymmetrization operator A_{12} can be divided into two parts $A_{12} = 1 + A'_{12}$, so the kernels are decomposed into sums of 'direct' and 'exchange' terms: $H(\mathbf{r},\mathbf{r}') = H_D(\mathbf{r},\mathbf{r}') + H_{ex}(\mathbf{r},\mathbf{r}')$ and $N(\mathbf{r},\mathbf{r}') = N_D(\mathbf{r},\mathbf{r}') + N_E(\mathbf{r},\mathbf{r}')$, where $N_D(\mathbf{r},\mathbf{r}') = \delta(\mathbf{r}-\mathbf{r}')$. The $H_{ex}(\mathbf{r},\mathbf{r}')$ term can also be simplified by splitting H of (1) into four parts,

$$H = H_1^{int} + H_2^{int} - \frac{\hbar^2}{2\mu}\Delta_{\mathbf{r}_{12}} + V_{12}, \tag{7}$$

where μ is the reduced mass of the clusters, H_i^{int} is the internal Hamiltonian of the cluster A_i, and

$$V_{12} = \sum_{i\epsilon A_1}\sum_{j\epsilon A_2}(V_{ij}^N + V_{ij}^C). \tag{8}$$

Using (7) we arrive at

$$H_D(\mathbf{r},\mathbf{r}') = \left[-\frac{\hbar^2}{2\mu}\Delta_{\mathbf{r}} + V_D(r) + E_1 + E_2\right]\delta(\mathbf{r}-\mathbf{r}'), \tag{9}$$

where the direct potential $V_D(r)$ is spherical and reads

$$V_D(r) = \left\langle \Phi_1^{int}\Phi_2^{int}\delta(\mathbf{r}_{12}-\mathbf{r})|V_{12}|\Phi_1^{int}\Phi_2^{int}\right\rangle \tag{10}$$

and E_i is the intrinsic energy of cluster A_i, i.e. $E_i = \langle\Phi_i^{int}|H_i^{int}|\Phi_i^{int}\rangle$. Taking into account (9), the integral equation (4) can be cast into an integro-differential equation form

$$\begin{aligned}\Big[-\frac{\hbar^2}{2\mu}\,\Delta_{\mathbf{r}} + V_D(r)\; g(\mathbf{r}) + \int d\mathbf{r}'\,(H_{ex}(\mathbf{r},\mathbf{r}') - EN_{ex}(\mathbf{r},\mathbf{r}'))\; g(\mathbf{r}') \\ = (E - E_1 - E_2)g(\mathbf{r}).\end{aligned} \tag{11}$$

Expanding $g(\mathbf{r})$ into spherical harmonics,

$$g(\mathbf{r}) = \frac{1}{r}\sum_{\ell m} g_\ell(r)\; Y_{\ell m}(\hat{r}), \tag{12}$$

we get the angular momentum projected many-body RGM wave function,

$$\Psi_{RGM}^{\ell} = A_{12}\left(\Phi_1^{int}\Phi_2^{int}\frac{1}{r}g_\ell(r)Y_{\ell m}(\hat{r})\right). \tag{13}$$

The angular momentum ℓ of the relative motion, by assumption, is a good quantum number as it is equal to the total angular momentum of the system. Making a partial wave expansion of the exchange parts of kernels (5) and (6) we get from (11)

$$\begin{aligned}[-\frac{\hbar^2}{2\mu}\frac{d^2}{dr^2} + V_D(r) + \frac{\hbar^2\ell(\ell+1)}{2\mu r^2}\; g_\ell(r) + \int_0^\infty dr'\left(H_{ex}^{\ell}(r,r')\right. \\ \left. - EN_{ex}^{\ell}(r,r')g_\ell(r')\right) = (E - E_1 - E_2)g_\ell(r).\end{aligned} \tag{14}$$

We should like to point out that the RGM equation of motion (11) or (14) refers to the relative motion wave function of the clusters containing the true dynamical variable $\mathbf{r}_{12}$. The final (11) or (14) form of the RGM equation of motion shows that the cluster-cluster interaction is energy dependent and non-local due to the

Pauli principle. The folding type direct potential term would describe the cluster-cluster interaction alone only if the antisymmetrization between A_1 and A_2 could be neglected.

In our case, the RGM reduces the many-body problem to a one-body Schrödinger-like equation (11). The microscopic character of the model, the composite structure of the clusters, appears through the kernel functions (5) and (6). In RGM calculations the most difficult task is to determine the kernel functions. It is out of our scope to enlarge upon this technical but important point. We mention that essentially two techniques have been developed; one that uses Jacobi coordinates in the evaluation of the matrix elements (5) and (6) and another that expresses the RGM kernels in terms of GCM kernels, which are to be given below.

2.2 The Generator-coordinate Method

An alternative formulation of the microscopic cluster model is the GCM. In this method the trial wave function is a continuous superposition of basis functions containing a specific non-dynamical continuous parameter. The starting point in the GCM is a two-center harmonic oscillator shell-model state

$$A_{12}\left(\Phi_1(\mathbf{X}_1-\varsigma_1)\Phi_2(\mathbf{X}_2-\varsigma_2)\right), \tag{15}$$

where $\mathbf{X}_1$ and $\mathbf{X}_2$ stand for $\mathbf{r}_1, ..., \mathbf{r}_{A_1}$ and $\mathbf{r}_{A_1+1}, ..., \mathbf{r}_{A_1+A_2}$, respectively, and the spin-isospin variables are suppressed. (e.g. $\mathbf{X}_1-\varsigma_1$ is to be understood as $\mathbf{r}_1-\varsigma_1, ..., \mathbf{r}_{A_1}-\varsigma_1, ...$)

In (15) $\Phi_i(\mathbf{X}_i-\varsigma_i)$ is a Slater determinant corresponding to the lowest-lying harmonic-oscillator shell-model configuration of the cluster A_i. The Slater determinant $\Phi_i(\mathbf{X}_i-\varsigma_i)$ is constructed of single-particle orbits which are centered at the position ς_i. For 1s-wave clusters these orbits are simple Gaussian functions

$$\left(\frac{\beta_i}{\pi}\right)^{3/4} \exp\left\{-\beta_i(\mathbf{r}-\varsigma_i)^2/2\right\}, \tag{16}$$

where $\beta_i = m w_i/\hbar$. Introducing relative coordinates, one can rewrite (15), apart from a trivial factor, as

$$A_{12}\left(\Phi_i^{int}\Phi_2^{int}\Gamma(\mathbf{R}_{cm}, \mathbf{r}_{12}, \mathbf{S}, \varsigma)\right), \tag{17}$$

where

$$\Gamma(\mathbf{R}_{cm}, \mathbf{r}_{12}, \mathbf{S}, \varsigma) = \exp\{-\bar{\alpha}(\mathbf{R}_{cm} - \mathbf{S})^2 - \bar{\beta}(\mathbf{R}_{cm} - \mathbf{S})(\mathbf{r}_{12} - \varsigma) - \bar{\gamma}(\mathbf{r}_{12} - \varsigma)^2\} \tag{18}$$

with $\mathbf{S} = (A_1\varsigma_1 + A_2\varsigma_2)/(A_1 + A_2)$, and $\varsigma = \varsigma_1 - \varsigma_2$, $\mathbf{R}_{cm} = \sum_{i=1}^{A} \mathbf{r}_i/(A_1 + A_2)$

$$\bar{\alpha} = (A_1\beta_1 + A_2\beta_2)/2, \tag{19}$$

$$\bar{\beta} = A_1 A_2(\beta_1 - \beta_2)/(A_1 + A_2), \tag{20}$$

$$\bar{\gamma} = A_1 A_2(A_1\beta_2 + A_2\beta_1)/2(A_1 + A_2)^2. \tag{21}$$

From Equations (17), (18) and (20), we see that when $\beta_1 \neq \beta_2$ the intrinsic and the center-of-mass motion are coupled. The Hamiltonian (1) requires the usage of translational invariant wave functions. Integrating over S in (17), we can get rid of the center-of-mass motion. The integration in (17) leads to the expression (apart from a constant factor)

$$\Psi(\varsigma) = A_{12}\left(\Phi_1^{int}\Phi_2^{int}\Gamma(\mathbf{r}_{12}, \varsigma, \gamma_o)\right), \tag{22}$$

where

$$\Gamma(\mathbf{r}_{12}, \varsigma, \gamma_o) = \exp\left\{-\gamma_o(\mathbf{r}_{12} - \varsigma)^2\right\}, \tag{23}$$

$$\gamma_o = A_1 A_2 \beta_1 \beta_2 / 2(A_1\beta_1 + A_2\beta_2). \tag{24}$$

The wave function (22) is considered to be a basis function in the GCM. The trial wave function now has the form

$$\Psi_{GCM} = \int d\varsigma f(\varsigma)\Psi(\varsigma). \tag{25}$$

The ς is called the generator coordinate, $\Psi(\varsigma)$ is the generating function, and the unknown $f(\varsigma)$ is the weight function. The latter is determined by the variational principle (3). Using the ansatz (25) in (3), we get the Griffin-Hill-Wheeler integral equation

$$\int d\varsigma' \left(h(\varsigma,\varsigma') - En(\varsigma,\varsigma')\right) f(\varsigma') = 0, \tag{26}$$

where the definitions of the kernels are

$$h(\varsigma,\varsigma') = \langle \Phi_1^{int}\Phi_2^{int}\Gamma(\mathbf{r}_{12},\varsigma,\gamma_o)|H|A_{12}\left(\Phi_1^{int}\Phi_2^{int}\Gamma(\mathbf{r}_{12},\varsigma',\gamma_o)\right)\rangle, \tag{27}$$

and

$$n(\varsigma,\varsigma') = \langle \Phi_1^{int}\Phi_2^{int}\Gamma(\mathbf{r}_{12},\varsigma,\gamma_o)|A_{12}\left(\Phi_1^{int}\Phi_2^{int}\Gamma(\mathbf{r}_{12},\varsigma',\gamma_o)\right)\rangle. \tag{28}$$

Note that the generator coordinate ς is a non-dynamical variable, and the weight function $f(\varsigma)$ has no direct physical interpretation. Looking at (22), (25) and (2) we can realize that the GCM wave function (25) corresponds to an RGM-type (2) ansatz with relative motion function

$$\tilde{g}(\mathbf{r}_{12}) = \int d\varsigma f(\varsigma)\Gamma(\mathbf{r}_{12},\varsigma,\gamma_o). \tag{29}$$

We mention that the inversion of (29), with some caution, is possible so the RGM and GCM are equivalent versions of the cluster model.

The GCM framework offers an advantage over the RGM. The kernels (27) and (28) can be calculated using the wave function of (15). Due to the fact that this function is a Slater determinant, the matrix elements of one- and two-body operators can be calculated in a straightforward manner. The RGM kernels can be obtained from GCM ones through various transformations.

Expanding, into spherical harmonics, the weight function and the locally peaked Gaussian (23), we obtain the angular momentum projected GCM wave function as

$$\Psi^{\ell}_{GCM} = \int_{o}^{\infty} d\varsigma f_\ell(\varsigma) A_{12}\left(\Phi_1^{int}\Phi_2^{int}\Gamma_\ell(r_{12},\varsigma,\gamma_o)\right), \tag{30}$$

where

$$\Gamma_\ell(r_{12},\varsigma,\gamma_o) = 4\pi r_{12}\varsigma i_\ell(2\gamma_o r_{12}\varsigma)\ \exp\{-\gamma_o(r_{12}^2+\varsigma^2)\} \tag{31}$$

and i_ℓ is the modified spherical Bessel function of the first kind. The integral equation (26) in the partial wave representation may be cast into the form

$$\int_{o}^{\infty} d\varsigma' \left(h_\ell(\varsigma,\varsigma') - En_\ell(\varsigma,\varsigma')\right) f_\ell(\varsigma') = 0, \tag{32}$$

where the kernels are defined by expansions like

$$h(\varsigma,\varsigma') = \frac{1}{\varsigma\varsigma'} \sum_{\ell m} h_\ell(\varsigma,\varsigma') Y_{\ell m}(\hat{\varsigma}) Y_{\ell m}(\hat{\varsigma}')^*. \tag{33}$$

The partial-wave representation of (29) is

$$\tilde{g}_\ell(r_{12}) = \int_{o}^{\infty} d\varsigma \Gamma_\ell(r_{12},\varsigma,\gamma_o) f_\ell(\varsigma). \tag{34}$$

3. Imposition of Gamow Asymptotics

3.1 Realization in the Generator-coordinate Method[3]

Equation (14) should be accompanied by one giving the asymptotic behavior of the relative motion function $g_\ell(r)$. The asymptotic form of $g_\ell(r)$ for the scattering of charged clusters is given by

$$g_\ell(r) \sim F_\ell(kr) + K_\ell G_\ell(kr), \tag{35}$$

where F_ℓ and G_ℓ denote the regular and irregular Coulomb functions, K_ℓ is the partial wave K-matrix, and $k = (2\mu(E - E_1 - E_2)/\hbar^2)^{1/2}$ is the wave number of the relative motion. At this point it is convenient to introduce the outgoing and incoming Coulomb waves with the definition

$$U_\ell^{(\pm)}(r) = G_\ell(r) \pm iF_\ell(r). \tag{36}$$

First we discuss a possible implementation of the general scattering asymptotics (35) in the framework of the GCM. The purely outgoing asymptotics of a Gamow state

$$\exp\{i(kr - \eta \ell n 2kr - \ell\pi/2 + \sigma_\ell)\} \tag{37}$$

clearly a special case of the general scattering asymptotics (35). In (37) σ_ℓ is the Coulomb phase shift and η is the Coulomb parameter, $\eta = Z_1 Z_2 e^2 \mu/\hbar^2 k$.

A number of procedures have been developed in the GCM to describe scattering wave functions [6]. Now we outline only de Takacsy's method, which has been used for the calculation of Gamow states, too. Let us divide the integration in (34) into two parts

$$\tilde{g}_\ell(r) = \int_0^{\varsigma_0} d\varsigma \Gamma_\ell(r,\varsigma,\gamma_o) f_\ell(\varsigma) + \int_{\varsigma_o}^{\infty} d\varsigma \Gamma_\ell(r,\varsigma,\gamma_o) f_\ell(\varsigma). \tag{38}$$

The first term of the r.h.s. of (38) is well approximated by a finite sum

$$\int_0^{\varsigma_0} d\varsigma \Gamma_\ell(r,\varsigma,\gamma_o) f_\ell(\varsigma) \approx \sum_{i=1}^{N} \Gamma_\ell(r,\varsigma_i,\gamma_o) f_\ell(\varsigma_i) W_i. \tag{39}$$

where W_i are the weights of a quadrature formula. We assume that the second term of the r.h.s. of (38) is responsible for the correct asymptotics prescribed in (35). The values of ς_o and N have to be chosen such that the function (39) may well describe the relative motion in the nuclear interaction region. The specific point now is to define the weight function so as to get the correct scattering asymptotics. For real k values, Friedrich, Hüsken and Weiguny gave a prescription and Fiebig and Weiguny extended this for complex k. They adopted the form

$$f_\ell^{(\pm)}(\varsigma) = \begin{cases} C \exp\{\pm i(k\varsigma - \eta \ell n 2k\varsigma - \ell\pi/2 + \sigma_\ell)\} \sum_{j=0}^{M} a_j/\varsigma^{j+1} & \varsigma \geq \varsigma_0 \\ 0 & \varsigma < \varsigma_0 \end{cases} \tag{40}$$

and determined the coefficients C, and a_j so that the expression

$$\int_0^{\infty} d\varsigma \Gamma_\ell(r,\varsigma,\gamma_o) f_\ell^{(\pm)}(\varsigma) \tag{41}$$

approximate the outgoing and incoming Coulomb waves $U_\ell^{(\pm)}(r)$ of (36). The relative error

$$\begin{aligned} &\mid \Big(U_\ell^{(\pm)}(r) - \int_0^{\infty} d\varsigma \Gamma_\ell(r,\varsigma,\gamma_o) f_\ell^{(\pm)}(\varsigma) \\ &\qquad / \exp\{\pm i(kr - \eta \ell n 2kr - \ell\pi/2 + \sigma_\ell)\} \mid \end{aligned} \tag{42}$$

was shown to go to zero as $1/r^{M+1}$.

All these considerations suggest that in the case of a scattering boundary condition (35) we may take for $g_\ell(r)$ the ansatz

$$g_\ell(r) = \sum_{i=1}^{N} f_{i,\ell}\Gamma_\ell(r,\varsigma_i,\gamma_o) + \int_0^\infty d\varsigma f_\ell^{as}(\varsigma)\Gamma_\ell(r,\varsigma,\gamma_o), \tag{43}$$

where $f_\ell^{as}(\varsigma)$ is an appropriate combination of $f_\ell^{(+)}(\varsigma)$ and $f_\ell^{(-)}(\varsigma)$ to get the asymptotics (35) instead of $U_\ell^{(+)}(r)$ and $U_\ell^{(-)}(r)$. The variational principle (3) leads to

$$\begin{aligned}&\sum_{j=1}^{N} \left(h_\ell(\varsigma_i,\varsigma_j) - En_\ell(\varsigma_i,\varsigma_j)\right) f_{j,\ell} \\ &= -\int_0^\infty d\varsigma \left(h_\ell(\varsigma_i,\varsigma) - En_\ell(\varsigma_i,\varsigma)\right) f_\ell^{as}(\varsigma).\end{aligned} \tag{44}$$

Here we have N equations but the number of unknowns are N+1 due to the fact that K_ℓ of (35) is also unknown. De Takacsy took a matching condition

$$f_{N,\ell} = f_\ell^{as}(\varsigma_N), \tag{45}$$

which is inspired by the continuity of $f_\ell(\varsigma)$. The N+1 simultaneous linear equations (44) and (45) can be uniquely solved, thus giving the value K_ℓ of the K-matrix for any energy. Since the value K_ℓ is related to the S-matrix by $S_\ell = (i - K_\ell)/(i + K_\ell)$, the resonance energy is defined by the condition $K_\ell = -i$. In numerical calculations, of course, the results have to be independent of the value ς_o and the discretization points ς_i. We have to mention that several questions have been raised concerning this approach, although it has proved to be successful in actual calculations. In other GCM approaches, (44) has a more sound foundation and the matching condition (45) is replaced by a new one.

3.1 Algebraic Version of the Resonating-group Model[4]

In this approach, the relative-motion part of the RGM wave function is expanded into a series of harmonic oscillator functions

$$\frac{1}{r} g_\ell(r) = \sum_{\vartheta=\vartheta_o}^{\infty} C_\vartheta^\ell \varphi_\vartheta^\ell(r), \tag{46}$$

where ϑ is the total number of oscillator quanta.

The summation in (46) starts from a specific value of ϑ_o because some $\varphi^{\ell}_{\vartheta}(r)$ states are Pauli-forbidden due to the indistinguishability of nucleons. Now the variational principle (3) leads to an infinite set of equations

$$\sum_{\vartheta'=\vartheta_o}^{\infty} \left(H^{\ell}_{\vartheta,\vartheta'} - E\delta_{\vartheta,\vartheta'}\right) C^{\ell}_{\vartheta'} = 0, \tag{47}$$

where

$$H^{\ell}_{\vartheta,\vartheta'} = \langle \Phi_1^{int}\Phi_2^{int}\phi^{\ell}_{\vartheta}(r_{12})Y_{\ell m}(\hat{r}_{12})|H|A_{12}\left(\Phi_1^{int}\Phi_2^{int}\phi^{\ell}_{\vartheta'}(r_{12})Y_{\ell m}(\hat{r}_{12})\right)\rangle. \tag{48}$$

Of course, in practical calculations the summation in (47) has to be cut at a finite ϑ_{max} value. The essential point in the algebraic version of the RGM is that the states with $\vartheta > \vartheta_{max}$ are also taken into account. The trick is that a prescribed (bound, scattering or resonant) asymptotic behavior of $g_\ell(r)$ implies an explicit form of C^{ℓ}_{ϑ} for large values of the quantum number ϑ. The asymptotic form of C^{ℓ}_{ϑ} will be denoted by $\bar{C}^{\ell}_{\vartheta}$. It was shown that the $\bar{C}^{\ell}_{\vartheta}$ coefficients obey an infinite set of equations,

$$-\frac{1}{2}[\vartheta(\vartheta+1) - \ell(\ell+1)]^{1/2}\bar{C}^{\ell}_{\vartheta-2} + \left[\vartheta + \frac{3}{2} + \left(\frac{m}{\mu\beta}\right)^{1/2} \frac{2\eta k}{(2\vartheta+3)^{1/2}} - \frac{k^2 m}{\mu\beta}\right]\bar{C}^{\ell}_{\vartheta}$$
$$-\frac{1}{2}\left[(\vartheta+2)(\vartheta+3) - \ell(\ell+1)\right]^{1/2}\bar{C}^{\ell}_{\vartheta+2} = 0 \;;\; \vartheta > n, \tag{49}$$

where n is a large value (~ 100). In the chargeless case ($\eta = 0$) (49) can be solved analytically. In the general case, within the limit of large ϑ, we can go over to a differential equation

$$\frac{\partial^2 \bar{C}^{\ell}}{\partial \rho_\vartheta^2} + \frac{1}{\rho_\vartheta}\frac{\partial \bar{C}^{\ell}}{\partial \rho_\vartheta} + \left[\frac{k^2 m}{\mu\beta} - \left(\frac{m}{\mu\beta}\right)^{1/2} 2\eta k - \frac{(\ell+\frac{1}{2})^2}{\rho_\vartheta^2}\right]\bar{C}^{\ell} = 0, \tag{50}$$

where $\rho_\vartheta = (2\vartheta+3)^{1/2}$ is considered to be a continuous variable. It was shown that the solution of (50),

$$\bar{C}^{\ell}_{\vartheta} = \frac{N_o}{\rho_\vartheta^{1/2}} W_{(-i\eta,\ell+\frac{1}{2})}\left\{-2ik\rho_\vartheta\left(\frac{m}{\mu\beta}\right)^{1/2}\right\} \tag{51}$$

corresponds to Gamow asymptotics (W is the Whittaker function and N_o is a normalization constant).

One can then write the ansatz

$$\sum_{\vartheta=\vartheta_o}^{n-2} C_\vartheta^\ell \varphi_\vartheta^\ell(r) + \sum_{\vartheta=n}^{N-2} C_\vartheta^\ell \varphi_\vartheta^\ell(r) + \sum_{\vartheta=N}^{\infty} \bar{C}_\vartheta^\ell \varphi_\vartheta^\ell(r), \tag{52}$$

where the coefficients C_ϑ^ℓ appearing in the first two terms of (52) are considered to be unknown; but, it is required that the coefficients of the second term fulfill (49). N can be chosen so large that $|H_{\vartheta,\vartheta'}^\ell| \approx 0$ if $\vartheta < n$ and $\vartheta' \geq N$. The third term of (52) ensures the correct Gamow asymptotics.

From the Schrödinger equation and from the (52) form of the wave function of the relative motion, a finite set of equations can be derived

$$\sum_{\vartheta'=\vartheta_o}^{N-2} (H_{\vartheta,\vartheta'}^\ell - E\delta_{\vartheta,\vartheta'})C_{\vartheta'}^\ell = 0, \qquad \vartheta < n; \tag{53}$$

$$\begin{aligned} -\frac{1}{2}[\vartheta(\vartheta+1) - \ell(\ell+1)]^{1/2} C_{\vartheta-2}^\ell + \left[\vartheta + \frac{3}{2} + \left(\frac{m}{\mu\beta}\right)^{1/2} \frac{2\eta k}{(2\vartheta+3)^{1/2}} - \frac{k^2 m}{\mu\beta}\right] C_\vartheta^\ell \\ -\frac{1}{2}[(\vartheta+2)(\vartheta+3) - \ell(\ell+1)]^{1/2} C_{\vartheta+2}^\ell = 0, \\ n \leq \vartheta \leq N-4; \end{aligned} \tag{54}$$

$$\begin{aligned} -\frac{1}{2}[(N-2)(N-1) - \ell(\ell+1)]^{1/2} C_{N-4}^\ell \\ + \left[N - \frac{1}{2} + \left(\frac{m}{\mu\beta}\right)^{1/2} \frac{2\eta k}{(2N-1)^{1/2}} - \frac{k^2 m}{\mu\beta}\right] C_{N-2}^\ell \\ = \frac{1}{2}[N(N+1) - \ell(\ell+1)]^{1/2} \bar{C}_N^\ell \ , \ \vartheta = N-2; \end{aligned} \tag{55}$$

$$C_{N-2}^\ell = \bar{C}_{N-2}^\ell. \tag{56}$$

The coefficients $C_{\vartheta_o}^\ell, C_{\vartheta_o+2}, \ldots, C_{N-2}^\ell$ can be calculated, from the inhomogeneous set of linear equations (53-55), for the arbitrary value of k. Then (56) gives a transcendental equation for the calculation of the resonance energy.

4. Complex Scaling in the Cluster Model

4.1 The Complex Scaling

In atomic and molecular physics the complex scaling (CS) method has become commonplace in the determination of positions and widths of resonances [7]. Since this method is still counted as a novelty in nuclear physics, we shortly summarize it for scattering from a local potential without mathematical rigour. The CS method is based on the so called 'ABC theorem ', which has a rigorous foundation [8]. Let us introduce the operator U_θ by the definition

$$(U_\theta f)(\mathbf{r}) = \exp\left\{i\frac{3}{2}\theta\right\} f(\mathbf{r}\exp\{i\theta\}), \quad 0 < \theta < \pi/2, \tag{57}$$

and transform the Hamiltonian

$$H(\mathbf{r}) = -\frac{\hbar^2}{2\mu}\Delta_\mathbf{r} + V(r) \tag{58}$$

into

$$H(\mathbf{r},\theta) = U_\theta H(\mathbf{r}) U_\theta^{-1} = -\frac{\hbar^2}{2\mu}\exp\{-i2\theta\}\Delta_\mathbf{r} + V(r\exp\{i\theta\}). \tag{59}$$

We mention that the interaction of the Hamiltonian (58) may generally be a non-local one. The 'ABC theorem' relates the spectrum of the non-hermition operator $H(\mathbf{r},\theta)$ to that of $H(\mathbf{r})$ for a broad class of potentials: (i) the bound-state eigenenergies of $H(\mathbf{r})$ remain eigenvalues of $H(\mathbf{r},\theta)$; (ii) the continuous spectrum of $H(\mathbf{r},\theta)$ is given by the expression $E\exp\{-i2\theta\}, 0 < E < \infty$ (i.e. the continuum of $H(\mathbf{r})$ is rotated down into the lower half of the complex energy plane); (iii) the complex poles E_{res} of the S-matrix become eigenvalues of $H(\mathbf{r},\theta)$ if $\theta > \frac{1}{2}|argE_{res}|$, and the corresponding eigenfunction is square-integrable; (iv) the resonant eigenvalues of $H(\mathbf{r},\theta)$ are independent of the value of θ.

Of course, the solution of the eigenvalue problem of the transformed Hamiltonian

$$H(\mathbf{r},\theta)\Psi(\mathbf{r},\theta) = E_\theta\Psi(\mathbf{r},\theta) \tag{60}$$

can only be obtained by using approximation methods. However, due to point (iii), we may determine resonance states using the standard methods for bound-state eigenvalue problems. Let us approximate $\Psi(\mathbf{r},\theta)$ by an expansion over a set of N linearly independent real square-integrable functions $\varphi_i(\mathbf{r})$, i = 1,...,N:

$$\Psi_N(\mathbf{r},\theta) = \sum_{i=1}^{N} C_i(\theta)\varphi_i(\mathbf{r}). \tag{61}$$

Using the generalized variational principle

$$\delta\left[\int d\mathbf{r}\bar{\Psi}_N(\mathbf{r},\theta)^* H(\mathbf{r},\theta)\Psi_N(\mathbf{r},\theta) / \int d\mathbf{r}\bar{\Psi}_N(\mathbf{r},\theta)^*\Psi_N(\mathbf{r},\theta)\right] = 0 \tag{62}$$

we get, for $C_i(\theta)$

$$\sum_{j=1}^{N}(H_{ij}(\theta) - EN_{ij})C_j(\theta) = 0, \tag{63}$$

where

$$\begin{aligned} H_{ij}(\theta) &= \int d\mathbf{r}\bar{\varphi}_i(\mathbf{r})^* H(\mathbf{r},\theta)\varphi_j(\mathbf{r}), \\ N_{ij} &= \int d\mathbf{r}\bar{\varphi}_i(\mathbf{r})^*\varphi_j(\mathbf{r}). \end{aligned} \tag{64}$$

In (62) and (64) the bar on the functions means that only their angular factors are to be complex conjugated. It is easy to verify that $H_{ij}(\theta)$ and N_{ij} may as well be expressed as matrix elements of the unrotated operators $H(\mathbf{r})$ and 1 between the back-rotated basis functions

$$\varphi_i(\mathbf{r},\theta) = U_\theta^{-1}\varphi_i(\mathbf{r}) = \exp\{-i\frac{3}{2}\theta\}\varphi_i(\mathbf{r}\exp\{-i\theta\}). \tag{65}$$

Due to the finite basis set approximation, it is expected that the 'ABC theorem' is fulfilled only approximately. For instance, the position of a resonance eigenvalue is not completely independent of θ. As a function of θ it moves along a trajectory in the complex energy plane. The best estimate for E_{res} is given by the particular θ value for which the rate of change with respect to θ is minimal. This procedure has been justified by the generalized virial theorem.

4.2 The Cluster Model and the Complex Scaling

The RGM equation (11) has the form of a Schrödinger equation with a complicated effective interaction. Because the interation in (11) is energy-dependent, we feel that the application of the CS method to the RGM equation goes beyond the mathematically proved range of the validity of the CS approach. Encouraged by the success of other attempts in which the sufficient conditions of the ABC theorem were not fulfilled either (e.g. in the description of the Stark effect), we tried to apply the CS method to the calculation of the resonant solution of the RGM equation.

We subjected the coordinate $\mathbf{r}_{12}$ to a complex scaling in (11). According to (65), we can then approximate the complex rotated relative motion function $g_\ell(r_{12},\theta)$ of a resonance state with the form

$$g_\ell(r_{12},\theta) = \sum_{i=1}^{N} f_{i,\ell}(\theta)\exp\{-i\frac{3}{2}\theta\}\Gamma_\ell(r_{12}\exp\{-i\theta\},\varsigma_i,\gamma) \tag{66}$$

or equivalently with

$$g_\ell(r_{12},\theta) = \sum_{i=1}^{N} f_{i,\ell}(\theta)\exp\{-i\frac{3}{2}\theta\}\Gamma_\ell(r_{12},\varsigma_i(\theta),\gamma(\theta)), \tag{67}$$

where

$$\gamma(\theta) = \gamma\exp\{-i2\theta\}, \varsigma_i(\theta) = \varsigma_i\exp\{i\theta\}, \tag{68}$$

and we can choose the value of γ freely.

Since now the resonance wave function is square-integrable, the second term of the r.h.s. of (43) can be omitted. Using the variational principle (62) on the subspace spanned by the functions (66), we get for the scaled Griffin-Hill-Wheeler equation

$$\sum_{j=1}^{N}(h^\ell_{ij}(\theta) - En^\ell_{ij}(\theta))f_{j,\ell}(\theta) = 0, \tag{69}$$

where $h^\ell_{ij}(\theta)$ and $n^\ell_{ij}(\theta)$ are the partial wave components of

$$\begin{aligned} &h_{ij}(\theta) = \\ &\langle\Phi_1^{int}\Phi_2^{int}\Gamma(\mathbf{r}_{12},\varsigma_i(\theta),\gamma(\theta))|H|A_{12}\left(\Phi_1^{int}\Phi_2^{int}\Gamma(\mathbf{r}_{12},\varsigma_j(\theta),\gamma(\theta))\right)\rangle, \end{aligned} \tag{70}$$

and

$$n_{ij}(\theta) = \langle\Phi_1^{int}\Phi_2^{int}\Gamma(\mathbf{r}_{12},\varsigma_i(\theta),\gamma(\theta))|A_{12}\left(\Phi_1^{int}\Phi_2^{int}\Gamma(\mathbf{r}_{12},\varsigma_j(\theta),\gamma(\theta))\right)\rangle, \tag{71}$$

respectively. Due to the momentum projection, in the kernels (27) and (28) however, the value of the parameter γ is fixed to γ_o of (24) so the GCM kernels (70) and (71) cannot be obtained from (27) and (28) by merely substituting γ_o and ς with $\gamma \exp\{-i2\theta\}$ and $\varsigma \exp\{i\theta\}$. Observing that $(\Re\gamma(\theta) < \gamma_o)$

$$\Gamma(\mathbf{r},\varsigma \exp\{i\theta\},\gamma(\theta)) = \left|\frac{\gamma_o^2}{\pi(\gamma_o - \gamma(\theta))}\right|^{3/2} \exp\left\{i\frac{3}{2}\, arg\, \frac{\gamma_o^2}{\gamma_o - \gamma(\theta)}\right\}$$
$$\int dt \exp\left\{-\frac{\gamma_o\gamma(\theta)}{\gamma_o - \gamma(\theta)}(\varsigma \exp\{-i\theta\} - \mathbf{t})^2\right\}\Gamma(\mathbf{r},\mathbf{t},\gamma_o), \tag{72}$$

we can derive the complex scaled kernels (70) and (71) from that of (27) and (28) by the help of a simple double integral transformation which can be carried out analytically.

5. Resonances in ^8Be and Summary

All reviewed methods of Sections 3 and 4 have been applied to the $(\alpha+\alpha)$ cluster model of ^8Be. The α cluster was described by one Slater determinant with $(1s)^4$ shell-model configuration. The size parameter of the single-particle orbit (1s) was chosen to be $\beta = 0.47301 fm^{-2}$. The Volkov 1 force [9] with Majorana exchange parameter m = 0.6, was used as the nuclear part of nucleon-nucleon potential, and the Coulomb interaction was taken into account.

Table I.

The complex energies $E_r - i\Gamma/2$ (MeV) of some resonances in ^8Be.

Model		$J^\pi = 0^+$	$J^\pi = 2^+$	$J^\pi = 4^+$
CS†		0.588-i0.121	3.068-i1.195	11.60-i3.52
Ref.4	Gamow	0.590-i0.120	3.340-i1.240	11.36-i2.47
	Scattering	0.590-i0.135	3.080-i1.495	11.70-i3.79
Ref.3	Gamow	—	3.160-i1.205	12.20-i3.88
	Scattering	—	3.160-i1.565	12.20-i4.27

† $\gamma = 0.32$ fm^{-2} and the zero of the energy scale is -53.41 MeV

The results for the lowest $J^\pi = 0^+, 2^+, 4^+$ resonance states of ^8Be are displayed in Table I. The three different methods give the same results for all states considered with accuracies well within the predicted widths of the resonances. The Figures show some details of the CS method as applied to the RGM equation.

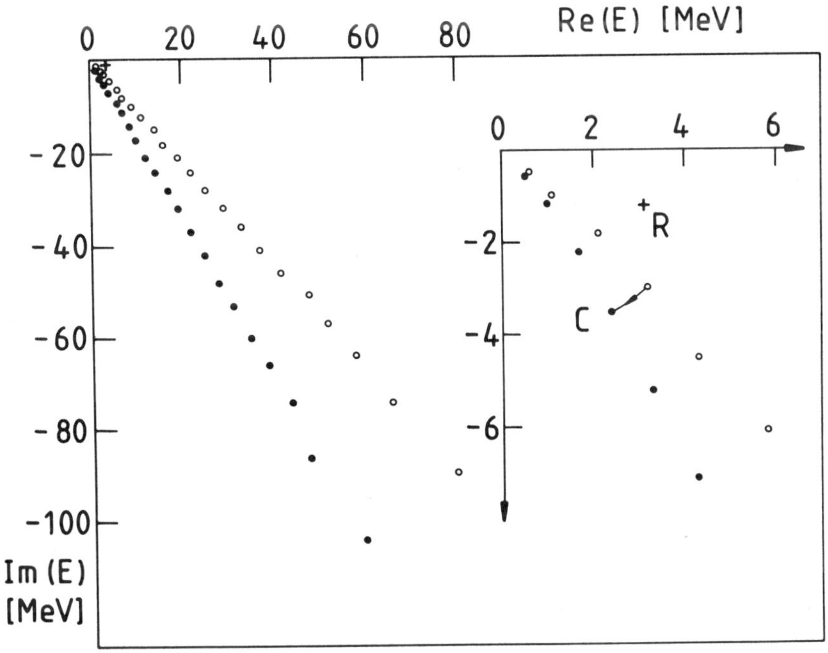

Figure 1.

The $J^{\pi} = 2^{+}$ energy values as given by equation (69) solved on a 24-term complex basis ($\gamma = 0.32 fm^{-2}$) with $\theta = 0.4(\circ)$ and $\theta = 0.5(\bullet)$; the point $+$ belongs to both θ values. The insert shows the area near the origin on an enlarged scale. (Taken from Reference[5].)

Figure 1 shows the complex energy values gained by (69) with two different θ values and with the discretization $\varsigma_i = 1fm, 2fm, ..., 24fm$. For each θ, 24 complex energy values are obtained. For a fixed θ, all of them but one are lined up on a straight line declining by 2θ from the real axis. The θ-stable value is to be regarded as a resonance. The other energies should be looked upon as a discretization of the

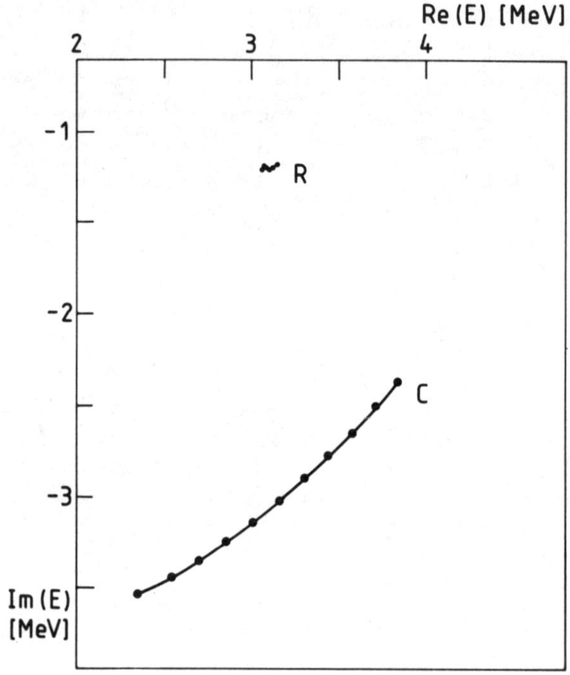

Figure 2.

The θ-trajectories of the resonance (R) and of a continuum point (C) of Figure 1. Both for R and C, θ varies from the right to the left between 0.3 and 0.5. (Taken from Reference[5])

rotated continuum. In Figure 2, the θ-trajectory of the resonance and a nearby continuum point are displayed. The continuum point is to be moved along a circular arc according to the ABC theorem. The small deviations from the straight lines and from the arc in Figures 1 and 2 are due to the finite basis set. For a precise determination of the energy of the resonance state in Figure 3, the relevant part of the θ trajectory of Figure 2 is thoroughly magnified. The resonance energy is taken to be the point where $dE/d\theta \approx 0$. Figures 1 and 2 illustrate that the

ABC theorem is fulfilled approximately and thereby justify the application of the CS method for resonance solutions of the RGM equation.

The advantages and disadvantages of the reviewed methods cannot be judged with full certainty, since only a few calculations have been carried out. However, one of the advantages of the CS method is obvious: the tedious construction of the correct asymptotics can be entirely avoided, and the bound state formalism of the GCM can be applied with minor modifications.

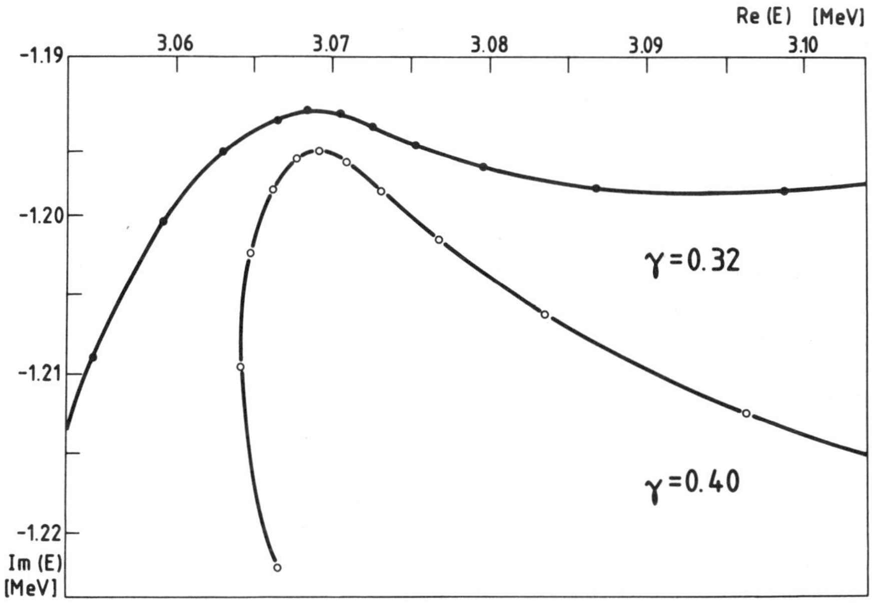

Figure 3.

A blownup plot of the θ-trajectory of R from Figure 2 (•) and the same with $\gamma = 0.4 fm^{-2}$(○).θ varies from 0.34 to 0.54 by steps 0.02. (Taken from Reference [5].)

We mention that in the microscopic cluster model other approaches have also been suggested for the direct determination of resonance states. In reference [10] a variational principle is introduced to calculate Gamow states and the corresponding formalism is developed in the GCM. In reference[11] a real stabilization method of the Hazi-Taylor-type is adopted to the GCM. Both methods have been applied only to a two-particle problem that is similar to the cluster model. Besides the fully microscopic RGM and GCM, extreme cluster models (which assume structureless clusters) are also useful in nuclear physics. Saito's orthogonality condition model (OCM) [12] is of special importance because it can be derived from the microscopic cluster model. Due to the structure of the OCM equation of motion it is difficult to determine resonance states. The only direct method applied [13] is based on the so called potential separable expansion technique [14] (also reviewed at this conference).

The RGM and GCM are very attractive approaches because they: (i) utilize the fundamental nucleon-nucleon interaction; and (ii) treat the Pauli-principle and the center-of-mass motion correctly. The RGM and GCM offer microscopic and dynamic descriptions of states of nuclei. Both methods have mainly been used for calculations of scattering and bound states of nuclei. The position and width of a resonance is usually determined by studying the phase shift. The disliked exponentially growing amplitude of the wave function of a resonance state can be handled. We can overcome the problem of non-square integrability, too, by redefining the inner product [15] or by introducing complex scaling [7]. We hope that we have shown that to calculate the resonance wave function in the framework of the cluster model is hardly more complicated than the usual description of bound and scattering states. We also hope that besides the determination of the position and the width of resonances, Gamow states will be invoked to calculate other physical quantities in the framework of the cluster model.

References

1. K. Wildermuth and W. McClure, *Cluster Representation of Nuclei*, Springer Tracts in Modern Physics Vol. 41, 1966; K. Wildermuth and Y. C. Tang, *A Unified Theory of the Nucleus*, Academic, New York, 1977; Microscopic Methods for the Interactions between Complex Nuclei, Prog. Theor. Phys. Suppl. **62**, (1977); Y. C. Tang, M. LeMere and D. R. Thompson, Phys. Rep. **47**, 167 (1978); H. Friedrich, Phys. Rep. **74**, 210 (1981).

2. Y. C. Tang, *Topics in Nuclear Physics*, ed. T. T. S. Kuo and S. S. M. Wong, Springer, Berlin, 1981, p. 572.

3. H. R. Fiebig and A. Weiguny, Z. Physics. **A279**, 275 (1976).

4. I. P. Okhrimenko, preprint ITP-84-166E, Kiev (1984); I. P. Okhrimenko, Few-Body Syst. **2**, 169 (1987).

5. A. T. Kruppa, R. G. Lovas and B. Gyarmati, Phys. Rev. **C37**, 383 (1988).

6. R. F. Barett, B. A. Robson and W. Tobocman, Rev. Mod. Phys. **55**, 155 (1983).

7. B. R. Junker, Adv. Atom. Molec. Phys. **18**, 207 (1982); W. P. Reinhardt, Ann. Rev. Phys. Chem. **33**, 223 (1982); Y. K. Ho, Phys. Rep. **99**, 1 (1983);

Proc. 1978 Sanibel Workshop Complex Scaling, Int. J. Quantum. Chem **14**, 343 (1978).

8. J. Aguilar and J. M. Combes, Commun. Math. Physics. **22**, 269 (1971); E. Balslev and J. M. Combes, ibid **22**, 280 (1971); B. Simon, ibid **2**, 1 (1972).

9. A. B. Volkov, Nucl. Phys. **74**, 33 (1965).

10. B. G. Giraud, M. V. Mihailović, R. G. Lovas and M. A. Nagarajan, Ann. Phys. (N.Y.) **140**, 29 (1982).

11. R. G. Lovas and M. A. Nagarajan, J. Phys. **A15**, 2383 (1982).

12. S. Saito, Prog. Theor. Phys. **41**, 705 (1969).

13. K. F. Pal, J. Phys. **A18**, 1665 (1985).

14. B. Gyarmati, A. T. Kruppa and J. Revai, Nucl. Phys. **A326**, 119 (1979); B. Gyarmati and A. T. Kruppa, Phys. Rev. **A33**, 2989 (1986); B. Gyarmati and A. T. Kruppa, Phys. Rev. **C34**, 95 (1986).

15. T. Berggren, Nucl. Phys. **A109**, 265 (1968); W. Romo, Nucl. Phys. **A116**, 618 (1968).

A NOTE ON THE CLUSTER MODEL AND COMPLEX SCALING

Erik Balslev

Matematisk Institut
Aarhus Universitet
Ny Munkegade. Bygning 530
DK-8000 AARHUS C
DENMARK

Abstract

In this note we indicate how to establish a rigorous basis for the calculation of nucleon resonances described in the preceding paper. The problem was mentioned to me by A. T. Kruppa and clarified through a discussion with him during the conference.

Let $H_0 = -\Delta$ on $\mathcal{D}_{H_0} = H^2(\mathbf{R}^n)$. Let V be a symmetric, H_0-compact, S_a-dilation-analytic operator in $\mathcal{H} = L_2(\mathbf{R}^n)$, where $S_a = \{\rho e^{i\varphi} | \rho > 0,\ |\varphi| < a\}$, $a > 0$.

Let N be a selfadjoint, compact operator on $\mathcal{H}$ such that $1 + N \geq \delta > 0$, and assume moreover that N is compact from $\mathcal{H}$ into $H^2(\mathbf{R}^n)$ and S_a-dilation-analytic as an operator in $\mathcal{C}\left(\mathcal{H}, H^2(\mathbf{R}^n)\right)$, extending through complex scaling to S_a as a $\mathcal{C}\left(\mathcal{H}, H^2(\mathbf{R}^n)\right)$-valued function $N(z)$.

Consider the following equation, where K is a real constant,

$$(H_0 + V + K)\Psi = E(1 + N)\Psi \tag{a2.1}$$

viewed as an equation for the resonance function Ψ associated with a resonance E. Formally we get, setting $(1 + N)^{\frac{1}{2}}\Psi = \Phi$,

$$(1 + N)^{-\frac{1}{2}}(H_0 + V + K)(1 + N)^{-\frac{1}{2}}\Phi = E\Phi \tag{a2.2}$$

In order to use the dilation-analytic theory, it is necessary to first give meaning to the operator on the l.h.s. of (a2.2). Note that $H^2(\mathbf{R}^n)$ is invariant under $N(z)$, and $N_1(z) = N(z)|H^2(\mathbf{R}^n)$ is compact on $H^2(\mathbf{R}^n)$. Moreover, if $N_1(z)f = \mu f$, $f \in \mathcal{H}$, $\mu \neq 0$, then $f \in H^2(\mathbf{R}^n)$; so, except for 0, the eigenvalues of $N_1(z)$ coincide with those of $N(z)$ and with the same eigenvectors. Let $(1+N)^{-\frac{1}{2}} = 1 + B$ and $(1+N_1)^{-\frac{1}{2}} = 1 + B_1$. Then B and B_1 are compact operators on $\mathcal{H}$ and $H^2(\mathbf{R}^n)$ respectively. It follows from the above that, apart from 0, B and B_1 have the same eigenvalues and eigenvectors, and $B_1 = B|H^2(\mathbf{R}^n)$. Moreover, B and B_1 are S_a-dilation-analytic, and $B_1(z) = B(z)|H^2(\mathbf{R}^n)$. Let us now consider the operator

$$\begin{aligned}\tilde{H} &= (1+B)(H_0+V+K)(1+B) \\ &= H_0 + K + V + BH_0 + H_0B + BV + VB + 2KB + BH_0B \\ &+ BVB + KB^2\end{aligned}$$

The complex-dilated operator $\tilde{H}(z)$ is given for $z \in S_a$ by

$$\begin{aligned}\tilde{H}(z) &= z^{-2}H_0 + K + V(z) + z^{-2}B(z)H_0 + z^{-2}H_0B(z) \\ &+ B(z)V(z) + V(z)B(z) \\ &+ 2KB(z) + z^{-2}B(z)H_0B(z) \\ &+ B(z)V(z)B(z) + KB^2(z)\end{aligned}$$

By the assumption on V and the compactness of $B(z)$ on $\mathcal{H}$, the terms $V(z)$, $z^{-2}B$ H_0, $B(z)V(z)$, $2KB(z)$ and $KB^2(z)$ are H_0-compact. By the above remarks, $B(z)$ $|H^2(\mathbf{R}^n) = B_1(z)$ is compact on $H^2(\mathbf{R}^n)$, hence the terms $z^{-2}H_0B(z)$, $V(z)B(z)$, $z^{-2}B(z)H_0B(z)$ and $B(z)V(z)B(z)$ are H_0-compact. Thus $\tilde{H}(z)$ is of the form

$$\tilde{H}(z) = z^{-2}H_0 + K + Q(z)$$

where $Q(z)$ is H_0-compact, and $\tilde{H}(z)$ is the dilation-analytic family associated with the operator

$$\tilde{H} = H_0 + K + Q, \quad Q = Q(1) .$$

We can now apply he standard dilation-analytic theory to define resonances of $(H_0 + K, \tilde{H})$. The essential spectrum of $\tilde{H}(z)$ is $K + e^{-2i\varphi}\overline{\mathbf{R}^+}$, and resonances are obtained for $0 < \varphi < a$ as discrete eigenvalues of $\tilde{H}(z)$ in the sector bounded by the half-lines $K + \mathbf{R}^+$ and $K + e^{-2i\varphi}\mathbf{R}^+$. The solution of (a2.1) is represented by the square-integrable eigenfunction of $\tilde{H}(z)$ corresponding to the discrete eigenvalue E (see Appendix 1).

In the case of the present work, N is an integral operator with an explicitly given kernel which is easily seen to satisfy the above conditions except that $1+N$ has the smallest eigenvalue 0. The finite-dimensional null space associated with this eigenvalue, however, is also null space of H_0+V+K. Using this fact, it should be possible to include this case in the theory, replacing $1+B$ by $(1+B)(1-P)$, where P is the orthogonal projection on the null space of $1+N$.

COMPLEX SCALING APPLIED TO TRAPPING OF ATOMS AND MOLECULES ON SOLID SURFACES

Nimrod Moiseyev

Department of Chemistry
Technion - Israel Institute of Technology
HAIFA 32000
ISRAEL

Abstract

The Complex coordinate method is applied to the calculation of the energy and lifetimes of predesorption states obtained by scattering of diatomic molecules from a flat surface and by scattering of atoms from a highly corrugated solid surface. A novel concept of complex quasi probability $Q(x)$ for atoms trapped on surfaces is described. The real part of $Q(x)$ is the probability to observe the absorbed atom at a certain point x on the corrugated surface. The quantity $ImQ(x)/|Q(x)-1|$ is suggested as a relative probability for desorption from a certain surface position x.

1. Introduction

The study of resonant phenomena in atomic and molecular scattering from surfaces, provides valuable information about adsorption- desorption kinetics in gas-surface processes [1-4]. The resonance states are obtained when an atom or a molecule is selectively adsorbed on the surface for a finite period of time. The lifetime of the adsorbed atom or molecule on the solid surface depends crucially on the structure of the surface. At sufficiently low surface temperatures and for small mass ratios between the trapped atom or molecule and the solid atoms, the scattering by the surface phonons can be neglected. In such a case only a *molecule* (and *not* an atom!) can be selectively trapped on a *flat* surface. The coupling between the internal rotational and vibrational modes of the trapped molecule with the molecule-surface vibrational modes allow energy transfer from the internal molecular modes to the molecular-surface mode. For sufficient large (intra molecular)→ (molecule/surface) energy transfer desorption occurs, and rotational (or vibrational) mediated selective adsorption states are obtained.

On a static surface, however, an atom can be selectively trapped only when the corrugation of the surface is sufficiently high. Due to the corrugation, there is a coupling between the motions of the trapped atom in the $\{x-y\}$ plane (i.e parallel to the solid surface plane) and in the $z-$direction (i.e. the coordinate perpendicular to the surface plane). For sufficiently large $(x,y) \rightarrow z$ energy transfer desorption occurs and resonance states of atoms at a corrugated surface are obtained. For a more detailed analysis of atomic and molecular scattering from surfaces see a recent review of Gerber [4].

In the next section we shall describe the complex coordinate method [5-7], which enables us to get accurate quantum energy levels and lifetimes of the trapped atom (or molecule) at the corrugated (or flat) solid surface *without* carrying out time dependent caculations. In section 3 the application of the complex coordinate method to the study of rotational predissociation and predesorption resonances will be described. By analyzing the complex rotated HD/Ag(111) resonance states we get that the trapped HD is an almost free rotor and therefore can be adequately described by classical dynamics[8]. Study of predesorption of CO from sodium chloride (100) surfaces by the complex coordinate method show that vibrational excitation of CO by an infrared laser does not have a strong effect on the resonance positions and widths. Consequently, the desorption of excited CO from NaCl(100) surfaces is a rotational predissociation phenomena [9]. In section 4 the application of the complex coordinate method to the study of resonances of atoms at a stepped surface is described. The $He/Cu(100)$ resonances are studied as an example. It is shown that there is a critical value of the corrugation amplitude, below which the trapping process is dominated by the coupling between adjacent diffraction channels and ***true bound localized states can be obtained as the corrugation strength parameter is increased***[10-11]. In section 5 a novel concept of complex quasi probability, $Q(x)$ for atoms trapped on surfaces is discribed. The real part of $Q(x)$ is the ***probability to observe the adsorbed atom at a certain point*** x on the corrugated surface. The quantity $ImQ(x)/|Q(x)-1|$ is suggested as a ***relative probability for desorption*** from a certain surface position x. For He/Cu(117) it was found that, most likely, the desorption process occurs around the middle of the terraces, where the probability density is small[12].

2. The Complex Coordinate Method - A Brief Review

The positions and widths (i.e. inverse lifetimes) of metastable states (selective adsorption or predesorption states in our studied case) are associated with the complex eigenvalues of the Schrödinger equation, such that:

$$\hat{H}\psi = (E_r - i\Gamma/2)\psi \tag{1}$$

where, $\psi(\vec{r}, E) \to \exp(i\sqrt{2E_{res}}r)$ as $r \to \infty$ if V(potential) $\to 0$ when $r \to \infty$ and $E_{res} = E = E_r$(position) $- i\Gamma/2$(width) [2]. Since $\psi(\vec{r}, E)$ approaches ∞ as $r \to \infty$ it does not belong to the Hermitian domain of $\hat{H}(\vec{r})$. This divergence of the eigenfunction is a necessary condition to have a non Hermitian Hamiltonian which supports complex eigenvalues. Resulting from the divergence of $|\psi_{res}|^2$ as $r \to \infty$, the number of particles in the coordinate space is not conserved at any given time.

This non physical behavior can be "corrected" by carrying out the following similarty transformation [13,14]

$$(\hat{S}\hat{H}\hat{S}^{-1})(\hat{S}\psi_{res}) = (E_r - i\Gamma/2)(\hat{S}\psi_{res}) \tag{3}$$

such that

$$\phi \equiv \hat{S}\psi_{res} \to 0 \text{ as } r \to \infty \tag{4}$$

and ϕ is a *square integrable function* (i.e. $\phi \in \mathcal{L}^2$). Since $\hat{S}$ changes the asymptotical (divergence) behavior of the resonance eigenfunction it can *not* be a unitary operator and $\hat{S}^{-1} \neq \hat{S}^{\dagger}$. There are several different possibilities for $\hat{S}$. In the complex coordinate method, as proposed by Balslev, Combes and Simon[15,16], $\hat{S}$ is a scaling operator such that

$$\hat{S}\psi(r) = \psi(e^{i\theta}r) \equiv \psi_\theta \tag{5}$$

and if θ is large enough $\psi_\theta \to 0$ as $r \to \infty$. Since the role of $\hat{S}$ is to "correct" the asymptotic behavior of ψ_{res}, Simon suggested to define as a piece-wise operator,

$$\hat{S}\psi(\vec{r}) = \psi(r) \text{ if } r \leq r_0 \tag{6}$$

$$\hat{S}\psi(\vec{r}) = \psi\{(r - r_0)e^{i\theta}\} \quad r > r_0$$

This procedure is known as Simon's exterior scaling method[17], which avoids the intrinsic non-analyticities of the potential by passing on the right-hand-side of the singularities points of the potential. Recently, with Hirschfelder, we proposed a new set of similarity transformations which enables one to select the "optimal" path in the complex coordinate plane which avoids the non-analyticities of the potential and provides the most stable resonance solution which can be obtained by a given number of basis functions[14].

The properties of the complex scaled Hamiltonian were studied in collaboration with Certain and Weinhold[18]. We derived the complex analog theorems to the variational, virial[19], hypervirial and Hellman-Feynman theorems which provide formal justification to the use of computational techniques developed for bound states. The resonances are associated with stationary solutions in the complex variational space for which

$$\frac{\partial E_r}{\partial \theta} = 0 \quad ; \quad \frac{\partial \Gamma}{\partial \theta} = 0 \tag{7}$$

and are represented by cusps in the θ trajectory calculations as it was proved by Friedland, Certain and myself.[20] More recently, Davidson, Engdahl and myself in a continuation of previous work of Froelich, Davidson and Brändas[21], derived new upper and lower bounds to the complex resonance eigenvalves[22].

The complex coordinate method has been successfuly applied to atomic autoionization resonances by Doolen and Nuttal[23]; to atoms in external field by Reinhardt[24]; to molecular autoionization resonances by Corcoran and myself[25]; to Feshbach predissociation resonances by Lefebvre[26] and by Moiseyev[27]; to positronium -hydrogen scattering by Ho; to Van der Waals complexes by Chu[28] and by Simons and Bacic[29], and to gas-surface scattering resonances which is the subject of this lecture.

3. Rotationally Mediated Selective Adsorption

These types of resonances are obtained resulting from the energy transfer from the excited rotational mode of the adsorbed molecule to the molecule-surface vibrational mode. The potential interaction of diatomic molecule with the solid surface is a function of two variables ($z-$the distance between the diatomic center of mass and the surface and $\alpha-$the angle between the molecular axis and a vector perpendicular to the surface plane) only if the following approximations are applicable:

(1) The interaction of the diatomic molecule with the surface phonons is negligible.

(2) The energy transfer from the motion of the adsorbed molecule in the $x-y$ plane (parallel to the surface plane) to the $z-$direction (the coordinate perpendicular to the surface plane) is very small and negligible.

(3) The precession of the diatomic around the normal to the surface is free (when the angle α between the molecular axis and the axis perpendicular to the surface is held fixed).

The condition (1) may be satisfied at sufficiently low surface temperatures. The two other conditions are satisfied in the two extreme cases:

Case 1: Interaction of a diatomic in its ground state (i.e. treated as a rigid rotor) with a flat surface (for example HD/Ag(111) interaction).

Case 2: A clearly preferable adsorption site on the surface which prevents the molecule from moving in the $x - y$ plane and where the potential interaction of the adsorbed diatomic molecule with the solid weakly affected by the precession of the diatomic along the normal to the surface (for example predesorption of CO from NaCl(100)).

Under the above approximations and within the Born-Oppenheimer type approximation (i.e. the potential depends parametrically on l-the diatomic bond length) the Hamiltonian is given by

$$\hat{H} = \frac{\hat{P}_z^2}{2\mu} + B_{rot}(l)\hat{j}^2(\alpha) + V(z, \alpha; l), \tag{8}$$

where B_{rot} is the rotational constant of the diatomic and j is the rotational angular momentum. The fact that the potential $V(z, \alpha; l)$ does not depend directly on the surface structure leads to the conservation of the magnetic quantum number, m, during the desorption process[8]. In such a case V can be expanded in terms of the Legendre polynomial series and the application of the complex coordinate method (CCM) is hence *very* similar to the application of the CCM to van der Waals rotational predissociation resonances [28,29,30].

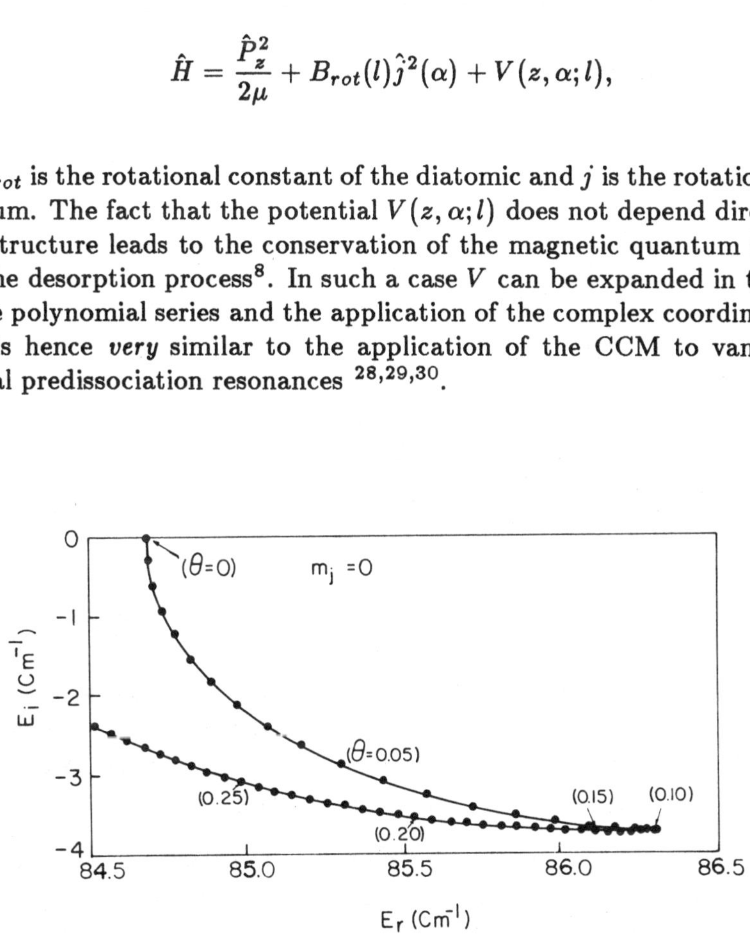

Figure 1

θ-trajectory calculations for surface scattering $HD - Ag(111)$ resonances.

In other words, the z-coordinate should be complex scaled in order to make the resonance eigenfunction square integrable as described in Section 2. Consequently, *after scaling z* by a complex factor, the *desorption states* can be expanded in terms of *square integrable* basis functions

$$\psi_{\text{desorption}} = \sum_{ij} C_{ij}(l)\phi_i(z)P_j(\cos\alpha), \tag{9}$$

where $\{P_j\}$ are Legendre polynomials, $\{\phi_i\}$ are a set of square integrable basis functions (harmonic oscillators in our case) and $\{C_{ij}\}$ are "variationally" obtained by solving the suitable complex secular equations (for the proof of the complex variational principle see Ref. 18:

$$\mathbf{H}_\theta \vec{C}_n = (E_r - iE_i)_n \vec{C}_n \tag{10}$$

where the matrix elements

$$\begin{aligned}(H_\theta)_{n,m} = (H_\theta)_{m,n} = \langle n = (i,j)|\hat{H}(ze^{i\theta},\alpha;l)|m = (i',j')\rangle = \\ = \langle \phi_i P_j|\hat{H}(ze^{i\theta},\alpha;l)|\phi_{i'}P_{j'}\rangle\end{aligned} \tag{11}$$

are given in Eqs(8-9) in Ref(8). The resonances are obtained for the optimal value of θ for which Eqs(7) are satisfied. An illustrative result of the θ-trajectory calculations is presented in Fig. 1 for surface scattering $HD/Ag(111)$ resonances and shows clearly a cusp in the $E_r - E_i$ plane for which Eqs(7) are satisfied. The resonance positions and widths obtaind from the $\theta-$trajectory calculations by Gerber, Elber, Maniv and myself[8] are summerized in Table I.

The population probabilities of the corresponding j states were found by projecting the resonance wave function obtained from Eq(10) onto the eigenfunctions of the $\hat{H}^{(0)}$ hamiltonian in which the coupling between different j states is neglected (i.e. $\langle \phi_i P_j|\hat{H}(ze^{i\theta},\alpha;l)|\phi_{i'}P_{j'}\rangle$ are taking as zeroes for $j \neq j'$).

As one can see from Table I the nine lowest resonance functions of $HD/Ag(111)$ are dominated by a single valve of j. *Therefore HD rotates almost freely above the $Ag(111)$ surface* and even for the interaction of light molecules such as hydrogen with the surface (where the quantum effects are expected to be large) the *resonance lifetime can be estimated from classical trajectory calculations.* See classical results presented in Ref(8).

Another case which was studied recently in collaboration with Ben-Ephraim, Folman and Heidberg[9] is the predesorption states of CO from Sodium Chloride. Infrared spectra, and spectral shifts of CO adsorbed and Sodium Salts were measured about twenty years ago[31]. It has been shown that the preferable adsorption

Table I

The rotational predissociation surface scattering resonanc positions, E_r, and widths, $-2E_i$, obtained form a basis set constructed of 50 harmonic oscillators. θ is the optimal rotation angle and $|C_j|^2, j = 0,1,2$ are population probabilities of the corresponding j state.

θ rad	E_r cm^{-1}	$-E_i$ cm^{-1}	$\|C\|^2_{j=0}$	$\|C\|^2_{j=1}$	$\|C\|^2_{j=2}$
0.140	36.669	6.442	0.26	**0.64**	0.10
0.235	53.843	1.219	0.03	**0.59**	0.38
0.230	70.423	6.443	0.06	**0.91**	0.03
0.105	86.314	3.728	0.08	**0.90**	0.02
0.075	123.240	4.274	0.09	0.29	**0.62**
0.235	183.171	3.209	0.03	0.04	**0.93**
0.215	227.122	2.603	0.02	0.02	**0.96**
0.020	254.981	2.987	0.20	0.12	**0.68**
0.230	271.040	3.271	0.01	0.01	**0.98**

site on the surface is above Na^+ and the approximations [2,3] mentioned above are applicable. Under these approximations the predesorption $CO/NaCl(100)$ positions and widths were calculated when the CO bond length was varied adiabaticaly. The CO/NaCl potential surface is described as a sum of dispersion, and electrostatic interactions (the detailed potential surface is given in Ref (9)).

In Fig. 2 we show the positions and widths of several low resonances (ordering by their positions) for different values of l_{co} (denoted by $\bullet, \circ, \triangle, \triangle$, and $\square$) (at $CO(v=1)$ l_{co} is changed only from 1.08 to 1.17). As one can see from Fig. 2 the $CO/NaCl$ predesorption positions and lifetimes (inverse lifetimes) are not much affected by the the value of l_{co}, and desorption of $CO(v=1)$ from $NaCl(100)$ surfaces is mainly a *rotational predissociation phenomena inspite of the fact that the adsorbed CO is vibrationally excited by an infrared laser.*

4. Trapping of Atoms at a Corrugated Surface

So far the complex coordinate method has been applied to model Hamiltonians for which the complex scaled resonance functions are completely localized in all directions. The case of atomic beam scattering from a corrugated surface is a

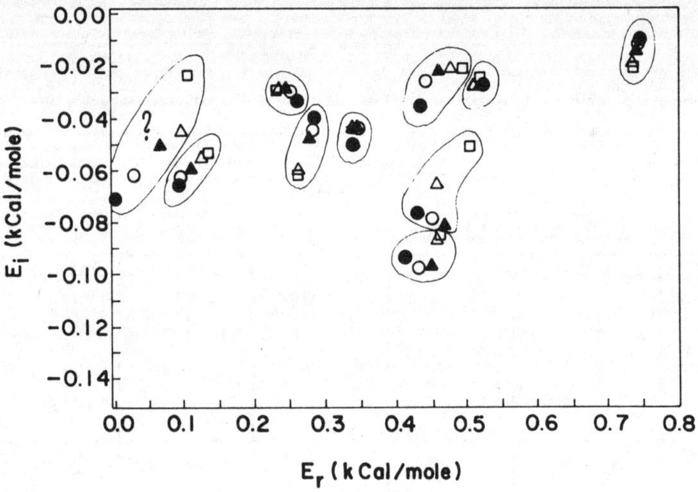

Figure 2

Predesorption positions positions and widths of CO above $NaCl(100)$ when the CO bond length is varied adiabatically. The question mark stands for unstable numerical solutions. The solutions which are enclosed by solid lines refer to the same predesorption states.

different one since the resonance functions are still extended in the $(x - y)$ plane (parallel to the surface plane) even after the scaling of the z-coordinate (perpendicular to the surface plane) by a complex factor. In collaboration with Maniv and Engdhal [10] we carried out the first successful application of the CCM to this special case. We considered a gas atom interaction with a statically one dimensional periodic solid surface. In such a case the atom-surface potential V satisfies the following condition,

$$V(x + a, z) = v(x, z) \tag{12}$$

where a is the lattice constant. Using the Bloch theorem and by expanding the eigenfunction of the Schrödinger equation in a Fourier series one can get that

$$\psi(x, ze^{i\theta}) = e^{ik_x x} \sum_{n=-\infty}^{+\infty} \phi_n(z) \exp(\frac{2n\pi ix}{a}) \tag{13}$$

where $k_x = k_o \sin\gamma$ is the component of the initial momentum in the $x-$direction, γ is the angle between the incoming atomic beam and the normal to the surface and $\phi_n(z)$ are square integrable functions (resulting from the complex scaling) which are associated with the $n-$th diffraction channel. Unlike the previous cases studied, here $\phi_n(n)$ is complex even when $\theta = 0$ resulting from the use of complex basis functions e.g. $\exp(2\pi nix/a)$) as a basis set. Therefore, the inner product of two complex eigenfunctions $(\psi_v, \psi_{v'}) = \delta'_{v,v}$ can not be taken as $\int \psi_v \psi_{v'} d\tau$ (which is the case when a real basis set is used) and not as the conventional scalar product $\int \psi^*_v \psi_{v'} d\tau$. (Since ψ_v and $\psi_{v'}$ are also complex resulting from the analytical continuation of the z coordinate into the complex plane). The inner product of two eigenfunctions of the complex scaled Hamiltonian are considered as

$$(\psi_v, \psi_{v'}) = \int \psi^L_v \psi_{v'} d\tau = \delta_{v,v'} \tag{14}$$

when

$$\begin{cases} \psi_{v'} = \sum_j c_{j,v'} \chi_j \\ \psi_v = \sum_j d_{j,v} \chi_j \end{cases}, \tag{15}$$

where $\{\chi_j\}$ are the basis functions (for example in our case $\chi_j = \phi_j(z) \exp(2j\pi ix)/a$) and $\{\phi_j\}$ are harmonic oscillator basis functions). The respectively linear parameters $\vec{c}$ and $\vec{d}$ are the *right* and the *left* eigenvectors of the Hamiltonian matrix,

$$H_{j'j}(\theta) = \langle \chi_{j'} | \hat{H}(x, ze^{i\theta}) | \chi_j \rangle,$$

such that

$$\mathbf{H}_\theta \vec{C}_{v'} = E_{v'} \vec{C}_{v'} \tag{16}$$

and,

$$\vec{d}^T_j \mathbf{H}_\theta = \tilde{E}_v \vec{d}^T_j \Rightarrow \mathbf{H}^T_\theta \vec{d}_j = \tilde{E}_v \vec{d}. \tag{17}$$

Since the determinant of a matrix is equal to that of its transpose, the eigenvalues of H_θ are the same as those of H^T_θ and $E_v = \tilde{E}_v$. In the case of complex scaled Hamiltonians, $\mathbf{H^T}(\theta) = \mathbf{H}^*(-\theta)$, and therefore $\vec{d}^*_v$ is an eigenvector of $\mathbf{H}(-\theta)$. The definition of thc inner product is necessary when one wishes to calculate complex expectation values of given dynamical operators. As an illustration example we studied the $He - Cu(110)$ interaction. The model potential is a corrugated Morse potential used by Armand and Manson [32],

$$V(x, z) = D[\frac{1}{v_0} e^{-2b(z-\phi(x))} - 2e^{-bz}], \tag{18}$$

where

$$\phi(x) = ha \cos(2\pi x/a) \tag{19}$$

$$v_0 = \{\text{the surface average of } \exp(2b\phi(x))\} = a^{-1} \int_{-\frac{a}{2}}^{\frac{a}{2}} \exp[2b\phi(x)] \tag{20}$$

and h is a dimensionless corrugation parameter. For the other parameter see Ref(10,32). The resonance eigenvalue $E_{res}(\gamma)$ is dependent on the incident angle γ through $kx = x_0 \sin\gamma$. For a given incident energy $E_0 = k_0^2$ the specular intensity is dominated by the pole factor I_{pole} of the S-matrix which is given by

$$I_{pole} = 1/|k_0^2 - E_{res}(\gamma)|^2. \tag{21}$$

In our numerical application we studied the resonance which is associated with the ground state of channel $n = 1$ embedded in the continuum of channel $n = 0$ (the channels are defined in Eq(13)). The resonance positions and widths, $E_{res} = E_r - i/2\Gamma$, as function of γ were calculated by the CCM. The scattering cross section is proportional to I_{pole} which is shown in Fig.3 and is in a good agreement with previous theoretical results[32].

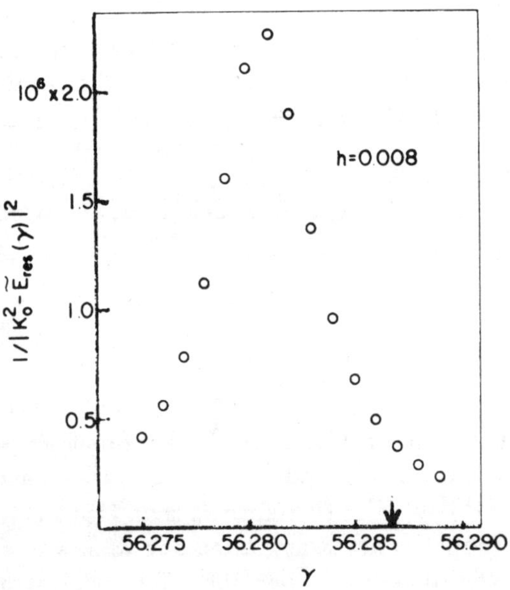

Figure 3

The S-matrix pole factor $1/|k_0^2 - E_{res}(\gamma)|^2$ as a function of incident angle for the system $He/Cu(110)$. $k_0 = 6.34\text{Å}^{-1}(E_0 = 21mev)$.

By varying the corrugation strength parameter h we obtained that below a critical value of $h \approx 0.05$, the trapping process is dominated by the coupling between adjacent diffraction channels. As h becomes larger than 0.05 the "bound" state of channel $n = 1$(the effective channel potentials are given by, $V(x = a/2, z) + (k_x + 2\pi n/a)^2$) gets below the threshold of the $n = 0$ channel. However,

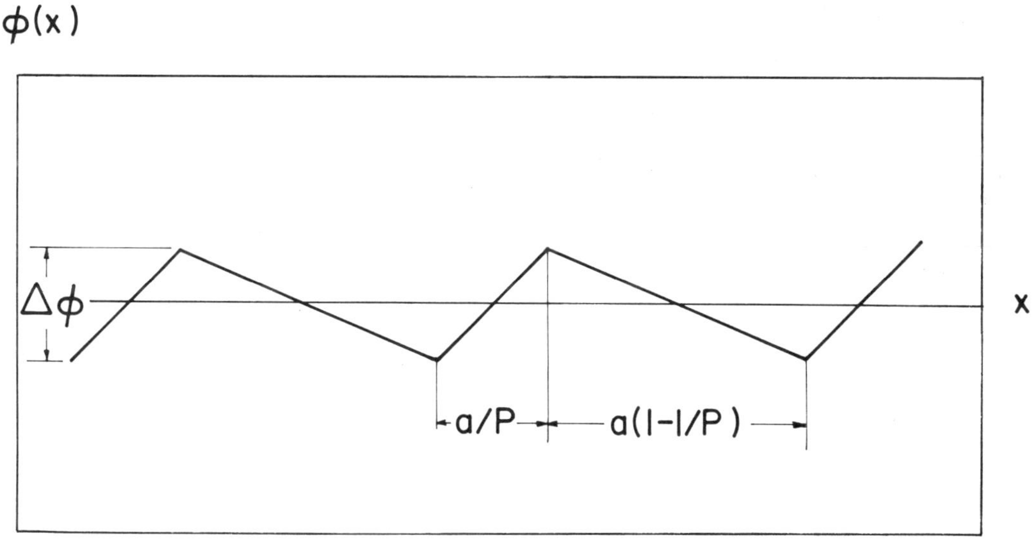

Figure 4

Periodic stepped structure surface, with a projected step length a/p, a terrace length $a(1-\frac{1}{p})$ along the x direction where a is the lattice constants.

it decays in time since it is still embedded in the continuum of the $n=-1$, $n=-2$, channels. Of course, if the corrugation amplitude h is sufficiently high the resonance state will become more localized and will have a longer life time on the surface as h is increased. This is the reverse of the observation at small corrugation where the lifetime of the selectively adsorbed atom on the surface becomes shorter as h is increased. Note that this result (i.e. larger $h \rightarrow$ larger width) is expected on the basis of the "zero order" approximation since the coupling between the two channels mixes the bound state of channel $n=1$ with the continuum of channel $n=0$ and by increasing h the coupling between the bound state and the continuum is increased too. With Engdahl[11], we have shown on the basis of infinite order perturbation theory that unlike the "new" molecular bound states which result from corrections to the Born Oppenheimer approximation, the existence of the new type of atom-surface bound state is not revealed by effective one dimensional potential surface.

5. Complex Quasi-Probability for Atoms Trapped on Stepped Surfaces

As discussed in Section 4 the inner product in the case of complex scaled Hamiltonians is defined by (ψ^L, ψ) where ψ^L is the eigenfunction of $\hat{H}^*(x, ze^{-i\theta})$

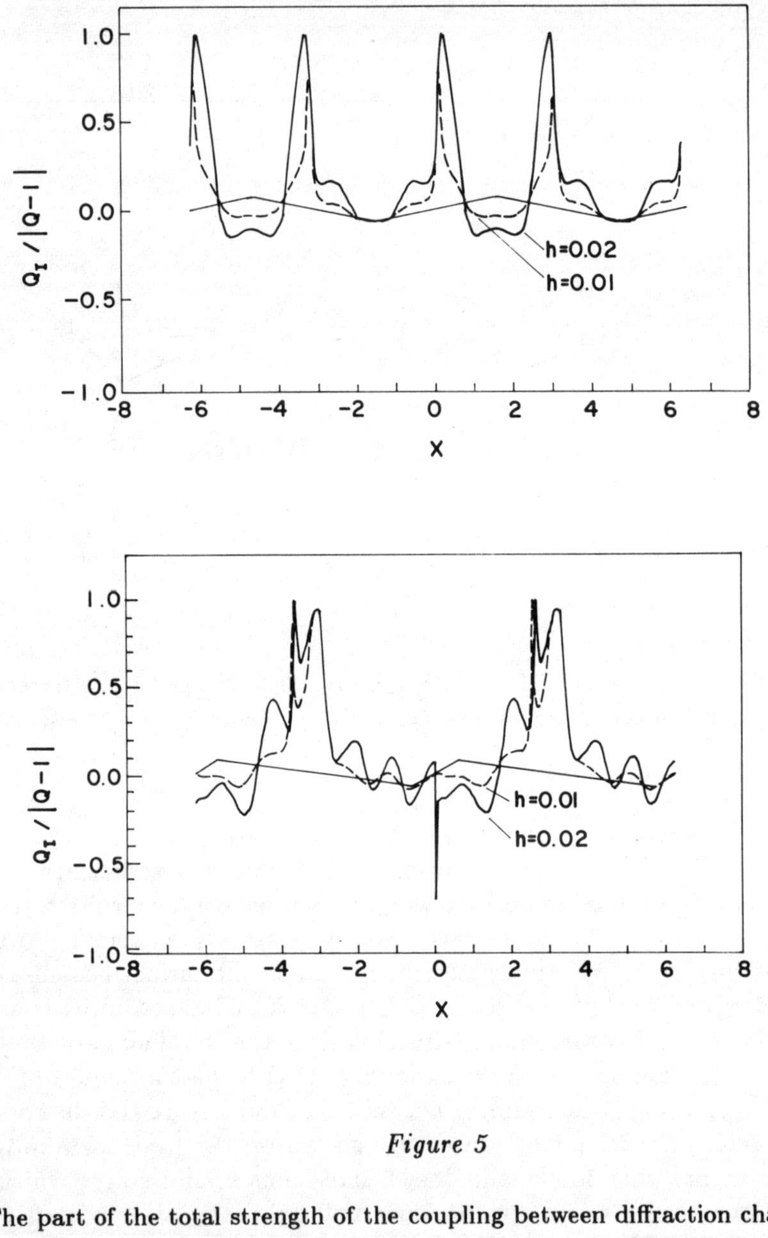

Figure 5

The part of the total strength of the coupling between diffraction channels which is associated with the instability of the trapped state as mesured locally by $Q_I(x)|Q(x)-1|$ for = 0.3 and Eq(3)=86.696°. (a) For Cu(110) surface (a=3.16Å and p=2 in Fig 3) (b) For Cu(117) surface (a=9.114Å and p=5.667 in Fig. 3)

and ψ is the eigenfunction of $\hat{H}(x, ze^{i\theta})$. Therefore, (ψ^L, ψ) gets complex values and can *not* be interpeted as the density probability. In collaboration with Engdahl

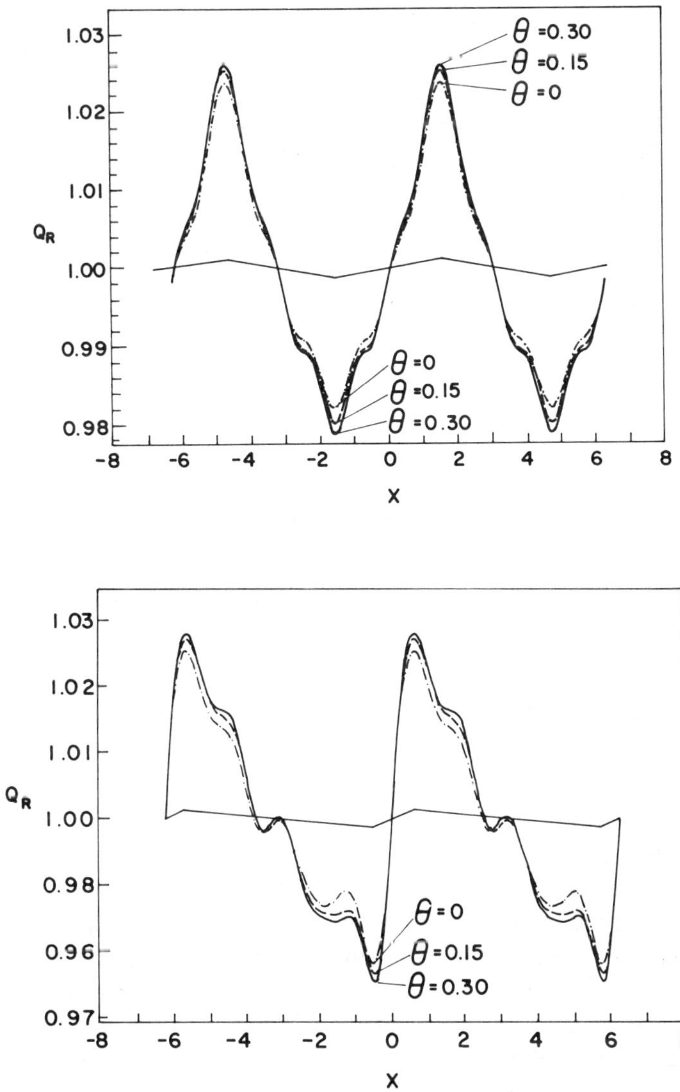

Figure 6

The real part of the complex quasi probability $Q(x)$ obtained for different values of the rotations angle θ. The value for the corrugation strength parameter is $h = 0.01$. **a)** For $Cu(110)$ surface. **b)** For $Cu(117)$ surface.

and Maniv[12] we proposed that when θ is taken as the smallest angle which is required to make ψ (or ψ_L) square integrable functions then the *real part of*

$$Q(x) = \int_{-\infty}^{+\infty} \psi^L \psi dz \tag{22}$$

is the probability of observing the adsorbed atom at a certain point x on the corrugated surface. The imaginary part of $Q(x)$ reflects the local instability of the trapped state on the surface and the quantity

$$\frac{\Im Q(x)}{|Q(x) - \langle Q \rangle|} \tag{23}$$

is loosely interpreted as a *relative probability for desorption from* a certain surface position x.

When an atom interacts with a stepped solid surface (see Fig. 4) then the complex quasi probability which is defined in Eq(22) is given by,

$$Q(x) = \sum_{\mu=-\infty}^{+\infty} q_\mu \exp(\frac{2\mu\pi i x}{a}) \tag{24}$$

where,

$$q_\mu = \sum_{n=-\infty}^{+\infty} \int_{-\infty}^{+\infty} \phi_n^L(z)\phi_n(z)dz \tag{25}$$

and $\{\phi\}$ are defined in Eq (13). In the absence of coupling between different scattering channels there is only one term, ϕ_n, in the series expansion of ψ presented in Eq (13) and therefore there is also only one term $\mu = 0$, in Eq(24). Consequently, for non corrugated surfaces (i.e. no coupling between different scattering channels) $Q(x) = 1$ for *any* value of x. The conclusion is clear:$Q(x) \neq 1$ as a result of the coupling between different channels and $|Q(x) - \langle Q \rangle|$ where $\langle Q(x) \rangle = 1$, is a local measure to the *total effective strength of the coupling between diffraction channels* $Q_I(x) \neq 0$, due to the coupling between different scattering channels and because of the rotation of the coordinate z into to the complex plane by θ (in order to make ψ a square integrable function). If θ is taken as the minimal angle required to make ψ square integrable then $\theta_{min} = 0$ for bound states and $Q_I(x) \neq 0$ only for *resonance* states and the *nonvanishing values of $|Q_I(x)|$ on the surface thus reflect the local instability of the surface resonance state in question and is related to the desorption process.*

Therefore, the ratio given in Eq(23) represent the *part* of the total effective strength of the coupling between diffraction channels which is "responsible" for the desorption process. Figs. 5 show this quantity for He trapped on $Cu(110)$ and on $Cu(117)$. It deviates significantly from zero around the middle of the terraces, where the probability density is small (see $Q(x)$ for $\theta = 0$ in Fig. 6). From Figs. 5 and 6 ($\theta = 0$) one can see that by increasing the corrugation amplitude the desorption regions are extended beyond the centers of the terraces, where the probabilty density is not small. In conclusion, we have shown that a complex quasi-probability density, obtained from *time-independent* calculations, provides valuable information about (time dependent) adsorption-desorption processes occurring on stepped solid surfaces.

References

1. H. Hoinnkes, Rev. Mod. Phys., **52**, 993 (1980).

2. V. Celli, Prog. Astronaut., 74, 50(1981); P. O. Goodman, ibid., **74**, (1981).

3. T. Engel and K. H. Rieder, Structural Studies of Surfaces. (Springer Berlin, 1982).

4. R. B. Gerber, Chem. Rev., **87**, 29 (1987) and references therin.

5. W. P. Reinhardt, Annu. Rev. Phys. Chem., **33**, 223 (1982).

6. B. K. Junker, Adv.At.Mol.Phys., **18**, 207 (1982).

7. N. Moiseyev, in lecture Notes in Physics, edited by S. Albeverio, L. S. Ferreira, and L. Streit (Springer, Berlin, 1984), Vol. **211**, p.235.

8. N. Moiseyev, T. Maniv, R. Elber, and R. B. Gerber, Mol.Phys., **55**, 1369 (1986).

9. A. Ben-Ephraim, M. Folman, J. Heidberg and N. Moiseyev, to be published

10. T. Maniv, E. Engdahl, and N. Moiseyev, J. Chem.Phys., **86**, 1048 (1987).

11. E. Engdahl and N. Moiseyev, Submitted for publication.

12. E. Engdahl, T. Maniv, and N. Moiseyev, Submitted for publication.

13. P.O. Löwdin, Florida Technical Note No. 803.

14. N. Moiseyev and J. O. Hirschfelder, J. Chem. Phys., **88**, 1063 (1988).

15. E. Balslev and J. M. Combes, Commun Math.Phys., **22**, 280 (1971).

16. B. Simon, Commun. Math. Phys., 27, 1 (1972), Ann. Math., **97**, 247 (1973).

17. B. Simon, Phys.Lett., **71A**, 211 (1979).

18. N. Moiseyev, P. R. Certain, and F. Weinhold, Mol.Phys., **36**, 1613 (1978).

19. E. Brändas and P. Froelich, Phys. Rev., **A16**, 2207 (1977).

20. N. Moiseyev, S. Friedland and P. R. Certain, J. Chem.Phys., **74**, 739 (1981).

21. P. Froelich, E. Davidson, and E. Brändas, Phys. Rev., **A28**, 2641 (1983).

22. E. R. Davidson, E. Engdahl, and N. Moiseyev, Phys. Rev., **A33**, 2436 (1986).

23. J. Nuttall, Bull. Am. Phys. Soc., **17**, 598 (1972); G. D. Doolen, M. Hidalgo, J. Nuttall and R. Stagat, in *Atomic Physics* (Eds. S. J. Smith and G. K. Walters, Plenum Press, New York, 1973), Vol. 3, see also Int. J. Quantum Chem. **14**, (1978).

24. W. P. Reinhardt, Int. J. Quantum Chem., **S10**, 359 (1976).

25. N. Moiseyev and C. T. Corcoran, Phys.Rev., **A20**, 814 (1979).

26. R. Lefebvre, Chem. Phys. Lett., **70**, 430 (1980).

27. N. Moiseyev, Int. J. Quantum Chem., **29**, 835 (1981).

28. S. I. Chu, J. Chem. Phys., **72**, 4772 (1980).

29. Z. Bacic and J. Simons, Int. J. Quantum Chem., **21**, 727 (1982).

30. N. Moiseyev and P. R. Certain, J. Phys. Chem., 89, 3853 (1985).

31. R. Gevirzman, Y. Kozirovski and M. Folman, Trans. Faraday Soc., 2206 (1969).

32. G. Armand and J. R.Manson, Surf. Sci., **119**, L299 (1982).

ON A THEOREM FOR COMPLEX SYMMETRIC MATRICES AND ITS RELEVANCE IN THE STUDY OF DECAY PHENOMENA

Charles E. Reid

Quantum Theory Project
Williamson Hall
University of Florida
Gainesville, Florida 32611
U.S.A

Erkki Brändas

Quantum Chemistry Group
Uppsala University
Box 518
S-751 20 Uppsala
SWEDEN

Abstract

The theorem that every square matrix is similar to a symmetric matrix is discussed and its relevance in the quantum mechanical formulation of decay phenomena is emphasized. An alternative proof of the theorem is derived. It is shown that the matrix $\mathrm{C}_n(0)$ of order n with elements of the first superdiagonal being 1 and all the remaining ones zero, is similar to $\mathbf{q}$ with the elements

$$q_{kl} = \exp(i\pi(k+l-2)/n)[\delta_{kl} - 1/n].$$

The possible occurence of the latter in connection with off-diagonal-longe-range-order (ODLRO) is indicated.

1. Introduction

In the quantum mechanical formulation of decay phenomena the complex scaling theory of Balslev and Combes[1,2] has led to new applications and theoretical development. In contrast to the standard description of time evolution generated by a self adjoint hamiltonian operator, the complex scaled hamiltonian occuring in the Fourier-Laplace transformed equation is no longer self adjoint or even normal.

To incorporate a generalized description into the well developed machinery of basis set selections and solutions of appropriate secular equations, a generalization of the traditional scalar product to the complex plane is necessary. For some recent accounts of complex scaling applications see Reinhardt[3] or Junker[4]. A simple extension of the scalar product in one dimension becomes[5] (requiring the integrand to be analytic)

$$\langle f|g\rangle_{C^\pm} = \int_{C^\pm} f^*(z^*)g(z)\,dz \tag{1}$$

where $C^\pm$ are complex paths in the upper or lower half-planes. One realization (for the radial scalar product) would be $z = \eta x$ where η is a complex scale factor and the integration interval is the positive real axis, $\mathcal{R}_+$. If f and g belong to the domain of complex rotations, i.e. the set $\mathcal{N}$ of analytic vectors associated with the dilation group[6] then one gets

$$\langle f|g\rangle_\eta = \int_{\eta\mathcal{R}_+} f^*(z^*)g(z)\,dz = \int_{\mathcal{R}_+} f^*(x)g(x)\,dx =$$

$$= \langle f|g\rangle = \int_{\mathcal{R}_+} [(\eta^*)^{\frac{1}{2}} f(\eta^* x)]^* \eta^{\frac{1}{2}} g(\eta x)\,d(\eta x) =$$

$$= \langle U(\eta^*)f|U(\eta)g\rangle = \langle f|U^\dagger(\eta^*)U(\eta)g\rangle \tag{2}$$

introducing the unbounded similitude $U(\eta)$ with domain $\mathcal{N}$ as

$$U(\eta)\phi(x) = \eta^{3N/2}\phi(\eta x) = \phi(\eta) \tag{3}$$

where 3N now are the number of degrees of freedom. We obtain

for all $\phi,\ \psi \in \mathcal{N}$

$$\langle\phi|\psi\rangle = \langle\phi|U^\dagger(\eta^*)U(\eta)\psi\rangle = \langle\phi(\eta^*)|\psi(\eta)\rangle \tag{4}$$

with $U(\eta)$ obeying the "star unitary" property

$$U^\dagger(\eta^*) = U^{-1}(\eta). \tag{5}$$

Similarly for a dilation analytic operator[1,2] H, one obtains

$$\langle\phi|H|\psi\rangle = \langle\phi|U^{\dagger}(\eta^*)U(\eta)HU^{-1}(\eta)U(\eta)\psi\rangle =$$

$$= \langle\phi(\eta^*)|H(\eta)\psi(\eta)\rangle \qquad (6)$$

with $H(\eta) = U(\eta)HU^{-1}(\eta)$.

The construction (3-6) leads to a matrix representation

$$\langle\phi_i(\eta^*)|H(\eta)\psi_j(\eta)\rangle = H_{ij}(\eta) \qquad (7)$$

which for $\phi_i^*(\eta^*) = \overline{\phi_i(\eta)} = \psi_i(\eta)$ implies that

$$H_{ij}(\eta) = H_{ji}(\eta) \qquad (8)$$

i.e. that the matrix H is *complex symmetric.* Such a construction employing a real basis was introduced earlier in connection with the derivation of a bivariational principle and the associated virial theorem[7].

Note that the C-product introduced by Moiseyev et. al.[8] also leads to a complex symmetric construction. However, rather than introducing a new bi-linear form, we have preferred to stay with the conventional sesquilinear form, since the latter involves the appropriate vector space and its dual.

Since complex symmetric constructions permeates the majority of complex scaling calculations (see e.g. Mishra et. al.[9] for some recent dilated propagator applications), it is obviously necessary to know that complex symmetric matrix representations do not invoke any constraints on the secular problem and associated projection operators referring to a given vector and its dual.

The theorem is seldom mentioned although it is more or less implicit in linear algebra treatments, see e.g. Löwdin[11]. For a more direct proof see Gantmacher[12], where it is explicitly shown that every square matrix is similar to a symmetric matrix and that every (complex) symmetric matrix is (complex) orthogonally similar to an explicitly given normal form.

For the benefit of the reader we give an alternative proof of the theorem. Although our proof to some extent parallels the Gantmacher construction, the explicit symmetric form is quite different. The connection between the latter and off-diagonal-long-range-order (ODLRO) in the sense of Yang[13] is briefly discussed as the main motivation for the present work.

2. Proof of the Theorem

Theorem: *Any matrix can be brought to complex symmetric form by a similarity transformation.*

Since any matrix is similar to a block-diagonal matrix in which the submatrices along the main diagonal are Jordan blocks[12], it suffices to show that any Jordan block can be transformed into a complex symmetric matrix. Moreover, since

$$\mathbf{C}_n(\lambda) = \begin{pmatrix} \lambda & 1 & 0 & 0 & \dots & \dots & 0 & 0 \\ 0 & \lambda & 1 & 0 & \dots & \dots & 0 & 0 \\ 0 & 0 & \lambda & 1 & \dots & \dots & 0 & 0 \\ 0 & 0 & 0 & \lambda & \dots & \dots & 0 & 0 \\ \vdots & \vdots & \vdots & \vdots & \ddots & & \vdots & \vdots \\ \vdots & \vdots & \vdots & \vdots & & \ddots & \vdots & \vdots \\ 0 & 0 & 0 & 0 & \dots & \dots & \lambda & 1 \\ 0 & 0 & 0 & 0 & \dots & \dots & 0 & \lambda \end{pmatrix}$$

can be written as $\lambda\mathbf{1} + \mathbf{C}_n(0)$, we may deal only with matrices of the form $\mathbf{C}_n(0)$. Given a matrix $\mathbf{A}$, we need to show that there is a matrix $\mathbf{B}$ such that $\mathbf{B}^{-1}\mathbf{A}\mathbf{B}$ is a symmetric matrix. This means that

$$\mathbf{B}^{-1}\mathbf{A}\mathbf{B} = (\mathbf{B}^{-1}\mathbf{A}\mathbf{B})^{\mathrm{T}} = \mathbf{B}^{\mathrm{T}}\mathbf{A}^{\mathrm{T}}(\mathbf{B}^{-1})^{\mathrm{T}} = \mathbf{B}^{\mathrm{T}}\mathbf{A}^{\mathrm{T}}(\mathbf{B}^{\mathrm{T}})^{-1}. \tag{9}$$

Multiplying this equation from the left by $\mathbf{B}$ and from the right by $\mathbf{B}^{\mathrm{T}}$ gives

$$\mathbf{A}\mathbf{B}\mathbf{B}^{\mathrm{T}} = \mathbf{B}\mathbf{B}^{\mathrm{T}}\mathbf{A}^{\mathrm{T}}.$$

Since $\mathbf{B}\mathbf{B}^{\mathrm{T}}$ is a symmetric matrix – call it $\mathbf{S}$ – we can write this as

$$\mathbf{A}\mathbf{S} = \mathbf{S}\mathbf{A}^{\mathrm{T}} = \mathbf{S}^{\mathrm{T}}\mathbf{A}^{\mathrm{T}} = (\mathbf{A}\mathbf{S})^{\mathrm{T}}$$

and so $\mathbf{A}\mathbf{S}$ is also symmetric. Thus if we can find a nonsingular symmetric matrix $\mathbf{S}$, factorizable into the form $\mathbf{B}\mathbf{B}^{\mathrm{T}}$, such that $\mathbf{A}\mathbf{S}$ is also symmetric, then the transformation (9) can be carried out.

Specializing at this point to the case where $\mathbf{A}$ is a Jordan block $\mathbf{C}_n(0)$, so that $a_{ij} = \delta_{i+1,j}$, we find

$$(AS)_{ij} = \sum_{k=1}^{n} a_{ik}s_{kj} = \sum_{k=1}^{n} \delta_{i+1,k}s_{kj} = \begin{cases} s_{i+1,j} & i < n; \\ 0 & i = n \end{cases}$$

Then, using the fact that both $\mathbf{A}\mathbf{S}$ and $\mathbf{S}$ are symmetric, we find

$$s_{i+1,j} = s_{j+1,i} = s_{i,j+1} \qquad (i \neq n \neq j) \tag{10}$$

while

$$s_{i+1,n} = 0. \tag{11}$$

Equation (10) shows that the elements of $\mathbf{S}$ depend only on the sum of the indices; that is, $\mathbf{S}$ is persymmetric. By (11) the only non-zero element in the last column of $\mathbf{S}$ is s_{1n}, and so all elements below and to the right of the secondary diagonal are zero. Those on the secondary diagonal must not vanish, since **S** must not be singular, while those above and to the left are arbitrary except for the requirement of persymmetry.

These requirements can be satisfied as follows: Let $\omega = \exp(\pi i/n)$, and

$$\mathbf{B} = \begin{pmatrix} 1 & \omega & \omega^2 & \dots & \omega^{n-1} \\ 1 & \omega^3 & \omega^6 & \dots & \omega^{3(n-1)} \\ \dots & \dots & \dots & \dots & \dots \\ 1 & \omega^{2n-1} & \omega^{2(2n-1)} & \dots & \omega^{(n-1)(2n-1)} \end{pmatrix}$$

That is, $b_{kl} = \omega^{(2k-1)(l-1)}$. Then

$$(BB^T)_{rs} = \sum_{l=1}^{n} b_{rl} b_{sl} = \sum_{l=1}^{n} \omega^{2(r+s-1)(l-1)} = \frac{1-\omega^{2n(r+s-1)}}{1-\omega^{2(r+s-1)}} \tag{12}$$

This depends only on $r+s$, leading to the required persymmetry. The numerator in equation (12) is zero for all r and s, but the denominator is zero only if $r+s = n+1$; that is, for the secondary diagonal. In this case each term of the sum is unity. Thus **S** has the form

$$\mathbf{S} = \begin{pmatrix} 0 & 0 & \dots & 0 & n \\ 0 & 0 & \dots & n & 0 \\ \dots & \dots & \dots & \dots & \dots \\ n & 0 & \dots & 0 & 0 \end{pmatrix}$$

As a specific example, for $n = 4$, $\omega = 2^{-\frac{1}{2}}(1+i)$, and $\mathbf{A} = \mathbf{C}_4(0)$

$$\mathbf{B}^{-1}\mathbf{AB} = \frac{1}{4}\begin{pmatrix} 3 & -\omega & -i & \omega^* \\ -\omega & 3i & \omega^* & 1 \\ -i & \omega^* & -3 & \omega \\ \omega^* & 1 & \omega & -3i \end{pmatrix}$$

Observation of this and other examples suggests that $\mathbf{B}^{-1}\mathbf{AB}$ is the sum of a persymmetric matrix and a diagonal matrix.

A proof of this follows:

If we denote the elements of **B** and its inverse by b_{kl} and α_{kl} respectively, and those of $\mathbf{B}^{-1}\mathbf{C}_n(0)\mathbf{B}$ by q_{kl}, we have

$$\sum_{k=1}^{n} \alpha_{jk} b_{kl} = \sum_{k=1}^{n} \alpha_{jk}\omega^{(2k-1)(l-1)} = \delta_{jl}$$

and

$$q_{jl} = \sum_{k=1}^{n-1} \alpha_{jk} b_{k+1,l} = \sum_{k=1}^{n-1} \alpha_{jk}\omega^{(2k+1)(l-1)}$$

Multiplying the first by $\omega^{2(l-1)}$ and comparing with the second gives

$$q_{jl} = \delta_{jl}\omega^{2(l-1)} - \alpha_{jn}\omega^{(2n+1)(l-1)} = \delta_{jl}\omega^{j+l-2} - \alpha_{jn}\omega^{l-1} \tag{13}$$

since $\omega^{2n} = 1$. Since **B** is the matrix associated with a Vandermonde determinant, we can use the standard expression for the inverse of this type of matrix[10] to find

$$\alpha_{jn} = \frac{(-1)^{j+n}\sigma_{n-j}(\omega,\omega^3,\ldots,\omega^{2n-3})}{\prod_{r=1}^{n-1}(\omega^{2n-1}-\omega^{2r-1})}$$

where σ_{n-j} is an elementary symmetric polynomial*. This can be evaluated by noting that its arguments, together with ω^{2n-1} $(= 1/\omega)$, are the roots of $x^n + 1 = 0$. Writing $-1/\omega^n$ for 1 and dividing out the undesired root $1/\omega$, we find

$$x^{n-1} + \frac{x^{n-2}}{\omega} + \frac{x^{n-3}}{\omega^2} + \ldots + \frac{1}{\omega^{n-1}} = 0 \tag{14}$$

Then the relation between the coefficients of an algebraic equation and the symmetric functions of its roots leads to the result that the numerator of α_{jn} is $\frac{1}{\omega^{n-j}} = -\omega^j$, since $\omega^n = -1$.

From the construction of equation (14) it is clear that the denominator in the expression for α_{jn} is the value that the left side takes when $x = \omega^{2n-1} = 1/\omega$. But these substitutions reduce each term to $-\omega$, so that the denominator is $-n\omega$, and $\alpha_{jn} = \omega^{j-1}/n$. Equation (13) then yields $q_{jl} = \omega^{j+l-2}(\delta_{jl} - 1/n)$.

The transformation with the matrix **B** as defined above is, of course, not unique, since any (complex) orthogonal transformation following this one will preserve the symmetry.

3. Conclusion

The matrix **q** constructed above looks rather "innocent" as it can be viewed as a unitary matrix with each element perturbed by $1/n$. Even if this perturbation is small (for large n) the consequences are of fundamental importance as **q** by construction corresponds to a Jordan block with Segré characteristic n.

It is interesting to note that such matrix representations may not be to unfamiliar in the area of complex deformations in general N-particle systems. For instance a large eigenvalue in the second order reduced density matrix, associated with Yang's assumption of ODLRO[13] may be analysed as follows. Consider a geminal of rank n expanded in a basis of $m = 2n$ orbitals, $(2n > N)$

$$g(1,2) = \sum_{i=1}^{n} g_i|i, i+n\rangle \tag{15}$$

* Generated by any of the expressions $\prod_{i=1}^n(x+\lambda_i) = \sum_{i=0}^n \sigma_{n-i}x^i$ respectively $\prod_{i=1}^n(1+\lambda_i t) = \sum_{i=0}^n \sigma_i t^i$ where really the symmetric functions should carry the arguments λ_i, for $i = 1\ldots n$.

where $|i, i+n\rangle$ is a (normalized) Slater determinant of the spin orbitals* i and $i+n$. In terms of the basis

$$[\{|i,i+n\rangle,\ 1 \le i \le n\},\quad \{|i,j\rangle,\ 1 \le i \le j \le 2n,\ i+n \neq j\}]$$

the second order reduced density matrix† $\Gamma^{(2)}(g^{N/2})$ assumes a very simple "box and tail" form, Coleman[14]. In the extreme case[14], the "box and tail" obtains a large eigenvalue λ_L with the remaining eigenvalues λ degenerate. The g_i in (15) become all equal to $1/\sqrt{n}$, and all $\lambda = N(N-2)/(4n(n-1))$. Thus $\Gamma^{(2)}(g^{N/2})$ writes

$$\Gamma^{(2)}(g^{N/2}) = \Gamma_L^{(2)} + \Gamma_D^{(2)} \tag{16}$$

with

$$\Gamma_L^{(2)} = |g\rangle\lambda_L\langle g|; \qquad \lambda_L = \frac{N}{2} - \frac{N(N-2)}{4n} \tag{17}$$

and

$$\Gamma_D^{(2)} = \lambda[\{\sum_{i,j} |i,i+n\rangle(\delta_{ij} - \frac{1}{n})\langle j,j+n|\} + \sum_{\substack{i<j \\ i+n\neq j}} |i,j\rangle\langle i,j|] \tag{18}.$$

Considering the part of $\Gamma_D^{(2)}$ that belongs to the "box", after application of $\exp(-\beta\hat{\mathcal{L}})$, where β is associated with the appropriate temperature of the system and bath and $\hat{\mathcal{L}}$ is the Liouville operator corresponding to the dynamics of the system exhibiting a "macroscopic eigenvalue" in the two-matrix, one obtains via the super operator formulation of Obcemea and Brändas[5] the expression (suppressing the "tail" part)

$$\exp(-\beta\hat{\mathcal{L}})\Gamma_D^{(2)} = \lambda \sum_{i,j} |i,i+n\rangle \exp(i\beta(\epsilon_i + \epsilon_j))(\delta_{ij} - \frac{1}{n})\langle j,j+n|. \tag{19}$$

In (19) we have assumed that the coherent pairs have the same energy and lifetimes $\tau_i = 2\hbar/\Gamma_i$, $\Gamma_i = 2\epsilon_i$. Assuming that the widths fulfil

$$\beta(\Gamma_i + \Gamma_j) = 2\pi(i+j-2)/n \tag{20}$$

---

* Explicitly, the geminal will have the form (for even N)

$$|g^{N/2}\rangle = g \wedge \ldots \wedge g = [S_{N/2}]^{-1} \sum_{\substack{j_r < j_{r+1} \\ 1 \le r \le N/2}} g_{j_1} \cdots g_{j_{N/2}} |\phi_{j_1}\phi_{j_1+n} \cdots \phi_{j_{N/2}}\phi_{j_{N/2}+n}\rangle$$

where $S_{N/2}$ is an appropriate normalization factor[17], given by a symmetric function of order $N/2$ with arguments $n_i = |g_i|^2$, $(i = 1 \ldots n)$.

† Please observe the difference between the second order reduced density matrices of Löwdin and Coleman:

$$\Gamma^{(2)} = \frac{N(N-1)}{2} D^2.$$

the two-matrix will exhibit a Jordan block with Segré characteristic n. For more details on applications to amorphous condensed matter, see references 16 and 17. Some additional comments on the appearence of Jordan blocks in physical problems as well as possible relations with quantum chaos are discussed in an accompanying article[17].

Acknowledgements

We are indebted to Dr. Nimrod Moiseyev for comments on Jordan Blocks and to M.Sc. Mikael Höghede for making the present TEX-file.

References

1. E. Balslev and J. Combes, Commun. Math. Phys **22**, 280 (1971).

2. B. Simon, Ann. Math. **97**, 247 (1973).

3. W. P. Reinhardt, Ann. Rev . Phys. Chem. **33**, 223 (1982).

4. B. R. Junker, Adv. Atomic Mol. Phys. **18**, 207 (1982).

5. E. J. Brändas, Proc. of Bielefeld Encounters in Phys. and Math.VI, "Resonances–Models and Phenomena", Lecture Notes in Phys., **211**, Springer Verlag, 307 (1984).

6. E. Nelson, Ann. Math, **70**, 572 (1959).

7. E. J. Brändas and P. Froelich, Phys. Rev. **A16**, 2207 (1977).

8. N. Moiseyev, P. R. Certain and F. Weinhold, Mol. Phys. **36**, 1631 (1978).

9. N. Mishra, H. A. Kurtz, O. Goscinski and Y. Öhrn, J. Chem. Phys. **79**, 1896 (1983).

10. A. C. Aitken, "Determinants and Matrices", Interscience Publ. Inc., N.Y., 118 (1951).

11. P. O. Löwdin, "Set theory and linear algebra–some mathematical tools to be used in quantum theory. Part II: Binary product spaces and their operators", Uppsala Techn. Note **509**, (1977).

12. F. R. Gantmacher, "The theory of matrices", Vol. II, Chelsea Publishing Company, New York (1959).

13. C. N. Yang, Rev. Mod. Phys. **34**, 694 (1963).

14. A. J. Coleman, Rev. Mod. Phys. **35**, 668 (1963); J. Math. Phys **6**, 1425 (1965).

15. C. H. Obcemea and E. J. Brändas, Ann. Phys. **151**, 383 (1983).

16. C. A. Chatzidimitriou-Dreismann and E. J. Brändas, Ber. Bunsenges. Phys. Chem. **92**, 549 (1988); **92**, 554 (1988).

17. E. J. Brändas and C. A. Chatzidimitriou-Dreismann, to appear in the proceedings of the symposium on "Resonances - The Unifying Route towards the Formulation of Dynamical Processes", Springer Verlag (1988).

CREATION OF LONG RANGE ORDER IN AMORPHOUS CONDENSED SYSTEMS

Erkki Brändas

Quantum Chemistry Group for Research in
Atomic, Molecular and Solid State Physics
Uppsala University
Box 518
S-751 20 Uppsala, Sweden

C. Aris Chatzidimitriou-Dreismann

Iwan N. Stranski-Institute for Physical and Theoretical Chemistry
Technical University of Berlin,
Strasse des 17 Juni 112
D-1000 Berlin 12
Federal Republic of Germany

Abstract

The second order density matrix $\Gamma^{(2)}$ of an N-body system, described by an AGP function, is considered. Certain theorems due to Yang, Sasaki and Coleman concerning the eigenvalue spectrum of $\Gamma^{(2)}$ and the possible appearance of coherent states, like e.g. BCS-states of superconductivity, are discussed in connection with Yang's concept of ODLRO ("off-diagonal long-range order"). The Complex Scaling Method (CSM) is introduced and some basic aspects of it are treated, like e.g. defining the appropriate scalar product, clarification of certain seemingly puzzling "inconsistencies" etc. Furthermore CSM is applied to the time evolution superoperator and the canonical density operator, and the connection with previous work on the mathematical foundation of Prigogine's theory of subdynamics is discussed. It is explicitly shown how the theoretical concepts and the mathematical framework are combined to demonstrate the appearance of a novel kind of organized forms, "coherent-dissipative structures", created spontaneously in condensed amorphous phases (like liquids at standard experimental conditions). The last chapter of this article is dealing with the application of the theoretical results into the physical context of molecular spectroscopic processes in condensed systems. Two experimental situations are considered in some detail: 1) far-infrared absorption in polar liquids, 2) photon-counting in steady-state luminescence. The theory allows for 1) the interpretation of the "anomalous" temperature dependence of far-infrared absorption bands in liquids, as well as 2) the appearance of coherent areas in luminescent solutions that are detected by the dynamically induced fluctuations (D-fluctuations) in the photoemission flux. The direct connection with experiments clearly demonstrates the physical content of the coherent-dissipative structures; it also illustrates the physical meaning of our novel theoretical approach to dynamical processes utilizing the application of the Complex Scaling Method in the framework of statistical mechanics.

1. Introduction

In this article we discuss a novel type of "off-diagonal long-range order" ODLRO as compared to Yang's fundamental analysis[1] of the possible existence of a macroscopically large eigenvalues of the fermionic two-matrix (or bosonic one-matrix), and the physical interpretation of this fact.

Since standard quantum theory in terms of self-adjoint operators are manifestly time reversible in character, there are obvious difficulties in providing satisfactory mappings of e.g. temperature dependent processes (which typically are irreversible) on strictly quantal formulations.

The latter, furthermore always deals with observables that are diagonalizable. It is hence also obvious that ODLRO in the meaning of Yang, i.e. giving a nonvanishing off-diagonal two-matrix (in the continuous coordinate representation) on account of the existence of the macroscopically large density matrix eigenvalue, seems to be the only way to exhibit such off-diagonal long-range order.

However, in recent years a new extension of quantum mechanics based on the celebrated theorems of dilatation analyticity Balslev and Combes[2], Simon[3] and van Winter[4] have led to a tremendous flow of applications in chemistry and physics, see e.g. this proceedings for relevant references. The reason for its popularity in physics and chemistry stems from the fact that one can treat a series of new, more general problems, see below, where the irreversibility problem is absolutely fundamental for the theoretical description, yet standard time reversible dynamics are nevertheless imbedded in the formalism.

In this extension of the formulation, which we will here refer to as CSM (the Complex Scaling Method) and which among other things incorporates the "terrible" Gamow waves into our hilbert space picture, save the inconvenience of a complex scaling transformation, one finds that irreducible non-diagonalizable blocks, so-called Jordan blocks, appear as a somewhat unwanted prize for the aforementioned convenience.

As is well known these Jordan blocks do not show up in standard treatments of self-adjoint operators, or in their associated hermitean matrix representations. However, in a dynamic formulation, provided by the Fourier-Laplace transformation of the time dependent Schrödinger (or Liouville) equation (within the CSM representation) which leads up to a time independent inhomogeneous equation these blocks may occur frequently on account of the general nature of the transformation.

The novelty of the theoretical formulation in this article stems from our ambition to incorporate CSM into the framework of statistical mechanics via complex scaling of the appropriate fundamental dynamical equations and/or the density operator, based on the time evolution superoperator and/or energy superoperator.

In connection with the study of the second order reduced (fermion) density matrix one may identify the appearing block in the density operator with a recently obtained (irreducible) Jordan block disguised as very "harmless" complex symmetric form[5]. This provides us with a different appearance of ODLRO, in comparison with Yang's original result.

In the last two chapters the consequences of this development is treated in quite some detail. In chapter 5, fundamental questions like the appearance of new *coherent-dissipative structures* in amorphous phases like a liquid, questions of microscopic irreversibility, EPR-correlations etc. are discussed in the light of the new irreducible patterns of the CSM theory. In the last chapter two *experimental contexts* are analysed in detail, i.e. 1) far-infrared absorption in polar liquids, and 2) the appearance of coherent areas in luminescent solutions that are detected by the dynamically induced fluctuations (D-fluctuations) in the photoemission flux.

Chapters 5 and 6 rely on the analysis of the *reduced second order density matrix* using the results due to Coleman[6], see e.g. chapter 2. Since these ideas are usually scattered in many various places we have brought together a self-contained section on *complex scaling* in chapter 3, and finally the appropriate *extension of CSM to subdynamics* in chapter 4. Even if some chapters are difficult to digest the first time, it is nevertheless possible to appreciate chapters 5 and 6 without a detailed understanding of the previous ones. In an oversimplified manner one might say that the present analysis becomes more comprehensible provided one might appreciate the close resemblance between equations (2.24) and (3.24).

2. Coherent States, ODLRO and Large Eigenvalues of $\Gamma^{(2)}(g^{N/2})$

In this chapter we will review some of the formulas and definitions needed for the development of a quantum theory of coherent-dissipative structures. We will try to make this section self-contained so that most relevant aspects of the present approach becomes easy to comprehend in its main features. We will start with the standard formulation.

First some definitions: The reduced density matrix of order p is defined as follows, see e.g. Löwdin[7], for N fermions

$$\Gamma^{(p)}(x_1 \ldots x_p | x'_1 \ldots x'_p) =$$

$$\binom{N}{p} \int \psi^*(x_1 \ldots x_p, x_{p+1} \ldots x_N) \psi(x'_1 \ldots x'_p, x_{p+1} \ldots x_N) \, dx_{p+1} \ldots dx_N \tag{2.1}$$

There exists several definitions with different normalizations:

$$\text{Löwdin}^7\text{: } \Gamma^{(p)};\ Tr\ \Gamma^{(p)} = \binom{N}{p} \tag{2.2}$$

$$\text{Coleman}^6\text{: } D^p;\ Tr\ D^p = 1 \tag{2.3}$$

$$\text{Yang}^1\text{: } \rho^{(p)};\ Tr\ \rho^{(p)} = p!\binom{N}{p} \tag{2.4}$$

We will use $\Gamma^{(p)}$ in what follows, and with a geminal we will mean a two particle (pair) function. Since the N-particle projection of the BCS-function[8] is an *antisymmetrized geminal power* (AGP) we will review the fundamentals of the AGP state. The interest in antisymmetrized geminal powers is related to the following two basic facts:

i) The two particle density operator $\Gamma^{(2)}(x_1, x_2 | x'_1, x'_2)$ has a simple sparse matrix representation, the "Box and Tail" [6,9].

ii) The AGP ansatz is the ground state for the polarization propagator[10,11,12].

There is also the generalization to GAGP states which we will not go into here. In addition to the usefulness of the polarization propagator technique in chemistry we will also mention the importance of the extreme case, see below and subsection 5.4.D, as a consequence of the assumption of equal a priori probability in the microcanonical ensemble as given by equilibrium (and quasiequilibrium) thermodynamics.

In this review we will focus on the first simplifying feature. Consider a geminal of rank s expanded in a basis of $r = 2s$ orthonormal spin orbitals, $(2s \geq N)$

$$g(1,2) = \sum_{i=1}^{s} g_i |\phi_i, \phi_{i+s}\rangle \tag{2.5}$$

where $|\phi_i, \phi_{i+s}\rangle$ is a (normalized) Slater determinant of the spin orbitals ϕ_i and ϕ_{i+s}. From g, one may construct the AGP as (using $N = 2m$ fermion states) the following antisymmetrized product

$$|g^{N/2}\rangle = [S_{N/2}]^{-1} \sum_{\substack{j_r < j_{r+1} \\ 1 \leq r \leq N/2}} g_{j_1} g_{j_2} \cdots g_{j_{N/2}} |\phi_{j_1} \phi_{j_1+s} \cdots \phi_{j_{N/2}} \phi_{j_{N/2}+s}\rangle \tag{2.6}$$

where the normalization integral $S_{N/2}$ is given by the symmetric sum

$$S_{N/2} = \sum_{\substack{j_r < j_{r+1} \\ 1 \leq r \leq N/2}} n_{j_1} n_{j_2} \ldots n_{j_{N/2}} \tag{2.7}$$

and $n_j = |g_j|^2$. The one-particle reduced density operator $\Gamma^{(1)}(g^{N/2})$ associated with the AGP is then written

$$\Gamma^{(1)}(g^{N/2}) = \sum_{j=1}^{s} \lambda_j^{(1)} \{|\phi_j\rangle\langle\phi_j| + |\phi_{j+s}\rangle\langle\phi_{j+s}|\} \tag{2.8}$$

with

$$Tr\ \Gamma^{(1)}(g^{N/2}) = N. \tag{2.9}$$

The eigenvalues $\lambda_i^{(1)}$ of $\Gamma^{(1)}(g)$ and $\Gamma^{(1)}(g^{N/2})$ are doubly degenerate with natural spin orbitals (NSO) ϕ_i and ϕ_{i+s}, given by

$$\lambda_i^{(1)} = [S_{N/2}]^{-1} n_i \frac{\partial S_{N/2}}{\partial n_i}. \tag{2.10}$$

Before we continue, one should point out that $\Gamma^{(p)}(g)$, $p = 1, 2$ are different albeit related. For instance

$$\Gamma^{(2)}(g) = |g\rangle\langle g| \tag{2.11}$$

$$Tr\ \Gamma^{(2)}(g) = 1 = \sum_{i=1}^{s} n_i \tag{2.12}$$

while

$$\Gamma^{(1)}(g) = \sum_{j=1}^{s} n_j \{|\phi_j\rangle\langle\phi_j| + |\phi_{j+s}\rangle\langle\phi_{j+s}|\}$$

$$Tr\ \Gamma^{(1)}(g) = 2.$$

Hence, while $\Gamma^{(1)}(g)$ and $\Gamma^{(1)}(g^{(N/2)})$ have the same NSO's, their occupation numbers are different as shown in (2.10) and below (2.7). The eigenvalues n_j of $\Gamma^{(1)}(g)$ are doubly degenerate as $\lambda_j^{(1)}$.

Although $|g\rangle$ is a natural geminal of $\Gamma^{(2)}(g)$ it is not in general an eigengeminal to $\Gamma^{(2)}(g^{N/2})$ (only in the extreme case, see below). However, the representation of $\Gamma^{(2)}(g^{N/2})$ in the basis

$$\{\{|\phi_i, \phi_{i+s}\rangle,\ 1 \leq i \leq s\}\ ,\quad \{|\phi_i, \phi_j\rangle,\ 1 \leq i \leq j \leq 2s;\ i+s \neq j\}\}$$

assumes a very simple "box and tail" form

$$\begin{pmatrix} b_{11} & b_{12} & \dots & b_{1s} & 0 & 0 & \dots & 0 & 0 \\ b_{21} & b_{22} & \dots & b_{2s} & 0 & 0 & \dots & 0 & 0 \\ \vdots & \vdots & \ddots & \vdots & \vdots & \vdots & & \vdots & \vdots \\ b_{s1} & b_{s2} & \dots & b_{ss} & 0 & 0 & \dots & 0 & 0 \\ 0 & 0 & \dots & 0 & t_{12} & 0 & \dots & 0 & 0 \\ 0 & 0 & \dots & 0 & 0 & t_{13} & \dots & 0 & 0 \\ \vdots & \vdots & & \vdots & \vdots & \vdots & \ddots & \vdots & \vdots \\ 0 & 0 & \dots & 0 & 0 & 0 & \dots & t_{(2s-2)2s} & 0 \\ 0 & 0 & \dots & 0 & 0 & 0 & \dots & 0 & t_{(2s-1)2s} \end{pmatrix} \tag{2.13}$$

The "box" $\mathbf{B}$, which has dimension s, has the elements

$$b_{j,j+s;k,k+s} \equiv b_{jk} = [S_{N/2}]^{-1} g_j g_k^* \frac{\partial^2 S_{N/2+1}}{\partial n_j \partial n_k};\quad 1 \leq j \neq k \leq s; \tag{2.14}$$

and

$$b_{j,j+s;j,j+s} \equiv b_{jj} = [S_{N/2}]^{-1} n_j \frac{\partial S_{N/2}}{\partial n_j} = \lambda_j^{(1)};\quad 1 \leq j \leq s \tag{2.15}$$

where in (2.14) $S_{N/2+1}$ is the symmetric function of order $N/2+1$. The "tail" $\mathbf{T}$, which has the dimension $2s(s-1)$ part of the matrix, has the elements

$$\begin{aligned} t_{ij} &= [S_{N/2}]^{-1} n_i n_j \frac{\partial^2 S_{N/2}}{\partial n_i \partial n_j} \\ &= \lambda_i^{(1)} \lambda_j^{(1)} + \frac{1}{2}\{n_j \frac{\partial \lambda_i^{(1)}}{\partial n_j} + n_i \frac{\partial \lambda_j^{(1)}}{\partial n_i}\} \\ & 1 \leq i \leq j \leq 2s;\ i+s \neq j \end{aligned} \tag{2.16}$$

Factorization properties of (2.14) and (2.16) have recently been discussed by Linderberg[13]and Goscinski[14], see also Bratoz[15]. For some recent AGP-calculations we refer to Ortiz[16] and Sangfeldt[17].

Instead of constructing the general BCS-function[8] for N goes to infinity we will take a closer look at the finite N case. The exact form of $\Gamma^{(2)}(g^{N/2})$ as reviewed here was thoroughly studied by Coleman[6,9] and furthermore investigated for the extreme case[9]. The upper bound on $\lambda^{(2)}$ was obtained by Yang[1] (for N even) and the existence of a large eigenvalue shown to be a consequence of the assumption of ODLRO[1]. The general case of N even and odd was studied by Sasaki[18] and the upper limit to $\lambda^{(2)}$ found to be

$$\lambda_i^{(2)} = \begin{cases} \leq (N-1)/2 & \text{if } N \text{ odd (Sasaki)} \; ; \\ < N/2 & \text{if } N \text{ even (Yang).} \end{cases} \tag{2.17}$$

Coleman introduced the following definition: $g^{N/2}$ (N even) is said to be extreme if all occupation numbers n_j of $\Gamma^{(1)}(g)$ are equal. Using (2.10), (2.14), (2.15) and (2.16) the "box and tail" becomes in the extreme case:

$$b_{jk} = b = \lambda^{(1)}\left(1 - \frac{N-2}{2(s-1)}\right) = \frac{N(2s-N)}{4s(s-1)}, \quad 1 \leq j \neq k \leq s \tag{2.18a}$$

$$b_{jj} = \lambda^{(1)} = \frac{N}{2s}, \quad 1 \leq j \leq s \tag{2.18b}$$

$$t_{ij} = \frac{N(N-2)}{4s(s-1)} = \lambda^{(1)} - b, \quad 1 \leq i \leq j \leq 2s; \; i+s \neq j. \tag{2.18c}$$

The characteristic equation for the "non-diagonal" part becomes

$$(\lambda^{(1)} + (s-1)b - \lambda^{(2)})(\lambda^{(1)} - b - \lambda^{(2)})^{(s-1)} = 0 \tag{2.19}$$

with the "large eigenvalue" given by

$$\lambda_1^{(2)} = \lambda_L^{(2)} = \lambda^{(1)} + (s-1)b = \frac{N}{2} - \frac{N(N-2)}{4s} \tag{2.20}$$

and an (s-1) degenerate eigenvalue

$$\lambda_j^{(2)} = \lambda_D^{(2)} = \lambda^{(1)} - b = \frac{N(N-2)}{4s(s-1)} \quad j = 2 \ldots s \tag{2.21}$$

Hence the "box and tail" becomes in the extreme case, a diagonal matrix with one eigenvalue $\lambda_1^{(2)} = \lambda_L^{(2)}$ non-degenerate and the remaining $(2s+1)(s-1)$ eigenvalues $\lambda_j^{(2)} = \lambda_D^{(2)}$, $j \neq 1$, degenerate*. In the extreme case $\Gamma^{(2)}(g^{N/2})$ thus writes

---

* Note that $s = N/2$ gives one degenerate eigenvalue 1 of order $\binom{N}{2}$. This is the

$$\Gamma^{(2)}(g^{N/2}) = \frac{N}{2s}\sum_{i=1}^{s}|\phi_i,\phi_{i+s}\rangle\langle\phi_i,\phi_{i+s}| + \frac{N(2s-N)}{4s(s-1)}\sum_{i\neq j}|\phi_i,\phi_{i+s}\rangle\langle\phi_j,\phi_{j+s}| +$$

$$+\frac{N(N-2)}{4s(s-1)}\sum_{\substack{i<j\\ i+s\neq j}}|\phi_i,\phi_j\rangle\langle\phi_i,\phi_j| = \Gamma_L^{(2)} + \Gamma_D^{(2)} \tag{2.22}$$

where

$$\Gamma_L^{(2)} = |\Phi_L\rangle\lambda_L^{(2)}\langle\Phi_L|; \quad \Phi_L = \sum_{i=1}^{s}\frac{1}{\sqrt{s}}|\phi_i,\phi_{i+s}\rangle = g \tag{2.23}$$

$$\lambda_L^{(2)} = \frac{N}{2} - \frac{N(N-2)}{4s}$$

and

$$\Gamma_D^{(2)} = \lambda_D^{(2)}\{\sum_{i,j}|\phi_i,\phi_{i+s}\rangle(\delta_{ij}-\frac{1}{s})\langle\phi_j,\phi_{j+s}| + \sum_{\substack{i<j\\ i+s\neq j}}|\phi_i,\phi_j\rangle\langle\phi_i,\phi_j|\} \tag{2.24}$$

$$\lambda_D^{(2)} = \frac{N(N-2)}{4s(s-1)}.$$

From (2.23) it is easily seen that if the basis $|\phi_i,\phi_{i+s}\rangle$ is localized in some sense the extreme geminal g is delocalized. Hence in the limit of large s this delocalization does not vanish because of the properties of the off-diagonal matrix elements. This limit obtains through the existence of a large eigenvalue, proportional to the number of particles, which also means that the off-diagonal parts of $\Gamma^{(2)}$ remains finite in the thermodynamic limit, i.e. $N \to \infty, \quad V_N \to \infty$ with $\frac{N}{V_N}$ is finite. For more details on the thermodynamic limit "lim therm" see e.g. Girardeau[19]. As we will see below, the degenerate eigenvalue $\lambda_D^{(2)} \to 0$ approaches 0 for large s, and this will lead to an interesting observation regarding the remaining part of the "box" after subtraction of the large "macroscopic" component.

Hartree-Fock independent particle model. In this case $\Gamma^{(2)}(g^{N/2})$ has a diagonal representation in the basis chosen here. Normally we will assume that $2s > N$. Note also that $\lambda_L^{(2)} = 0$ for $s = N/2 - 1$ which gives $\lambda_D^{(2)} = \frac{N}{N-4}$. This is obviously unphysical, but the possibility of such a condensation off the real axis exists in the CSM framework, see subsection 5.4.E

From (2.22) we obtain as a check

$$Tr\ \Gamma^{(2)}(g^{N/2}) = \frac{N}{2} + 2s(s-1)\frac{N(N-2)}{4s(s-1)} = \binom{N}{2} \quad (2.25)$$

and from (2.23) and (2.24) that

$$Tr\ \Gamma^{(2)}(g^{N/2}) = Tr\ \Gamma_L^{(2)} + Tr\ \Gamma_D^{(2)} =$$

$$= \frac{N}{2} - \frac{N(N-2)}{4s} + (2s+1)(s-1)\frac{N(s-2)}{4s(s-1)} = \binom{N}{2}. \quad (2.26)$$

Before continuing we note that if the rank s of the geminal g becomes infinite $\lambda_L^{(2)}$ approaches $N/2$ and the degenerate eigenvalue $\lambda_D^{(2)}$ goes to zero. The degeneracy also becomes infinite, but in such a way that the trace relations (2.25) and (2.26) should hold in the limit.

The results quoted here can be strictly formulated into a theorem of Coleman[9]:

"The geminal g is an eigenfunction of $\Gamma^{(2)}(g^{N/2})$ with a non-vanishing eigenvalue if and only if g is of extreme type, i.e. the eigenvalues of $\Gamma^{(1)}(g)$ are all equal."

Since Yang's assumption of ODLRO leads to a large eigenvalue of the 2-matrix, flux quantization, and the Meissner effect, it is obvious that the extreme condition for g is a necessary condition for ODLRO. However, one may wonder whether this is a sufficient condition also, i.e. whether the extreme condition for g is sufficient for ODLRO. As we will see in following chapters also another type of off-diagonal order will appear although not corresponding to the macroscopically large eigenvalue.

To complete our analysis we will now evaluate one and two-body expectation values in the extreme type situation. For $\sum_{i=1}^{N}\hat{h}_i$ one obtains

$$Tr\ \{\hat{h}\Gamma^{(1)}(g^{N/2})\} = \frac{N}{2s}\sum_{i=1}^{s}\{\langle\phi_i|\hat{h}|\phi_i\rangle + \langle\phi_{i+s}|\hat{h}|\phi_{i+s}\rangle\} \quad (2.27)$$

which if ϕ_i and ϕ_{i+s} have identical spatial parts but opposite spins reduces to (if $\hat{h}$ is spin independent)

$$Tr\ \{\hat{h}\Gamma^{(1)}(g^{N/2})\} = \frac{N}{2}\langle\phi|\hat{h}|\phi\rangle. \quad (2.28)$$

For a two-electron property $\sum_{i<j}\hat{v}_{ij}$ one obtains similarly

$$Tr\ \{\hat{v}\Gamma^{(2)}(g^{N/2})\} = Tr\ \{\hat{v}\Gamma_L^{(2)}\} + Tr\ \{\hat{v}\Gamma_D^{(2)}\} \quad (2.29)$$

with

$$Tr\ \{\hat{v}\Gamma_L^{(2)}\} = \frac{1}{s}[\frac{N}{2} - \frac{N(N-2)}{4s}] \sum_{i=1}^{s} \sum_{j=1}^{s} \langle \phi_i, \phi_{i+s} | \hat{v} | \phi_j, \phi_{j+s} \rangle = \lambda_L^{(2)} \langle g | \hat{v} | g \rangle \quad (2.30a)$$

and

$$Tr\ \{\hat{v}\Gamma_D^{(2)}\} =$$

$$= [\frac{N(N-2)}{4s(s-1)}] \{ \sum_{i=1}^{s} \langle \phi_i, \phi_{i+s} | \hat{v} | \phi_i, \phi_{i+s} \rangle - \langle g | \hat{v} | g \rangle + \sum_{\substack{i<j \\ i+s \neq j}} \langle \phi_i, \phi_j | \hat{v} | \phi_i, \phi_j \rangle \}. \quad (2.30b)$$

In the singlet geminal case it reduces to (after summation over spin)

$$Tr\ \{\hat{v}\Gamma_L^{(2)}\} = \frac{1}{s}[\frac{N}{2} - \frac{N(N-2)}{4s}] \sum_{i=1}^{s} \sum_{j=1}^{s} \langle \phi_i, \phi_i | \hat{v} | \phi_j, \phi_j \rangle = \lambda_L^{(2)} \langle g | g \rangle \quad (2.31a)$$

and

$$Tr\ \{\hat{v}\Gamma_D^{(2)}\} = [\frac{N(N-2)}{4s(s-1)}] \{ \sum_{i=1}^{s} \langle \phi_i, \phi_i | \hat{v} | \phi_i, \phi_i \rangle - \frac{1}{s} \sum_{i=1}^{s} \sum_{j=1}^{s} \langle \phi_i, \phi_i | \hat{v} | \phi_j, \phi_j \rangle \} +$$

$$+ \sum_{1 \leq i < j \leq s} \{ 4 \langle \phi_i, \phi_j | \hat{v} | \phi_j, \phi_i \rangle - 2 \langle \phi_j, \phi_i | \hat{v} | \phi_i, \phi_j \rangle \}. \quad (2.31b)$$

The formulas above are applicable to the general extreme case, both when there exists a large eigenvalue and in other cases of interest, see later chapters. In particular we will see how ODLRO may appear from a complex dilation, see chapters 3 and 4, of the first part of equation (2.24).

3. The Complex Scaling Method (CSM), Similarity (Non-unitary) Transformations and Complex Symmetric Forms

3.1. The Complex Scaling Method, CSM

Since the method of complex scaling, or complex rotation, plays such a fundamental role in our present study we will give a short review of some of its more puzzling aspects. It is not improper to consider the deformation of the coordinates into complex valued degrees of freedom as an attempt to analytically continue quantum mechanics beyond its conventional domain of a time reversible microscopic formulation into a valid theory of a more general nature, where irreversibility is naturally imbedded in the dynamics. The meaning of this somewhat cryptic statement will be clarified in more detail below.

Let us first introduce some mathematical concepts needed for the actual extension. Although our technique emanates from the pioneering work of Balslev and Combes[2], Simon[3] and van Winter[4], we will here follow the domain characterization of Brändas[20] in order to arrive at a transformation theory that not only contains standard quantum mechanics, but also leads to the desired extensions, i.e. including mappings onto a contractive (semigroup) evolution, cf. Kumičák and Brändas[21], and the concomitant approach to equilibrium. As we will see in accompanying sections, this has fundamental consequences also for the properties of the CSM derived canonical density matrix.

It is customary to introduce the N-body hamiltonian as $H = T + V$, where T is the kinetic energy operator and V the interaction potential. Since H is an unbounded operator, albeit bounded from below, we need to restrict the domain somewhat. Let $\mathcal{H}$ denote the hilbert space, complete with respect to the standard norm of square integrability, i.e. $\Phi \in \mathcal{H}$ with $\|\Phi\|_{L^2} < \infty$. Then the domain of H is given by

$$\mathcal{D}(H) = \{\Phi \in \mathcal{H},\ \ H\Phi \in \mathcal{H}\} \tag{3.1}$$

We will basically consider interactions such that the unboundedness of H arises entirely from the kinetic energy operator T. The potential V will be further specified below, however at the present stage it will be sufficient to treat the case with $\mathcal{D}(H) = \mathcal{D}(T)$.

A key quantity in the CSM formulation is the scaling operator

$$U(\theta) = \exp(iA\theta) \tag{3.2}$$

where

$$A = \frac{1}{2}\sum_{k=1}^{N}[\vec{p}_k\vec{x}_k + \vec{x}_k\vec{p}_k] \tag{3.3}$$

is the generator of the scaling transformation, θ is a parameter, which may be real or complex and $\vec{x}_k$ and $\vec{p}_k$ are the coordinate and momentum of particle k. If $\theta \in R$, then U is a unitary operator, defining the dilation group, with $\mathcal{D}(U) = \mathcal{H}$, which effectuates the scaling

$$U(\theta)\Phi(x_1,\ldots,x_N) = \exp\left(\frac{3N}{2}\theta\right)\Phi(e^{\theta}x_1,\ldots,e^{\theta}x_N) \tag{3.4}$$

By considering complex scalings, i.e. $U(\eta)$ with $\eta = |\eta|e^{i\theta}$ (note that our parameter in U now refers to the whole analytic parameter η rather than the dilation group parameter θ) Balslev and Combes[2] proved that for certain classes of hamiltonians, i.e. so-called dilatation analytic hamiltonians, the continuous spectrum changed under the complex deformation and became rotated down -2θ in the complex energy plane, thereby opening up sectors on the "unphysical" Riemann sheet, where possible finite dimensional "resonance eigenvalues" would appear. They also demonstrated the non-existence of singularly continuous spectra in these cases, as well as invariance of the exposed spectrum (including bound states) to variations of θ.

Without going into too many mathematical details, we will thus consider $\eta = |\eta|e^{i\theta}$, with $0 \leq \theta < \theta_0$. Generally θ_0 depends on the potential, for instance in the Coulomb case no limit is invoked, while in some other cases, as will be seen below it is natural to have $\theta_0 \leq \frac{\pi}{4} - \delta$. We also introduce Ω as

$$\Omega = \{\eta, |\arg(\eta)| < \theta_0\} \tag{3.5}$$

Furthermore we make the decomposition $\Omega = \Omega^+ \cup \Omega^- \cup R$, where the real axis $R = R^+ \cup R^- \cup \{0\}$ partitions Ω into its upper and lower parts in the complex plane. In what follows we will assume that the interaction V is a sum of two body interactions V_{ij} such that V_{ij} is Δ_{ij} -compact in $L^2(R^3)$, se e.g. reference 2 for details, and furthermore that $V_{ij}(\eta) = U_{ij}(\eta)V_{ij}U_{ij}^{-1}(\eta)$, $\eta \in R^+$, has a compact analytic extension to Ω^+. The definition of the dilatation analytic family of operators $H(\eta)$ is then given by

$$H(\eta) = U(\eta)HU^{-1}(\eta), \quad \eta \in R^+ \tag{3.6a}$$

and

$$H(\eta) = \eta^{-2}T + V(\eta), \quad \eta \in \Omega^+ \tag{3.6b}$$

In the expression above it is notable that $H(\eta)$ is obtained in two steps. First via a unitary transformation to a scaled representation and thereafter followed by an analytic continuation to Ω^+. Although this is a mathematically rigorous procedure it is preferable, as we will see below, to consider the unbounded similarity transformation $U(\eta)$, $\eta \in \Omega^+$, directly. We will then introduce the domain $\mathcal{N}(\Omega)$ as the subset or Nelson class[22]

$$\mathcal{N}(\Omega) = \{\Phi, \Phi \in \mathcal{H};\ U(\eta)\Phi \in \mathcal{H};\ \eta \in \Omega\} \tag{3.7}$$

We furthermore complete the space with respect to the norm

$$\int_{-\theta_0}^{+\theta_0} \|U(\eta)\Phi\|_{L^2}^2 d\theta = \|\Phi\|_{N_{\theta_0}}^2 \tag{3.8}$$

Since we also want the first and second partial derivatives to satisfy (3.8) it is natural to define the spaces $\mathcal{N}_{\theta_0}^{(i)}$, $i = 0, 1, 2$, analogously. We are now in the position to make an alternative definition of the analytic family $H(\eta)$, see reference 20, i.e.

$$H(\eta) = U(\eta)HU^{-1}(\eta);\ \mathcal{D}(UHU^{-1}) = \mathcal{N}_{\theta_0}^{(2)} \tag{3.9a}$$

and

$$H(\eta) = \eta^{-2}T + V(\eta);\ \mathcal{D}(UHU^{-1}) \to \mathcal{D}(T) \tag{3.9b}$$

Note that $\mathcal{N}_{\theta_0}^{(i)}, i = 0, 1, 2$ are hilbert spaces, which allow $H(\eta)$ to be interpreted as a similarity transformation of the self-adjoint unscaled operator H. We also note that $U(\eta)$ exhibits the "star-unitary" property

$$U^\dagger(\eta^*) = U^{-1}(\eta) \tag{3.10}$$

which can be easily proven from

$$\langle\Phi|\Psi\rangle = \langle\Phi(\eta^\star)|\Psi(\eta)\rangle = \langle\Phi|U^\dagger(\eta^\star)U(\eta)|\Psi\rangle \tag{3.11}$$

which holds for all $\Phi,\Psi \in \mathcal{N}_{\theta_0}$.

In the definition presented above, leading of course to the same analytic family $H(\eta)$, we can see that the two steps involved in equation (3.9) are different in comparison to (3.6). Here the first step consists of restricting $\mathcal{H}$ to a smaller domain $\mathcal{N}_{\theta_0}^{(2)}$ for which the scaling $U(\eta)$ is defined for all η such that $\arg(\eta) < \theta_0$. After making the complex rotation, i.e. changing the parameter θ in e^θ from real to complex, one completes the Nelson subset (dense in $\mathcal{H}$) to $\mathcal{D}(T)$), which means that completion is made with respect to the "standard" L^2-norm for the functions and its first and second partial derivatives. Before ending this subsection we want to emphasize that the present definition is particularly useful in connection with time evolution and associated Lyapunov converters[21]. Other uses of this formulation will be stressed in the next subsection.

3.2. Similarity Transformations and Complex Symmetric Forms

One can understand the abstract transformations of the previous subsection in a very simple way. Consider e.g. the integral

$$\mathrm{I}(b) = \int_0^{+\infty} e^{-br} dr = \frac{1}{b} \tag{3.12}$$

whose simple analytic form trivially allows analytic continuation to negative values of b. In the latter case the integral is not convergent even if $\mathrm{I}(b) = \frac{1}{b}$ for all $b\neq 0$. If $b\neq 0$ is complex (or negative) so that the integrand in (3.12) does not vanish for $r \to \infty$, then one may consider the trick of defining a complex integration path with θ sufficiently large so that

$$\mathrm{I} = \lim_{R\to\infty} \int_0^{Re^{i\theta}} e^{-br} dr = \frac{1}{b} \tag{3.13}$$

The complex path can hence be used to explicitly compute a numerical value of the integral I even when its analytical form is unknown.

When evaluating matrix elements, using a complex deformation as in (3.13) a word of warning should be issued. For instance if one does not realize that the spaces $\mathcal{N}_{\theta_0}^{(i)}$, $i = 0, 1, 2$, have a finer topology than $\mathcal{H}$ then paradoxical situations may occur. Consider e.g. the resolvent $(H - \lambda I)^{-1}$, which for $\lambda^\star \neq \lambda$ is a bounded operator in $\mathcal{H}$. *This is not true in* $\mathcal{N}_{\theta_0}$. We will exemplify this as follows (we are indebted to B. Nagel[23] for this example). In momentum space we obtain with $H_0 = k^2$; $\Phi = \pi^{-\frac{1}{4}} e^{-\frac{k^2}{2}}$, $\eta = |\eta| e^{i\theta}$, $\theta_0 \leq \frac{\pi}{4} - \delta$ the following result

$$(U(\eta)\Phi) = \pi^{-\frac{1}{4}} \frac{1}{\sqrt{\eta}} e^{-\frac{k^2}{2\eta^2}} \tag{3.14}$$

and

$$\|U(\eta)\Phi\|^2_{L^2} = \frac{1}{\sqrt{\cos(2\theta)}} \tag{3.15a}$$

$$\|U(\eta)\Phi\|^2_{N_{\theta_0}} = \int_0^{\frac{\pi}{2}} \frac{d\theta}{\sqrt{\cos\theta}} < \infty \tag{3.15b}$$

Since $0 < \theta_0 \leq \frac{\pi}{4} - \delta$ then $k^n e^{-\frac{k^2}{2}} \in \mathcal{N}_{\theta_0}$ for all n, and hence $\Phi \in \mathcal{N}^{(\infty)}_{\theta_0}$. Now it is easy to show that if $-2\theta_0 < \arg\lambda < 0$, then $\chi = (H_0 - \lambda I)^{-1} e^{-\frac{k^2}{2}} \notin \mathcal{N}_{\theta_0}$ one obtains

$$(U(\eta)\chi)(k) = \frac{1}{\sqrt{\eta}}(\eta^{-2}k^2 - \lambda)^{-1} e^{-\frac{k^2}{2\eta^2}} \tag{3.16}$$

and

$$\|\chi\|^2_{L^2} = 2(\cos 2\theta)^{\frac{3}{2}} \int_0^\infty \frac{e^{-k^2}dk}{(k^2 - \lambda^* \cos 2\theta e^{-2i\theta})(k^2 - \lambda \cos 2\theta e^{2i\theta})} =$$

$$= [\lambda \cos 2\theta e^{2i\theta} = k_0^2 e^{2i\epsilon};\ \ k_0 = |\lambda|^{\frac{1}{2}}(\cos 2\theta)^{\frac{1}{2}}] = \tag{3.17}$$

$$= 2(\cos 2\theta)^{\frac{3}{2}} \int_0^\infty \frac{e^{-k^2}dk}{|k + k_0 e^{i\epsilon}|^2 (k - k_0 e^{i\epsilon})(k - k_0 e^{-i\epsilon})}$$

It is obvious from (3.17) above that the divergence of the integral as $\epsilon \to 0$ determines whether $\|\chi\|^2_{N_{\theta_0}}$ is finite or not. For small ϵ one finds that

$$\|\chi\|^2_{L} \sim \frac{e^{-k^2}}{4k_0^2} \int_{-\delta}^{\delta} \frac{dx}{(x - i\epsilon k_0)(x + i\epsilon k_0)}$$
$$\sim \pi \frac{e^{-k_0^2}}{4k_0^3|\epsilon|} \tag{3.18}$$

which since $\int_{-\delta}^{\delta} \frac{d\epsilon}{|\epsilon|} = \infty$ implies that $\chi \notin \mathcal{N}_{\theta_0}$.

This simple example clarifies the following somewhat surprizing situation. In accordance with (3.11) we write the following matrix element with respect to the operator H, utilizing (3.9) and (3.10), with $\Phi, \Psi \in \mathcal{N}^{(2)}_{\theta_0}$, such that

$$\langle\Phi|H|\Psi\rangle = \langle\Phi(\eta^*)|H(\eta)|\Psi(\eta)\rangle \tag{3.19}$$

holds. It is important to note that $H(\eta)\Psi(\eta) = U(\eta)HU^{-1}(\eta)U(\eta)\Psi = U(\eta)(H\Psi)$ are all valid operations since $\Psi \in \mathcal{N}^{(2)}_{\theta_0}$ implies $H\Psi \in \mathcal{N}_{\theta_0}$ and hence (3.19) reduces to (3.11).

The invariance (3.19) shows that the matrix element is not altered by the complex deformation, if Ψ, Φ are in the right domain, see above. However, compare e.g. (3.12) and (3.13), the scaled integral to the right in (3.19) may have a finite value for a suitable η even when the unscaled integral to the left in (3.19) may not exist. Hence the right hand side of (3.19) exists for a more general class of functions than those that are L^2 in its unscaled form or more precisely those that belong to $\mathcal{N}_{\theta_0}^{(2)}$. Thus we see that complex scaling allows us to extend quantum mechanics beyond its conventional domain in that it assigns a well defined meaning to vectors and matrix elements with respect to operators (resolvents), requiring in analogy with (3.19) the existence of the scaled representation for some nonreal η.

Before we continue, we want to emphasize that we have implicitly assumed that $V\Psi \in \mathcal{N}_{\theta_0}$ for $\Psi \in \mathcal{N}_{\theta_0}^{(2)}$. Even if this may be a restriction on V in general one can circumvent this problem by considering more general deformations like e.g. exterior scaling etc. These are, however, technical points that will not alter our general considerations and conclusions as well as physical interpretation.

With the preceeding development in mind, it is now easy to avoid the following paradoxical situation. If we employ (3.19) with H replaced by the resolvent operator one may ask whether the following relation with $\eta \in \Omega^+$, $\lambda^* \neq \lambda$

$$\langle \Phi | (H - \lambda I)^{-1} | \Psi \rangle = \langle \Phi(\eta^*) | (H(\eta) - \lambda I)^{-1} | \Psi(\eta) \rangle \tag{3.20}$$

holds for all $\Phi, \Psi \in \mathcal{N}_{\theta_0}$. The puzzle consists in that the left hand side of (3.20) is always finite for λ complex, while the right hand side may become infinite if λ is a complex eigenvalue of the complex rotated operator $H(\eta)$, which is a possibility according to the Balslev-Combes theorem[2]. Now we can return to our analysis carried out in (3.18). For instance with the choice $H = H_0 = k^2$ and $\Psi = \Phi$ with $-2\theta_0 < \arg\lambda < 0$ it was demonstrated that $\chi = (H - \lambda I)^{-1}\Phi \notin \mathcal{N}_{\theta_0}$ and hence (3.11) (with χ replacing Ψ) is *not* applicable and the equality (3.20) is *not* valid. Nevertheless the scaled matrix element in (3.20) exists for e.g. $2\theta_0 > \arg\lambda > 0$, and is equal to the unscaled integral and more importantly has a meromorphic continuation into the sector given by $-2\theta_0 < \arg\lambda < 0$.

For simplicity we have displayed a simple situation with one threshold at zero, but the generalization to the general manybody situation should be obvious.

It is also clear that we have considered the hilbert spaces $\mathcal{N}_{\theta_0}^{(i)}$, $i = 0, 1, 2$ in order to find convenient domains for the unbounded (in $\mathcal{H}$) similitude $U(\eta)$, $\eta \in \Omega$. After appropriate deformations completion with respect to the standard L^2 norm is made. In this manner we arrive at the formal eigenvalue relation ($\eta = |\eta| e^{i\theta}$)

$$H(\eta)\Psi(\eta) = \varepsilon(\eta)\Psi(\eta) \quad \eta \in \Omega^+ \tag{3.21}$$

with $\arg\eta$ sufficiently large to uncover the resonance ε. The conjugate of equation (3.21) becomes

$$\overline{H(\eta)}\ \overline{\Psi(\eta)} = \overline{\varepsilon(\eta)}\ \overline{\Psi(\eta)} \quad \eta \in \Omega^- \tag{3.22}$$

with the involution $\overline{A(\eta)} = A^\star(\eta^\star)$ being introduced to include also (complex) optical potentials. Obviously (3.21) and (3.22) in conjunction with (3.19) motivates the construction

$$\langle\overline{\Psi(\eta^\star)}|H(\eta)|\Psi(\eta)\rangle = \varepsilon(\eta)\langle\overline{\Psi(\eta^\star)}|\Psi(\eta)\rangle \tag{3.23}$$

from which stationary variational principles can be derived in almost the same fashion as in ordinary quantum mechanics[24,25] , with the important distinction that the extremum property of the principle has been lost. Note also that if H satisfies $H^\star = H$ then there is no restriction *from the variational point of view* to assume that $\Psi^\star = \Psi$ [26] and then the construction (3.23) defining a trial $\varepsilon(\eta)$ based on a trial $\Psi(\eta)$ and corresponding $\overline{\Psi(\eta^\star)} = \Psi(\eta)^*$ becomes complex symmetric. However, in many cases it may be useful to analyse the wavefunction in terms of nonreal spherical harmonics. One might then leave the angular momentum eigenfunctions as they are and consider each angular block in a complex symmetric representation even if the corresponding term in the wavefunction representation contains the nonreal component $Y_{l,m}(\vartheta,\varphi)$.

Since η sometimes are used primarily as a numerical convergence factor we may replace the explicit η-dependence for the quantity A by replacing $A(\eta)$ by A^c. Hence we write

$$H^c\Psi^c = \varepsilon\Psi^c \tag{3.21$'$}$$

and for the complex conjugate equation

$$H^{c\star}\Psi^{c\star} = \varepsilon^\star\Psi^{c\star} \tag{3.22$'$}$$

and

$$\langle\Psi^{c\star}|H^c|\Psi^c\rangle = \varepsilon\langle\Psi^{c\star}|\Psi^c\rangle \tag{3.23$'$}$$

Even if this simple notation is very appealing in that almost any standard quantum mechanical technique can be taken over provided it is appropriately modified, one should note that it results in a formulation that goes beyond conventional quantum mechanics "on the real axis". The most direct consequence is, as mentioned, the appearance of complex resonance eigenvalues and associated Gamow vectors[27]. A closer study of the full generalized spectral properties of the complex deformed problem[28] shows that these eigenstates essentially deflate the (generalized) spectral density, giving in an asymptotic sense a decomposition of the continuum into resonances and background. This has the important consequence that each Gamow vector represents a well-defined section of the continous spectrum associated with the unscaled self-adjoint problem. In other words one can say that *the Gamow representation condensates an infinite dimensional hilbert space associated with a particular spectral part of the continuum into a finite dimensional linear space of suitable Gamow vectors.* This condensation will play a fundamental role in the later section of this study. The prize for this convenience is the appearance of so-called Jordan blocks, see e.g. Reid and Brändas[5].

If H in (3.21) and (3.22) has the property that it is real for $\eta \in R^+$ then it follows, as we have pointed out, that the construction (3.23) will lead to complex symmetric forms. Furthermore it is a general theorem that *any matrix can be brought to a complex symmetric form by a similarity transformation*, see e.g. Gantmacher[29]. Even if this is always possible in general it may not be the most convenient construction in every problem. Nevertheless, it will be an important representation in our subdynamical formulation as will be presented here.

In the preceeding article[5] a very simple complex symmetric form was found, i.e.

$$q_{kl} = (\delta_{kl} - \frac{1}{s}) \exp(i\pi \frac{k+l-2}{s}), \tag{3.24}$$

where s is the dimensionality of the Jordan block and

$$1 \leq k, l \leq s. \tag{3.25}$$

with the property that the matrix in (3.24) is <u>similar</u> to

$$\begin{pmatrix} 0 & 1 & 0 & \dots & 0 \\ \vdots & \ddots & \ddots & \dots & \vdots \\ & & \ddots & \ddots & 0 \\ \vdots & & & \ddots & 1 \\ 0 & \dots & & \dots & 0 \end{pmatrix} \tag{3.26}$$

Before we continue to demonstrate the relevance of this simple form in connection with the quantum formulation of the appearance of dissipative structures, we will first make a short digression into the appropriate superoperator extension. Note, however, the close resemblance between the first part of (2.24) and (3.24).

4. Subdynamics in the Light of CSM

In order to set the frame for a dilatation analytic extension of a subdynamical superoperator formulation we will introduce the following quantities. As will be seen below, this leads directly to the relevant evolution superoperator and the corresponding Boltzmann operator.

To allow for different choices of representation we define the superoperator $\hat{P}$ by

$$\hat{P} = A \;\rangle\langle \;+\; \rangle\langle\; B \tag{4.1}$$

where the choice of A and B will be made below and $\rangle\langle$ is short for the "bra" and "ket" components of the density matrix. With this notation it is straight-forward to show that

$$e^{\hat{P}} = e^A \;\rangle\langle\; e^B \tag{4.2a}$$

and

$$\partial\rho = dA\rho + \rho dB \tag{4.2b}$$

There are basically two choices of A and B and $\hat{P}$ that correspond to physically meaningful subdynamical formulations.

1. The first choice of operators concerns time evolution, see e.g. Obcemea and Brändas[30]. One obtains $\hat{P} = -i\hat{L}t, \quad A = -iHt = B^\dagger$ which leads to

$$e^{-i\hat{L}t}\ \rangle\langle = |e^{-iHt}\ \rangle\langle e^{-iHt}\ | \tag{4.3a}$$

and

$$i\frac{\partial \varrho}{\partial t} = \hat{L}\varrho \tag{4.3b}$$

In CSM using the construction[30] one gets with

$$\varrho^c_{kl} = |\Psi^c_k\rangle\langle\Psi^c_l| \tag{4.4}$$

and

$$H^c\Psi^c_k = \varepsilon_k\Psi^c_k \tag{4.5}$$

with

$$\varepsilon_k = E_k - i\epsilon_k \tag{4.6}$$

the following eigenvalue relation

$$e^{-i\hat{L}^c t}\varrho^c_{kl} = e^{\{-i(E_k-E_l)-(\epsilon_k+\epsilon_l)\}t}\varrho^c_{kl} \tag{4.7}$$

This choice is good in that (4.7) contains energy differences, while the widths are added. Note, however, that ϱ in (4.4) for $k = l$ is *not* a projector, i.e. $\varrho^2 \neq \varrho$, which makes the interpretation fundamentally different as compared to the one based on the analytically continued fundamental unit, see below.

2. The second choice concerns the Boltzmann factor containing $\beta = \frac{1}{k_B T}$, where k_B is the Boltzmann constant and T the absolute temperature. Here the representation obtains from $\hat{P} = -\beta\hat{L}$ and $A = -\beta H = B$ satisfying

$$-\frac{\partial\rho}{\partial\beta} = \hat{L}_B\rho \tag{4.8}$$

with the eigenvalue relation

$$e^{-\beta\hat{L}^c_B}|\Psi^c_k\rangle\langle\Psi^{c*}_l| = e^{-\beta\{(E_k+E_l)-i(\epsilon_k+\epsilon_l)\}}|\Psi^c_k\rangle\langle\Psi^{c*}_l| \tag{4.9}$$

Note that $\rho^c_{kk} = |\Psi^c_k\rangle\langle\Psi^{c*}_k|$ here *is a (nonselfadjoint) projector* i.e. $\rho^{c2} = \rho^c$, while $\rho^{c\dagger} \neq \rho^c$. In addition to this feature we also see that *both widths and energies are* *added* in (4.9). In what follows we will for convenience let $\beta \rightarrow \frac{\beta}{2}$ so that our formulation agrees with the standard definition of the Boltzmann factor, i.e.

$$e^{-\frac{\beta}{2}\hat{L}^c_B}\rho^c_{kk} = e^{-\beta\varepsilon_k}\rho^c_{kk} \tag{4.10}$$

Note that $\hat{L}$ in (4.3b) is defined by

$$\hat{L} = H\ \rangle\langle\ -\ \rangle\langle\ H \tag{4.11}$$

while $\hat{L}_B$ in (4.8) is given by

$$\hat{L}_B = H\ \rangle\langle\ +\ \rangle\langle\ H \tag{4.12}$$

We also see the difference between ϱ and ρ in the two formulations above. In the undilated formalism they are of course identical.

For additional comments on the usefulness of the analytically continued projector in connection with variational principles and related virial theorems, see Obcemea, Froelich and Brändas[31].

As we have stated at the end of the previous chapter, complex scaling leads to a certain asymptotic decoupling of the localized resonance state from the background. This is also very easy to visualize from the physical point of view. By scaling the coordinates so that the appropriate outgoing Gamow waves, fulfilling the Siegert condition[32] become square integrable, means that the environment will be shielded from the system in a manner precisely given by the deformation. The incorporation of complex scaling into the present superoperator picture therefore gives both in a physical as well as mathematical sense a valid subdynamical picture. The consequences of this "philosophy" will be explicitly explored below.

Although complex scaling exhibits the above mentioned localization properties, it is important to remember that irreducible non-diagonal structures, so called Jordan blocks, may appear in contrast to the conventional formulation "on the real axis". Here this will be a desired "complication" since it will in fact be associated with a new coherence pattern of quantum statistical origin.

It is important to realize that these irreducible structures occur in the second order density operator, and not on the Schrödinger level. It was immediately observed, see e.g. references[33,34], that Jordan blocks could appear in connection with complex scaling. It was further pointed out by Moiseyev and Friedland[35] that there always exists (if the kinetic and potential operators do not commute) at least one rotation angle θ so that the complex rotated matrix hamiltonian exhibits a Jordan block with Segrè characteristic larger than unity. Another line of thought emanates from the suggestion of Marcus, see e.g. Moiseyev[36], that many avoided crossings in the energy plane would be an indication of Quantum Chaos. That such situations can occur is not unusual, see e.g recent studies of the so-called

Zel'dovich phenomenon[37]. In connection with this discussion we suggest that the present ocurrence of irreducible blocks in the density matrix, as will be studied and analysed below, will shed additional light on the idea of Quantum Chaos, if it exists at all as a meaningful concept.

5. Emergence of Coherent-Dissipative Structures

5.1. Introductory Remarks

The following considerations propose a treatment of the spontaneous emergence of coherent-dissipative structures in amorphous phases, like a liquid, from first principles. In the formal context of kinetic equations and, in general, of other phenomenological theories, the appearance of coherent (or cooperative) effects is nowadays not an uncommon feature. Within the microscopic level of description of natural phenomena, however, the appearance of these structures must be considered to be surprising, owing to the well known reversible character of the fundamental equations of motion (classical as well as quantal). In fact, one expects even intuitively that time reversible mechanical laws cannot account for the appearance of qualitatively new forms of matter, as e.g. highly organized coherent structures, starting from a disordered form, as e.g. a liquid being in the so-called thermal equilibrium.

The matter in thermal equilibrium, however, is not totally "disordered". As standard statistical mechanics tells us, there exist correlation patterns that obey the BBGKY (Bogoliubov-Born-Green-Kirkwood-Yvon) hierarchy (classical or quantal) and may have a very large size. These patterns appear due to the thermal motion of the molecules (or atoms) in the condensed system, and thus correlations are continuously created and destroyed. The "distribution" of these correlation patterns is represented by the reduced distribution functions that fulfill the BBGKY hierarchy ; see Balescu[38]. An elementary excitation (e.g., absorption of a photon by a quantum system) that may occur in the equilibrated macroscopic system can be traced as follows. All the correlation patterns which contain the excited quantum system, e.g. the molecule, become disturbed and thus they begin to "relax", showing a time dependence that is different from the one they had at thermal equilibrium. One should remember, however, that the BBGKY hierarchy is still valid, although some of the reduced distribution functions have now become explicitly time dependent[38].

The following observation is crucial in the present context and should be stressed. During the actual relaxation time of the excited quantum system or (the molecule), the motion of all the other molecules that belong to the same correlation pattern (as the excited one) are also affected. In other terms, the dynamical behaviour of all these molecules has been changed, even if they do not interact with each other directly, and although only one molecule has been initially disturbed.

It should be reminded that the considerations above are also valid in classical statistical mechanics. Thus they ought to be looked upon as illustrative. The analysis to be presented in the following sections, however, are quantum mechanical in nature, without a classical analogue. They reveal the possibility that "self-organized" structures may appear, after excitation of some of the elementary quantum systems that constitute the macroscopic condensed system. It should also be pointed out that the size of these structures is primarily not determined by the range of the intermolecular interactions (as, e.g. the classical and thus illustrative viewpoint might suggest). The possibility of spontaneously created Jordan blocks in the density matrix, which represent the coherent-dissipative structures, appears naturally within the generalized quantum mechanical formalism used above.

Moreover it will also become clear that (after complex scaling) the appearance of the irreducible non-diagonalizable Jordan blocks, in the second order reduced density matrix corresponding to the whole system, represents a purely dynamical phenomenon that cannot occur within "standard" quantum mechanics, i.e. in the case of isolated quantum systems treated via the time independent Schrödinger equation. In fact we saw that the quantization conditions that lead to the appearance of the Jordan blocks are directly connected with the "life times" that the Complex Scaling Method generically assigns to the constituents of the Jordan block. In a more physical language we may say that in a space that consists of states with infinite lifetime, there can be no coherent-dissipative structures (of the kind described below).

It might also be interesting to mention the following similarity between the coherent-dissipative structures under consideration and the well known dissipative structures of the Brussels school[39,40,41] appearing in phenomenological thermodynamics of systems far from equilibrium. In both cases, organized forms of matter appear only after external supply of energy to the (initially) equilibrated, unstructured macroscopic system; these organized forms, moreover, die out, if the external energy supply stops; furthermore, their size cannot fall below a critical value.

5.2. On Microscopic Irreversibility

The discovery of the Second Law of classical thermodynamics was and still is in conflict with the microscopic mechanical theory by the fact that time symmetry is "broken" in the nature. The well known "contradiction" between "reversible" microscopic dynamics (classical as well as quantal) and "irreversible" thermodynamical processes still represents, undoubtedly, one of the most challenging problems of modern science.

This contradiction touches upon the meaning of the physical entity known as "time". This is by no means "only a philosophical" problem, since e.g. this is experienced by every chemist considering the time development of a chemical reaction. Many scientists are therefore not willing to accept Einstein's view that, "irreversibility is only an illusion, a subjective impression, coming from exceptional initial conditions"; cf. Prigogine[41], p.203. Therefore, can irreversibility be founded on the microscopic level of description of natural phenomena? It is obvious that this question is intimately connected also with the appearance of coherent-dissipative structures as given by the theoretical framework of this paper and hence has a direct bearing on the aforementioned fundamental problem.

Here one should also stress the work of Prigogine and coworkers concerning the microscopic foundation of time and entropy as observables represented through superoperators; for more references see, e.g., Prigogine[41] and George[42] and references cited therein. It is clear that this work contrasts with the "Boltzmann-Gibbs" traditional viewpoint, i.e. that the entropy is some functional of the distribution function (or the density operator) in phase space.

However, every attempt to introduce irreversibility on the microscopic level is inevitably coupled to the question: What are the direct and experimentally accessible new theoretical predictions that cannot satisfactorily be predicted and explained via standard "reversible" microscopic theory, be it classical or quantum mechanical? Formulated more emphatically, what would be the deciding reason (this should of course be intimately connected with the experimental situation) that would force us to accept that a "deep revision" of the well established time reversible theory is absolutely inevitable? It is well known that the majority of the scientific community is convinced that, thus far, these questions have not been answered in a convincing manner. Motivated by these remarks we will therefore consider (see chapter 6) the connection between the present coherent-dissipative structures as formulated in the subdynamical CSM approach and two real experiments explicitly.

The formalism presented and interpreted in the following sections of this chapter do indeed make contact with the problem of the precise "microscopic foundation" of irreversibility; cf. also chapter 4. This is a simple consequence of the property attributed to the complex scaling method as being a model for e.g. resonance phenomena, i.e. being devised to deal with irreversible phenomena (decay of molecular quantum states, etc.) starting from first principles; cf. references cited in chapter 3. In what follows, we will not make further mentioning of this problem. Instead, our interest will be focused on spectroscopic applications, see chapter 6, as well as the application of CSM to the canonical ensemble formalism of statistical mechanics, see present chapter. Nevertheless we emphasize again the underlying conceptual connection between the present work and previous treatments based on Prigogine's subdynamics[30] and star-unitary transformations, see e.g. Dreismann[43,44].

5.3. Spontaneous Creation of Coherent-Dissipative Structures in Amorphous Condensed Systems: Formal Derivations

5.3.A. Jordan Blocks and the Reid-Brändas Corollary

As starting point for the present derivation, the following algebraic theorem may be considered:

Any square matrix can be brought to complex symmetric form by a similarity transformation; c.f. e.g. Gantmacher[29].

Of particular interest for our purposes, however, is the following corollary[5], which is also used explicitly in the proof of the above theorem as given by Reid and Brändas[5] :

The $s \times s$ matrix (s: a positive integer) $\mathbf{C}_s(0)$ defined by

$$\begin{pmatrix} 0 & 1 & 0 & \dots & 0 \\ \vdots & \ddots & \ddots & \dots & \vdots \\ & & \ddots & \ddots & 0 \\ \vdots & & & \ddots & 1 \\ 0 & \dots & & \dots & 0 \end{pmatrix} \tag{5.1}$$

is similar to the symmetric $s \times s$ matrix $\mathbf{q}$ with matrix elements

$$q_{kl} = (\delta_{kl} - \frac{1}{s}) \exp(i\pi \frac{k+l-2}{s}), \tag{5.2}$$

with

$$1 \leq k,l \leq s. \tag{5.3}$$

Now it is easily seen that the range (5.3) of the integer variables k and l is not unique. This follows from the proof of this corollary. Namely, the explicit form (5.2) of $\mathbf{q}$ can be rewritten as

$$q_{kl} = (\delta_{kl} - \frac{1}{s}) \omega^{k+l-2} \tag{5.2a}$$

with $\omega \equiv \exp(i\pi/s)$, which implies

$$\omega^s = -1. \tag{5.4}$$

From the derivation presented in the corollary[5], however, it becomes clear that the condition (5.4) is the only crucial one used in the proof. Thus the replacement

$$\omega \to \omega' \equiv \omega \exp(i\frac{\pi}{s}2K), \tag{5.5}$$

where K is an arbitrary (negative or positive) integer, preserves the result (5.2). In other words: The matrix elements of $\mathbf{q}$ may be given by equation (5.2), with the "generalized" range

$$k,l = K+1, \dots, K+s \tag{5.3a}$$

(where K is any integer) for the integer variables k and l which appear in the exponent of ω, in (5.2a). Thus we have found that the $s \times s$ matrix $\mathbf{q}$ corresponds to the special Jordan block $\mathbf{C}_s(0)$ with Segrè characteristic s. This just means that there exists no similarity transformation (e.g., unitary, star-unitary, orthogonal, etc.) that is able to diagonalize the matrix $\mathbf{q}$. As the more general form of a Jordan block corresponding to $\mathbf{C}_s(0)$ is given by $\mathbf{C}_s(\lambda) = \lambda\mathbf{1} + \mathbf{C}_s(0)$, where $\mathbf{1}$ is the unit $s \times s$ matrix and λ a (real) constant, we see that $\mathbf{C}_s(0)$ and a fortiori $\mathbf{q}$ have eigenvalues and traces (of all orders) equal to zero. These remarks will play an important role by the physical interpretation of the formal results of this chapter; see below, and chapter 6.

5.3.B. On the Second Order Density Matrix

The second point in the present derivations concerns the second order density matrix $\Gamma^{(2)}(g^{N/2})$ connected with the AGP function $|g^{(N/2)}\rangle$, N being the number of "fermions"; see chapter 2 for a detailed account. From equation (2.13) we find that the matrix $\Gamma^{(2)}(g^{N/2})$ takes a simple "box and tail" form in the basis

$$\{\{|i, i+s\rangle,\ 1 \le i \le s\},\ \ \{|i,j\rangle,\ j \ne i+s,\ 1 \le i,j \le 2s\}\} \tag{5.6}$$

as has originally been proved by Coleman[9]. (Here and in the following we introduce the shorthand notation $|i\rangle$ instead of $|\phi_i\rangle$ used in chapter 2, except when we want to make the identification more explicit.) To begin with we recall that the rank s of the geminal

$$|g(1,2)\rangle = \sum_{i=1}^{s} g_i |\phi_i, \phi_{i+s}\rangle \tag{5.7}$$

is assumed to be finite, and that one has $s > N/2$. The AGP function $|g^{(N/2)}\rangle$ is called extreme if the eigenvalues of the first order reduced matrix $\Gamma^{(1)}(g)$ are equal. From a theorem[9] it follows that the eigenvalues of $\Gamma^{(1)}(g^{(N/2)})$ are equal, so that in the the AGP $|g^{(N/2)}\rangle$, equation (2.6), may be regarded as corresponding to a system of N "particles" equally distributed over $2s$ orbitals ($2s > N$).

In the following we consider extreme AGP's exclusively. As stated in chapter 2, in this case $|g\rangle$ is an eigenfuncion of the matrix $\Gamma^{(2)}(g^{N/2})$, which corresponds to the large eigenvalue $\lambda_L^{(2)}$ of the matrix $\Gamma^{(2)}(g^{N/2})$, cf. chapter 2. The remaining eigenvalues are equal and constitute a $(2s+1)(s-1)$ degenerate eigenvalue $\lambda_D^{(2)}$ (which is small if $2s/N \gg 1$). The matrix $\Gamma^{(2)}(g^{N/2})$ can therefore be decomposed as follows:

$$\Gamma^{(2)} = \Gamma^{(2)}(g^{N/2}) = \Gamma^{(2)}_{\text{Box}} + \Gamma^{(2)}_{\text{Tail}} = (\Gamma^{(2)}_L + \Gamma^{(2)}_S) + \Gamma^{(2)}_{\text{Tail}}, \tag{5.8}$$

where the "large box-part" is given by

$$\Gamma_L^{(2)} = |g\rangle\lambda_L^{(2)}\langle g|, \quad |g\rangle = \frac{1}{\sqrt{s}}\sum_i |\phi_i, \phi_{i+s}\rangle, \tag{5.9}$$

the "small box part" is given by

$$\Gamma_S^{(2)} = \lambda_D^{(2)} \sum_{i,j} |\phi_i, \phi_{i+s}\rangle (\delta_{ij} - \frac{1}{s}) \langle \phi_j, \phi_{j+s}|, \tag{5.10}$$

and the "tail part" is given by

$$\Gamma_{\text{Tail}}^{(2)} = \lambda_D^{(2)} \{ \sum_{\substack{i<j \\ i+s \neq j}} |\phi_i, \phi_j\rangle \langle \phi_i, \phi_j| \}. \tag{5.11}$$

Full details are presented in chapter 2, cf. also Coleman[9].

5.3.C. Density Matrix of the Canonical Ensemble and Complex Scaling

In this section we implement the concept of the canonical ensemble into the formalism. This is, for our purposes, physically necessary because of the fact that condensed quantum systems cannot be properly treated as an ensemble of isolated small systems. It should also be observed that the concept of the canonical density operator is confined, in a strict sense, to physical systems being in thermal equilibrium[38]. Nevertheless, this concept has also found a wide applicability to systems being in quasiequilibrium and, more general, in a steady state. For the density operator ρ we have in general

$$\rho = \sum_{k,l} |\Psi_k\rangle \rho_{kl} \langle \Psi_l| \tag{5.12}$$

where $\{|\Psi_k\rangle\}$ is a set of appropriate (many-particle) states. In the special case where these states are eigenfunctions of the system hamiltonian H, i.e. $H\Psi_k = E_k\Psi_k$, one defines the <u>canonical</u> density operator through the well known formula

$$\rho_{\text{can}}^{\text{d}} = \frac{1}{Z} \sum_k |\Psi_k\rangle e^{-\beta E_k} \langle \Psi_k| \tag{5.13}$$

where $\beta = \frac{1}{k_B T}$ (k_B: Boltzmann constant) and Z is an appropriate normalization constant. (The superscript d stays for "diagonal", see below.) In the conventional representation, i.e. before applying CSM, H is of course self-adjoint and E_k is real. Thus one sees immediately that the characteristic Boltzmann factor $e^{-\beta E_k}$ can be obtained by the action of the superoperator with $\frac{1}{Z}e^{(-\beta \hat{H})}$ with $\hat{H} = \frac{1}{2}\hat{L}_B$ (see (4.12)) on the operator $\rho_{kk} = |\Psi_k\rangle\langle\Psi_k|$, as the following equations show, see e.g. (4.10)

$$e^{-\beta \hat{H}} \rho_{kk} = e^{-\beta E_k} \rho_{kk} \tag{5.14}$$

In this equation, $\hat{H}$ is the energy superoperator, see e.g. Prigogine[41]. This quantity is a direct generalization of the conventional hamiltonian. For the proof of (5.14) see chapter 4, where it is simply seen that the superoperator $\hat{H}$ is also self-adjoint, since H is self-adjoint. The physical meaning of the superoperator in (5.14) is naturally that it produces the well known Boltzmann factors. The factorization property displayed in chapter 4 and showed in (5.14) can also be applied to the general ket-bra operators $\rho_{kl} = |\Psi_k\rangle\langle\Psi_l|, \quad k \neq l$, in a consistent manner producing purely quantum mechanical off-diagonal contributions to the density operator, see Eq. (5.12), i.e. correlations of pure quantum mechanical character (cf. Primas[45], d'Espagnat[46], Wheeler and Zurek[47], and the corresponding remarks in chapter 6). We thus obtain

$$e^{-\beta\hat{H}}\rho_{kl} = e^{-\frac{\beta}{2}(E_k+E_l)}\rho_{kl} \tag{5.15}$$

from which one obtains the off-diagonal terms that correspond to (5.13), i.e. dividing the exponential factor in (5.15) by Z. From this it follows that:

(i) the density operator is self-adjoint, i.e. $\rho^\dagger = \rho$

(ii) the correct diagonal elements are proportional to $e^{-\beta E_k}$.

Let us summarize the situation in the following way and let

$$\rho_{\mathrm{m}} \equiv \sum_{k,l} |\Psi_k\rangle c_{kl}\langle\Psi_l| \tag{5.16}$$

with

$$c_{kk} \equiv \text{const.}, \quad \text{all } k, \tag{5.17a}$$

and

$$c_{kl} = c^\star_{lk}, \quad (l \neq k) \tag{5.17b}$$

be a "non-thermalized" density operator. (Clearly, this operator corresponds to a microcanonical-type ensemble.) The canonical density operator ρ_{can} is then obtained through

$$\rho_{\mathrm{can}} = \frac{1}{Z}e^{-\beta\hat{H}}\rho_{\mathrm{m}} \tag{5.18}$$

It follows that by disregarding the off-diagonal elements of ρ_{can} one obtains the very well known "diagonal" part as given in Eq. (5.13), which - per definition - contains no information about EPR-correlations between different states.

Thus far no complex scaling has been considered. Now we proceed to the extension of the above results in the framework of complex scaled quantities. This is easily done here, because of the simple analytic form of the "Boltzmann" superoperator. Thus we need in principle only to know how to scale the hamiltonian H. In the following we make use of some results, see e.g. chapters 3 and 4, or Brändas[20]. There is also a recent very pedagogical review by Löwdin[48] on the Complex Scaling Method that we recommend:

(i) The complex scaled hamiltonian H^c is given by a similarity transformation

$$H^c \equiv UHU^{-1} \tag{5.19}$$

where the operator U, is the unbounded dilation operator fulfilling the condition

$$\left(U^{\dagger}\right)^{-1} = U^{\star} \tag{5.20}$$

in the appropriate domain, see chapter 3.

(ii) The hamiltonian H^c is complex symmetric in the same domain i.e.

$$(H^c)^{\dagger} = (H^c)^{\star} \tag{5.21}$$

This simple property is of particular importance in the context under consideration, because it allows us to construct the complex scaled canonical operator ρ^c_{can} in a very simple way. Due to the simple analytic form of the exponential operator one finds immediately

$$\left[e^{-\beta H}\right]^c = e^{-\beta H^c} \tag{5.22}$$

Using (4.5) and (4.6) where

$$E_k, \epsilon_k = \text{real} \tag{5.23a}$$

and

$$\epsilon_k \geq 0 \tag{5.23b}$$

one finds

$$e^{-\frac{\beta}{2}H^c}|\Psi^c_k\rangle\langle\Psi^{c\star}_l|e^{-\frac{\beta}{2}H^c} \equiv \omega_{kl}|\Psi^c_k\rangle\langle\Psi^{c\star}_l| \tag{5.24a}$$

with ω_{kl} given by (4.9) up to a factor $\frac{1}{2}$, i.e.

$$\omega_{kl} = e^{-\frac{\beta}{2}\{(E_k+E_l)-i(\epsilon_k+\epsilon_l)\}} \tag{5.24b}$$

This proves that the density operator being here under consideration

$$\rho^c \equiv \sum_{k,l}|\Psi^c_k\rangle\omega_{kl}\langle\Psi^{c\star}_l| \tag{5.25}$$

is trivially complex symmetric, i.e.

$$\rho^{c\dagger} = \rho^{c\star} \tag{5.26}$$

and this is in agreement with (5.21), since ρ^c is per definition an operator function of H^c, cf. also Obcemea *et al.*[31]. We also see easily that all the results concerning the complex scaled quantities coalesce to "unscaled" results in the limit

$$\epsilon_k \to 0 \tag{5.27}$$

if one uses the accompanying formal transitions:

$$H^c \to H \tag{5.28}$$

and

$$\Psi_k^c \to \Psi_k \tag{5.29}$$

This remark may be considered as an additional check that the above considerations concerning the complex scaled extension of the canonical density operator are physically meaningful.

Finally we will formulate the analogue of equation (5.18) in the CSM representation. The density operator of a micro-canonical ensemble ρ_{m}^c may be represented as

$$\rho_{\mathrm{m}}^c = \sum_{k,l} |\Psi_k^c\rangle c_{kl} \langle \Psi_l^{c*}| \tag{5.30}$$

with $c_{kk} = \mathrm{const}$, all k, and $c_{kl} = c_{lk}$, which indeed is a complex symmetric operator. The corresponding complex scaled canonical density operator ρ_{can}^c is then obtained through the relation

$$\rho_{\mathrm{can}}^c = \frac{1}{Z} e^{-\frac{\beta}{2}H^c} \rho_{\mathrm{m}}^c e^{-\frac{\beta}{2}H^c} \quad . \tag{5.31a}$$

From the consideration above one finds that this operator is also complex symmetric. From (5.24) and (5.30) we obtain

$$\rho_{\mathrm{can}}^c = \frac{1}{Z} \sum_{k,l} c_{kl} \omega_{kl} |\Psi_k^c\rangle \langle \Psi_l^{c*}| \tag{5.31b}$$

where ω_{kl} is given by (5.24b) above, which gives the explicit form of the canonical operator needed in the derivations below.

5.3.D. Coherent-Dissipative Structures on the Microscopic Level of Description

We proceed now to present the formal conditions that accompany the spontaneous creation of microscopic coherent-dissipative structures. The physical discussion of this formal derivation will be presented in the next section; for further illustration within the experimental context we refer to chapter 6.

The three main formulas that are needed in the following are given by (5.2), (5.10) and (5.31b) i.e.

$$q_{kl} = (\delta_{kl} - \frac{1}{s}) \exp(i\pi \frac{k+l-2}{s}), \tag{5.2}$$

and

$$\Gamma_S^{(2)} = \lambda_D^{(2)} \sum_{i,j} |\phi_i, \phi_{i+s}\rangle(\delta_{ij} - \frac{1}{s})\langle\phi_j, \phi_{j+s}|, \tag{5.10}$$

as well as

$$\rho_{\text{can}}^c = \frac{1}{Z} \sum_{k,l} c_{kl}\omega_{kl} |\Psi_k^c\rangle\langle\Psi_l^{c*}| \tag{5.31b}$$

Comparing Eq. (5.10) with (5.30) we see the following formal correspondence. Applying CSM to the "small box-part" in (5.10), utilizing the replacements $|k, k+s\rangle \to |(k, k+s)^c\rangle$, is nothing but a special case of the complex scaled "microcanonical" operator ρ_{m}^c with the coefficients

$$c_{kl} \equiv \lambda_D^{(2)}(\delta_{kl} - \frac{1}{s}) \tag{5.32}$$

In what follows, the absolute values of the numbers c_{kl} (representing quantum correlations, see above) do not obey any restriction, and thus - if desired - they may also be considered to be arbitrarily small. We can immediately see that $\Gamma_S^{(2)^c}$ is also complex symmetric, as one would expect. We are now ready to ask what happens to $\Gamma_S^{(2)^c}$ if the systems of interest are subject to the canonical ensemble formalism of statistical mechanics. With the considerations of the previous section in mind, we just make use of equation (5.31a). The "part of the ensemble" defined by (5.10) is then subject to the transformation

$$\gamma \equiv \frac{1}{Z} e^{-\frac{\beta}{2}H^c} \Gamma_S^{(2)^c} e^{-\frac{\beta}{2}H^c} \tag{5.33}$$

Here, H^c represents the complex scaled second order reduced hamiltonian[49]. Using the explicit result (5.31b) and the additional assumption that the (real) energies of all the pairs $|(k, k+s)^c\rangle$ are equal, i.e.

$$E_k = E, \quad (k = 1, 2, \ldots s) \tag{5.34}$$

we immediately obtain

$$\gamma \equiv \sum_{k,l} \gamma_{kl} |(k, k+s)^c\rangle\langle(l, l+s)^{c*}| \tag{5.35a}$$

with

$$\gamma_{kl} = \frac{\lambda_D^{(2)}}{Z} e^{-\beta E}(\delta_{kl} - \frac{1}{s}) e^{i\frac{\beta}{2}(\epsilon_k + \epsilon_l)} \tag{5.35b}$$

We thus make the following crucial observation. The matrix elements γ_{kl} of the operator γ, Eq. (5.35), are formally very similar to the matrix elements of the matrix q, Eq. (5.2). More precisely, if we require the validity of the "quantization" conditions

$$\pi \frac{k-1}{s} = \frac{\beta}{2}\epsilon_k, \quad (k = 1, 2, \ldots s) \tag{5.36}$$

we obtain the important equality

$$\gamma_{kl} = \frac{1}{Z}\lambda_D^{(2)} e^{-\beta E} q_{kl} \equiv \text{const} \cdot q_{kl} \tag{5.37}$$

In other terms if the "widths" ϵ_k and the energies E_k of the complex scaled pairs $|(k, k+s)^c\rangle$ fulfill the conditions (5.36) and (5.34), then the matrix elements of the canonical density operator γ constitute a Jordan block $\mathbf{C}_s(0)$ (due to the Reid-Brändas corollary, see subsection 5.3.A). The density operator γ represents the coherent-dissipative structures, i.e., the cooperative (or: synergetic) phenomenon of interest. This remark concludes the desired formal derivations. Further considerations and the physical context of the formalism are presented in section 5.4.

5.4. Spontaneous Creation of Coherent-Dissipative Structures. Further Derivations and Physical Discussion

In this section, we present some remarks concerning the physical interpretation of the previous theory. It will become clear that many formal concepts and formal steps in the above derivations have a physical meaning which may be considered as "natural", in the light of the concrete experimental context of the next chapter. Additionally, the minimal size of the coherent-dissipative structures will be formally derived in general terms. Some remark on the differences between the physical nature of these structures and the well-known superconducting BCS-states (which, of course, represent typical coherent structures) may contribute to the clarification. The following considerations will also enable us to apply the formalism to concrete spectroscopic experiments, see chapter 6.

5.4.A. Spectroscopic Transitions and the Basis Set S

As we intend to apply the above formalism to spectroscopic experiments, we will interpret the units $|\phi_i, \phi_{i+s}\rangle$, or $|i, i+s\rangle$ from now on, appearing in the box-part of the second order density matrix $\Gamma^{(2)}(g^{N/2})$ in the following manner: $|i\rangle$ and $|i+s\rangle$ represent the quantum states of a microscopic, molecular system "before" and "after" a spectroscopic transition "happens", respectively. Or, in other words, these kets represent two possible quantum states of a microscopic system that characterize a spectroscopic transition, i.e., the emission or the absorption of one photon (for examples, see chapter 6).

One should observe that these two states must be considered to be entangled, even with the state of the contributing photon, at least for times of the order of the "transition time" of the considered microscopic system; cf. d'Espagnat[46], Primas[45,50] and Wheeler and Zurek[47].

Thus it appears to be natural to represent a spectroscopic transition through a unit (or "pair") $|i, i+s\rangle$, which in the present physical context points out the fact that the two possible states $|i\rangle$ and $|i+s\rangle$ are intrinsically connected via

Einstein-Podolsky-Rosen (EPR) correlations; cf., eg., the above references. As is well-known, these correlations are of pure quantum mechanical nature and thus they have no classical analogue.

In more conventional terms (for chemists) one can also say that, "during the time" the spectroscopic transition "takes place", the quantum system has to be described by some linear superposition of $|i\rangle$ and $|i+s\rangle$. (This is possible, however, only if the degrees of freedom of the electromagnetic field are not considered explicitly – as usual in the field of "linear" spectroscopy- and if the microscopic system, say the molecule, is isolated "during" the transition time.)

5.4.B. On Fermionic Degrees of Freedom, and the Units $|i, i+s\rangle$

The absorption and emission of photons by a microscopic quantum system (e.g., a molecule) is caused by the interaction of some relevant electronic degrees of freedom of the system with some other degrees of freedom of the electromagnetic field acting on the system. As is well known, in the physical context of "linear" molecular spectroscopy (i.e., by sufficiently weak external fields) the semiclassical (or: Dirac-) form

$$H_I = -\vec{D} \cdot \vec{E}(t) \tag{5.38}$$

for the interaction hamiltonian yields excellent results; here $\vec{D}$ is the electronic dipole moment operator of the system, and $\vec{E}(t)$ is a classical function, or a c-number, that represents the electric component of the field[51]. (Of course, this operator describes dipole-allowed transitions, like those discussed in chapter 6.) In a oversimplified manner (which, however is in common use between experimentalists) one can say that "an electron undergoes a transition between two orbitals", and this is accompanied by the emission (or absorption) of a photon. In a more detailed (and thus realistic) description one needs of course to study the electronic many-body system introducing "correlation" for each individual molecular state taking part in the emission (or absorption). Hence the correlation features manifest themselves in the associated well-known properties (see also chaptcs 2 and 3) of the one- and two-matrix. Thus it is immediately clear that fermionic degrees of freedom are involved in the build-up of the pairs $|i, i+s\rangle$, as they have been interpreted in the previous subsection. To be a little more specific, let us consider, as an example, the case where (mainly) one electron is "involved" in the interaction with the field; then we have

$$\vec{D} = e\vec{r} \ , \tag{5.39}$$

the symbols are self-explanatory. The entanglement of the two particle states represented by $|i, i+s\rangle$ is consequently intrinsically connected with the fact that the many-body correlations lead to a two-body density matrix representing in general non-integral pair- and one-particle occupations. In any case it is clear that the degrees of freedom connected with the present emission or absorption process of the microscopic quantum system have manifestly fermionic origin even though they are coupled (paired) into (quasi) bosonic rotational or vibrational degrees of freedom in accordance with the spectroscopic processes considered. It is hence obvious that the unit $|i, i+s\rangle$ which is associated with two electronic and thus fermionic degrees of freedom should be viewed as one boson-type degree of freedom, and when we refer to the number of degrees of freedom below, we will always mean the latter type. Let us also point out the –more or less– obvious fact that, the properties of the density matrix $\Gamma^{(2)}$ being presented in chapter 2 (or subsection 5.3.B) are valid because of the fermionic character of the single-particle states $|i\rangle$ and not (as erroneously it might be believed) because these states may be occupied by distinguishable, different particles that, moreover, also happen to be fermions. (The latter point of view is intuitively motivated by the build-up of Cooper pairs from "individual" electrons in the BCS theory of superconductivity; see subsection 5.4.G. To avoid confusion, we emphasize here that the presently used concept of "paired degrees of freedom" is not identical with the concept of "particle pairs".)

5.4.C. On the Spectral Resolution and the Degeneracy Condition $E_k = E$

In subsection 5.3.D we have used the condition $E_k = \text{const.}$, for all $k = 1, \ldots s$. This means that the (real) energies of all the units that constitute the aforementioned Jordan block are equal. Moreover, it suffies to assume that

$$|E_k - E| \leq \Delta E, \quad k = 1, \ldots s. \tag{5.34'}$$

This condition is, of course, more realistic, and it permits us to recognize its physical meaning easily: In any realistic spectroscopic experiment, the energy of the detected transitions can only be measured up to a finite accuracy which is given by the so-called spectral resolution of the apparatus. It should be observed that all the detected spectroscopic transitions which fulfill condition (5.34′) are then indistinguishable in the sense that there is no possibility to attribute each of them to some individual quantum system (e.g., molecule) of the spectroscopic specimen; for more details and examples, see chapter 6.

5.4.D. On Geminals of Extreme Type

The spectroscopic transitions fulfilling the condition (5.34′) are indistiguishable, as discussed above. Hence it appears to be "natural" to assume that the N/2 considered spectroscopic transitions be equally "distributed" over all the possible units $|k, k+s\rangle$. This assumption is in line with the basic ideas of statistical mechanics of equilibrium (and quasi-equilibrium); it corresponds to the best "unbiased choice" that conforms with equation (5.34′). We should point out, however, that this remark may illustrate the "extremity" of the AGP function $|g^{(N/2)}\rangle$ in the present physical context; but it does not explain (or give a clear physical insight into) the fact that the considered coherent-dissipative structures are connected –as presented in section 5.3– with functions of that type. (Almost the same remark holds true for the case of superconductivity, cf. Coleman[9,52].)

5.4.E. On the Finite Rank of $\Gamma^{(2)}$ and the Associated Dissipative Phenomena

In the derivations presented in section 5.3, we considered the case where the index s is finite exclusively. In other words, the dimension of the physical space under consideration is assumed to be finite. Therefore one might wonder why this restriction does not prevent the treatment of irreversible (or: dissipative) phenomena as those being under consideration in this article. Indeed, as it has been pointed out by many authors (e.g., cf. Prigogine[41] and references cited therein) a necessary condition to break the reversible character of the fundamental equations of motion (Schrödinger or Liouville-von Neumann equation) is that the physical system under consideration has infinitely many degrees of freedom. The clarification of this apparent "contradiction" is more or less trivial, and it is given by the physical meaning of the concept of the "resonance" of the complex scaling method. Namely, as an example, an isolated resonance that may be revealed after complex scaling, sometimes is found to be associated with a real, measured "bump" in some observed continuous spectrum. This very well known fact just shows that a space of infinite dimension, in some conventional quantum representation, can be represented even by an one-dimensional space after the appropriate complex scaling; cf. Reed and Simon[53].

In the context of section 5.3, we showed how coherent-dissipative structures can appear after complex scaling of the density operator. Remember that, prior to the application of the complex scaling, no cooperative phenomenon (of the kind of the considered coherent-dissipative structures) has been reported. These remarks are also in line with the fact that we considered the canonical ensemble associated with the (second order) density matrix; see subsection 5.3.C. This means physically that any degree of freedom of the canonical ensemble should be considered to be "coupled" with infinitely many degrees of freedom of the thermal bath. (Remember that the use of the canonical ensemble formalism presupposes the existence of a thermal bath, which is assumed to be weakly coupled with any system of the ensemble; cf., e.g. Balescu[38].) In the light of the CSM, one can say that the "lifetimes" associated with the imaginary energies $i\epsilon_k$, Eq. (4.6) or (5.23), are due to those "couplings". The above comments may also be used in order to clarify the appearance of a contractive evolution operator $[\exp(-i\hat{L}t)]^c$, i.e. a Lyapunov converter, after complex scaling; cf. Obcemea and Brändas[30].

5.4.F. Minimal Size of the Coherent-Dissipative Structures. Derivation and Physical Interpretation

In this subsection we proceed in the completion of the formalism by determining the "size" the coherent-dissipative structures can have. This is a crucial task, because of its importance for the direct confrontation of the theory with real spectroscopic experiments, see chapter 6. In the previous section we found the "quantization" conditions

$$\pi\frac{k-1}{s} = \frac{\beta}{2}\epsilon_k, \quad (k = 1, 2, \dots s) \tag{5.36}$$

which are necessary for the appearance of Jordan blocks of Segrè characteristic s. (As is well known, these blocks are not diagonizable, and – as a consequence– the scaled pairs $|(k, k+s)^c\rangle$ constituting them, appear to "act cooperatively"; see below.) Of particular importance for the further derivation is the above proved relation (5.3a) concerning the range of the indices k and l of q_{kl} and thus also of the number k of the left-hand-side (lhs) of Eq. (5.36): It has already been shown that

$$k = K+1, \dots, K+s \tag{5.3a}$$

where K is an arbitrary integer (positive of negative); see subsection 5.3.A. Here the following points should be observed. Any choice of the value of K affects the magnitude of the lhs of (5.36). On the other hand, the values of the index k of the quantities ϵ_k do not have any physical significance; they may always be renumbered in order to achieve the conventional numbering $\epsilon_1, \dots, \epsilon_s$, which just corresponds to the numbering $1 \leq k, l \leq s$ of the rows and columns of the $s \times s$ matrix q_{kl}. Therefore, the above condition may be rewritten as

$$\pi\frac{k-1}{s} = \frac{\beta}{2}\epsilon_{k-K}, \quad (k = K+1, \dots K+s) \tag{5.40}$$

Now we proceed in the determination of the physically meaningful range of k (and K) through the following considerations:

(1) As we already know from the general theory of CSM, the "widths" ϵ_k, equation (5.23), should be positive real quantities. (This physically means that one is dealing with decaying states.) Thus, from (5.40) it follows the condition

$$K \geq 1 \tag{5.41}$$

(2) Secondly, we require that no "pair lifetime"

$$\tau_k = \frac{\hbar}{2\epsilon_k} \tag{5.42}$$

be greater than the characteristic relaxation time τ_{rel} of the relaxation of a quantum system, which can be studied in a specific spectroscopic experiment by standard methods. (E.g., in the case of far-infrared absorption one can measure the rotational relaxation time τ_{rot} of a dipolar molecule, whereas in the case of fluorescence one can measure the molecular fluorescence decay time τ_{F} ; cf. chapter 6. These two quantities represent two examples for τ_{rel}.) This requirement thus physically means that all the pairs constituting the considered Jordan block will "disappear" (through decay) after a time of the order of the experimentally measured time interval $\tau_{\rm rel}$. Or, in other terms, the "lifetime" of the coherent-dissipative structure is not longer than the quantum relaxation process which creates this structure; see also below. This requirement is formally represented through:

$$\tau_{\rm rel} = \tau_1 = \frac{\hbar}{2\epsilon_1} = \frac{\hbar\beta}{4\pi}\frac{s}{K} \tag{5.43}$$

as it follows immediatelly from equations (5.40) and (5.3a). We know from (5.40) that $K \geq 1$. Thus we arrive at the following crucial point:

(3) The choice

$$K = 1, \tag{5.41a}$$

which restricts condition (5.41), also guarantees that the Segrè characteristic of the considered Jordan block becomes minimal! Namely one sees easily that any other choice, i.e. $K > 1$, yields a numerical value for s which is proportional to K, whereas the <u>Segrè characteristic</u> of the considered Jordan block coincides with the number of the pairs constituting the Jordan block, which is always equal to s, see (5.3a).

Let us summarize the above results: The "quantization" conditions that lead to the creation of Jordan blocks may be formulated as

$$\frac{1}{2}\beta\epsilon_k = \pi\frac{k}{s}, \quad k = 1, \ldots s \tag{5.44}$$

or, alternatively,

$$\tau_k = \frac{\hbar\beta}{4\pi}\frac{s}{k}, \quad k = 1, \ldots s \tag{5.45}$$

see equations (5.36), (5.40) and (5.43). The additional equirement

$$\tau_{\rm rel} = \frac{\hbar\beta}{4\pi}\frac{s}{1}, \tag{5.46}$$

guarantees that –physically speaking– all the "members" (i.e. pairs) of the Jordan block, and therefore also the Jordan block itself, decay within a time of the order of the (measurable) relaxation time of the microscopic quantum systems under consideration. Also we have seen that the same conditions which led us to the equations (5.44) up to (5.46) determine, at the same time, the minimal "size" $s_{\rm min}$ that the Jordan block can achieve:

$$s_{\rm min} = \frac{4\pi}{\hbar\beta}\tau_{\rm rel} \tag{5.47}$$

These results are needed explicitly for the treatment of the two spectroscopic experiments which are presented and discussed in chapter 6.

5.4.G. Further Remarks on the Physical Interpretation of the Formalism.

In this subsection we represent additional comments and remarks that may illustrate the above formalism and also help to visualize its physical significance. The two real spectroscopic experiments to be discussed in the next chapter will then illustrate the formalism furthermore.

(1) We found in subsection 5.3.D that the complex scaled part $\gamma \equiv (\Gamma^{(2)}_{S,\mathrm{can}})^c$ of the "thermalized" or: canonical second order density matrix, see eq. (5.33), is similar to a Jordan block, if the appropriate "quantization" conditions are fulfilled. As is well known, however, a Jordan block is characterized by its non-diagonalizability. Hence, there exists no unitary, star-unitary, restricted similarity, or another (fully arbitrary) similarity transformation that diagonalizes γ! Thus the units, or pairs, that constitute the considered part γ of the full density matrix $\Gamma^{(2)}$ act cooperativelly and create a new "unit" which we called a coherent-dissipative structure. In other terms we say that –in the case under consideration– all the units $|(k,k+s)^c\rangle$ for $k = 1,\dots s$, coalesce and act "as a whole". (Here, and in the following parts of this article, we will suppress the upper index "c" on the complex-scaled quantities, if there is no danger for confusion.) In more physical terms we may say that, in the considered case, one cannot attribute a probability to the "particular" spectroscopic transition represented by $|k,k+s\rangle\langle(k,k+s)^*|$ just because of the fact that now this particular transition does not exist (and this holds for all $k = 1,\dots s$). Namely, the facts that (i) $\mathrm{Tr}\gamma = 0$, and (ii) that physically meaningful occupation probabilities must be non-negative, $p_k \geq 0$, imply $p_k \equiv 0$, for all k.

(2) It has been mentioned that the units $|k,k+s\rangle$ (before and after complex scaling) are considered to represent a possible spectroscopic transition between the "unperturbed states" $|k\rangle$ and $|k+s\rangle$ of one quantum system. All these units $(k = 1,\dots s)$ have the same (real) energy, $|E_k - E| \leq \Delta E$. Thus all these units are multiplied by the same Boltzmann factor, in the formalism of the canonical ensemble; cf. subsection 5.3.D. Now, one quantum system (or more of them) may appear to be delocalized over the space of states

$$S_E \equiv \{|k,k+s\rangle;\quad E_k = E;\;\; k = 1,\dots s\} \tag{5.48}$$

As is well known, the complex scaling method assigns "widths" ϵ_k to that units. These widths are considered to determine the "life times" (or decay times) of the elements of S_E through the relation $\tau_k = \frac{\hbar}{2\epsilon_k}$. These "life times", however, cannot be identified with possible real decay (or relaxation) times of an excited quantum system, because, in the present case, all the elements of S_E act cooperatively, and thus the whole Jordan block represents the above decay of a single quantum system. The requirement $\tau_{\mathrm{rel}} = \tau_1$, Eq. (5.43), which can be reformulated as $\tau_{\mathrm{rel}} \geq \tau_k$ (for all $k = 1,\dots s$) may be interpreted in physical terms as the condition that the coherent-dissipative structure associated with the Jordan block over S_E does decay with the time constant τ_{rel}. From the above remarks it follows that there does not exist a physical interpretation of each individual τ_k in terms of measurable quantities.

(3) Further clarification of the physical content of the above theory may now be achieved by pointing out some conceptual differences between our formalism and the well known BCS theory (or better, model) of superconductivity. It has been proved by Blatt[54] and repeatedly pointed out by Coleman[6,9,55] that the BCS ground state ansatz is "equivalent" to an AGP ansatz. As is well known, the formation of Cooper pairs of electrons is accompanied by a negative energy, which –in conventional treatments– is considered to be the "main necessary physical condition" for the formation of the superconducting state. But, as e.g. Coleman[6] and Yang[1] have pointed out, the most fundamental aspects of the phenomenon of superconductivity ought to be presented mainly in connection with the wave function of this state. As an example, Coleman found that AGP's of extreme type do represent the cases where "coherent pairing" is most intense. In this context, moreover, it is interesting to observe that the explicit form of the interaction hamiltonian does not appear in the fundamental papers of Yang and Coleman cited above. (The reason for this is just that the full hamiltonian is not needed in detail, in order to describe many of the main aspects of superconductivity!)

The above remarks may be used to stress the following point: In the case of superconductivity, the ground state is represented by the part $\Gamma_L^{(2)}$, Eq. (5.9), of the total density matrix, where the geminal $|g\rangle$ is extreme. This physical state is clearly associated with a pure quantum state which, additionally, has the lowest possible energy of all the states of the system. In contrast to this, a coherent-dissipative structure is not represented by a pure quantum state (and therefore we do not call it a "state" but a "structure"). Furthermore, the "pairing" $|k, k+s\rangle$ has here a totally different physical meaning (see above), and it is associated with a positive (real) energy E. This shows that the coherent part $\Gamma_L^{(2)}$ is associated with a very high energy (being of the order of $E\lambda_L^{(2)}$,), and thus there is a vanishing probability at quasi-equilibrium for the occupancy of this state. (Remember that, in the case of superconductivity, the "opposite" is the case; cf. chapter 2.) Thus the condensed system may occupy the "non-coherent" part (which is of no interest in the present context), or the "small-box part" $\Gamma_S^{(2)}$, which, after complex scaling, exhibits a qualitatively new form of cooperative (or, synergetic) behavior, which is formally represented by the Jordan block of order s. Please also remember that the appearance of the Jordan block is immediately connected with the phase factors $e^{-\frac{i\beta}{2}(\epsilon_k+\epsilon_l)}$, cf. Eqs. (5.31), which the CSM introduces in the formalism of the second order density matrix.

In more physical terms, one may say that the spontaneous creation of coherent-dissipative structures is due to the dynamical process of quantum transitions, which, as already has been mentioned in section 5.1, do disturb the equilibrium (quantum or classical) correlation patterns of the quantum systems in the condensed macroscopic system. Clearly, these structures are not caused by "noise", and they ought to be considered to belong to the field of quantum dynamics, rather than to the theory of stochastic processes. It also should be pointed out that the coherent-dissipative structures always have a finite lifetime. In contrast to this fact, a superconducting state has an infinite lifetime. Another (rather formal) feature that might be worth mentioning is the following: In the case of BCS states, it holds that the occupation number of any involved pair state is equal to one. In contrast to this, in the present case there is no specification of the expectation value of the occupancy of the units that create the Jordan block. From the above considerations we conclude that the coherent-dissipative structures and the superconducting states represent qualitatively different organized forms of matter.

(4) At this stage we would like to mention that there seems to be a similarity between the coherent-dissipative structures and the very well known dissipative structures of Prigogine and coworkers: Both cannot have a "size" smaller than a critical one; both cannot continue to exist, if "external energy supply" stops. On the other hand, it should be stressed that Prigogine's dissipative structures are concepts of phenomenological thermodynamics, whereas the coherent-dissipative structures are concepts of microscopic theory. Thus the following speculation may be mentioned: The formalism of coherent-dissipative structures may represent the framework in which the phenomenological dissipative structures can be established "from first principles".

(5) As we mentioned in chapter 2, the extreme condition on the AGP under consideration is a necessary condition for the appearance of Yang's concept of off-diagonal long-range order (ODLRO). We also have shown in subsection 5.3.D that the appearance of coherent-dissipative structures is intimately connected with the existence of off-diagonal terms in $\Gamma_S^{(2)}$. In this context, we would like to stress that no intuitively appealing explanation (i.e., an explanation in classical mechanical terms) of the emergence of coherent-dissipative structures is possible. Namely, as Yang[1] points out: "Since off-diagonal elements (of the density matrix) have no classical analog, the off-diagonal long-range order ... is a quantum phenomenon not describable in classical mechanical terms".

6. Applications to Condensed Matter Spectroscopy

In this chapter we will make contact with two real experiments. In section 6.1 we will discuss a far-infrared absorption experiment in liquids and show the difficulties interpreting it from first principles. Then we will discuss the physical context of this experiment in the light of the aforementioned theory of coherent-dissipative structures. It will be shown that now all the above "difficulties" disappear totally. In section 6.2 we will make contact with a recent experiment in luminescence emission spectroscopy, in the visible spectral range. Some particular aspects of this experiment are presented and shown to contradict the predictions of the "orthodox" theory. Once more, the analysis of this experiment in the light of the present theory will help us to clarify some physical reasons for the detected cooperative effect known as "D-fluctuation". It should be observed that these considerations also represent the first application of the CSM to realistic condensed matter problems.

6.1. On the Far-Infrared Absorption in Liquids

6.1.A. Anomalous Temperature Dependence of FIR Absorption Bands in Liquids

An anomalous "red shift" of the far-infrared (FIR) absorption band of dilute solutions of acetonitrile (CH_3CN) in n-heptane with increasing temperature was first observed by van der Elsken and coworkers[56,57]. The experiment shows that the whole FIR absorption band shifts continously to lower frequences with increasing temperature; cf. fig. 1. As these authors pointed out: "This now is contrary to what one could expect for a pure rotation band of a gas and is also in disagreement with existing theories for rotational bands in liquids" see e.g. Kroon *et al.*[56].

As a matter of fact, this effect seems to contradict the validity of the Boltzmann distribution law and/or the validity of the equipartition theorem of (classical as well as quantum) statistical mechanics. This effect, namely, indicates that, with increasing temperature of the liquid, the lower energy levels of the hamiltonian representing the absorbing system become more populated at the expense of the population of the higher energy levels. It should here be pointed out that we are considering a typical absorption experiment where the absorbing system always remains in (quasi-)equilibrium because of the fact that the external field is very weak. The absorption of FIR radiation is supposed to be due to the "rotations" of the dipolar molecules (with dipole moment $\vec{m}$), which couple with the FIR field. This coupling is represented by the interaction hamiltonian

$$H_I = -\sum_j \vec{m}_j \cdot \vec{E}(t) \tag{6.1}$$

where $\vec{E}(t)$ is the electric component of the FIR field; see, e.g., Gordon[58].

It is also easily seen that this effect is of generic origin, i.e., independent of the following questions:

1) What is the precise definition of the hamiltonian of an absorbing quantum system.

2) How strong is the rotational-translational coupling for the dipolar molecules appearing in eq. (6.1).

3) What changes show the density and/or the viscosity of the liquid with increasing temperature, etc., etc...

Namely, as long as no "pathological" situations (like, e.g. a phase transition) are

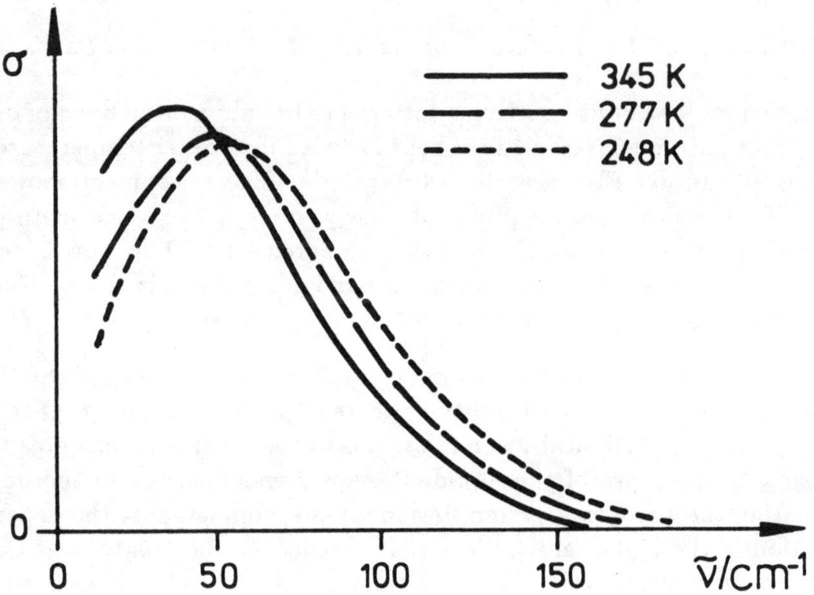

Figure 1

Schematic representation of the effect of the temperature dependence of the absorption cross section of CH_3CN dissolved in n-heptane (dilute solution, 0.4%). For details see Kroon *et al.*[56], cf. also Lippert *et al.*[59].

considered, the total hamiltonian H is just

$$H = H_0 + H_I \tag{6.2}$$

where H_0 represents the isolated system (i.e., the spectroscopic specimen when the external field is switched off). After some reflexions one sees that none of the above questions can reveal a reason that could explain the aforementioned effect. This reasoning is also confirmed by the fact that further experimental work carried out by different research groups proved that the above "anomalous" effect is <u>independent</u> of (i) the concentration of the solution and (ii) the physical and/or chemical nature of the solvent. Moreover it has also been found that this effect appears in the FIR absorption bands of many other molecules (in dilute solution or even in neat phase), like, e.g., $CHCl_3$, CH_2Cl_2 and C_6H_5Cl; for references, cf. Lippert *et al.*[59].

A nearer analysis of the physical parameters and conditions underlying the molecular motion in the liquid and the FIR absorption, based on the well known energy-time uncertainty relation, has shown that this absorption cannot be just considered to be given by the "rotational" motion of the polar molecules (to be thought of, if necessary, as "dressed" with some other molecules in their environment), Lippert *et al.*[59]. Here it is not possible to consider the FIR absorption spectra as an incoherent superposition of many "transitions" between (quasi-rotational) energy levels of the hamiltonian (6.2). To prove this, it is sufficient to notice that the thermal energy $\frac{1}{2}k_BT$ corresponds to about 100 cm^{-1}, at room temperature, and, at the same time, the main part of the considered spectra (cf. fig. 1) is situated between, say, 10 cm^{-1} and 100 cm^{-1}. Thus many of the elementary transitions $\Delta E_{if} = E_f - E_i$ that, at first glance, are believed to constitute the FIR spectrum incoherently, are strongly correlated, the correlations being created by the thermal disturbance. On these physical grounds, furthermore, it has been shown that the off-diagonal part of the density operator (in the representation of the hamiltonian of the absorbing quantum molecular systems) plays the main role by the FIR absorption process under consideration[59]. This result was obtained by considering the validity of the standard linear response treatment of the FIR absorption (see e.g., Berne and Harp[60]) in the light of van Vliet's revisited linear response theory[61,62].

For illustration it might be helpful to formulate the aforementioned anomalous effect in the "time domain", too. This is easily done with the aid of the standard fluctuation-dissipation theorem (FDT), cf. Berne and Harp[60]. The FDT permits, under certain conditions, the calculation of the time autocorrelation function for the dipole operator $\vec{m}(t)$. Then the measured temperature dependence of the FIR bands can be expressed as follows: with increasing temperatur, the rotational relaxation time of the absorbing quantum systems increases too! This statement also questions the applicability of the standard FDT in the physical context under consideration, Lippert *et al.*[59].

6.1.B. Coherent-Dissipative Structures and FIR Absorption

As we already know from introductory quantum theory, the absorption of radiation occurs in "quantized form". Expressions of the kind "a molecule absorbs an energy quantum $h\nu$" are in common use. Thus sometimes it may happen that one is inclined to consider the absorption of each FIR photon as a spacially localized process; i.e., one tacitly assumes that the photon energy $h\nu$ becomes absorbed in some well defined volume element ΔV being given by the size of one absorbing molecule (because of the "fact" that the photon becomes absorbed by one molecule).

In the light of quantum mechanics, however, the above remarks should not be taken seriously, because they are in contradiction with the standard uncertainty relations. Namely, it is easily seen that photons of a field with wavelength $\lambda_0 \pm \Delta\lambda$ cannot be thought of to be confined in a volume element with linear dimension considerably smaller than $\Delta\lambda$. In the experiment under consideration, cf. fig. 1, it is reasonable to put $\Delta\tilde{\nu} \approx \tilde{\nu}_0 \approx 40$ cm^{-1} and thus $\Delta\lambda \approx \lambda_0 \approx 1/40$-cm $= 2.5 \cdot 10^{-2}$ cm . But in a volume with $\Delta V \approx (2.5 \cdot 10^{-2})^3$ cm^3 there are approximately $N \sim 10^{16}$ molecules of the liquid. Thus the dynamics of every one elementary absorption process should be considered to depend – in principle – on a huge number of degrees of freedom, rather than on the degrees of freedom of one molecule only. These remarks clearly confirm the corresponding considerations of the last section, where we pointed out that the FIR absorption bands under consideration are not simply incoherent superpositions of many, very closely spaced elementary transitions (between eigenstates of the hamiltonian H_0).

After these illustrative considerations, we proceed to show what are the consequences of the theory presented in the previous chapters, within the physical context of FIR absorption.

Let us consider a molecular absorbing system (or, quantum system). This system may be a dipolar molecule, like for instance CH_3CN. Now owing to the strong interactions between the roto-translational degrees of freedom, it seems to be more reasonable to take a dipolar molecule together with a sufficient large number of other molecules in its environment, as the definition of an appropriate quantum system. Of course there are many such (dipolar) molecules distributed throughout the liquid. Let the possible states of these quantum systems be $|i\rangle$, $i = 1, \ldots s$. The whole spectroscopic specimen (that is the macroscopic system consisting of a large number of dipolar molecules including environmental molecules) is assumed to be in thermal equilibrium (or, in quasi-equilibrium).

Now we will consider the introduction of a weak FIR field (remember that $E(t)$ in equation (6.1) is a c-number). As usual in "linear" spectroscopy, one can assume that this field is monochromatic: $\vec{E}(t) = \vec{E}_0 \cos\omega t$. Thus, all the present FIR photons have the same energy $\hbar\omega$. An elementary spectroscopic transition consists in the absorption of one photon by some quantum system. Thus to each $|i\rangle$ one can associate an excited state $|i+s\rangle$ that might be considered to follow from $|i\rangle$ "after the absorption of one FIR photon".

In the light of recent quantum theoretical developments dealing with (i) the quantum theory of measurement,(ii) Bell's inequalities and (iii) the experimental verification of the existence of non-local Einstein-Podolsky-Rosen (EPR) correlations (cf. d'Espagnat[46], Primas[45], and Wheeler and Zurek[47]), however, one should be inclined to consider the states $|i+s\rangle$, $|i\rangle$, and the photonic state as being <u>entangled</u> "during the absorption".

Thus it appears to be natural to represent the absorption process under consideration by the set of pairs:

$$\mathcal{S} = \{|i, i+s\rangle, \quad i = 1, \ldots s\}. \tag{6.3}$$

At this stage of consideration, moreover, it is also natural to speak, let us say, about the probability per unit time that a quantum system undergoes a transition (represented by an element of $\mathcal{S}$) owing to the absorption of one FIR photon.

We stress again that the unit represented by $|i, i+s\rangle$ involves pairs of fermionic degrees of freedom. This is, as we have seen, due to the fact that the dipole moment operator $\vec{m}$ appearing in the coupling hamiltionian (6.1) represents an electronic distribution within each molecule of the dissipative system, whose size will be determined below; furthermore, the "absorption" is effectuated by the coupling which is given (in the semiclassical dipole-dipole approximation) by (6.1)*. The above interpretation of the pairs constituting the set $\mathcal{S}$ shows that each pair is associated with the same (real) energy $E = \hbar\omega$. This remark is related to the corresponding assumption of chapter 5 that led to equation (5.34) and thus to the emergence of the Jordan blocks in the CSM-transformed density matrix; see chapter 5.

Now we proceed to the determination of the minimal dimensionality s_{min} of the space $\mathcal{S}$, equation (6.3), whose elements represent physically possible spectroscopic transitions which are due to FIR absorption.

As stated in chapter 5, we can obtain an estimate of s_{min} by the following reasoning: Let τ_{rot} be the well-known rotational relaxation time of a quantum system. This relaxation time is a characteristic "molecular" parameter of the FIR absorption, in the context of all the conventional theories and/or models describing this absorption process, e.g. cf. Gordon[58], or Berne and Harp[60]. Hence, in the conventional "one-body" point of view one considers τ_{rot} to be the characteristic decay time of the time autocorrelation function $< \vec{m}(0) \cdot \vec{m}(t) >$ of the molecular dipole moment $\vec{m}$, see equation (6.1). In chapter 5, however, we saw that the "one-body" relaxation processes can create Jordan blocks in the CSM-transformed second-order density matrix of the condensed system. The "quantization condition" for this creation was found to be (see equation (5.45))

$$\tau_i = \frac{\hbar}{4\pi k_B T}\frac{s}{i}, \tag{6.4}$$

where the variable i takes the natural range

$$i = 1, 2, \ldots s. \tag{6.5}$$

---

* Of course the real dipole moment operator may have many-body contributions, and the real coupling should show the quantum character of the perturbing field. Nevertheless these remarks do not affect the present considerations.

In this case the above mentioned "obvious" meaning of the elements of S is lost. Here all the members of S should be considered to show a cooperative or coherent behaviour. In other terms, all the "elementary transitions" contained in S become strongly correlated, thus defining a coherent-dissipative structure. Hence in the present case it is physically meaningless to ask "what is the probability that a quantum system undergoes a transition represented by the specific pair $|i, i+s\rangle$".

From (6.4) we see that the longest "life time" is given by $\tau_1 = (\hbar/4\pi k_B T)s$. Thus we observe that for times t fulfilling the condition

$$t > \tau_1, \tag{6.6}$$

the aforementioned coherent-dissipative structure may be considered to have decayed, because all the units $|i, i+s\rangle$ that constitute this structure have "life times" shorter than τ_1. Therefore it appears to be natural to identify the "one-body" quantity τ_{rot} with the longest of all the "life times" characterizing the Jordan block (that is, the coherent structure) under consideration:

$$\tau_1 = \tau_{rot}. \tag{6.7}$$

In other words, condition (6.7) tells that the life time of the coherent-dissipative structure, which "accompanies" the absorption of one FIR photon by some quantum system, cannot be longer than the duration of the rotational relaxation process which (in the light of the conventional fluctuation-dissipation theorem, cf. section 6.1.A) is intimately connected with the considered FIR absorption.

This condition, moreover, also permits the determination of the aforementioned minimal value s_{min} that the dimension of the coherent-dissipative structure can take. From (6.4) and (6.7) we obtain

$$s_{min} = \frac{4\pi k_B T}{\hbar}\tau_{rot}. \tag{6.8}$$

(One sees immediately that the condition $\tau_{rot} = \tau_i$ with $i > 1$, instead of (6.7), implies a value for the parameter s which is greater than that of the equation (6.8).) At room temperature, $T = 300K$, we have $\hbar/4\pi k_B T \approx 2 \cdot 10^{-15}s$; for the value

$$\tau_{rot} \approx 10^{-12}s$$

being typical for molecular liquids, we thus obtain:

$$s_{min} \approx 500. \tag{6.9}$$

This result states that the coherent structures created by the dynamics of the considered absorption process "couple" at least 500 elementary transitions which in conventional theory are assumed to be dynamically independent.

We are now in the position to interpret illustratively the "anomalous" temperature dependence of the FIR absorption bands (mentioned in section 6.1.A) from "first principles":

An increase of the temperature causes an increasing number of "collisions" per unit time between the molecular absorbing systems (or quantum systems). Therefore the size of the correlation patterns can also grow, and the same holds true for the size of the coherent-dissipative structures which cannot be smaller than a critical size characterized by s_{min}; see (6.8). Now remember that the FIR absorption under consideration is governed by the interaction hamiltonian given by equation (6.1). Interpreting this interaction classically, one sees that the interaction energy (and thus also the FIR absorption rate) depends on the reorientations of the (molecular) dipole moments $\vec{m}_j$. Thus in our case it is clear that, if coherent-dissipative structures are apparent in the liquid, the FIR absorption is caused by the reorientations of these structures, rather than by the reorientations of the individual molecules. In this case, an increase of the temperature can cause an increase of the size of these structures, which plainly means that these structures also become "heavier" and thus "rotate" slower. But, a slower reorientational motion (that is, a larger rotational relaxation time) is expected to cause a red shifted FIR absorption spectrum, just as the experimental results substantiate!

Thus we see that the FIR-band temperature dependence under consideration (see section 6.1.A) appears to be "anomalous" only in the context of the conventional theories which describe the absorption process – more or less – in the "one-body" point of view. Moreover, the existence of coherent-dissipative structures in the liquid does explain the observed effect from first principles.

6.1.C. Supplementary Remarks

(1) The succesfull theoretical treatment of the aforementioned "anomalous" temperature dependence of FIR absorption bands of liquids clearly states that the conventional point of view which considers each (caged) dipolar molecule as individual uncorrelated quantum (or classical) systems for the description of dynamical processes in condensed matter is physically unjustified. In more specific terms, one may say that the micro-dynamical process under consideration is due to the "cooperative" behaviour of many quantum systems that acts as a "whole" for the absorption of one FIR photon. Once again we stress that this "cooperation" between molecular systems follows from first principles of quantum mechanics and CSM (see the previous chapters), and is entirely due to the non-diagonal part of the density operator of the condensed system.

(2) It may also be illustrative to express the weakness of the theoretical treatment of the FIR anomaly within the language of conventional theories. As an example, let us consider the treatment of FIR absorption within first-order perturbation theory, cf. Gordon[58]. Here the irradiation energy loss (that is, the absorption

rate), $(-\dot{E}_{rad})$, is assumed to be given by the ansatz:

$$-\dot{E}_{rad} = \sum_{f,i} \hbar\omega_{fi} \cdot P_{f\leftarrow i} \cdot \rho_i. \tag{6.10}$$

Here $\hbar\omega_{fi}$ is the energy for the transition $|i\rangle \Rightarrow |f\rangle$, $P_{f\leftarrow i}$ is the corresponding transition rate (which is assumed to be given by the so-called Golden Rule) and ρ_i is the probability of finding the absorbing quantum system in the initial state $|i\rangle$. (Parenthetically it should be stressed that this ansatz has the logical status of an axiom, within the theory reviewed by Gordon[58].) But it is immediately obvious that this ansatz neglects the off-diagonal parts of the relevant density operator, ρ, totally. Namely, the above "probabilities" ρ_i are nothing but the diagonal parts (ρ_{ii}) of the density operator! Thus, this ansatz cannot contain information about a possible ODLRO. (The physical reasons showing the importance of the non-diagonal parts of ρ, in the context of FIR absorption, have been discussed by Lippert *et al.*[59].) In more physical terms: the ansatz (6.10) describes the absorption process just as an incoherent superposition of all the possible elementary transitions $|i\rangle \Longrightarrow |f\rangle$.

(3) As mentioned above, the considered experiment indicates the existence of the predicted coherent-dissipative structures in the liquid. From this, we may infer the following conclusions : The concept of individually existing molecules (in the liquid) cannot be maintained, in the physical context under consideration; the "molecules" loose their individualities since they "act cooperatively" and create novel irreducible, indivisible "individuals", i.e. the coherent-dissipative structures. At the risk of over-simplifying one could say, in classical-mechanical terms, that correlation patterns rather than molecules are physically meaningful concepts (in the context under consideration). A qualitatively very similar conclusion has recently been drawn by Hertz[63], within the physical context of the H^+ mobility in aqueous solutions.

(4) It should be reminded that the process of creation of coherent-dissipative structures, as discussed above, differs substantially from the physical processes leading to superconductivity or superfluid states. Nevertheless, these structures, too, cannot be considered to be localized in the configurational space. The actual geometrical dimensions of the coherent-dissipative structures have not been estimated thus far, because this problem is not intimately connected with the clarification of the physical problem under consideration (for an example of such an estimate, see section 6.2.B below.).

(5) Additionally we would like to mention that there may also be other cooperative phenomena in liquids, which could be observable in different spectral regions, e.g. in the middle infrared; here vibrational modes become excited, too. In an article by Michel and Lippert[64], there have been reported experimental observations of cooperative phenomena in liquid/solid CH_3CN that seem to have their origin in a

coupling of molecular oscillations and that lead to the formation of non-stationary waves (or modes) in configurational space.

(6) Using the scientific language which is born out by the CSM investigation, one might say that the photon with energy $\hbar\omega$ will be absorbed as a consequence of the resonance structure (or manifold) displayed by the Jordan block appearing as a possible complex deformation of the two-matrix connected with the set S, equation (6.3).

(7) In many instances, repeatedly referred to in the previous development, we have tried to illustrate the entity "coherent-dissipative structure" using the term "correlation pattern". Therefore, to prevent possible confusion and misunderstanding, it is necessary to appreciate that coherent-dissipative structures are intimately (albeit not in the "BCS" sense) connected with Yang's ODLRO in that the corresponding density operator cannot be diagonalized and thus have no classical-mechanical analogue (cf. chapter 5).

6.2. On the Steady-state Photoemission

6.2.A. D-Fluctuations in the Luminiscence

Recently, the basic concepts of Prigogine's star-unitary transformation theory (cf. Prigogine[41] or George[42] and references cited therein) have been formally analyzed and shown to be applicable into the physical context of luminiscence (and in particular, of fluorescence) spectroscopy in condensed systems (cf. Dreismann[43,44] and references cited therein). A photon-counting spectroscopic experiment[43] has been proposed and carried out[65,66,67] that demonstrates the predicted applicability of Prigogine's theory within the considered physical context.

The present theoretical analysis have led to the prediction that dynamically induced fluctuations (D-fluctuations) may appear in in the photo-emission flux of luminescent condensed matter, even under "perfect" steady-state excitation conditions[43]. This novel kind of fluctuations has nothing to do with "stochastic" or "random" noise, but it is due to the true time evolution of the photoemitting quantum systems which are coupled with the many degrees of freedom of the matter in their environment. We have also given an estimate of the characteristic time of these D-fluctuations from first principles although their "amplitude" is not predictable. This characteristic time is of the order of magnitude of $10^{-6}s$, for a typical fluorescence transition in the visible spectral region:

$$\tau_D \approx 10^{-6}s.$$

This prediction, however, appears to be very surprising for the following reasons:

a) The excited states of many substances (like, e.g., the aromatic π-electron systems) normally show a natural fluorescence decay time, i.e. the life time of the fluorescing state, of the order: $\tau_F \approx 10^{-8}s$. As is well known, τ_F is connected with the temperature independent Einstein A-coefficient[68]. Moreover, the photoemission is considered as a spontaneous transition related to "one-body" dynamics – i.e., it is mainly of intramolecular character, cf. e.g. Birks[69]. Of course, it is well-known that the spectral distribution of light often depends on the environment of the emitting quantum systems, e.g., on the nature of the solvent, in the case of dissolved aromatic compounds; cf. Birks[69], Lippert[70,71,72] and references cited therein. But the time characteristics of the photemission are believed to be governed by "statistics" rather than by "dynamics".

b) Thus the photoemission taking place in condensed matter is widely believed to be governed by the "statistics of independent events", which, in the case under consideration, is given by the well known Poisson statistics. Here, we are furthermore dealing with conventional luminescence experiments and standard instrumentation and/or physical parameters: There is a steady state, broad band excitation (i.e. the exciting irradiation has a constant intensity and is given by some discharge lamp); the spectral resolution of the luminescence is $\Delta\lambda \geq 1\ \mathrm{cm}^{-1}$; the time resolution for investigations of photon statistics is $\Delta t > 10^{-9}s$.

c) D-fluctuations were predicted to be frequency dependent, see e.g. Dreis mann[43,44], because they are due to cooperative processes between spectrally different spectroscopic transitions. This prediction, however, is clearly in contrast to all known conventional considerations of luminiscence transitions, in the physical context under consideration. In the light of standard theory (cf., e.g. Birks[69]) one believes that the thermal motion destroys all possible "coherences" between emitting quantum systems. To show this, one usually refers to the characteristic time $\hbar/k_BT$ which is of the order $10^{-14}s$ at room temperature, and thus:

$$\tau_F \gg \hbar/k_BT. \tag{6.11}$$

This relation is then interpreted in the sense that the thermal disturbance acts very often (say, $10^{-8}s/10^{-14}s = 10^6$ times) before the spectroscopic photo-emitting transition occurs, thus destroying all conceivable correlations and/or memory effects between different quantum systems*.

Nevertheless, the hitherto existing experimental results strongly indicate a certain "cooperative" behaviour of vibronic transitions (depending on wave length) with a characteristic time of the predicted order of magnitude ($\approx 10^{-6}s$). This characteristic time differs substantially from the one that quantum beats between vibronic transitions would exhibit†. In the following we will see that the spontaneously created coherent-dissipative structures, as predicted in the previous chapters of this

* It should be mentioned that this point of view is widely accepted to be "generally valid" in all standard spectroscopic fields; cf. the similar remarks concerning the FIR absorption, in the previous sections.

† E.g., a typical vibrational mode with $\Delta\tilde{\nu} \approx 1000\ \mathrm{cm}^{-1}$ would give a theoretical "beat" frequency of the order $10^{14}s^{-1}$.

article, are able to explain some aspects of the D-fluctuations. The framework of the present theoretical considerations is given by the CSM, and is therefore – at first sight – independent of the star-unitary transformation theory, which has been previously used in the prediction of those fluctuations. Thus it is not surprising that it was not possible, thus far, to find the aforementioned characteristic time within the present formalism. Nevertheless, the extension of the luminescent coherent-dissipative structures (even in the configurational space) can now be estimated and compared to the size of the coherence area that the experimental findings support.

6.2.B. Coherent-Dissipative Structures and D-fluctuations in the Luminiscence of Condensed Systems

As mentioned above, even in the framework of classical statistical mechanics it is well known that correlations between "distant" molecules (in a condensed system) exist; see section 5.1. Here, as an example, let us consider a dilute solution (say 10^{-5} mol/l) of a luminescent (in particular - a fluorescent) compound in some inert solvent, at room temperature. This solution constitutes the spectroscopic specimen.

Now we can try to visualize a possible mechanism that might create correlations in the photoemission flux, by the following "classical" considerations: The thermal "bombardement" of any luminescent quantum system of the molecules of the surrounding can also be considered to disturb slightly its electronic charge distribution, and thus also its coupling operator with the electromagnetic field. Thus it is possible that the existing correlations between "distant" excited quantum systems (which do not interact directly with another) may imply some correlation between their deexcitation times, too. The quantitative study of this mechanism requires the detailed knowledge of the higher-order distribution functions entering into the BBGKY hierarchy (cf. section 5.1.). This knowledge, however, does not exist, as the BBGKY hierarchy is usually only studied up to the second or third order; cf. Balescu[38].

In the following, we will proceed to make direct contact between the theory leading to coherent-dissipative structures and the aforementioned luminescence experiment. For this, let $|i\rangle$, $i = 1, \ldots s$, represent a possible electronic ground state of a luminescent quantum system *. Let also $|i+s\rangle$ represent the electronically excited state that creates $|i\rangle$ after a photon emission $\hbar\omega$. Thus, the pair $|i, i+s\rangle$ represents the indicated spectroscopic transition, and we can schematically write:

$$|i, i+s\rangle \implies |i\rangle + |photon\rangle. \tag{6.12}$$

In the light of EPR-correlations, cf. section 6.1.B, the pair $|i, i+s\rangle$ may also

* Remember that usually this "quantum system" is not identical with one "molecule".

be considered to represent the entanglement between $|i\rangle$, $|i+s\rangle$, and the photon. Thus we can define the set

$$S = \{|i, i+s\rangle,\ i = 1, \ldots s\}, \tag{6.13}$$

whose elements have the previously mentioned obvious physical meaning. Hence it is also natural – at this stage of consideration (!) – to speak, say, about the probability that a specific quantum system undergoes one of the transitions associated with an element of S.

The next step in our considerations is very similar to that taken in the case of FIR absorption, see section 6.1.B. We observe that there exists a minimal dimensionality, s_{min}, of S, so that coherent-dissipative structures over S can emerge only if $s \geq s_{min}$. The reasoning is the same as in the FIR absorption study, with the only difference that the relevant "one-body" characteristic relaxation time is now given by the radiative life time τ_F. Thus, instead of (6.7), we require here:

$$\tau_1 = \tau_F. \tag{6.14}$$

This condition thus guarantees that the "lifetime" of the coherent-dissipative structure (which is associated with the created Jordan block over the set S) is not larger than the duration of the deexcitation process (which is the very physical reason for the emergence of that structure). Hence, if coherent-dissipative structures are apparent, we can write schematically:

$$|Jordan\ block\rangle \Longrightarrow |ground\ state\rangle + |photon\rangle\ , \tag{6.15}$$

for every one measured photon that has originated from some quantum system belonging to the structure represented by the Jordan block.

For $T = 300K$ and the typical value (for a fluorescent transition in the visible spectral range) $\tau_F \approx 10^{-8}s$, we obtain:

$$s_{min} = \frac{4\pi k_B T}{\hbar}\tau_F \approx 5 \cdot 10^6. \tag{6.16}$$

In the present case, the "one-body" relaxation time τ_F is independent of temperature.

This remark points out a significant conceptual difference between (6.16) and the corresponding formula (6.8) in the FIR absorption case. In the latter case, the rotational relaxation time τ_{rot} does depend on temperature, because of the fact that the rotational relaxation is owing to thermal motion of the molecules.

The considerations above have shown that, if coherent structures do exist in the spectroscopic specimen, some of the emitted photons are delocalized in the physical space represented by the set S. This "delocalization" is characterized by equation

(6.16), and it appears to be a product of a "one-body" physical parameter (τ_F) and a factor ($4\pi k_B T/\hbar$) that represents a kind of "coupling" between many elementary transitions (that is, "one-body" processes). Now let us try to estimate some typical (i.e., not the minimal) spacial size of the coherent-dissipative structures. The last remark enforce us to ask the following question: How small can the volume V be, in which an emitted (and measured!) photon can be thought of as being localized at the instant of time of its emission from a quantum system? At first glance the answer seems to be that V should be equal to the volume that the emitting quantum system (say, a molecule) occupies, e.g. $V = (10\ \text{Å})^3$. Nevertheless, this appears to be an erroneous estimate, since it does not take into account the quantal uncertainty relations. The physically meaningful answer to the above question should be given by the following consideration: Let the emitted light be measured with a spectral resolution $\Delta\lambda$. From the uncertainty relation

$$\Delta p \cdot \Delta x \geq \hbar/2, \tag{6.17}$$

we obtain with $|p| = \hbar|k|$ and $|k| = 2\pi/\lambda$ that

$$|\Delta p| = \hbar 2\pi \Delta\lambda/\lambda^2, \tag{6.18}$$

and thus

$$\Delta x \geq \frac{1}{4\pi}\frac{\lambda^2}{\Delta\lambda}. \tag{6.19}$$

For $\lambda \approx 500\ nm$ and a typical resolution of $\Delta\lambda \approx 1\ nm$ we obtain

$$\Delta x \geq 2 \cdot 10^{-3}\ \text{cm}\ . \tag{6.20}$$

From this we infer that each measured photon is "delocalized" in the instant of its creation in a volume V with:

$$V \geq V_{min} \approx (\Delta x)^3, \tag{6.21}$$

and thus $V \gg (10\text{Å})^3$. It should be observed that V_{min} has been determined in the standard way, i.e. without taking into account the existence of coherent-dissipative structures. Now let us proceed and consider a coherent-dissipative structure with dimensionality s_{min}. We assume that the measured photon has been emitted from this structure. In order to obtain a crude estimate, it seems natural to identify the volume V_{min}, equation (6.21), with the volume that can be thought of to be associated with every element $|i, i+s\rangle$ of S^*. If now s_{min} cubes, each of them with volume V_{min}, are thought of to be arranged is such a way as to touch one-another in order to build up a large cube, then the large cube will have the linear dimension

$$d \geq (s_{min} \cdot V_{min})^{1/3} = (s_{min})^{1/3}\Delta x. \tag{6.22}$$

---

* The fact that V_{min} contains a huge number of fluorescent molecules does not invalidate the present estimation.

The numerical values of equations (6.16) and (6.20) yield

$$d \geq d_{min} \approx 3 \cdot 10^{-1} \text{ cm} . \tag{6.23}$$

This linear dimension might also be considered as an "upper bound" for the spatial range of the coherent-dissipative structure with dimensionality s_{min}. This appears to be the case if one assumes that the elementary volumes V_{min} may as well "overlap", thus giving a smaller linear extension than d.

Of course, the experimental investigation of the last situation requires a much more refined instrumentation and a more precise measurement than those reported in the aforementioned references (see section 6.2.A). Nevertheless, the estimate (6.23) corresponds to a linear dimension of a spectroscopic specimen that can be easily arranged in a real experiment. Indeed, this was also the case in the previously mentioned experiment concerning the detection of D-fluctuations.

It is interesting to note that the present numerical estimate can also be achieved independently from the physical context of coherent-dissipative structures. This can be shown to follow from the formalism being represented by the D-transformation, Dreismann[43,44], in the following way: As was mentioned in section 6.1.A, the typical characteristic time of the D-fluctuations is $\tau_D \approx 10^{-6}s$. We also mentioned that the excitation of a molecule also (of course) disturbs the equilibrium correlation patterns to which it contributes. Let us assume that the typical velocity of the propagation of that disturbance is of the order of the velocity of sound† , v_s. Typically one has in liquids that $v_s \approx 10^3 \ m \cdot s^{-1}$. Thus, the above disturbance propagates at most a distance

$$d_D \equiv \tau_D \cdot v_s \approx 10^{-1} \text{ cm} . \tag{6.24}$$

during a time interval $\Delta t = \tau_D$. In other words: If D-fluctuations do exist, then a time-resolved measurement (like those reported in the aforementioned references) made in the time scale given by τ_D would also indicate that the coherence area of these fluctuations has the linear dimension d_D. In the light of these crude estimates, it is indeed surprising that

$$d \approx d_D$$

as we obtain from the last two equations! Further clarification of this important matter would be achieved, if we could succeed in determining some characteristic "fluctuation time" of the coherent-dissipative structures within the above CSM formalism which yielded the present spatial estimate. We conclude that the above estimated value for the spatial extension of the coherent-dissipative structures is of the same order as the size of the coherence area that is shown to exist through

† For experimentalists, this assumption appears to be "natural".

the detection of D-fluctuations. A preliminary account of this work was given recently[73].

Acknowledgements

We are indebted to M.Sc. Mikael Höghede for contributing to an earlier version of this TEX-file. C. A. C.-D. wish to thank the members of the Quantum Chemistry Department (University of Uppsala) for elucidative discussions, and the Deutsche Forschungsgemeinschaft and the Fonds der Chemischen Industrie for financial support. We are also indebted to the Swedish Natural Science Research Council for financial contributions.

References

1. C. N. Yang, Rev. Mod. Phys. **34**, 694 (1963).
2. E. Balslev and J. M. Combes, Commun. Math. Phys. **22**, 280 (1971).
3. B. Simon, Ann. Math. **97**, 247 (1973).
4. C. van Winter, J. Math. Anal. **47**, 633 (1974).
5. C. E. Reid and E. J. Brändas, "On a Theorem for complex symmetric matrices and its relevance in the study of decay phenomena", this volume.
6. A. J. Coleman, Rev. Mod. Phys. **35**, 668 (1963).
7. P. O. Löwdin, Phys. Rev. **97**, 1474 (1955).
8. J. Bardeen, L. N. Cooper and J. N. Schrieffer, Phys. Rev. **108**, 1175 (1957).
9. A. J. Coleman, J. Math. Phys. **6**, 1425 (1965).
10. J. Linderberg and Y. öhrn, Int. J. Quant. Chem. **12**, 161 (1977).
11. B. Weiner and O. Goscinski, Phys. Rev. **A22**, 2374 (1980).
12. B. Weiner and O. Goscinski, Phys. Rev. **A27**, 57 (1983).
13. J. Linderberg, Israel J. Chem. **19**, 93 (1980).
14. O. Goscinski, Int. J. Quant. Chem. **S16**, 591 (1982).
15. S. Bratoz and Ph. Durand, J. Chem. Phys. **43**, 2670 (1965).

16. J. V. Ortiz, B. Weiner and Y. öhrn, Int. J. Quant. Chem. **S15**, 113 (1981).

17. E. Sangfeldt, O. Goscinski, N. Elander and H. Kurtz, Int. J. Quant. Chem. **S15**, 133 (1981).

18. F. Sasaki, Phys. Rev. **138**, B 1338 (1965).

19. M. D. Girardeau, Proceedings of Conference on "Reduced Density Matrices with Applications to Physical and Chemical Systems", held at Queen's University, Kingston, during August 28 -September 1, 1967; Queens Papers on Pure and Applied Mathematics, No 11 (1968) p.111.

20. E. Brändas, Int. J. Quantum Chem. **S20**, 119 (1986).

21. J. Kumičák and E. Brändas, Int. J. Quant. Chem. **32**, 669 (1987).

22. E. Nelson, Ann. Math. 70, 572 (1959).

23. B. Nagel, Private communication (1986).

24. D. A. Micha and E. Brändas, J. Chem. Phys. **55**, 4792 (1971).

25. E. Brändas and P. Froelich, Phys. Rev. **A16**, 2207 (1977).

26. E. Brändas, J. Mol. Spectry. **27**, 236 (1968).

27. G. Gamow, Z. Phys. **51**, 204 (1928).

28. E. Engdahl, E. Brändas, M. Rittby and N. Elander, J. Math. Phys. **27**, 2629 (1986).

29. F. R. Gantmacher, "The theory of matrices", Vol. II, Chelsea Publishing Company, New York (1959).

30. Ch. Obcemea and E. Brändas, Ann. Phys. **151**, 383 (1983).

31. Ch. Obcemea, P. Froelich and E. J. Brändas, Int. J. Quantum Chem. **S15**, 695 (1981).

32. A. J. F. Siegert, Phys. Rev. **56**, 750 (1939).

33. E. Brändas and P. Froelich, Int. J. Quantum Chem. **13**, 619 (1978).

34. N. Moiseyev, P. R. Certain and F. Weinhold, Mol. Phys. **36**, 1631 (1978).

35. N. Moiseyev and S. Friedland, Phys. Rev. **A22**, 618 (1980).

36. N. Moiseyev, Lecture Notes in Physics. Vol. **256**, 122 (1985).

37. A. J. Engelmann, M. A. Natiello, M. Höghede, E. Engdahl and E. Brändas,

Int. J. Quantum Chem. **31**, 841 (1987).

38. R. Balescu, Equilibrium and Non-Equilibrium Statistical Mechanics (Wiley-Interscience, New York, 1975).

39. P. Glansdorff and I. Prigogine, Thermodynamic theory of structure, stability, and fluctuations (Wiley-Interscience, New York, 1971).

40. G. Nicolis and I. Prigogine, Self-organization in non-equilibrium systems (Wiley-Interscience, New York, 1977).

41. I. Prigogine, From Being To Becoming (W. H. Freeman and Company, San Francisco, 1980); I. Prigogine and T. Y. Petrosky, Physica **147A**, 33 (1988).

42. C. George, F. Mayné and I. Prigogine, Adv. Chem. Phys. **61**, 223 (1985).

43. C. A. Chatzidimitriou-Dreismann, Int. J. Quantum Chem. **23**, 1505 (1983).

44. C. A. Chatzidimitriou-Dreismann, Int. J. Quantum Chem. **S19**, 369 (1986).

45. H. Primas, Chemistry, Quantum Mechanics and Reductionism, Springer Verlag, Berlin (1983).

46. B. d'Espagnat, Conceptual Foundations of Quantum Mechanics, Benjamin, London (1976); Physics Reports, **110**, 201 (1984).

47. J. A. Wheeler and W. H. Zurek (editors), Quantum Theory and Measurement, Princeton Univ., Princeton (1983).

48. P.-O. Löwdin, "On the change of Spectra Associated with Unbounded Similarity Transformations of a Many-Particle Hamiltonian and the Occurrence of Resonance States in the Method of Complex Scaling", Adv. Quantum Chem. (1988) in press.

49. T. K. Lim, Chem. Phys. Letters **4**, 521 (1970).

50. H. Primas, "Contextual quantum objects and their ontic interpretation", in: Proceedings of "The Copenhagen Interpretation 60 Years after the Como Lecture", World Scientific (1987); in press.

51. P. A. M. Dirac, The Principles of Quantum Mechanics, (4th edition), Oxford University Press, Oxford (1958).

52. A. J. Coleman, in "Quantum Statistics and the Many-Body Problem (Eds. S. B. Trickey, W. P. Kirk, and J. W. Dufty, Plenum Press, New York and London, 1975) p.239.

53. M. Reed and B. Simon, Methods of Modern Theoretical Physics (IV: Analysis of Operators), Academic Press, New York, 1978.

54. J. M. Blatt, Prog. Theoret. Phys. (Kyoto) **23**, 447 (1960).

55. A. J. Coleman, Phys. Rev. Lett. **13**, 406 (1964).

56. S. G. Kroon and J. van der Elsken, Chem. Phys. Letters **1**, 285 (1967).

57. R. M. van Aalst, J. van der Elsken, D. Frenkel and G. H. Wegdam, Faraday Disc. Chem. Soc. **6**, 94 (1972).

58. R. G. Gordon, Adv. Magnetic Resonance **3**, 1 (1968).

59. E. Lippert, C. A. Chatzidimitriou-Dreismann and K.-H. Naumann, Adv. Chem. Phys. **57**, 311 (1984); (see chapter 3).

60. B. J. Berne and G. D. Harp, Adv. Chem. Phys. **17**, 63 (1970).

61. K. M. van Vliet, J. Math. Phys. **19**, 1345 (1978).

62. K. M. van Vliet, J. Math. Phys. **20**, 2573 (1979).

63. H. G. Hertz, Chem. Scripta **27**, 479 (1987).

64. H. Michel and E. Lippert, in "Organic Liquids" (Eds. A. D. Buckingham, E. Lippert and S. Bratos, Wiley, New York, 1978) p.297.

65. C. A. Chatzidimitriou-Dreismann, J. Chem. Phys. **80**, 561 (1984).

66. J. Degen, H. H. Schmidtke, and C. A. Chatzidimitriou-Dreismann, Theoret. Chim. Acta **67**, 37 (1985).

67. J. Degen, J. Phys. C: Solid State Physics **20**, 3765 (1987); Ber. Bunsenges. Phys. Chem. **91**, 1215 (1987).

68. Th. Förster, Fluoreszenz organischer Verbindungen, Vandenhoeck & Ruprecht, Göttingen (1951).

69. J. B. Birks, Photophysics of Aromatic Molecules, Wiley, London (1970).

70. E. Lippert, in: "Optische Anregung organischer Systeme", W. Foerst (ed.), pp. 342-355, Verlag Chemie, Weinheim (1964).

71. E. Lippert, Acc. Chem. Research **3**, 74 (1970).

72. E. Lippert, W. Rettig, V. Bonačić-Koutecký, F. Heisel, and J. A. Miehé, Adv. Chem. Phys. **68**, 1 (1987).

73. C. A. Chatzidimitriou-Dreismann and E. J. Brändas, Ber. Bunsenges. Phys. Chem. **92**, 549 (1988); *ibid* **92**, 554 (1988).

THE LERTORPET SYMPOSIUM VIEW ON A GENERALIZED INNER PRODUCT

edited by

Nils Elander

Manne Siegbahn Institute of Physics
Frescativägen 24
S-104 05 Stockholm 50
SWEDEN
and

Erkki Brändas

Quantum Chemistry Group
Uppsala University
Box 518
S-751 20 Uppsala
SWEDEN

Abstract

In the following sections we have collected the discussions and viewpoints regarding the generalized scalar (or inner) product as they developed during the symposium week.

1. Introduction

Resonant states in the present context are represented by Gamow waves. They have the property of diverging with increasing values of the radial coordinate. This leads to obvious problems when one wants to define inner products over such functions, and this is therefore one of the central problems of this symposium. Since several methods can be used to tame these terrible Gamow waves it was then quite natural that the corresponding definition of the generalized inner product was brought up by the "grass roots". Several evening discussions started spontaneously. The result of them was that one finally agreed that essentially all definitions led to the same result but of course from different starting points. Since this problem has been a very central topic in several publications, albeit still felt to be an open question, it was suggested that the outcome of our discussions should be published.

The paper flows naturally as the contribution from the different spokesmen. The regularization was presented by the pioneer, Tore Berggren, followed by Borbála Gyarmati and Tamás Vertse in their extension of the regularization idea to a contour integration in the complex plane, see sections 1-3. The complex scaling as seen from a mathematical point of view is discussed in particular by Bernard Helffer in his paper in these proceedings, comparing different resonance notions. We will return to this aspect in the summary, section 6. The scalar product and the associated turn over rule is reviewed from a functional topological angle, in section 4, by the two organizers Brändas and Elander. The C-product introduced by Nimrod Moiseyev, Philip Certain and Frank Weinhold is finally presented by the former in section 5 and will be extended by him for the most general case when the Hamiltonian matrix is constructed from complex basis basis functions. A summary is given at the end.

2. *Tore Berggren* : The regularization method I

The products are defined using a regularization procedure based on the convergence factor $\exp\{-\varepsilon r^p\}$ where p was chosen to be 2 in the first paper (see e.g. Nucl. Phys., **A109**, 265 (1968)). The success of this regularization depends on the asymptotic behaviour of the wave functions in the product which restricts the allowable combinations in as manner to be described below. It was also desirable that the generlized inner product should be as similar as possible to the ordinary Hilbert space inner product: It should be linear in one factor (in quantum mechanical usage, particularly the Dirac bra-c-ket notation, the second factor) and antilinear in the other factor.

The integral

$$I(\varepsilon,K)=\int_0^\infty \exp\{-\varepsilon r^2\}\exp\{iKr\}dr=\frac{1}{\sqrt{\varepsilon}}\exp\{-K^2/4\varepsilon\}\int_{\frac{-iK}{2\sqrt{\varepsilon}}}^{\infty}\exp\{-t^2\}dt \tag{1}$$

exists for all $\varepsilon>0$ but has a finite limit for $\varepsilon\to 0$ only if either $\Im K<0$ or else $\Re K^2>0$, i.e.

$$|\Re K|>|\Im K| \tag{2}$$

Thus if we take two solutions $u_1(r)$ and $u_2(r)$ to the radial Schrödinger equation having the asymptotic behaviour

$$u_1(r) \sim \exp\{ik_1 r\} \quad (3a)$$

and

$$u_2(r) \sim \exp\{ik_2 r\} \quad (3b)$$

then the product $u_1(r)^* u_2(r)$ will behave as

$$u_1(r)^* u_2(r) \sim \exp\{+i(-k_1^* + k_2)r\} \quad (4)$$

To ensure that

$$|\Re\{-k_1^* + k_2\}| > |\Im\{-k_1^* + k_2\}| \quad (5)$$

for any combination of k_1 and k_2 we may stipulate that

$$\Re k_1 < 0 \ ; \ -\Im k_1 < -\Re k_1; \text{(incoming wave solution)} \quad (6a)$$

and

$$\Re k_2 < 0 \ ; \ -\Im k_2 < -\Re k_2; \text{(outgoing wave solution)} \quad (6b)$$

or *vice versa*. This restricts $u_1(r)$ and $u_2(r)$ to be selected from *different* function spaces (which may be considered to be the duals of each other) corresponding to asymptotically outgoing ($\Re k > 0$) or asymptotically incoming ($\Re k < 0$) solutions. Thus if we take the factor that stands in the linear position of the product from the set of outgoing solutions, we must take the other factor from the set of incoming solutions and *vice versa*. It is useful to distinguish the solutions having incoming waves asymptotically by means of a tilde (˜) or a similar typographical mark. Thus we have the regularizable possibilities

$$\langle \tilde{u}_1 | u_2 \rangle = \lim_{\varepsilon \to 0} \int \exp\{-\varepsilon r^2\} \tilde{u}_1(r)^* u_2(r) dr \quad (7a)$$

$$\langle u_2 | \tilde{u}_1 \rangle = \lim_{\varepsilon \to 0} \int \exp\{-\varepsilon r^2\} u_2(r)^* \tilde{u}_1(r) dr \quad (7b)$$

both with k_1 outgoing and k_2 incoming wave vectors.

Within this scheme it is prohibited to form products like $\langle u_1 | u_2 \rangle$ or $\langle \tilde{u}_1 | \tilde{u}_2 \rangle$. It may well be that stronger, but then less regular, methods of regularization enable one to assign non-infinite values also to such inner products. A consistent application of such a procedure should lead to orthogonality relations of the form (for complex energy eigen states only)

$$\langle u_i | u_j \rangle = \langle \tilde{u}_i | \tilde{u}_j \rangle = 0 \quad (8)$$

because we want $\langle u_i |$ to be the adjoint of $| u_i \rangle$, and if

$$H_r | u_i \rangle = W_i | u_i \rangle \quad (9)$$

then also

$$\langle u_i | H_r = \langle u_i | W_i^* \tag{10}$$

Here H_r is the appropriate radial Hamiltonian, which for real potentials is a real operator. (One consequence of the regularization is to eliminate all surface contributions to a matrix element that would contradict Hermiticity.) The stated orthogonality is an immediate consequence of the last two equations and the nonreality of W_i.

3. *Borbála Gyarmati and Tamás Vertse* : The regularization method II

Having read the pioneering works of Berggren (Nucl. Phys.**A109**, 265 (1968)) and Romo (Nucl. Phys.**A116**, 617 (1968)) we became interested in making calculations by using Gamow functions in nuclear physics problems. For example, we tried to describe isobaric analogue resonances in a very much simplified way having only one bound neutron and a resonant proton (Gamow) wave function. Our potential was on the other hand a realistic one. It was the sum of a finite- range nuclear potential and an infinite-range Coulomb potential, the latter being the Coulomb potential of a sharp edged sphere with homogenous charge distribution. The numerical integration of the radial Schrödinger equation and the calculation of the complex energy eigenvalue of the resonance was done by generalizing a bound state procedure. But it was hard to find ways of normalizing the solution. We tried the methods available at that time and found none of them suitable for numerical calculations. The use of convergence factors required the functions to be normalized up to very large distances; the analytic continuation on the other hand would require the knowledge of the indefinite integral of the squared outgoing Coulomb wave: $O_\ell(kr) = G_\ell(kr) + \imath F_\ell(kr)$ in closed form. It was even not clear that the Gamow function could be normalized in the presence of a Coulomb potential of infinite range. Therefore we studied the normalization of Gamow functions (Nucl. Phys., **A160**,523 (1971)). In the contribution by Vertse et al. in this book a section is devoted to this question. This part does accordingly not need to be repeated here and we therefore restrict ourselves to a few comments.

Within the range (a) of the nuclear potential, which had a Saxon-Woods plus spin-orbit force form, we calculated the contribution to the norm by numerical integration and the only question was how to calculate the contribution of the asymptotic region where the solution oscillates with exponentially growing amplitude. It is important to realize that the asymptotic solution is dependent on the product kr and oscillations could be damped if we allowed r to be also complex beyond the point a, i.e. rotate r with an angle $\varphi > 0$

$$r = a + |r - a| \exp(\imath\varphi) \tag{11}$$

If $\pi - \psi > \varphi > -\arg(k) \equiv -\psi > 0$, then $\frac{3\pi}{2} > \theta \equiv \varphi + \psi + \frac{\pi}{2} > \frac{\pi}{2}$ therefore the asymptotic solution goes to zero

$$\lim_{r\to\infty} O_\ell(k|r|e^{\imath\varphi}) \propto \lim_{r\to\infty} \exp[|k||r|(\cos\theta + \imath\sin\theta)] = 0 \tag{12}$$

We showed in (Nucl. Phys., **A160**,523 (1971)) that $\exp(-\varepsilon r^n) = \exp(-\varepsilon |r|^n e^{in\varphi})$ is a convergence factor for complex r, if $|\varphi| < \frac{\pi}{2n}(n > 1)$. We can thus replace the norm with the convergence factor along the real axis with the norm of the rotated outgoing solution without convergence factor, whenever we use $-\psi < \varphi < \frac{\pi}{2n}$.

$$\lim_{\varepsilon\to 0} \int_a^\infty \exp(-\varepsilon r^n) O_\ell^2(kr)\, dr = \int_a^\infty O_\ell^2(ke^{i\varphi}x)\, d(xe^{i\varphi}) \qquad (13)$$

This is because of the integrand of the right hand side falls exponentially with r and the integral does exist without the convergence factor. The proof is simple and is based on the Cauchy theorem. The right hand side of the equation does not depend on n which causes the result to be independent of the form of the convergence factor that we used.

Using the asymptotic form of the Coulomb functions we were able to prove that the Gamow functions are normalizable even in the presence of a Coulomb potential of infinite range. For neutrons the simple form of the Riccati-Hankel functions made it possible for us to give the integral in closed form. From this form one can immediately recognize that the procedure given by Romo, i.e. to calculate the integral in the upper half of the $k-$plane and continue the result analytically to the appropriate value (i.e. the value of k corresponding to a resonance pole of the resolvent or equivalently the Gamow wave) in the lower half of the $k-$plane, leads to the same result as that obtained from the use of a convergence factor or another suitable complex transformation. Via the complex scaling (11) we have in fact rotated the external Hamiltonian $T + V_{coul}(r)$ to complex r beyond (a) and used its solution in the integral of the right hand side of eq.(13). This procedure, which later became known as *exterior complex scaling* is very much suited for numerical work. Since the integrand goes to zero for large radial distance its contribution becomes negligible beyond a certain range like in the case of a bound state wave function.

As a summary we would like to stress the following points:

1. Any convergence factor $\exp(-\varepsilon r^n)$ with $1 < n < \frac{1}{2}\pi/\arctan(\gamma/\kappa)$ leads to the same limit $(k = \kappa - i\gamma)$

2. This limit coincides with the one reached by integrating at k, $\Im k > 0$ and in the result substituting k with k_0, $\Im k_0 < 0$ (k_0 is the resonance pole).

3. The norm for short-ranged potentials is given in closed form.

4. The existence of the norm for short-ranged + Coulomb potentials is proven.

5. The method used is contour integration in the complex r-plane (equivalent to exterior complex scaling). The practical consequence is that the limiting

procedure

$$\lim_{\varepsilon \to 0} \int \exp\{-\varepsilon r^2\} \ldots, dr \tag{14}$$

can be avoided.

4. *Erkki Brändas and Nils Elander* **: The complex scaling method and the turn over rule**

The extensions of the regularization theory of Berggren by Gyarmati and Vertse do in a sense precede the (exterior) complex dilation methods, and appeared independently of the rigorous mathematical theory. For the details of this development we refer to the original work of Balslev and Combes (Commun. Math. Phys. **22**, 280 (1971)), Simon(Ann. Math. **97**, 247 (1973)) and van Winter (J. Math. Anal. **47**, 633 (1974)). Our contribution to this presentation will rest on previous work by Brändas and Froelich (Phys. Rev.**A16**, 2207 (1977) including the discussion of generalized variation principles and the virial theorem, Brändas (Int. J. Quantum Chem. **S20**, 119 (1986)) and Kumičák and Brändas (Int. J. Quant. Chem. **32**, 669 (1987)) as well as the paper by Brändas and Dreismann in this book.

Despite the historic comments given above, the utilization of complex paths to define convenient integration procedures has existed since the last century, and particular applications to theoretical physics were made already by Sommerfeld and later by Hartree. More recent pre-CSM exploitations of these ideas were advocated by A. M. Dykhne and A. V. Chaplik, Sov. Phys. JETP **13**, 1002 (1961) and by J. Nuttall and H. L. Cohen, Phys. Rev. **188**, 1542 (1969). The exterior scaling procedure was anticipated by C. A. Nicolaides and D. R. Beck, Phys. Lett. **65**, 11 (1978), see also previous section, well before the work of B. Simon, Phys. Lett. **71**, 211 (1979), and J. D. Morgan and B. Simon J. Phys. **B14**, L167 (1981) and before recent rigorous proofs appeared, see papers by Helffer and Sjöstrand in this book. Note also the beautiful theory of generalized functions of Carleman at the end of the forties.

The CSM is, from our point of view, not only a way to extract direct information on physical processes out of higher order Riemann sheets. The strength of this method is also that it (rigourously) enables the direct calculation of scattering information in terms of Green's functions, spectral densities or S-matrix poles and their respective residues. In general it could be quite dangerous to give these poles any direct physical interpretation even if it is sometimes rather tempting. Instead these poles and residues get their meaning through parameters (c.f. Fano line shapes and the Breit-Wigner ansatz which are discussed by Rittby et al. in this book) in expansions which can be used to go back to the physical world on the real axis, see Rittby et al. and Elander et al. in this book.

From these points of view it is thus neccessary that all generalizations and extensions "beyond conventional" real axis quantum mechanics can be brought back to the real physical world.

For completencss we will collect some of the main points in this section, although they also appear elsewhere. The key to the theory is the complex scaling operator

$$U(\theta) = \exp(iA\theta) \tag{15}$$

where A is the generator of the scaling transformation and θ is a real or complex parameter. If $\theta \in R$, then U is a unitary operator , defining the dilation group, with $\mathcal{D}(U) = \mathcal{H}$ i.e. the full hilbert space, which effectuates the scaling

$$U(\theta)\Phi(x_1,\ldots,x_N) = \exp\left(\frac{3N}{2}\theta\right)\Phi(e^{\theta}x_1,\ldots,e^{\theta}x_N) \tag{16}$$

The idea behind CSM is to consider an analytical continued operator $H(\eta)$ which is obtained from a non-deformed, self-adjoint operator $H(1)$ via a similarity transformation. The general theory of Balselv and Combes deals wiht an an N-body Schrödinger Hamiltonian $H = T + V$. The domain of this Hamiltonian is defined by that of the kinetic energy term T and the interaction potential term V. We need to introduce the domain to η

$$\Omega = \{\eta, |\arg(\eta)| < \theta_{crit}\} \tag{17}$$

where θ_{crit} is defined by the potential V. Following Brändas (1986) we can now introduce the domain $\mathcal{N}(\Omega)$ of $U(\theta)$ as a subset dense of the standard hilbert space $\mathcal{H}$ usually called the Nelson class (E. Nelson, Ann. Math. **70**, 572 (1959)).

$$\mathcal{N}(\Omega) = \{\Phi, \Phi \in \mathcal{H};\;\; U(\eta)\Phi \in \mathcal{H};\;\; \eta \in \Omega\} \tag{18}$$

which then is completed with respect to the norm

$$\int_{-\theta_0}^{+\theta_0} \|U(\eta)\Phi\|^2_{\mathrm{L}^2} d\theta = \|\Phi\|^2_{\mathrm{N}_{\theta_0}} \tag{19}$$

to a hilbert space $\mathcal{N}_{\theta_0}$ The form of T as a sum second partial derivatives necessitates analogous definition of the spaces $\mathcal{N}^{(i)}_{\theta_0}$, $i = 0,1,2$. The potential V naturally requires a domain such that

$$\mathcal{N}_V(\Omega) = \{\Phi, \Phi \in \mathcal{H};\;\; U(\eta)V\Phi \in \mathcal{H};\;\; \eta \in \Omega\} \tag{20}$$

In analogy with (19) we define $\mathcal{N}^{(V)}_{\theta_0}$ as given by all Φ such that $V\Phi \in \mathcal{N}_{\theta_0}$. The domain of $U(\eta)$ can now be specified as

$$\mathcal{N}_H = \mathcal{N}^{(0)}_{\theta_0} \cap \mathcal{N}^{(1)}_{\theta_0} \cap \mathcal{N}^{(2)}_{\theta_0} \cap \mathcal{N}^{(V)}_{\theta_0}$$

It is also clear that we have considered the hilbert spaces $\mathcal{N}^{(i)}_{\theta_0}$, $i = 0,1,2$ in order to find convenient domains for the unbounded (in $\mathcal{H}$) similitude $U(\eta)$, $\eta \in$

Ω. After appropriate deformations completion with respect to the standard L^2 norm is made. In this manner we arrive at the formal eigenvalue relation ($\eta = |\eta|e^{i\theta}$)

$$H(\eta)\Psi(\eta) = \varepsilon(\eta)\Psi(\eta) \quad \eta \in \Omega^+ \tag{21}$$

with $\arg\eta$ sufficiently large to uncover the resonance ε. The conjugate of equation (21) becomes

$$\overline{H(\eta)}\ \overline{\Psi(\eta)} = \overline{\varepsilon(\eta)}\ \overline{\Psi(\eta)} \quad \eta \in \Omega^- \tag{22}$$

with the involution $\overline{A(\eta)} = A^*(\eta^*)$ being introduced to include also (complex) optical potentials. Obviously (21) and (22) motivates the construction

$$\langle\overline{\Psi(\eta^*)}|H(\eta)|\Psi(\eta)\rangle = \varepsilon(\eta)\langle\overline{\Psi(\eta^*)}|\Psi(\eta)\rangle \tag{23}$$

from which stationary variational principles can be derived in almost the same fashion as in ordinary quantum mechanics , with the important distinction that the extremum property of the principle has been lost. Note also that if H satisfies $H^* = H$ then there is no restriction *from the variational point of view* to assume that $\Psi^* = \Psi$ and then the construction (23) defining a trial $\varepsilon(\eta)$ based on a trial $\Psi(\eta)$ and corresponding $\overline{\Psi(\eta^*)} = \Psi(\eta)^*$ becomes complex symmetric. However, in many cases it may be useful to analyse the wavefunction in terms of nonreal spherical harmonics. One might then leave the angular momentum eigenfunctions as they are and consider each angular block in a complex symmetric representation even if the corresponding term in the wavefunction representation contains the nonreal component $Y_{l,m}(\vartheta, \varphi)$.

Even if this simple notation is very appealing in that almost any standard quantum mechanical technique can be taken over provided it is appropriately modified, one should note that it results in a formulation that goes beyond conventional quantum mechanics "on the real axis". The most direct consequence is, as mentioned, the appearance of complex resonance eigenvalues and associated Gamow vectors. A closer study of the full generalized spectral properties of the complex deformed problem (E. Engdahl et al. J. Math. Phys. **27**, 2629 (1986)) shows that these eigenstates essentially deflate the (generalized) spectral density, giving in an asymptotic sense a decomposition of the continuum into resonances and background. This has the important consequence that each Gamow vector represents a well-defined section of the continous spectrum associated with the unscaled self-adjoint problem. In other words one can say that *the Gamow representation condensates an infinite dimensional hilbert space associated with a particular spectral part of the continuum into a finite dimensional linear space of suitable Gamow vectors.*

Our generalized inner product is naturally associated with the source term in the poisson and poisson like equations which may be used to generate coulomb and exchange densites in selfconsistent-field schemes as discussed by Elander et al. in this book.

As a summary we give a few basic facts and comparisons with the previous method.

1. In contrast to the traditional theory of hilbert space, ($\mathcal{H}$), where the dual space, appearing in the bra-position, is identical (via completion) to $\mathcal{H}$ or appropriate domain, *this is not so in CSM.*

2. In contrast to the "convergence factor" regularization schemes, the hamiltonian in CSM *does not exhibit hermitian symmetry.*

3. A turn over rule can be defined provided the domains of the operators in question are explicitly given.

4. The new (semi)-norm can be zero for non-trivial vectors in $\mathcal{H}$. (In a deeper sense this leads to the occurrence of so called Jordan blocks in CSM).

5. Orthogonality is replaced by bi-orthogonality between Gamow states or resonances and bound state wave vectors

6. The scalar product between functions out of the same space (in the bra-c-ket positions) *cannot be defined in CSM.*

7. Using Gamow waves, satisfying certain eigenvalue relations, the CSM construction to assign a given value to a particular matrix element becomes equivalent with the regularization procedure of Berggren.

8. Any finite matrix representation of an (dilation analytic) operator can be brought to complex symmetric form by a similarity transformation, see below. If the (dilation analytic) hamiltonian is real and selfadjoint before applying the CSM, then the corresponding matrix representation becomes trivially complex symmetric, if the bra-basis is the complex conjugate of the ket-basis. See particularly the contribution by Reid and Brändas in this book.

Although the appearance of a complex symmetric form seems to impose a certain structure on the actual matrix representation this invokes no restriction on the formulation. As mentioned under point 8, there exists a general theorem, see e.g. F. R. Gantmacher, "The theory of matrices", Vol.II, Chelsea Publishing Company, New York (1959), that reads *any matrix can be transformed to complex symmetric form by a similarity transformation.* Despite the convenience and simplicity of complex symmetric matrices the next remark shows that they are not always the most natural choice.

5. *Nimrod Moiseyev* : The C-product

The properties of the complex scaled Hamiltonian were studied in collaboration with Certain and Weinhold, Mol. Phys., **36** 1613 (1978). In particular we derived a complex analogue to the variational theorem based on a *new* scalar product where the complex conjugation in the second position (corresponding to the bra) was not carried out. This led to the notion of the so called C-product. This also had the attractive feature of leading to complex symmetric matrix representations in a very natural way. Even if, as has been emphasized here, every matrix can be brought to such a complex symmetric form, this may not be the most advantageous or natural construct for every problem. For example, in the case of van der Waals predissociation resonance complex basis functions are commonly used in solving the suitable closed coupled equations. We will briefly consider another case below. For details we refer to the proceedings contribution and references therein.

We have considered, at this workshop, a gas atom interaction with a statically one dimensional periodic solid surface. In such a case the atom-surface potential V satisfies the following condition,

$$V(x+a,z) = v(x,z)$$

where a is the lattice constant. Using the Bloch theorem expanding the eigenfunction of the Schrödinger equation in a Fourier series one obtains

$$\psi(x, ze^{i\theta}) = e^{ik_x x} \sum_{n=-\infty}^{+\infty} \phi_n(z) \exp(\frac{2n\pi ix}{a})$$

where $k_x = k_o \sin\gamma$ is the component of the initial momentum in the $x-$direction, γ is the angle between the incoming atomic beam and the normal to the surface and $\phi_n(z)$ are square integrable functions (resulting from the complex scaling) which are associated with the $n-$th diffraction channel. Unlike earlier studies $\phi_n(n)$ is complex even when $\theta = 0$ resulting from the use of complex basis functions e.g. $\exp(2\pi nix/a)$) as a basis set. Therefore, the inner product of two complex eigenfunctions $(\psi_v, \psi_{v'}) = \delta'_{v,v}$ can not be taken as $\int \psi_v \psi_{v'} d\tau$ (which is the case when a real basis set is used) and not as the conventional scalar product $\int \psi_v^* \psi_{v'} d\tau$ (Since ψ_v and $\psi_{v'}$ are also complex resulting from the analytical continuation of the z coordinate into the complex plane). The *new* inner product (although still in a C-product language, since no complex conjugation is taken) is given by the bi-linear form of two eigenfunctions of the complex scaled Hamiltonian considered as

$$(\psi_v, \psi_{v'}) = \int \psi_v^L \psi_{v'} d\tau = \delta_{v,v'}$$

when

$$\begin{cases} \psi_{v'} = \sum_j c_{j,v'} \chi_j \\ \psi_v^L = \sum_j d_{j,v} \chi_j \end{cases},$$

where $\{\chi_j\}$ are the basis functions (for example in our case $\chi_j = g_j(z)\exp[\imath x(k_x + 2\bar{v}j\pi\imath x/a)]$ and $\{g_j\}$ are harmonic oscillator basis functions). The respectively linear parameters $\vec{c}$ and $\vec{d}$ are the *right* and the *left* eigenvectors of the Hamiltonian matrix,

$$H_{j'j}(\theta) = \langle\chi_{j'}|\hat{H}(x, ze^{i\theta})|\chi_j\rangle,$$

such that

$$\mathbf{H}_\theta \vec{C}_{v'} = E_{v'}\vec{C}_{v'}$$

and,

$$\vec{d}_j^T \mathbf{H}_\theta = \tilde{E}_v \vec{d}_j^T \Rightarrow \mathbf{H}_\theta^T \vec{d}_j = \tilde{E}_v \vec{d}.$$

Note that we here use the variable θ explicitely instead of η as above. Of course $\arg\eta = \theta$.

Since the determinant of a matrix is equal to that of its transpose, the eigenvalues of H_θ are the same as those of H_θ^T and $E_v = \tilde{E}_v$. In the case of complex scaled Hamiltonians, $\mathbf{H^T}(\theta) = \mathbf{H}^*(-\theta)$, and therefore $\vec{d}_v^*$ is an eigenvector of $\mathbf{H}(-\theta)$. The definition of the inner product is necessary in this case, particularly when one wishes to calculate complex expectation values of given dynamical operators, as well as the corresponding scattering cross section.

6. Summary

It is obvious that the various notions as well as the particular circumstances in which the actual resonance formulations appear may lead to confusion. As can be seen from the present discussion there should, however, be no misunderstanding regarding the extension of analytic expansion techniques and corresponding matrix element constructions to involve also complex diverging Gamow waves.

It is interesting to see how parallel developments in neighbouring fields have emerged almost unperturbed through the years and yet one finds that the C-product language and the time-reversed notion of the pioneers in nuclear physics are factually quite related. It is also interesting to note that a similar conclusion occurs also in the mathematical comparison between various resonance definitions, see e.g. Helffer's contribution to this symposium.

In our earlier Weyl theory studies, see for instance Engdahl et al. J. Math. Phys., **27**, 2629 (1986), it follows furthermore from a residue study of the classical Green's function associated with a second order differential operator that the bilinear form obtains naturally during the contour integration. Once again it should be pointed out that this is not in contradiction with the last comment (before the summary) of this editorial. The representation, as analysed here, can always be made complex symmetric. Yet the matrix element that corresponds to a physical quantity, e.g. a surface cross section, may nevertheless be derived from an expression that does not "subscribe" to complex symmetry. The appropriate way to interpret complex (semi) norms from a satisfactory physical point of view has not been fully investigated at the time for the meeting, and is left as a future discussion topic.

Subject Index

A

B

C

D

E

F

G

H

I

J

K

L

M

N

O

P

T

U

V

W

Z

Lecture Notes in Mathematics

Vol. 1236: Stochastic Partial Differential Equations and Applications. Proceedings, 1985. Edited by G. Da Prato and L. Tubaro. V, 257 pages. 1987.

Vol. 1237: Rational Approximation and its Applications in Mathematics and Physics. Proceedings, 1985. Edited by J. Gilewicz, M. Pindor and W. Siemaszko. XII, 350 pages. 1987.

Vol. 1250: Stochastic Processes – Mathematics and Physics II. Proceedings 1985. Edited by S. Albeverio, Ph. Blanchard and L. Streit. VI, 359 pages. 1987.

Vol. 1251: Differential Geometric Methods in Mathematical Physics. Proceedings, 1985. Edited by P.L. García and A. Pérez-Rendón. VII, 300 pages. 1987.

Vol. 1255: Differential Geometry and Differential Equations. Proceedings, 1985. Edited by C. Gu, M. Berger and R.L. Bryant. XII, 243 pages. 1987.

Vol. 1256: Pseudo-Differential Operators. Proceedings, 1986. Edited by H.O. Cordes, B. Gramsch and H. Widom. X, 479 pages. 1987.

Vol. 1258: J. Weidmann, Spectral Theory of Ordinary Differential Operators. VI, 303 pages. 1987.

Vol. 1260: N.H. Pavel, Nonlinear Evolution Operators and Semigroups. VI, 285 pages. 1987.

Vol. 1263: V.L. Hansen (Ed.), Differential Geometry. Proceedings, 1985. XI, 288 pages. 1987.

Vol. 1265: W. Van Assche, Asymptotics for Orthogonal Polynomials. VI, 201 pages. 1987.

Vol. 1267: J. Lindenstrauss, V.D. Milman (Eds.), Geometrical Aspects of Functional Analysis. Seminar. VII, 212 pages. 1987.

Vol. 1269: M. Shiota, Nash Manifolds. VI, 223 pages. 1987.

Vol. 1270: C. Carasso, P.-A. Raviart, D. Serre (Eds.), Nonlinear Hyperbolic Problems. Proceedings, 1986. XV, 341 pages. 1987.

Vol. 1272: M.S. Livšic, L.L. Waksman, Commuting Nonselfadjoint Operators in Hilbert Space. III, 115 pages. 1987.

Vol. 1273: G.-M. Greuel, G. Trautmann (Eds.), Singularities, Representation of Algebras, and Vector Bundles. Proceedings, 1985. XIV, 383 pages. 1987.

Vol. 1275: C.A. Berenstein (Ed.), Complex Analysis I. Proceedings, 1985–86. XV, 331 pages. 1987.

Vol. 1276: C.A. Berenstein (Ed.), Complex Analysis II. Proceedings, 1985–86. IX, 320 pages. 1987.

Vol. 1277: C.A. Berenstein (Ed.), Complex Analysis III. Proceedings, 1985–86. X, 350 pages. 1987.

Vol. 1283: S. Mardešić, J. Segal (Eds.), Geometric Topology and Shape Theory. Proceedings, 1986. V, 261 pages. 1987.

Vol. 1285: I.W. Knowles, Y. Saitō (Eds.), Differential Equations and Mathematical Physics. Proceedings, 1986. XVI, 499 pages. 1987.

Vol. 1287: E.B. Saff (Ed.), Approximation Theory, Tampa. Proceedings, 1985–1986. V, 228 pages. 1987.

Vol. 1288: Yu. L. Rodin, Generalized Analytic Functions on Riemann Surfaces. V, 128 pages, 1987.

Vol. 1294: M. Queffélec, Substitution Dynamical Systems – Spectral Analysis. XIII, 240 pages. 1987.

Vol. 1299: S. Watanabe, Yu.V. Prokhorov (Eds.), Probability Theory and Mathematical Statistics. Proceedings, 1986. VIII, 589 pages. 1988.

Vol. 1300: G.B. Seligman, Constructions of Lie Algebras and their Modules. VI, 190 pages. 1988.

Vol. 1302: M. Cwikel, J. Peetre, Y. Sagher, H. Wallin (Eds.), Function Spaces and Applications. Proceedings, 1986. VI, 445 pages. 1988.

Vol. 1303: L. Accardi, W. von Waldenfels (Eds.), Quantum Probability and Applications III. Proceedings, 1987. VI, 373 pages. 1988.

Lecture Notes in Physics

Vol. 303: P. Breitenlohner, D. Maison, K. Sibold (Eds.), Renormalization of Quantum Field Theories with Non-linear Field Transformations. Proceedings, 1987. VI, 239 pages. 1988.

Vol. 304: R. Prud'homme, Fluides hétérogènes et réactifs: écoulements et transferts. VIII, 200 pages. 1988.

Vol. 305: K. Nomoto (Ed.), Atmospheric Diagnostics of Stellar Evolution: Chemical Peculiarity, Mass Loss, and Explosion. Proceedings, 1987. XIV, 468 pages. 1988.

Vol. 306: L. Blitz, F.J. Lockman (Eds.), The Outer Galaxy. Proceedings, 1987. IX, 291 pages. 1988.

Vol. 307: H.R. Miller, P.J. Wiita (Eds.), Active Galactic Nuclei. Proceedings, 1987, XI, 438 pages. 1988.

Vol. 308: H. Bacry, Localizability and Space in Quantum Physics. VII, 81 pages. 1988.

Vol. 309: P.E. Wagner, G. Vali (Eds.), Atmospheric Aerosols and Nucleation. Proceedings, 1988. XVIII, 729 pages. 1988.

Vol. 310: W.C. Seitter, H.W. Duerbeck, M. Tacke (Eds.), Large-Scale Structures in the Universe – Observational and Analytical Methods. Proceedings, 1987. II, 335 pages. 1988.

Vol. 311: P.J.M. Bongaarts, R. Martini (Eds.), Complex Differential Geometry and Supermanifolds in Strings and Fields. Proceedings, 1987. V, 252 pages. 1988.

Vol. 312: J.S. Feldman, Th.R. Hurd, L. Rosen, "QED: A Proof of Renormalizability." VII, 176 pages. 1988.

Vol. 313: H.-D. Doebner, T.D. Palev, J.D. Hennig (Eds.), Group Theoretical Methods in Physics. Proceedings, 1987. XI, 599 pages. 1988.

Vol. 314: L. Peliti, A. Vulpiani (Eds.), Measures of Complexity. Proceedings, 1987. VII, 150 pages. 1988.

Vol. 315: R.L. Dickman, R.L. Snell, J.S. Young (Eds.), Molecular Clouds in the Milky Way and External Galaxies. Proceedings, 1987. XVI, 475 pages. 1988.

Vol. 316: W. Kundt (Ed.), Supernova Shells and Their Birth Events. Proceedings, 1988. VIII, 253 pages. 1988.

Vol. 317: C. Signorini, S. Skorka, P. Spolaore, A. Vitturi (Eds.), Heavy Ion Interactions Around the Coulomb Barrier. Proceedings, 1988. X, 329 pages. 1988.

Vol. 318: B. Mercier, An Introduction to the Numerical Analysis of Spectral Methods. V, 154 pages. 1989.

Vol. 319: L. Garrido (Ed.), Far from Equilibrium Phase Transitions. Proceedings, 1988. VIII, 340 pages. 1988.

Vol. 320: D. Coles (Ed.), Perspectives in Fluid Mechanics. Proceedings, 1985. VII, 207 pages. 1988.

Vol. 321: J. Pitowsky, Quantum Probability – Quantum Logic. IX, 209 pages. 1989.

Vol. 322: M. Schlichenmaier, An Introduction to Riemann Surfaces, Algebraic Curves and Moduli Spaces. XIII, 148 pages. 1989.

Vol. 323: D.L. Dwoyer, M.Y. Hussaini, R.G. Voigt (Eds.), 11th International Conference on Numerical Methods in Fluid Dynamics. XIII, 622 pages. 1989.

Vol. 324: P. Exner, P. Šeba (Eds.), Applications of Self-Adjoint Extensions in Quantum Physics. Proceedings, 1987. VIII, 273 pages. 1989.

Vol. 325: E. Brändas, N. Elander (Eds.), Resonances, Proceedings, 1987. XVIII, 564 pages. 1989.